INTRODUCTORY OPERATIONAL AMPLIFIERS AND LINEAR ICs

Theory and Experimentation

ROBERT F. COUGHLIN
ROBERT S. VILLANUCCI

WENTWORTH INSTITUTE OF TECHNOLOGY

PRENTICE HALL, Englewood Cliffs, New Jersey 07632

Library of Congress Cataloging-in-Publication Data

Coughlin, Robert F.
Introductory operational amplifiers and linear ICs : theory and experimentation / Robert F. Coughlin, Robert S. Villanucci.
p. cm.
Includes index.
ISBN 0-13-477514-7
1. Operational amplifiers. 2. Linear integrated circuits.
I. Villanucci, Robert S., II. Title.
TK7871.58.06C67 1989 89-8412
621.39'5--dc20 CIP

Editorial/production supervision and interior design: **Lillian Glennon**
Cover design: **Wanda Lubelska**
Manufacturing buyer: **Dave Dickey**

Printed in the United States of America

10 9 8 7 6 5 4 3 2 1

ISBN 0-13-477514-7

Prentice-Hall International (UK) Limited, *London*
Prentice-Hall of Australia Pty. Limited, *Sydney*
Prentice-Hall Canada Inc., *Toronto*
Prentice-Hall Hispanoamericana, S.A., *Mexico*
Prentice-Hall of India Private Limited, *New Delhi*
Prentice-Hall of Japan, Inc., *Tokyo*
Simon & Schuster Asia Pte. Ltd., *Singapore*
Editora Prentice-Hall do Brasil, Ltda., *Rio de Janeiro*

TO:

SAREH,

SHEVAUN,

RALPH, and

NORMA

Contents

Preface

It is easy to solve many practical problems by using that most powerful tool, the operational amplifier (op amp). The original op amps have spawned an array of specialized and general-purpose devices. Some have one or more characteristics that have been optimized to perform one particular set of tasks, better, faster, and at lower cost. Others have been designed to be incorporated into more complex linear integrated circuits. These modern, powerful, application-specific devices can be used to easily solve many more difficult problems.

This book is written to show in the simplest way, the fastest, and most modern techniques for using operational amplifiers as well as the more complex linear integrated circuits (ICs). Every circuit drawing has pin numbers plus component values, and all op amps or ICs are clearly identified. Each circuit works. Each circuit teaches a principle of operation or solves a class of practical problems, and can be used as a laboratory exercise.

A formal laboratory exercise is provided at the end of each chapter. Each lab exercise gives a step-by-step procedure to show the student how to complete the learning objectives with maximum efficiency. All exercises have been class tested. They are associated directly with the text to reinforce learning and help bridge the gap between reading about a task and actually doing it.

Each laboratory exercise contains a recommended list of parts that are industry standards. In addition, the student should have access to the following:

1. Dual-trace oscilloscope, 10 MHz or better, with probes
2. Digital multimeter
3. Analog ac voltmeter for chapter 10
4. Function generator (1 MHz)
5. Adjustable regulated 0-to 30-V dc power supply; current capacity should exceed 100 mA
6. Breadboarding system with a ±15-V power supply at 50 mA and +5V at 500 mA

With these instruments the reader can gain experience with the principles presented in each chapter and observe the solutions to real problems.

Calculus is *not* required in this text but basic algebra is used throughout. Derivations have purposely been omitted and where appropriate, design and analysis equations are presented in their simplest form. Since we choose to eliminate derivations, we use the space gained to show how easy it is to work with the equations by providing clear example problems. For instructors who wish key derivations, laboratory advice, and further tutorials, a separate *Instructor's Resource Manual* is available.

ORGANIZATION OF THE TEXT

This book consists of 12 chapters, providing more than enough material for a one-semester course. The first four chapters show how the op amp and advanced ICs are used to monitor or compare voltages and provide on/off control. Chapter 1 introduces the op amp and its characteristics and identifies some of its practical limitations. Chapter 2 shows how op amps can act as voltage comparators. The AD584 pin-programmable precision voltage reference is introduced as well as the LM311 precision voltage comparator.

Chapter 3 shows how voltage comparators are used to make such circuits as a portable TTL logic probe, window detector, crowbar protection for power supplies, and a light-column voltmeter with the LM3914 IC. In Chapter 4, positive feedback is added to the basic comparator. The result is a class of circuits that perform on/off control with a measure of noise immunity. One example is an electronic equivalent of the familiar wall thermostat.

Chapter 5 introduces negative feedback to show how easy it is to make, analyze, and design amplifier circuits. Applications include the programmable gain amplifier (PGA) and a high-resistance voltmeter.

Chapter 6 extends the benefits of negative feedback to illustrate mathematical operations with op amps and linear ICs. Addition, subtraction, averaging, and scaling are done with an op amp. An AD534 analog multiplier IC completes the math hierachy, with multiplication, division, squares, and square roots. We have avoided integration, derivatives, and differential equations because analog technology in this field has long been obsolete. Chapter 7 further applies the principles of negative feedback to show how op-amp converters (current to voltage, voltage to current, and current to current) are used to make a variety of useful instrumentation circuits. Examples include (1) a universal polarity-insensitive ac-dc voltmeter complete with function and range switches, (2) applications with temperature- and light-sensitive transducers, and (3) semiconductor device testers.

Chapter 8 demonstrates practical measurement techniques that can be solved only by differential measurements using instrumentation amplifiers such as the powerful INA11. Applications include extracting heart-beat signals from body noise, and measuring strain, stress, and weight.

Chapter 9 explains the simplest, most current way to make a variety of inexpensive and also precision timers and signal generators or oscillators. A brief sampling of applications is presented for that ubiquitous

IC—the 555. Simple low-cost square- and triangle-wave generators are then introduced. The highlight of this chapter is the latest technology on how to make a precision sine-wave generator whose frequency can be precisely and widely adjusted using only a single resistor! Its amplitude can be precisely adjusted independent of frequency, and vice versa. As a bonus, we show how to use the new AD639 universal trigonometric function generator to construct a precision sine-wave generator.

The most useful basic active filters are given in Chapter 1. We have chosen circuits that are simple to design. They are also the easiest to build and tune precisely. The notch filter design example uses a few more parts but is the most practical way to accomplish the job with precision.

Finally, Chapters 11 and 12 teach the use of digital-to-analog converters and analog-to-digital converters using real hardware. They show how to connect DACs and ADCs to microprocessors. In addition, circuits have been included to show you how to test or operate microprocessor-compatible DACs and ADCs without a working microprocessor or computer.

COMPLETE LIST OF DEVICES COVERED IN THIS BOOK

Operational Amplifier

A741C	General-purpose op amp
LM301	General-purpose op amp
LM358	Dual single-supply, op amp
OP-07	Ultralow-offset-voltage op amp
LF411A	FET input op amp
LF442A	Dual low-power op amp
TLO-81	JFET input op amp
CA3140	MOSFET input op amp

Linear Integrated Circuits

LM311	Voltage comparator
AD584	Pin-programmable precision voltage reference
7805	Three-terminal voltage regulator
LM3914	Dot/bar display driver
HI-508A	Analog switch
AD534	Four-quadrant analog multiplier
AD590	Two-terminal IC temperature transducer
INA105	Precision differential amplifier
INA110	Precision instrumentation amplifier
NE555	IC timer
AD630	Modulator/demodulator
AD639	Universal trigonometric function generator
DAC-08	Digital-to-analog converter
AD558	Eight-bit microprocessor-compatible D/A converter
AD670	Analog-to-digital converter

Other Devices

HDSP-4830	Ten-element bar-graph array
C107B	Silicon-controlled rectifier (SCR)
IN914	Silicon diode
MRD500	Photodiode
CA4029	CMOS presettable up/down counter

Acknowledgments

For an abundance of constructive suggestions, we thank our colleague and old friend Professor Fred Driscoll, as well as Professors John Marchand, William Megow, Ronald J. Young, and Dominic Giampetro, who gave such enthusiastic support to this project. In particular, we thank Mrs. Phyllis Wolff, who expertly prepared the manuscript and cheerfully made the inevitable revisions. She is a fine human being who helps others. So are, and so do, Annette and Barbara.

Robert F. Coughlin

Robert S. Villanucci

Wentworth Institute of Technology
Boston, Massachusetts
1990

Characteristics and Specifications of Operational Amplifiers

1

LEARNING OBJECTIVES

Upon completion of this chapter on the characteristics and specifications of operational amplifiers, you will be able to:

- Determine what individual circuits are required to build a general-purpose operational amplifier.
- Draw the schematic symbol and name each terminal of an op amp.
- Identify the various package styles used to house op amps and then to select the most appropriate one for your application.
- Correctly apply dc power to an op amp.
- Select the correct value of a load resistor for the output terminal of an op amp.
- Estimate the output current of an op amp under various load conditions.
- Determine, from the device data sheet, both differential input resistance and single-ended output resistance.
- Determine, from the data sheet, input bias current, input offset current, and input offset voltage.
- Predict the maximum signal frequency that an amplifier can process.
- Predict how fast the output voltage of an op amp can change for a specific input voltage change.

1-0 INTRODUCTION

Operational amplifiers are complex electronic circuits packaged in convenient cases. These units, together with only a few support components, can be used to construct an extraordinary variety of useful circuits. One of the principal advantages of operational amplifiers is the ease with which they can be used.

We begin the study of operational amplifiers by learning the "language" of the device. This will require us first to define and understand those common symbols and terms that are associated with, and used to explain, operational amplifiers. As we are introduced to this new language, many of the important characteristics and specifications of operational amplifiers are presented. These specifications, obtained from the device's data sheet, are used throughout the book.

Our study begins with a functional diagram of the basic general-purpose operational amplifier. The associated circuitry within the package is presented only briefly to illustrate the type of electronics required to produce an operational amplifier. Paramount in our introduction to the operational amplifier is an understanding of the circuit symbol, available packaging, dc power requirements, and capabilities of the output terminal to deliver current and voltage. We also study the operational amplifier's input current and voltage specifications as well as its frequency response characteristics and other selected features.

1-1 OPERATIONAL AMPLIFIER

The typical general-purpose operational amplifier is a solid-state device with very high voltage gain or amplification capabilities. It is a direct-coupled device able to accept either dc or ac input signals and process these voltages to produce a variety of both linear and nonlinear outputs. The name "operational amplifier," or simply *op amp,* is obtained from early applications where it performed mathematical *operations* such as addition, subtraction, integration, multiplication, and division. Today, use of the op amp has expanded to the point where it has become the basic "building block" for applications such as voltage comparators, current or voltage regulators, active filters, oscillators, timers, and many others.

An operational amplifier is broadly classified as a linear integrated-circuit, high-gain differential amplifier. Presently, the op amp is inexpensively fabricated on a silicon wafer where all required electronics are part of a single *chip.* The chip is then packaged in a suitable case for safety and ease of use. As a result of this type of manufacturing, the device is called a monolithic ("single-stone") IC.

1-1.1 Block Diagram of a Basic Op-Amp Circuit

General-purpose op amps are multistate amplifier systems. As can be seen from the functional circuit representation of Fig. 1-1, the basic op amp consists of an *input stage* with two input terminals, an *output stage* with one output terminal, and an *intermediate stage* to properly connect the output signal of the first stage to the input terminal of the output stage.

Dc power is obtained from a bipolar (or dual) supply and applied to the op amp's external power supply terminals. Dc power is thus connected to each internal stage of the op amp.

Depending on the application, input signal $V_{(+)}$ and $V_{(-)}$ can be applied to either input terminal or simultaneously to both inputs. The

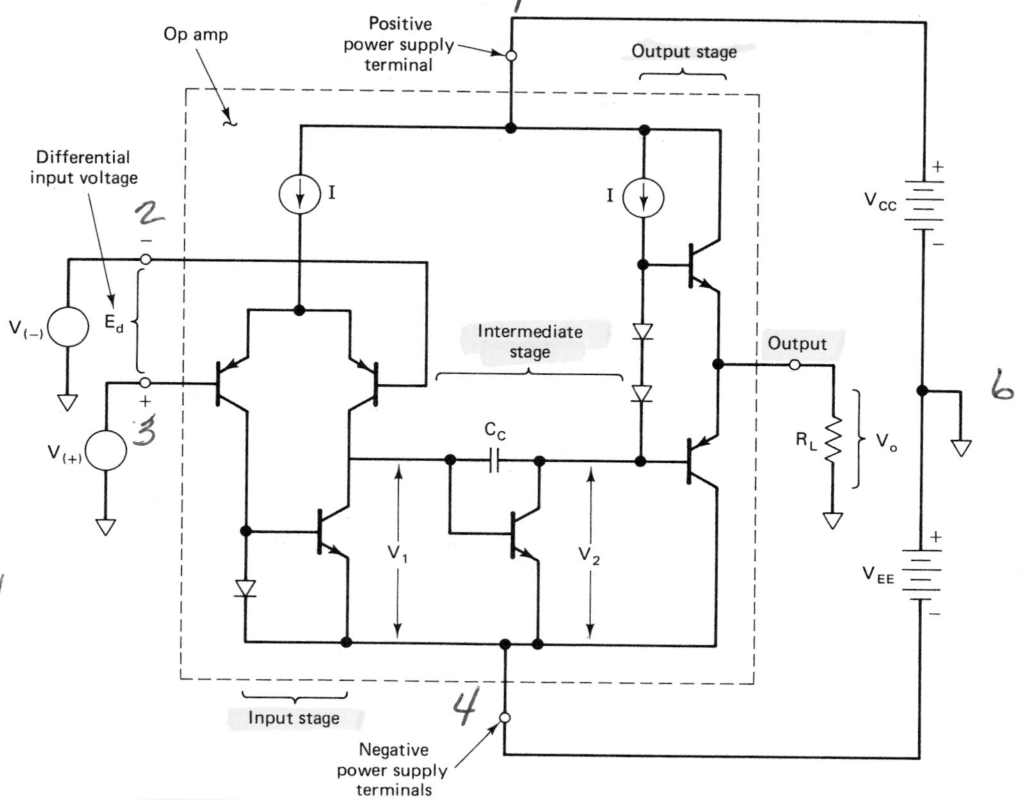

Figure 1-1 Basic functional block diagram of a general-purpose operational amplifier.

resulting output voltage is monitored across the load resistor R_L connected to the op amp's output terminal. The output voltage will depend on the input signals and characteristics of the op amp.

1-1.2 Differential Amplifier Input Stage

The input stage of the op amp in Fig. 1-1 is called a *differential amplifier.* It has very high input resistance characteristics as well as large voltage gain. When input signals $V_{(+)}$ and $V_{(-)}$ are applied, the *difference voltage* E_d* is amplified by the considerable gain of this stage and appears as the output voltage V_1.

1-1.3 Level-Shifter Intermediate Stage

Signal voltage V_1 at the output of the differential amplifier is directly coupled to the input of the intermediate level-shifter stage. This stage performs two major functions. First, it shifts the dc voltage level at the output of the differential amplifier to a value required to bias the output stage. Second, the level shifter allows input signal V_1 to pass nearly unaltered and become input signal V_2 for the output stage.

*Throughout this book we use the term E_d to identify the differential voltage that appears between the input terminals of an op amp.

1-1.4 Push-Pull Output Stage

The signal voltage V_2 at the output of the level-shifter circuit is coupled directly into the output stage. The most common output stage is a PNP/NPN push-pull transistor configuration. Use of a push-pull circuit as the final stage allows the op amp to have a very low output resistance. As shown in Fig. 1-1, load resistor R_L is connected to the output terminal to develop output voltage V_o.

What has been presented to this point is basic information on the internal architecture of an op amp. The actual circuitry is more complicated.

1-2 ELEMENTS OF AN OPERATIONAL AMPLIFIER

1-2.1 Op-Amp Symbol

The op-amp symbol in Fig. 1-2 is a triangle that points in the direction of signal flow. This component has a *part identification number* (PIN) placed within the triangular symbol. The PIN refers to a particular op amp with specific characteristics. The 741C op amp illustrated here is a general-purpose op amp that is used throughout the book for illustrative purposes.

The op amp may also be coded on a circuit schematic with a *reference designator* such as U1, IC 101, and so on. Its PIN is then placed beside the reference designator in the parts list of the circuit schematic.

1-2.2 Op-Amp Terminals

All op amps have at least five terminals, as shown in Fig. 1-2: (1) The positive power supply terminal V_{CC} or $+V$ at pin 7, (2) the negative power supply terminal V_{EE} or $-V$ at pin 4, (3) output pin 6, (4) the inverting ($-$) input terminal at pin 2, and (5) the noninverting (+) input terminal at pin 3. Some general-purpose op amps have additional specialized terminals. (The pins above refer to the mini-DIP case discussed in the following section.)

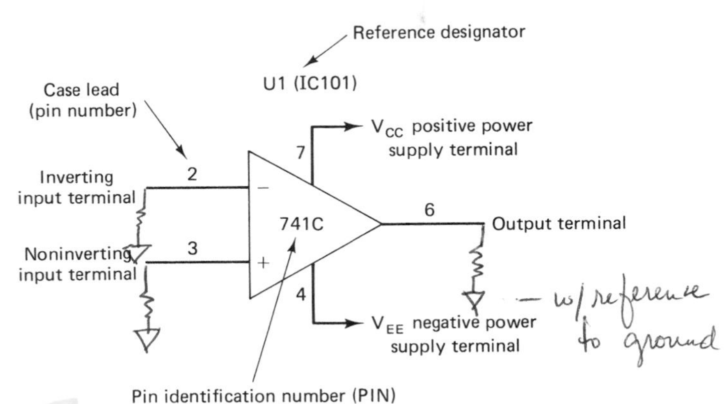

Figure 1-2 Schematic symbol and lead identification for a general-purpose operational amplifier (mini-DIP).

1-2.3 Op-Amp Packages

The op amp is fabricated on a silicon chip and packaged in a suitable case. Three of the most common types of packages are those shown in Fig. 1-3.

The *metal can* package (Fig. 1-3a) is popular where space is at a premium. The silicon chip is bonded inside the package to the bottom or *sealing plane* of the case. External leads extend outward from the sealing plane to facilitate circuit interconnections. Internally, fine-gauge wires are welded to the silicon chip at the appropriate circuit points and connected to the external leads. Since the sealing plane and cover of this package are constructed of metal, heat generated by currents flowing through the silicon chip is quickly dissipated by the package. For this reason, the metal can package is used for large power devices.

Metal can packages are available with 3, 5, 8, 10, and 12 leads. A small metal *tab* is placed adjacent to one lead on the package as a lead designator. For example, in Fig. 1-3a, the tab represents pin 10 and the leads are numbered counterclockwise from the tab.

Figure 1-3b illustrates the *dual-in-line* (DIP) package. This case, molded from either plastic or ceramic, is currently the most widely available and most popular package configuration. The reason for its wide acceptance is ruggedness, ease of breadboarding, and adaptability to printed circuit board packaging.

The DIP package is available with 8 pins, called the *mini*-DIP, for low-density circuitry, as well as 14- and 16-pin dual-in-line packages.

As viewed from the top down, a small notch or dot is placed on one end of the DIP case to identify pin 1. As with the metal case, the pin designation is obtained by counting counterclockwise from pin 1.

A surface-mounted technology-style package (SMT) is shown in Fig. 1-3c. These packages provide a higher circuit density for a package of a given size. Additionally, SMTs have lower noise and improved frequency response characteristics.

SMT components are available in (1) *plastic lead chip carriers* (PPCCs), (2) *small outline integrated circuits* (SOICs), and (3) *leadless ceramic chip carriers* (LCCCs).

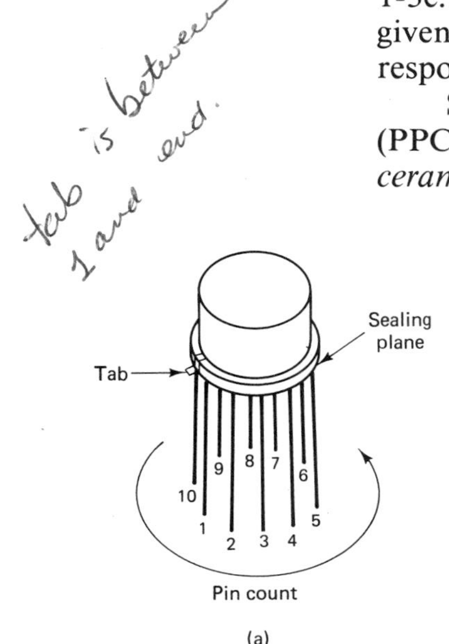

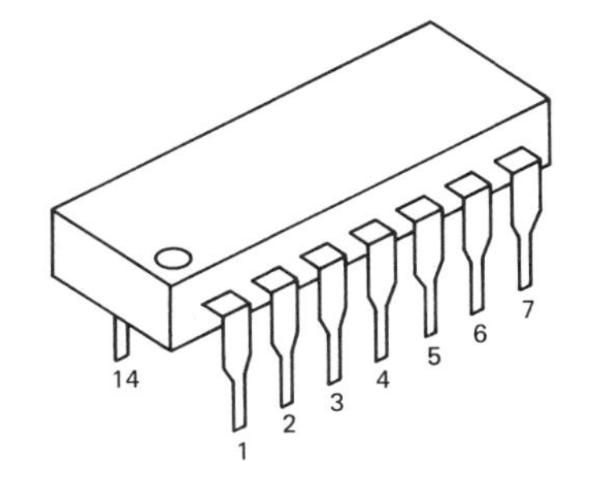

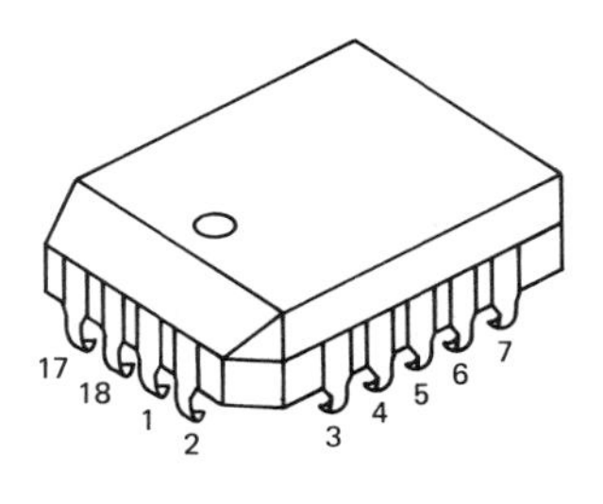

Figure 1-3 Three popular packages used to house integrated circuits: (a) 10-pin TO-5 style metal can type of IC package; (b) 14-pin dual-in-line plastic package (DIP); (c) 18-pin rectangular plastic lead chip carrier (PLCC).

SMTs are available in a variety of pin configurations. A small notch or dot identifies pin 1 as viewed from the top; the entire pin designation is obtained by counting counterclockwise from pin 1.

1-2.4 Op-Amp Packaging Code

Useful information can be obtained from the *packaging code* stamped on the case by the manufacturer. Figure 1-4 shows an 8-pin mini-DIP package with a typical packaging code used by Motorola. (*Note:* Other manufacturers use somewhat different packaging codes.)

The top row of the code gives the following information: first, the letter code for the manufacturer; in this case MC indicates that Motorola is the manufacturer. This is followed by the part identification number (PIN) to identify the device, here a 1741 (or 741) general-purpose linear integrated-circuit op amp. The PIN is followed by the letter P, to indicate a plastic 8-pin mini-DIP package, and C, to mean that this device will operate reliably over the *commercial* temperature range 0 to 70°C.

A second line of the package code, if present, provides a manufacturer's date code. The first two digits indicate the year and the second two digits indicate the week of fabrication. The example of Fig. 1-4 shows that this device was fabricated in the thirty-sixth week of 1984.

1-2.5 Characteristics of the Ideal Op Amp

An ideal op amp is shown in Fig. 1-5. When a dc voltage (V_{CC}, V_{EE}) is applied to the ideal op-amp, zero supply current is drawn into the device. No power is wasted by the op amp and therefore the temperature of an ideal device never increases.

Both input terminals have an ideal input resistance, R_{in}, of infinite ohms. As a result, the input terminals of an ideal op amp draw no input current I_{in}. Since $I_{in} = 0\text{A}$, the op amp can be connected directly to any circuit without fear of circuit loading.

The ideal op amp has a single output terminal with an output resistance R_o equal to zero ohms. If R_o were to equal zero ohms, any load resistor value could be connected to the op amp and large output currents I_o would be possible. The ideal op amp also has infinite gain (A_{OL}). The input signal, E_d, is amplified by the infinite gain of the ideal op amp to produce an output voltage V_o.

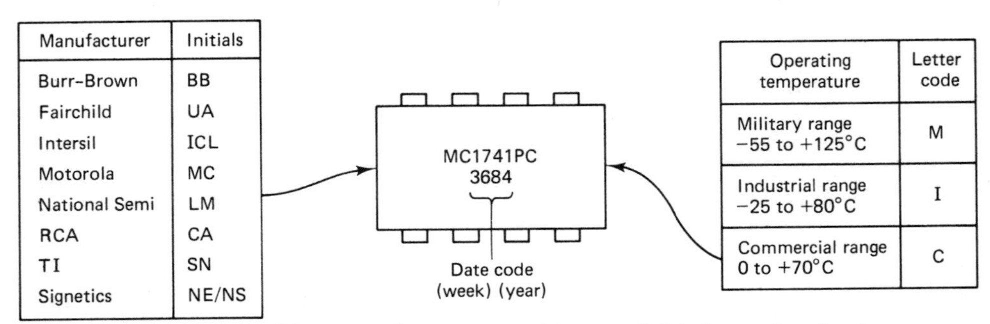

Figure 1-4 The manufacturer provides useful information in the form of a packaging code stamped on the case of the IC.

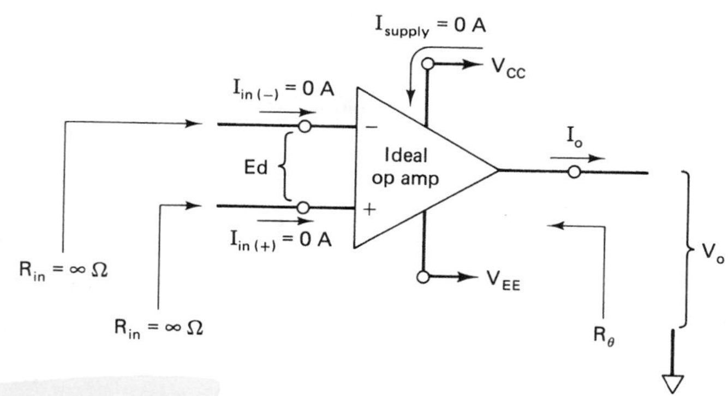

Figure 1-5 Some characteristics and specifications of an ideal operational amplifier.

The ideal op amp is linear and can amplify signals from dc (0 Hz) to frequencies approaching ∞ Hz. The ideal op amp does not exist, but the remarkable performance of modern op amps approaches these ideals in many respects. In the remaining sections of this chapter a practical 741 op amp will be discussed.

1-3 POWERING THE OP AMP

1-3.1 Bipolar Power Supply

A bipolar power supply is created by using two separate dc voltage sources. As shown in Fig. 1-6, the positive terminal of the V_{EE} supply is connected internally to the negative terminal of the V_{CC} supply. This terminal is brought out external to the supply and called the *power supply common* (PSC). The PSC terminal is designated on a circuit schematic with an inverted triangular (ground) symbol. This "reference" terminal is therefore the *point with respect to which all voltages are measured* in an op-amp circuit. V_{CC} and V_{EE} are measured with respect to the power supply common terminal.

1-3.2 Power Supply Quality

Power supplies used for op-amp circuits should be well regulated. For general-purpose experimentation, the typical supply should have a bipolar output of ±15 V as in Fig. 1-6. It should be capable of delivering

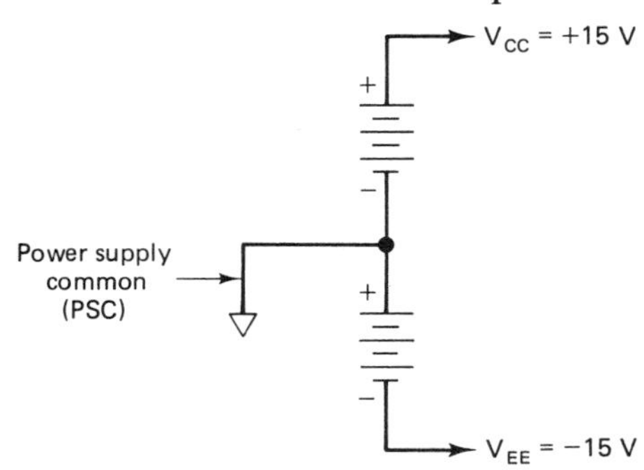

Figure 1-6 A bipolar power supply is represented by two dc voltage sources connected in series aiding. Power supply common (PSC) is brought out as the reference terminal.

currents up to 100 mA. Other common terminal voltages, suitable for most op-amp circuits, are ±12 V or two 9-V batteries. The maximum specified power supply voltage for a commercial-grade 741 op amp is ±18 V.

1-3.3 Applying DC Power to the Op Amp

Figure 1-7a shows the power supply terminals of the 741 op amp properly connected with pin 7 wired to the V_{CC} supply and pin 4 wired to the V_{EE} supply. A simplified representation of the power supply connections is shown in Fig. 1-7b. Each power supply terminal is labeled with the polarity and magnitude of voltage to be applied. Op-amp schematics are also drawn with no power supply terminals as in Fig. 1-7c. While this "shorthand" form is often used to reduce the schematics complexity, the power supply connections are *always* implied and must be wired. A good rule to remember is to connect the power supply terminals first before any other wiring is done.

1-3.4 Power Supply Current and Power Consumption at Idle

When a power supply is first connected to the 741 op amp, the device draws a supply current I_{supply} into pin 7, to establish bias currents for the internal transistors. See Fig. 1-8. Typically, the 741's supply current is 1.7 mA. If no other terminals of the device are being used, 1.7 mA is the supply current required at "idle" (Fig. 1-8).

To obtain the power consumption of the op amp at idle, multiply the total supply voltage (30 V) by the supply current:

$$\begin{aligned} P_{\text{idle}} &= V_{\text{supply}} \times I_{\text{supply}} \\ &= (15\text{ V} + 15\text{ V})(1.7\text{ mA}) = 51\text{ mW} \end{aligned} \tag{1-1}$$

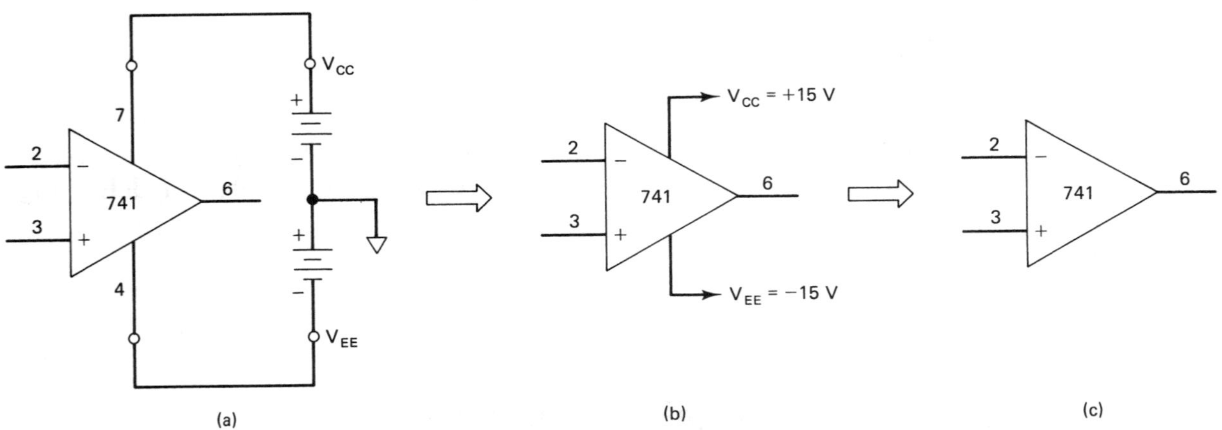

Figure 1-7 Various methods of showing the power supply terminals of the op amp connected to the dc power supply: (a) power supply connections shown completely; (b) simplified representation; (c) power supply connections not shown.

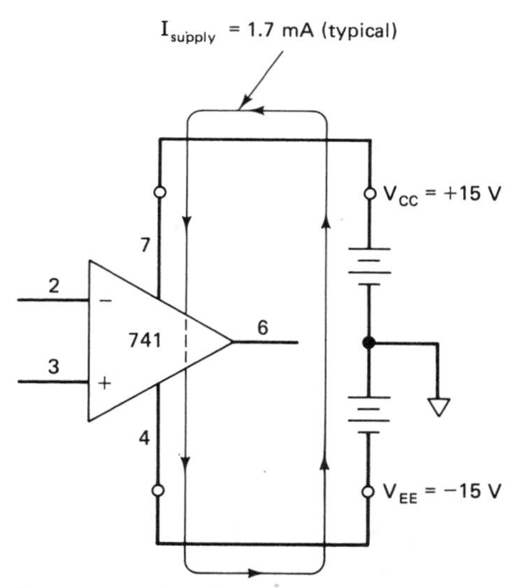

Figure 1-8 As soon as the dc power supply is connected to the op amp a supply current is drawn from it to bias the internal circuitry of the op amp.

All of this power is converted directly into heat within the chip, and the device package will warm slightly.

The 741C's data sheet gives a *maximum power dissipation* specification for the 8-pin mini-DIP package as 0.93 W. Since 51 mW of power is required for biasing, about 5% of the package's power capability is required at idle. For a large system with 100 or more op amps mounted to a printed circuit board, the idle power consumption would exceed 5 W.

1-4 OUTPUT TERMINAL OF THE OP AMP

1-4.1 Output Current Direction

A load on the op amp is modeled by R_L in Fig. 1-9a. One end of R_L is connected to the output terminal 6 of the op amp. The other end goes to the power supply common (PSC) or common terminal identified by a ground symbol. When V_o goes positive with respect to ground in Fig. 1-9a, the output terminal of the op-amp *sources* current I_o to the load. This load current is supplied by the positive supply V_{CC}. Internally, the op amp sources a current to the load when the NPN transistor of the push-pull output stage is conducting (see also Fig. 1-1).

When V_o goes negative as in Fig. 1-9b, the output terminal of the op amp *"sinks"* a current I_o from the load. This load current is supplied by the negative supply V_{EE}. Internally, the op amp sinks a current from the load when the PNP transistor of the push-pull output stage is conducting. The polarity of voltage and direction of the current of the output terminal are under the control of the inputs, as discussed in Section 1-6.4.

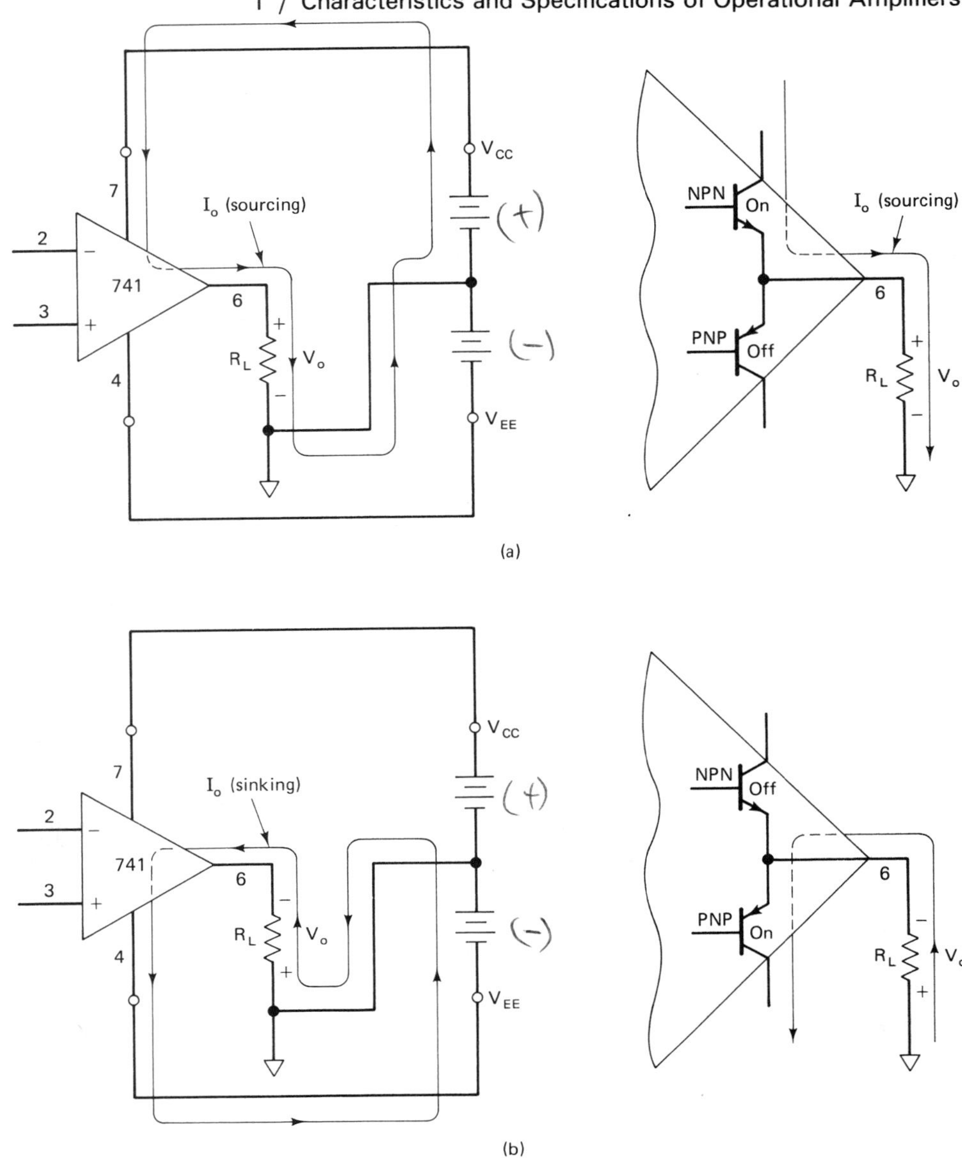

Figure 1-9 Op amp output currents. (a) The output current I_o is furnished by V_{CC} if the op-amp output stage NPN transistor is active. (b) If the PNP transistor is active, output current is furnished by V_{EE} and enters pint 6. In (a), output current is sourced by the op amp; in (b), output current is sinked by the op amp.

1-4.2 Output Voltage Limits

Maximum positive and negative voltage limits of the output are shown as $\pm V_{\text{sat}}$ in Fig. 1-10. V_o can never go above the positive supply voltage V_{CC} or below the negative supply voltage V_{EE}. These primary limits are called the *power supply rails.*

When V_o is positive, the NPN transistor in Fig. 1-9a must have at least 2 V across it to remain on. Therefore, the output terminal cannot be driven closer than about 2 V to the positive supply rail. This upper limit is called *positive saturation* or $+V_{\text{sat}}$. For $V_{CC} = 15$ V, $+V_{o(\max)} \simeq +V_{\text{sat}} = +13$ V.

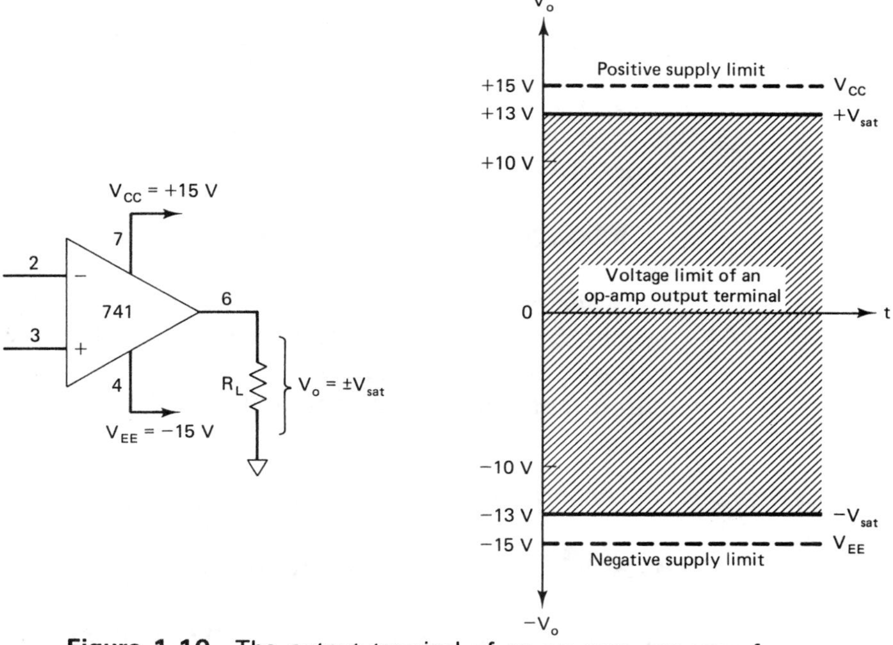

Figure 1-10 The output terminal of an op amp can vary from a maximum lower limit of $-V_{sat}$ to a maximum upper limit of $+V_{sat}$. For power supplies of -15 V, V_O can swing from -13 V to $+13$ V.

When V_o is negative, the PNP transistor in Fig. 1-9b must have at least 2 V across it to remain on. The output terminal cannot be driven closer than about 2 V to the negative supply rail. This lower limit is called *negative saturation* or $-V_{sat}$. For $V_{EE} = -15$ V, $V_{o(max)} \simeq -V_{sat} = -13$ V.

1-4.3 Output Current Limit

Manufacturers of general-purpose op amps such as the 741 guarantee output voltage swings of ± 12 V (min) into loads of 10 kΩ or greater. If R_L is an open circuit, the output current of the op amp is given by

$$I_o = I_L = \frac{\pm V_{o(\min)}}{R_L} = \frac{\pm 12 \text{ V}}{\infty\ \Omega} = 0 \text{ A} \tag{1-2}$$

Since the op amp has to deliver no output current, it dissipates only idle power of about 50 mW. If R_L is changed to 10 kΩ,

$$I_o = I_L = \frac{\pm 12 \text{ V}}{10 \text{ k}\Omega} = \pm 1.2 \text{ mA}$$

The output terminal will then source or sink 1.2 mA of current. This is the recommended value of output current for best performance. As a general rule:

Always select, when possible, a load resistor of 10 kΩ or larger for the output of an op amp.

When R_L is reduced to 2 kΩ, the manufacturer will only guarantee an output voltage swing of ±10 V (minimum). I_o is then ±10 V/2 kΩ = ±5 mA.

An output of ±5 mA is the maximum that a 741 op amp should supply. As a general rule:

Never select a load resistor below 2 kΩ for the output of an op amp.

The output voltage levels cited above are *minimum* values from the manufacturer's data sheet. The output of the op amp is an almost ideal voltage source for load resistance above 2 kΩ. V_o and R_L determine the output current; V_o does not change as I_o changes.

1-5 SHORT-CIRCUIT PROTECTION AT THE OUTPUT

1-5.1 Short-Circuit Current

Normally, when a 0-Ω load or short circuit is connected to the output of a device, large currents are quickly generated. These excessive currents overheat the device and can destroy it. This is not the case with the 741 op amp. Internal circuitry has been provided to monitor and limit the output current to a safe preset value called the *short-circuit current* I_{SC}. For the 741 op amp I_{SC} is limited to 25 mA (actually, 15 to 30 mA, depending on the device).

When the op amp is in "short-circuit protection," it acts as a constant-current source. I_{SC} is independent of the value of load resistor. To learn what values of load resistor will cause I_{SC} to flow, refer to the plot of V_o versus R_L in Fig. 1-11. For small values of load resistors, V_o is set by I_{SC} and R_L. V_o is determined by

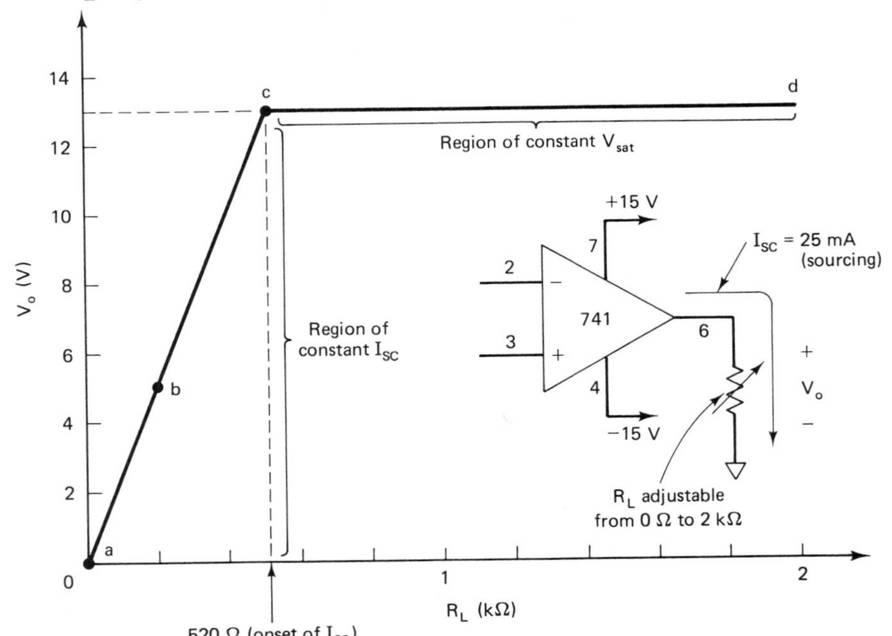

Figure 1-11 Plot of how V_o changes with small values of R_L. When R_L is between 0 Ω (short circuit) and 520 Ω, V_o is limited to the value that I_{SC} can develop. When R_L is above 520 Ω V_o is limited by V_{sat}.

$$V_o = I_{SC}R_L \tag{1-3}$$

Adjust R_L to 0 Ω (short circuit), then V_o = (25 mA) (0 Ω) = 0 V. As shown by point *a* in Fig. 1-11, readjust R_L to say 200 Ω, V_o = (25 mA) (200 Ω) = 5 V (point *b*). The output voltage V_o is being limited by both I_{SC} and the small values of R_L. If R_L is set to 520 Ω, V_o = (25 mA) (520 Ω) = 13 V (point *c*). Therefore, any load resistor whose value is smaller than 520 Ω will initiate the short-circuit protection circuitry of the 741 op amp. Loads from 0 to 520 Ω are shown by line *ac*. Loads above 520 Ω up to 2 kΩ are shown by line *cd* in Fig. 1-11. The output now acts like a constant-voltage source equal to +13 V.

1-5.2 Power Consumption under Short-Circuit Conditions

Figure 1-12a illustrates a short-circuit load (R_L = 0 Ω). Output current is limited to I_{SC} = 25 mA. Since V_o = 0 V, the entire V_{CC} supply voltage will appear across the NPN output transistor. The power dissipated by the op amp in short-circuit protection is

$$P_{D(SC)} = V_{CE} \times I_{SC} = V_{CC}I_{SC}$$
$$= 15\text{ V} \times 25\text{ mA} = 375\text{ mW} \qquad \text{from Section 1-3.4} \tag{1-4}$$

The remaining internal circuitry of the op amp requires about 50 mW of power. The total power dissipation of the package under short-circuit conditions is then

$$P_{D(\text{total})} = P_{D(SC)} + P_{\text{idle}}$$
$$= 375\text{ mW} + 50\text{ mW} = 425\text{ mW} \tag{1-5}$$

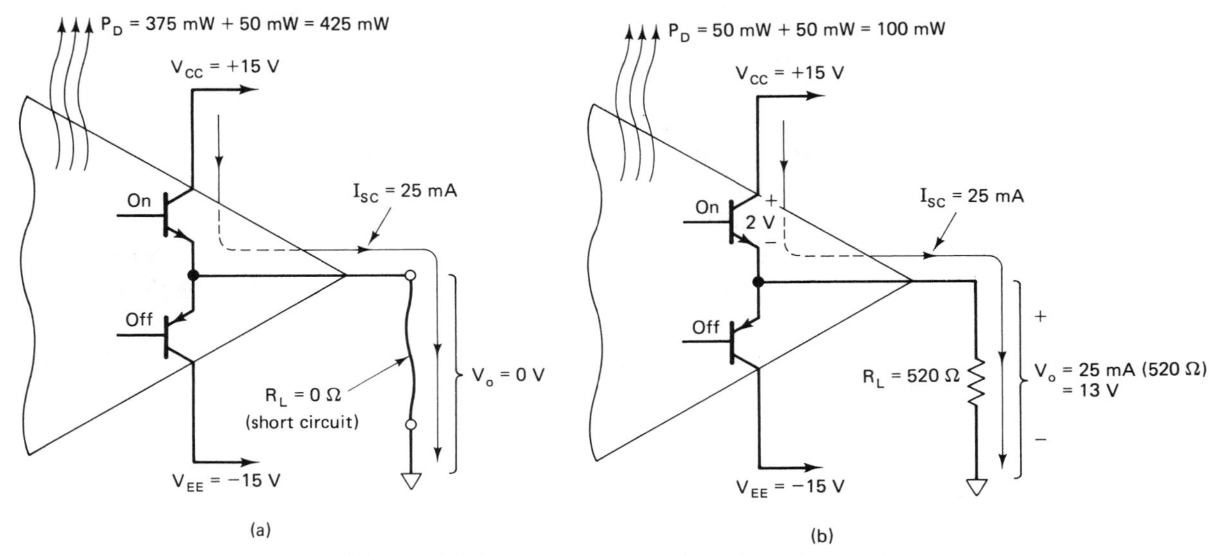

Figure 1-12 An op amp will dissipate less power and operate at a lower temperature as the value of load resistor is increased. (a) The power dissipated by the op amp is 425 mW when the output is shorted; (b) the power dissipated by the op amp is reduced to 100 mW when R_L is increased to 520 Ω.

Since the package is capable of dissipating 0.93 W, the device is protected from destruction.

R_L is changed to 520 Ω in Fig. 1-12b (see also Fig. 1-11). To find op-amp power, first determine V_o.

$$V_o = I_{SC}R_L = (25 \text{ mA})(520\ \Omega) = +13 \text{ V}$$

Solve for V_{CE} internal from

$$\begin{aligned} V_{CE} &= V_{CC} - V_o \\ &= 15 \text{ V} - 13 \text{ V} = 2 \text{ V} \end{aligned} \tag{1-6}$$

The power being dissipated can be found as $P_{D(SC)} = (2 \text{ V})(25 \text{ mA}) = 50 \text{ mW}$ and the total power dissipated by the package will be $P_{D(\text{total})} = 50 \text{ mW} + 50 \text{ mW} = 100 \text{ mW}$. The op amp will operate cooler since it dissipates less power.

1-6 *INPUT TERMINALS*

1-6.1 Differential Input Resistance

The differential input resistance R_i of the 741 op amp is specified by the manufacturer sheet at 2 MΩ. R_i is the resistance measured between either the inverting or the noninverting input terminal, with the other terminal connected to ground. An R_i = 2 MΩ may seem quite large, but op amps with FET inputs (such as analog devices, AD548) exhibit differential input resistances of 1 teraohm ($1 \times 10^{12}\ \Omega$).

1-6.2 Input Bias Currents and Input Offset Voltage

Input bias current I_{bias} is defined as the average value of currents flowing into both inverting $I_{B(-)}$ and noninverting $I_{B(+)}$ input terminals. Typical values of $I_{\text{bias}} = (I_{B(-)} + I_{B(+)})/2 = 80$ nA for the 741 op amp. This magnitude, although small, can be greatly improved. Selection of the AD548 BIFET op amp results in an input bias current of 5 pA = (5×10^{-12} A). As a general rule, the inputs of an op amp draw negligible (zero) current.

An input offset voltage V_{io} exists in all op amps. V_{io} is unavoidable and results from the manufacturer's inability to exactly match input transistors. V_{io} ranges from ±2 to ±6mV for the 741. The smaller the value of V_{io} the better the input transistor matching. An OP-07A op amp has a V_{io} as low as $\pm 10 \mu$V.

V_{io} is modeled as a small dc voltage source in series with the non-inverting input terminal of the op amp. Since V_{io} can be either a positive or negative voltage there is no way of knowing what polarity effect it will have on the output.

1-6.3 Input Voltage Limits

The inputs of a general-purpose op amp such as the 741 must never be brought closer than about 2 V to either power supply rail. For a ±15-V supply, the inputs must be restricted to below +13 V and above −13 V. Exceeding these limits will inhibit proper operation and can result in device damage.

1-6.4 Differential Input Voltage

Op amps amplify the differential input voltage E_d. E_d is the *difference* voltage between the noninverting (+) and the inverting (−) inputs. Both input terminals can have voltages that are positive (above ground), negative (below ground), or straddle ground potential. This is not important. What is important is the difference voltage between them, E_d can be determined from

$$E_d = V_{(+)} - V_{(-)} \tag{1-7}$$

$V_{(+)}$ is the voltage measured at the noninverting input with *respect to ground.* $V_{(-)}$ is the voltage measured at the inverting input *with respect to ground.*

EXAMPLE 1-1

Calculate the differential input voltages E_d in Fig. 1-13.

SOLUTION: (a) From Eq. (1-7) and Fig. 1-13a,

$$E_d = V_{(+)} - V_{(-)} = (5.1\ \text{V}) - (5.0\ \text{V}) = +0.1\ \text{V}$$

The noninverting input is *above* (more positive than) the inverting input. Hence the sign of E_d is (+).

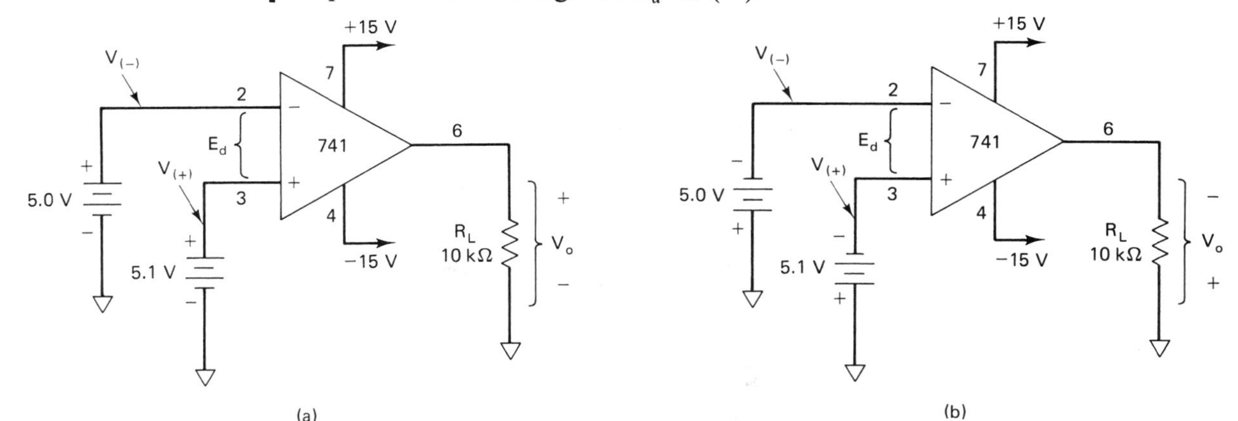

Figure 1-13 The output voltage polarity is set by the polarity on E_d. (a) If the differential input voltage E_d is positive, the output voltage V_o will be positive; (b) if the differential input voltage E_d is negative, the output voltage V_o will be negative.

(b) From Eq. (1-7) and Fig. 1-13b:

$$E_d = V_{(+)} - V_{(-)} = (-5.1\text{ V}) - (-5.0\text{ V}) = -0.1\text{ V}$$

The noninverting input is *below* (more negative than) the inverting input. Hence the sign of E_d is (−).

The following can be concluded from Example 1-1:

1. E_d is positive if the noninverting input terminal is *above* (more positive than) the inverting input.
2. E_d is negative if the noninverting input terminal is *below* (more negative than) the inverting input.

1-7 OPEN-LOOP VOLTAGE GAIN

1-7.1 Definition of A_{OL}

Manufacturers' data sheets define the open-loop voltage gain of A_{OL} of an op amp as the *large-signal voltage gain.* For the 741, the value of A_{OL} is given typically as 200,000.

An op amp amplifies the differential input voltage E_d by its open-loop gain A_{OL} to produce the single-ended output voltage V_o, where

$$V_o = (A_{OL}) \times E_d \tag{1-8}$$

When using Eq. (1-8) to predict the value of V_o, never trust your *mathematically* correct answer. There are two practical limitations imposed by the output terminal that override the math:

1. V_o can never exceed a maximum value of $\pm V\text{sat} = \pm 13$ V, (± 15 V power supply).
2. I_o cannot exceed its short-circuit output current $I_{SC} = 25$ mA.

These principles are illustrated by the following examples.

1-7.2 Open-Loop Gain and Output Voltage

The polarity of V_o depends on the polarity of E_d in Eq. (1-8) and E_d is found from Eq. (1-7). We show how to calculate V_o using Fig. 1-13 in the following examples.

EXAMPLE 1-2

Find V_o for both circuits of Fig. 1-13.

SOLUTION: (a) From Example 1-1a, $E_d = +0.1$ V for Fig. 1-13a, V_o is found from Eq. (1-8): $V_o = (A_{OL})\,E_d = (200{,}000)\,(0.1\text{ V}) = +20{,}000$ V. It is actually impossible for V_o to reach +20,000 V, as Eq. (1-8) predicts mathematically. V_o can only rise to a limit of $+V_{sat} = 13$ V ($V_{CC} = +15$ V on Fig. 1.13a).

(b) From Example 1-1b, $E_d = -0.1$ V. Using Eq. (1-8), we obtain

$$V_o = (A_{OL})\, E_d = (200{,}000)\,(-0.1 \text{ V}). = -20{,}000 \text{ V}$$

V_o is again limited at saturation to a voltage $-V_{sat} = -13$ V.

1-7.3 Short-Circuit Current and Output Voltage

The second practical limitation on output voltage V_o (see Section 1-7.1) is that I_o cannot exceed I_{SC}. $I_{SC} = +25$ mA for the 741 op amp and is assumed to exist if R_L is below 520 Ω.

EXAMPLE 1-3

Find V_o in Fig. 1-13 (for both circuits) if R_L is replaced by a 100-Ω resistor.

SOLUTION: (a) From Example 1-2a, $E_d = +0.1$ V and V_o heads for $+V_{sat}$. Since $R_L = 100\Omega$ (less than 520 Ω) use Eq. (1-8) to find V_o. $V_o = I_{SC}R_L = (25 \text{ mA})100\ \Omega = +2.5$ V. The op amp is sourcing I_{SC}.

(b) From example 1-2b, $E_d = -0.1$ V and V_o heads for $-V_{sat}$. Since $R_L = 100\ \Omega$ (less than 520 Ω) $V_o = I_{SC}\, R_L = (-25 \text{ mA})\ 100\ \Omega = -2.5$ V. The op amp is sinking I_{SC}.

We conclude from Example 1-3 that the output voltage V_o will be at $\pm V_{sat}$ if R_L is above 520 Ω and that V_o will be limited by Eq. (1-8) if R_L is below 520 Ω.

1-8 FREQUENCY RESPONSE CHARACTERISTICS

The ideal op amp of Section 1-2.5 should have an infinite open-loop gain and amplify equally signal frequencies from 0 Hz (dc) to ∞ Hz. As shown in Fig. 1-14, a practical op amp like the 741 has neither $A_{OL} = \infty$ nor equal amplification for all frequencies. Figure 1-14 is taken from the 741 data sheet.

At very low frequencies (approaching dc), open-loop gain is constant at 200,000. However, as input signal frequencies increase, above 5 Hz, gain decreases at a constant rate until the gain is 1 or 0 dB at f_T, the *unity-gain frequency*. Since the low-frequency end of the response curve is 0 Hz (dc), f_T defines the *small-signal unity-gain bandwidth* of the 741 and is equal to 1 MHz.

The product of voltage gain A_{OL} and frequency is a constant for frequencies above 5 Hz and is given by

$$f_T = (\text{unity-gain bandwidth}) = A_{OL}\, f \tag{1-9}$$

where f is the input signal frequency.

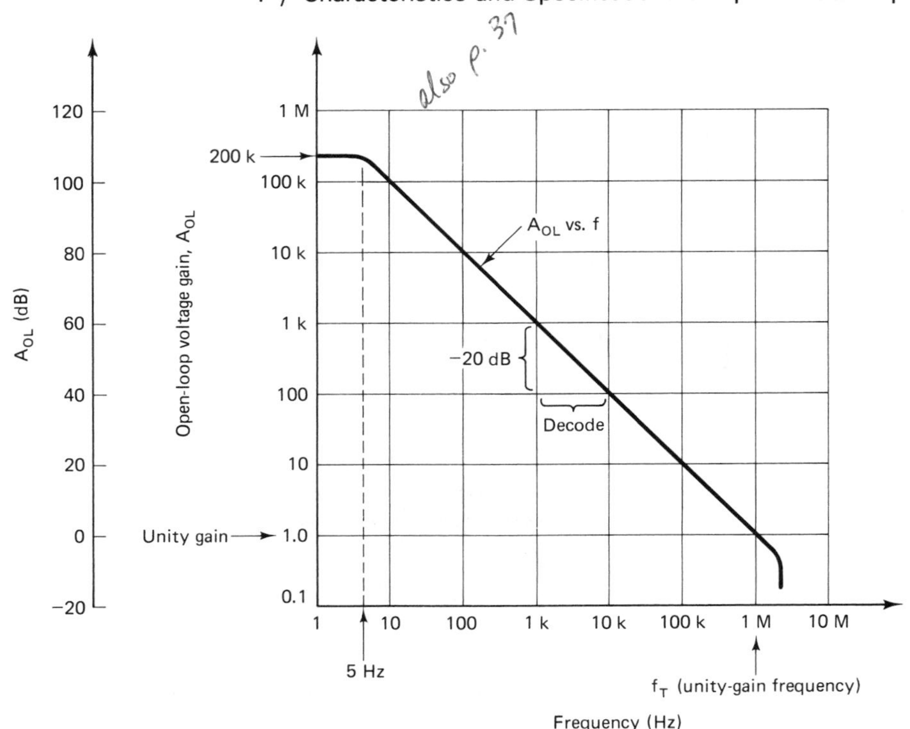

Figure 1-14 Open-loop voltage gain (A_{OL}) versus frequency for the 741 general-purpose op amp.

EXAMPLE 1-4

Find the open-loop voltage gain of a 741 op amp if the input signal frequency, f, is (a) 100 kHz; (b) 10 kHz; (c) 1kHz.

SOLUTION: Locate $f_T = 1$ MHz where $A_{OL} = 1$ in Fig. 1-14. Substitute into Eq. (1-9) $f_T = 1 \text{ MHz} = A_{OL} \times f$. Thus Eq. (1-9) can be written as $A_{OL} = 1 \text{ MHz}/f$. Then

$$\text{(a) } A_{OL} = \frac{1 \text{ MHz}}{100 \text{ kHz}} = 10$$

$$\text{(b) } A_{OL} = \frac{1 \text{ MHz}}{10 \text{ kHz}} = 100$$

$$\text{(c) } A_{OL} = \frac{1 \text{ MHz}}{1 \text{ kHz}} = 1000$$

Example 1-4 shows that as frequency *increases* from 1 to 10 to 100 kHz, A_{OL} *decreases* or rolls off from 1000 to 100 to 10. An increase of frequency by a factor of 10 is called a *decade.* For each decade increase in frequency, gain drops by a factor of 10 or −20 dB. For these reasons, the frequency response (A_{OL} versus f) of an op amp is said to roll off by −20 dB per decade.

For high-frequency applications, select an op amp with as large a value of f_T as possible. Newer op amps such as the LF411 or the CA3140 have values of f_T of 4.0 MHz and 4.5 MHz, respectively. Select an OP-17 with an f_T = 30 MHz or an HA2542 with an f_T = 45 MHz for even higher-frequency applications. All of the devices above have the same basic pin pattern as those of the industrial standard 741.

1-9 SLEW RATE

Slew rate, SR, is the speed at which the output voltage of an op amp can change in response to an abrupt change of input voltage. SR has units of volts per microsecond and is given by

$$\text{SR} = \frac{\Delta V_o}{\Delta T} \tag{1-10}$$

The maximum slew rate of the 741 is typically 0.5 V/μs and is shown graphically in Fig. 1-15. Distortion will result if the output is asked to change faster than the maximum slew rate of the op amp.

EXAMPLE 1-5

Find the time for V_o to change by 10 V if the slew rate of an op amp is 1 V/μs.

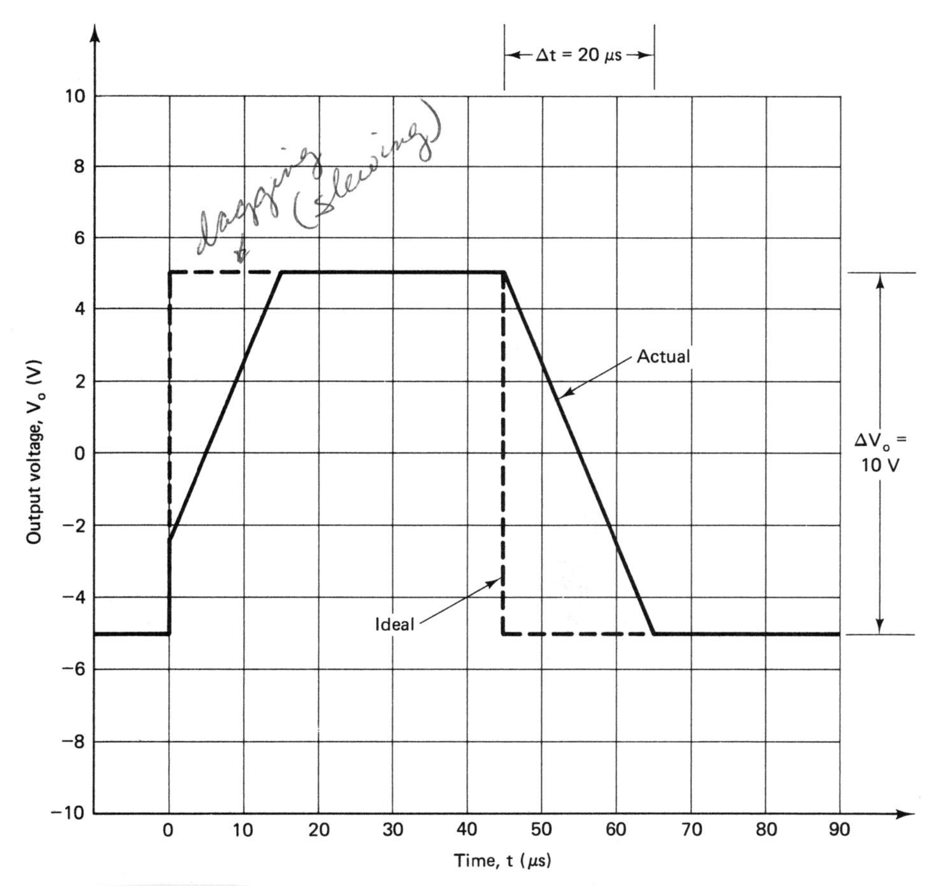

Figure 1-15 The output terminal of the 741 op amp can change no faster than 0.5 V/μs. It takes 20 μs for the output voltage to change 10 V.

SOLUTION: Rewriting Eq. (1-10) gives us

$$\Delta T = \frac{\Delta V_o}{\text{SR}} = \frac{10\text{ V}}{1\text{ V}/\mu\text{s}} = 10\ \mu\text{s}$$

For high-speed applications, select an op amp with improved slew rate characteristics. The OP-17 op amp has a very high maximum slew rate of 60 V/μs, while the HA2542 has a slew rate of as high as 375 V/μs. In Chapter 2 the op amp is used as a voltage comparator in practical applications. The op amp will operate open loop and should have a high slew rate. For this reason we use the 301 op amp, which is a fast version of the 741, having a slew rate of 10 V/μs.

End of Chapter Exercises

Name: ______________________

Date: __________ **Grade:** ___

A. FILL-IN THE BLANKS

Fill in the blanks with the best answer.

1. The op amp's function block diagram shows three stages; __________, __________, and __________ .

2. The op amp symbol is a __________ that points in the direction of signal flow.

3. A general purpose op amp such as the __________ has a minimum of __________ terminals.

4. An ideal op amp has __________ input resistance and __________ output resistance.

5. General purpose op amps use __________ power supplies whose voltage is measured with respect to power supply common.

6. Differential input voltage E_d is the difference between __________ and __________ . Both are measured with respect to __________ .

7. Idle power drawn by a 741 has a typical value of __________ .

8. The op amp limits output current to a maximum of __________ .

9. Output voltage depends on both __________ and __________ .

10. The maximum values for output voltage are limited to values called __________ and __________ .

B. TRUE/FALSE

Fill in **T** if the statement is true, and **F** if any part of the statement is false.

1. ______ When R_L equals 200Ω, $I_L = V_o/R_L$.
2. ______ When R_L equals 200Ω, $V_o = I_{SC} \times R_L$.
3. ______ When $E_d = 0.1$V and $\pm$V = 15 v, output V_o goes to 20,000 V.
4. ______ When $E_d = -0.1$V and $\pm$V = $\pm$15V, $V_o = -13$V.
5. ______ When the output of an op-amp is short circuited it is heated by an idle power of 50mW plus a load current power of $I_{SC} \times V_{CC}$.

C. CIRCLE THE CORRECT ANSWER

Circle the correct answer for each statement.

1. Typical bias currents for the 741 are (50 nA, 5nA).
2. Op amps with FET inputs have (higher, lower) input resistance than bipolar junction transistor inputs.
3. The op amp's output is a (single-ended, differential) voltage.
4. Input voltage at the non-inverting input should never go (above V_{CC}, below V_{EE}).
5. Open loop gain of an op amp (decreases, increases) by 20 dB as frequency is decreased by a decade.

D. MATCHING EXERCISE

Match the name or symbol in column **A** with the statement that matches best in column **B**.

	Column A		Column B
1. ______	13 V	a.	Open loop gain
2. ______	50 mW	b.	I_{SC}
3. ______	500 nA	c.	$+V_{sat}$
4. ______	25 mA	d.	I_{bias}
5. ______	A_{OL}	e.	Idle power

PROBLEMS

1-1. **(a)** Name the three basic stages internal to a general-purpose op amp.
(b) Describe briefly the function of each amplifier stage.

1-2. Draw the symbol for an op amp and label each terminal.

1-3. **(a)** Name the three package styles popular for housing op amps.
(b) Which of the styles named in part (a) is chosen most often, and why?

1-4. Name the three types of surface-mounted packages mentioned in this chapter and give the letter designation of each.

1-5. The package code on Fig. 1-4 is replaced with the following MC1301PM/1277. List
(a) the manufacturer,
(b) the PIN, and the
(c) usable temperature range.
(d) Indicate the week and year of manufacture.

1-6. The following questions apply to an ideal op amp.
(a) What is R_{in}?
(b) How much current do the input terminals require?
(c) What is R_o?
(d) What is the voltage gain A_{OL}?

1-7. What is the maximum power supply voltage that can be connected to a commercial-grade 741?

1-8. Determine the power dissipated by a 741C op amp at idle if the bipolar supply used is
(a) ± 9 V;
(b) ± 12 V;
(c) ± 18 V. In each case assume that I_{supply} remains constant at 1.7 mA.

1-9. The output voltage V_o is a negative value.
(a) Is the op amp sinking or sourcing current?
(b) Which push-pull transistor in the output stage of the op amp is not conducting?

1-10. Repeat Problem 1-9 for a positive value of V_o.

1-11. Estimate the value of both $+V_{sat}$ and $-V_{sat}$ if the power supply for a 741C op amp is
(a) ± 9 V;
(b) ± 12 V;
(c) ± 18 V.

1-12. **(a)** What is the short-circuit current for a 741 op amp?
(b) Will I_{SC} flow if $R_L = 300\ \Omega$?
(c) What will V_o equal if $R_L = 100\ \Omega$?

1-13. What is the value of R_L that will just initiate short-circuit protection for the 741 if the dc supply is ± 15 V?

1-14. Determine the total power dissipated by a 741 op amp in short-circuit protection when R_L is
(a) 500 Ω;
(b) 300 Ω;
(c) 100 Ω.
(d) Does the temperature of the op amp increase or decrease as R_L is lowered?

1-15. What is the differential input resistance R_i of a 741 op amp?

1-16. Assume that power to 741 op amp is $V_{CC} = 12$ V and $V_{EE} = -12$ V. What is the maximum value of input signal that can be applied to either input terminal?

1-17. Determine both the magnitude and polarity of E_d for Fig. 1-13a if $V_{(+)} = +1$ V and $V_{(-)}$ is
(a) 0 V;
(b) +15 V;
(c) −0.6 V.

1-18. Repeat Problem 1-17 if $V_{(+)}$ is changed to −0.3 V.

1-19. Solve for the output voltage V_o for each of the values of E_d in Problems 1-17 and 1-18. Assume that a 741 op amp is powered by a ±18-V supply.

1-20. Solve for V_o for each of the values of E_d in Problems 1-17 and 1-18. Assume that a 741 op amp is used with a load of 150 Ω.

1-21. Use Fig. 1-14 to determine A_{OL} if the input signal frequency is
(a) 100 Hz;
(b) 10 Hz.

1-22 Find the time for V_o to change from +6 V to −6 V if the slew rate of the op amp is
(a) 10 V/μs;
(b) 60 V/μs;
(c) 375 V/μs.

Laboratory Exercise 1

Name: ____________________

Date: __________ **Grade:** __

OP-AMP CHARACTERISTICS

OBJECTIVES: Upon completion of this laboratory exercise on the characteristics of op amps, you will be able to (1) interpret the manufacturer's packaging code; (2) measure $\pm V_{sat}$; (3) measure both I_{load} and I_{supply}; (4) measure short-circuit current; (5) predict both the magnitude and polarity of V_o; and (6) measure slew rate.

REFERENCES: Chapter 1 and 741 Data Sheet

PARTS LIST

1	741 op amp	1	100-Ω $\frac{1}{2}$-W resistor
1	301 op amp	4	10-k Ω resistors

Procedure A: Manufacturer's Package Code

1. Refer to the packaging code stamped on your 741 op amp and the data sheet provided as an appendix. Answer the following questions:

(a) Name the device manufacturer. __________ .
(b) Give the exact PIN (part identification number).

__________ .

(c) What letters/numbers are used to identify the case style you are

using? __________ .
(d) What is the operating temperature range of your device?

__________ .

(e) When was your device manufactured?

Week __________ Year __________ .

2. The terminal numbers shown on Fig. L1-1 are for a 741 op amp packaged in an 8-pin mini-DIP case. If the device you are using is

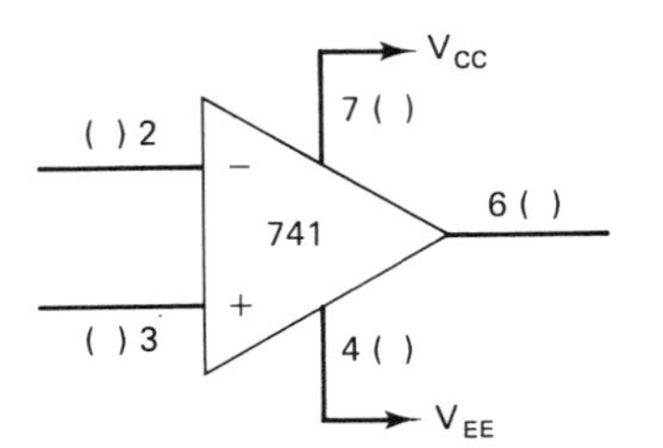

Figure L1-1 Pin numbers.

packaged in a different-style case, use the appropriate data sheet and provide the correct terminals for your device in the parentheses supplied beside each pin.

Procedure B: Measuring $\pm V_{sat}$

3. Wire the circuit shown in Fig. L1-2. Measure the power supply voltages. V_{CC} = ________ , V_{EE} = ________ . (*Note:* Since these voltages are measured with respect to ground, be sure to include the correct polarity.) Also measure the total voltage between points x and y. $V_{CC} + |V_{EE}|$ = ________ .

4. Estimate both $\pm V_{sat}$ using the power supply values obtained in step 3.

5. Measure $V_o = +V_{sat}$ in Fig. L1-2. $+V_{sat}$ = ________ . To measure $-V_{sat}$, first remove the wire connecting pin 3 to point x in Fig. L1-2 and wire pin 3 to point y. Measure $V_o = -V_{sat}$ = ________ . Do the estimated and measured values compare favorably? ________ . Explain. ________________

6. Short both input terminals together and connect them to ground as shown in Fig. L1-3. From the data sheets for a 741, A_{OL} = 200,000 (200 k) as a typical value and E_d in Fig. L1-3 should be 0 volts.

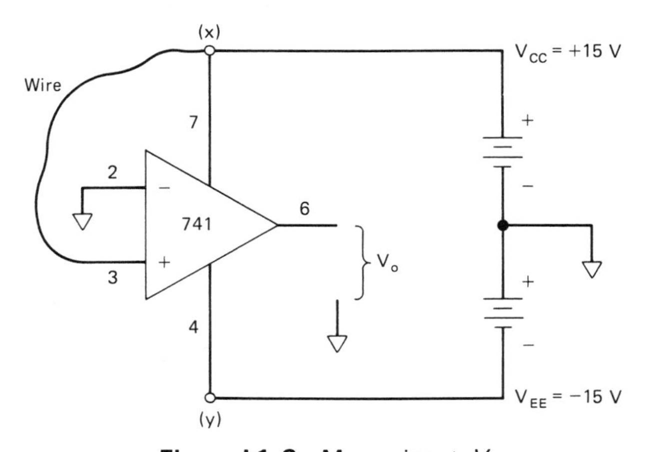

Figure L1-2 Measuring $\pm V_{sat}$.

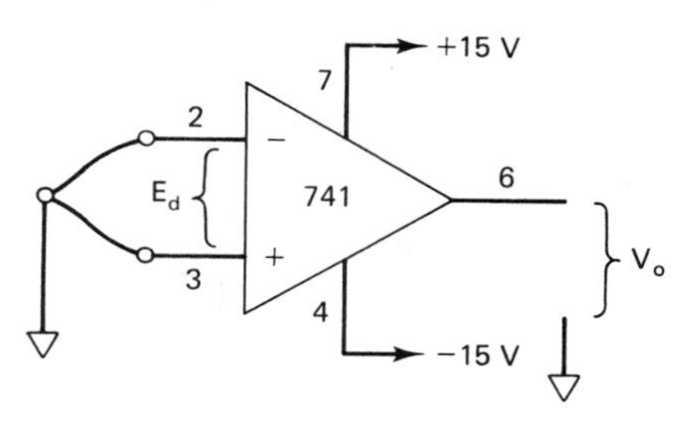

Figure L1-3

Solve for V_o using the equation $V_o = A_{OL}E_d$. Theoretical V_o = __________. Measure and record the output voltage. V_o = __________. Explain why V_o is at $+V_{sat}$ (or $-V_{sat}$) and not at 0V? ______________________________.

Procedure C: Measuring Both I_{load} and I_{supply}

7. Wire the circuit shown in Fig. L1-4. Obtain the typical value of supply current from the manufacturer's data sheet, at the end of this lab. $I_{supply(typ)}$ = __________. Measure $I_{supply(+)}$ at points a–a' and $I_{supply(-)}$ at points b–b'. $I_{supply(+)}$ = __________, $I_{supply(-)}$ = __________. For a device to be within its specifications, the measured value of I_{supply} should be *equal to* or *less than* the data sheet value. Is your device within "spec"? __________.

8. Figure L1-5 illustrates the op amp "sourcing" I_L current where the following equation is true: $I_{supply(+)} = I_L + I_{supply(-)}$. Wire Fig. L1-5 with $R_L = 10\ k\Omega$ and the input terminals connected as shown. Measure I_L

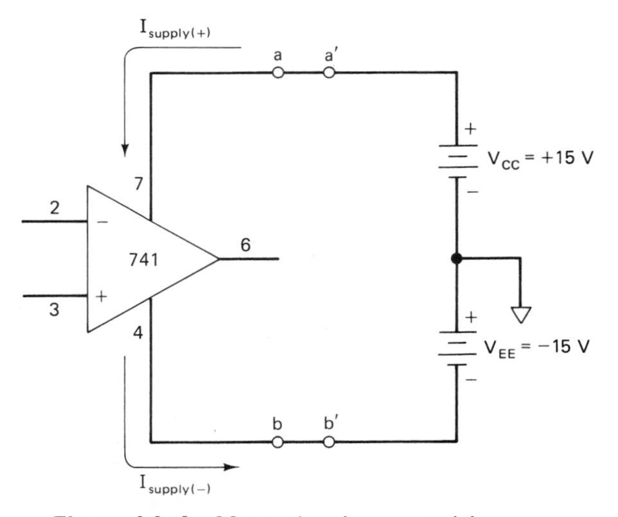

Figure L1-4 Measuring $I_{supply\ (+)}$ and $I_{supply\ (-)}$.

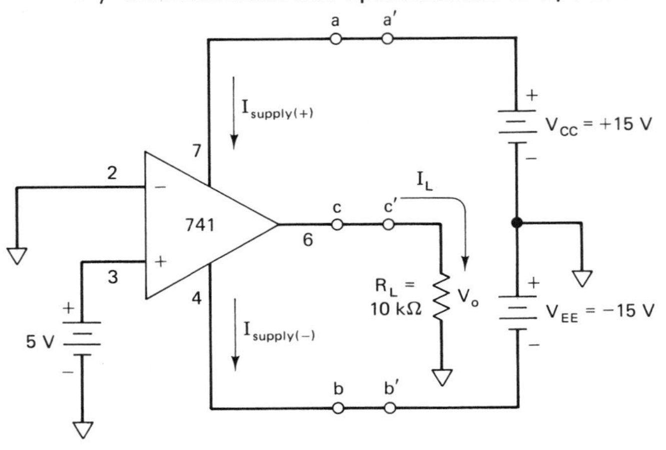

Figure L1-5 "Sourcing" I_L.

at c–c'. I_L = __________. Calculate $I_{supply(+)}$ using the measured value of $I_{supply(-)}$ from step 7. $I_{supply(+)}$ = __________. Measure $I_{supply(+)}$ at points a–a'. $I_{supply(+)}$ = __________. Do the calculated and measured values of $I_{supply(+)}$ compare favorably? __________. The output terminal of the op amp is "sourcing" current. Is V_o at $+V_{sat}$? __________.

9. Figure L1-6 illustrates the op amp "sinking" I_L current where the following equation is true: $I_{supply(-)} = I_L + I_{supply(+)}$. Wire Fig. L1-6 and again measure I_L at points c–c'. I_L = __________. Calculate $I_{supply(-)}$ using the measured value of $I_{supply(+)}$ from step 7. $I_{supply(-)} = I_L + I_{supply(+)}$ __________. Measure $I_{supply(-)}$ at points b–b'. $I_{supply(-)}$

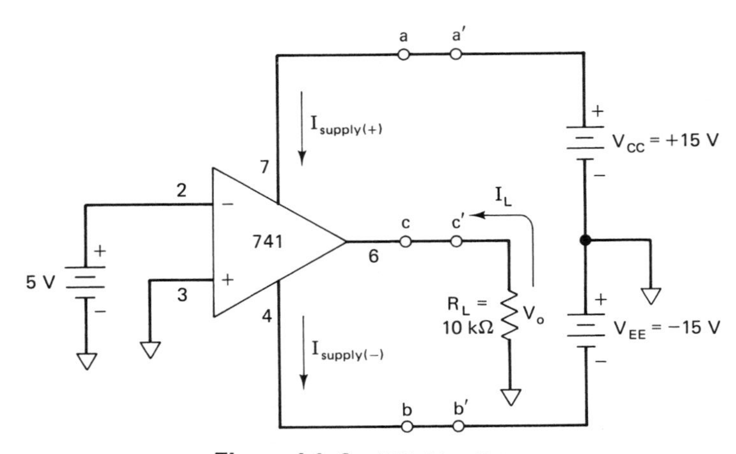

Figure L1-6 "Sinking" I_L.

= __________ . Do the results compare favorably? __________ .

When the op amp is "sinking" current, is V_o at $\pm V_{sat}$? __________ .

Procedure D: Short-Circuit Current Measurements

10. Obtain the typical value of short-circuit current I_{SC} for your device from the data sheet. $+I_{SC}$ = __________; $-I_{SC}$ = __________ .

11. Figure L1-7 illustrates the op amp wired to deliver positive I_{SC}. Measure $+I_{SC}$ at points c–c' and show the direction of the current flow on Fig. L1-7. $+I_{SC}$ = + __________ . Does this measured value compare favorably with the data sheet value? __________ .

12. To measure $-I_{SC}$, remove the wire from pin 3 and V_{CC} and connect pin 3 to V_{EE}. Measure $-I_{SC}$ at points c–c'. $-I_{SC}$ = − __________ . Does the minus sign indicate a change in current direction? __________ .

Procedure E: Predicting the Magnitude and Polarity of V_o

13. Both the magnitude and polarity of V_o are controlled by the input terminals of the op amp. The magnitude of $V_o = A_{OL}E_d$ and the polarity of V_o depend on the polarity of E_d, where $E_d = V_{(+)} - V_{(-)}$. To demonstrate how the input terminals control V_o, start by building Fig. L1-8. This circuit is a simple resistor divider network to establish two input voltages V_1 and V_2 of *approximately* +5 V and −5 V, respectively.

14. Measure both V_1 and V_2 with respect to ground, and record their values. V_1 = __________ ; V_2 = __________ . Use these voltages as inputs to the next three op-amp circuits as well as the equations above to predict that V_o can never exceed the practical limits of $\pm V_{sat}$.

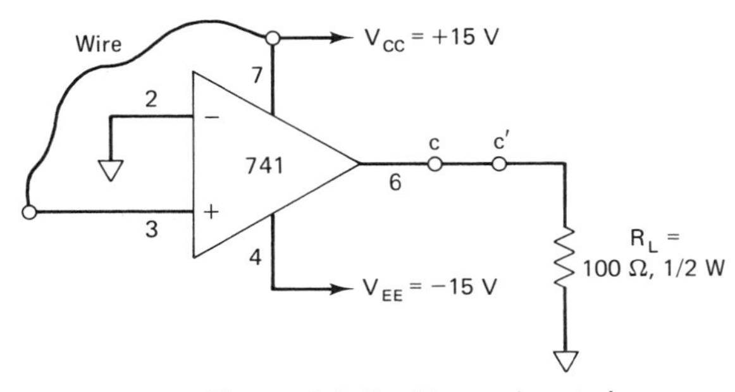

Figure L1-7 Measuring $+I_{SC}$.

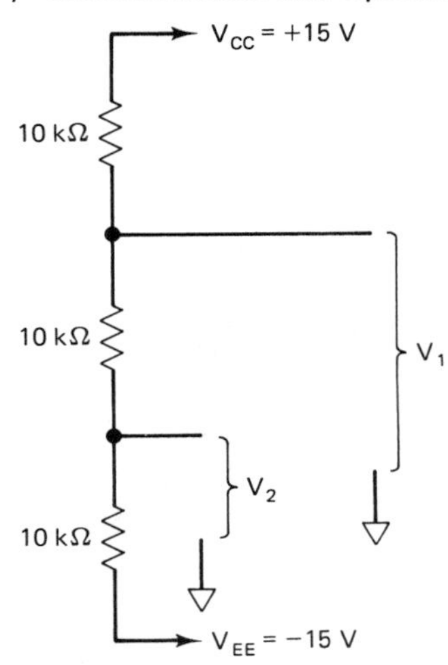

Figure L1-8 Input voltages.

15. Wire Figure L1-9. Ground the (+) input terminal and connect V_2 (from Fig. L1-8) to the (−) input terminal. Calculate both the magnitude and polarity of E_d and V_o. $E_d = V_{(+)} - V_{(-)} =$ __________ ; $V_o = A_{OL}E_d =$ __________ . Will V_o be at $\pm V_{sat}$? __________ .

16. Measure and record the magnitude and polarity of V_o. $V_o =$ __________ .

17. Wire Fig. L1-10. Again ground the (+) input and connect V_1 (from Fig. L1-8) to the (−) input. Again calculate both the magnitude and polarity of E_d and V_o __________ .

18. Measure and record the magnitude and polarity of V_o. $V_o =$ __________ .

19. Wire Fig. L1-11 to observe the effect of two input voltages applied simultaneously to the op amp. Use Fig. L1-8 to apply V_1 to the (+) input and V_2 to the (−) input terminals. Calculate both the magnitude and polarity of E_d and V_o. $E_d = V_{(+)} - V_{(-)} =$ __________ ; $V_o = A_{OL}E_d =$ __________ .

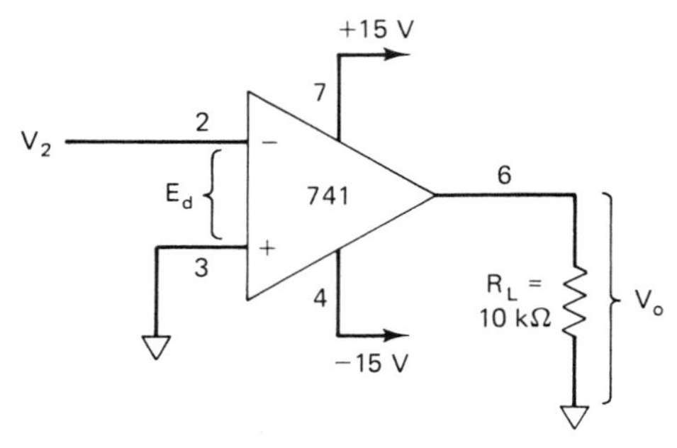

Figure L1-9

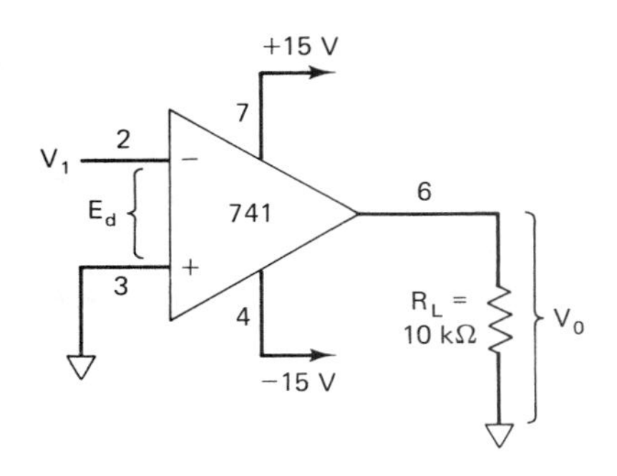

Figure L1-10

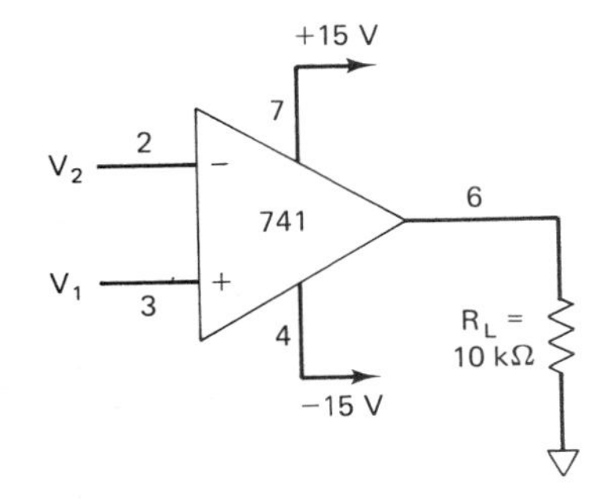

Figure L1-11

20. Measure and record the magnitude and polarity of V_o. V_o = ___________.

Procedure F: Measuring the Slew Rate

21. The circuit shown in Fig. L1-12 will be used to measure the slew rate of both the 741 and 301 op amps. Wire Fig. L1-12 as shown. Set $V_i = \pm 5$ V (peak) [10 V (p-p)] at a square wave test frequency of 10 kHz. Connect channel 1 of the CRO (cathode-ray oscilloscope) to the output terminal, monitoring V_o. Switch the CRO to 5 V/div with a time base of 10μs/div (see Fig. L1-13). Obtain the waveshape displayed in Fig. L1-13.

22. Measure Δt, the time for ΔV_o to change a total of 26 V. Compute the value of slew rate from SR = $\Delta V_o/\Delta t$ in units of $V/\mu s$. SR

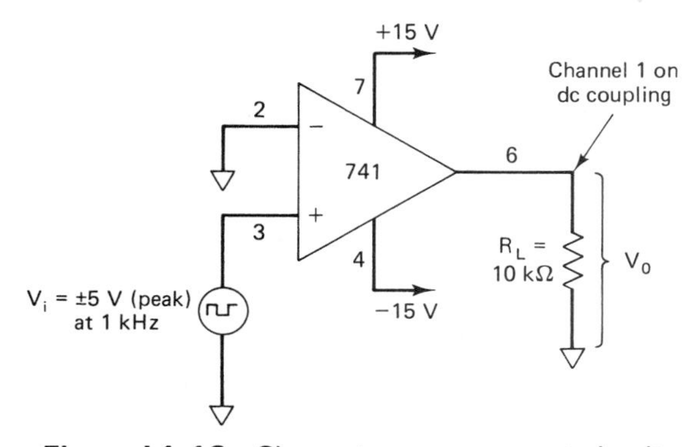

Figure L1-12 Slew rate measurement circuit.

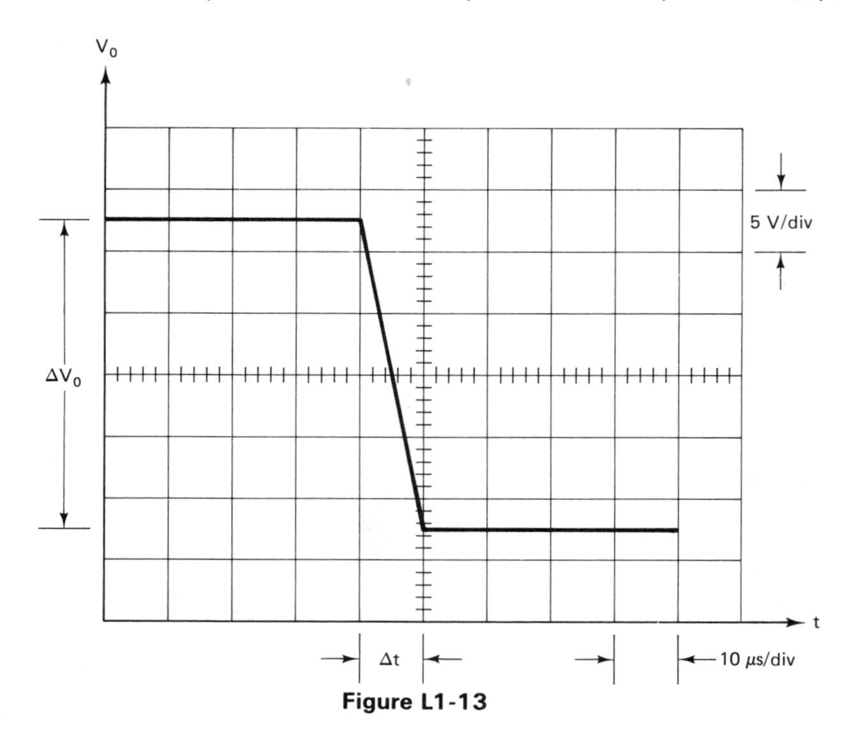

Figure L1-13

= __________ . V/μs. What is the data sheet value of SR for the 741 op amp? SR = __________ .

23. Replace the op amp in Fig. L1-12 with the faster 301 that has a slew-rate value of 10 V/μs. Maintain all conditions identical to the previous test except change the CRO time base to 1μs/div. Sketch, in Fig. L1-14, the waveform to measure slew rate.

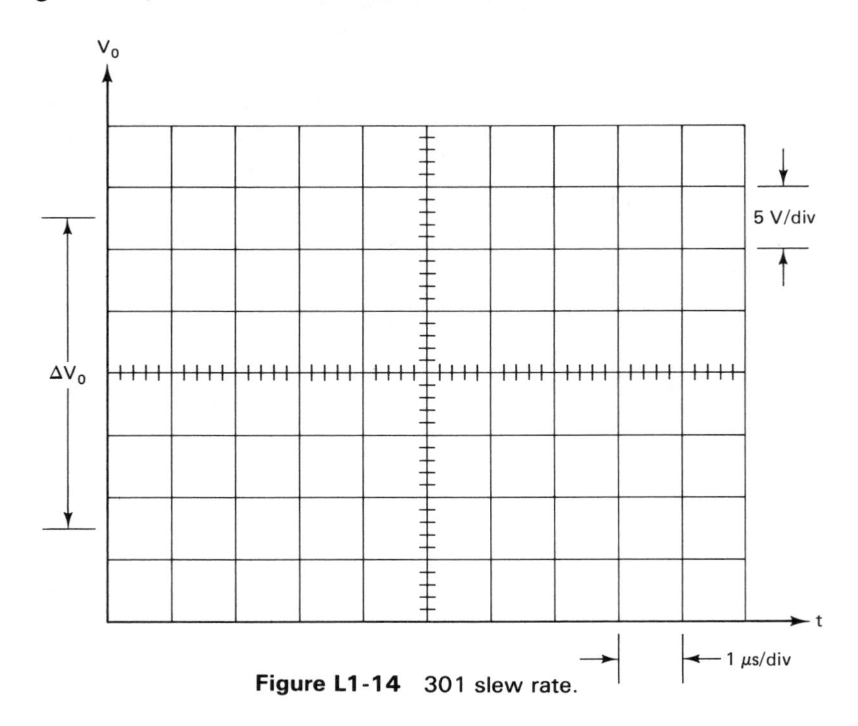

Figure L1-14 301 slew rate.

24. Measure the time Δt for ΔV_o to change a total of 10 V. Compute the value of slew rate. SR. = __________ .

25. Does this measured value compare favorably with the manufacturer's specification? __________ .

CONCLUSION:

Low Cost High Accuracy IC Op Amps

AD741 SERIES

FEATURES

Precision Input Characteristics

- Low V_{os}: 0.5mV max (L)
- Low V_{os} Drift: 5μV/°C max (L)
- Low I_b: 50nA max (L)
- Low I_{os}: 5nA max (L)
- High CMRR: 90dB min (K, L)

High Output Capability

- A_{ol} = 25,000 min, 1kΩ load (J, S) T_{min} to T_{max}
- V_o = ±10V min, 1kΩ load (J, S)

AD741 SERIES FUNCTIONAL DIAGRAMS

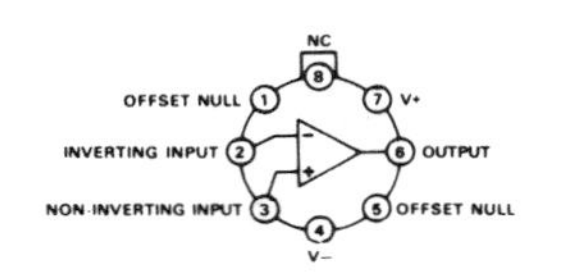

TO-99
TOP VIEW

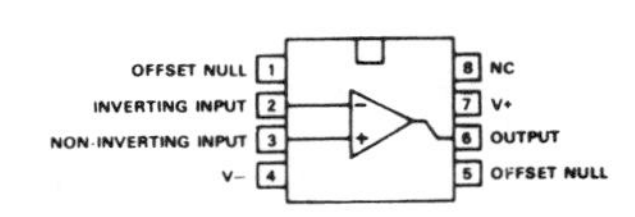

8-PIN MINI DIP
TOP VIEW

GENERAL DESCRIPTION

The Analog Devices AD741 series are high performance monolithic operational amplifiers. All the devices feature full short circuit protection and internal compensation.

The Analog Devices AD741J, AD741K, AD741L and AD741S are specially tested and selected versions of the standard AD741 operational amplifier. Improved processing and additional electrical testing guarantee the user precision performance at a very low cost. The AD741J, K and L substantially increase overall accuracy over the standard AD741C by providing maximum limits on offset voltage drift and significantly reducing the errors due to offset voltage, bias current, offset current, voltage gain, power supply rejection, and common mode rejection. For example, the AD741L features maximum offset voltage drift of 5μV/°C, offset voltage of 0.5mV max, offset current of 5nA max, bias current of 50nA max, and a CMRR of 90dB min. The AD741S offers guaranteed performance over the extended temperature range of −55°C to +125°C, with max offset voltage drift of 15μV/°C, max offset voltage of 4mV, max offset current of 25nA, and a minimum CMRR of 80dB.

HIGH OUTPUT CAPABILITY

Both the AD741J and AD741S offer the user the additional advantages of high guaranteed output current and gain at low values of load impedance. The AD741J guarantees a minimum gain of 25,000 swinging ±10V into a 1kΩ load from 0 to +70°C. The AD741S guarantees a minimum gain of 25,000 swinging ±10V into a 1kΩ load from −55°C to +125°C.

All devices feature full short circuit protection, high gain, high common mode range, and internal compensation. The AD741J, K and L are specified for operation from 0 to +70°C, and are available in both the TO-99 and mini-DIP packages. The AD741S is specified for operation from −55°C to +125°C, and is available in the TO-99 package.

SPECIFICATIONS (typical @ +25°C and ±15V dc, unless otherwise specified)

Model	AD741C Min	AD741C Typ	AD741C Max	AD741 Min	AD741 Typ	AD741 Max	AD741J Min	AD741J Typ	AD741J Max	Units
OPEN LOOP GAIN										
R_L = 1kΩ, V_O = ±10V							50,000	200,000		V/V
R_L = 2kΩ, V_O = ±10V	20,000	200,000		50,000	200,000					V/V
T_A = min to max R_L = 2kΩ	15,000			25,000			25,000			V/V
OUTPUT CHARACTERISTICS										
Voltage @ R_L = 1kΩ, T_A = min to max							±10	±13		V
Voltage @ R_L = 2kΩ, T_A = min to max	±10	±13		±10	±13					V
Short Circuit Current		25			25			25		mA
FREQUENCY RESPONSE										
Unity Gain, Small Signal		1			1			1		MHz
Full Power Response		10			10			10		kHz
Slew Rate		0.5			0.5			0.5		V/μs
Transient Response (Unity Gain)										
Rise Time $C_L \leqslant$ 10V p-p		0.3			0.3			0.3		μs
Overshoot		5.0			5.0			5.0		%
INPUT OFFSET VOLTAGE										
Initial, $R_S \leqslant$ 10kΩ, Adj. to Zero		1.0	6.0		1.0	5.0		1.0	3.0	mV
T_A = min to max		1.0	7.5		1.0	6.0			4.0	mV
Average vs. Temperature (Untrimmed)									20	μV/°C
vs. Supply, T_A = min to max								30	100	μV/V
INPUT OFFSET CURRENT										
Initial		20	200		20	200		5	50	nA
T_A = min to max		40	300		85	500			100	nA
Average vs. Temperature								0.1		nA/°C
INPUT BIAS CURRENT										
Initial		80	500		80	500		40	200	nA
T_A = min to max		120	800		300	1,500			400	nA
Average vs. Temperature								0.6		nA/°C
INPUT IMPEDANCE DIFFERENTIAL	0.3	2.0		0.3	2.0			1.0		MΩ
INPUT VOLTAGE RANGE[1]										
Differential, max Safe									±30	V
Common Mode, max Safe	±12	±13		±12	±13			±15		V
Common Mode Rejection, R_S = ⩽ 10kΩ, T_A = min to max, V_{IN} = ±12V	70	90		70	90		80	90		dB
POWER SUPPLY										
Rated Performance		±15			±15			±15		V
Operating							±5		±18	V
Power Supply Rejection Ratio		30	150		30	150				μV/V
Quiescent Current		1.7	2.8		1.7	2.8		2.2	3.3	mA
Power Consumption		50	85		50	85		50	85	mW
T_A = min					60	100				mW
T_A = max					45	75				mW
TEMPERATURE RANGE										
Operating Rated Performance	0		+70	-55		+125	0		+70	°C
Storage	-65		+150	-65		+150	-65		+150	°C

NOTES

[1] For supply voltages less than ±15V, the absolute maximum input voltage is equal to the supply voltage.

Specifications subject to change without notice.

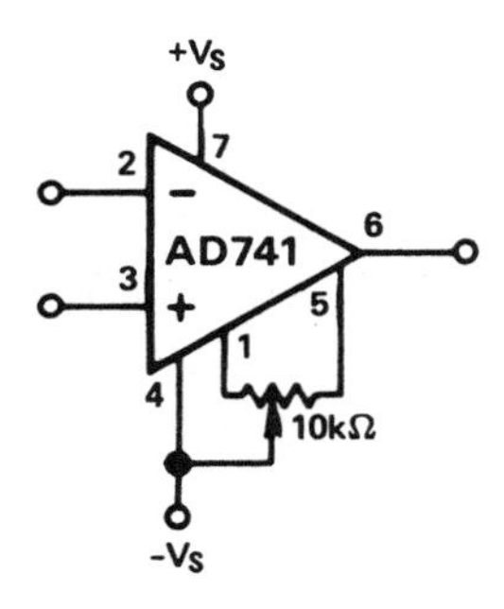

Standard Nulling Offset Circuit

Model	AD741K Min	AD741K Typ	AD741K Max	AD741L Min	AD741L Typ	AD741L Max	AD741S Min	AD741S Typ	AD741S Max	Units
OPEN LOOP GAIN										
$R_L = 1k\Omega, V_O = \pm 10V$							50,000	200,000		V/V
$R_L = 2k\Omega, V_O = \pm 10V$	50,000	200,000		50,000	200,000					V/V
T_A = min to max $R_L = 2k\Omega$	25,000			25,000			25,000			V/V
OUTPUT CHARACTERISTICS										
Voltage @ $R_L = 1k\Omega$, T_A = min to max										V
Voltage @ $R_L = 2k\Omega$, T_A = min to max	±10	±13		±10	±13		±10	±13		V
Short Circuit Current		25			25			25		mA
FREQUENCY RESPONSE										
Unity Gain, Small Signal		1			1			1		MHz
Full Power Response		10			10			10		kHz
Slew Rate		0.5			0.5			0.5		V/μs
Transient Respone (Unity Gain)										
Rise Time		0.3			0.3			0.3		μs
Overshoot		5.0			5.0			5.0		%
INPUT OFFSET VOLTAGE										
Initial, $R_S \leqslant 10k\Omega$, Adj. to Zero		0.5	2.0		0.2	0.5		1.0	2	mV
T_A = min to max			3.0			1.0			4	mV
Average vs. Temperature (Untrimmed)		6.0	15.0		2.0	5.0		6.0	15.0	μV/°C
vs. Supply, T_A = min to max		5	15.0		5	15.0		30	100	μV/V
INPUT OFFSET CURRENT										
Initial		2	10		2	5		2	10	nA
T_A = min to max			15			10			25	nA
Average vs. Temperature		0.02	0.2		0.02	0.1		0.1	0.25	nA/°C
INPUT BIAS CURRENT										
Initial		30	75		30	50		30	75	nA
T_A = min to max			120			100			250	nA
Average vs. Temperature		0.6	1.5		0.6	1.0		0.6	2.0	nA/°C
INPUT IMPEDANCE DIFFERENTIAL		2			2			2		MΩ
INPUT VOLTAGE RANGE[1]										
Differential, max Safe		±30			±30			±30		V
Common Mode max Safe		±15			±15			±15		V
Common Mode Rejection,										
$R_S \leqslant 10k\Omega$, T_A = min to max										
$V_{IN} = \pm 12V$	90	100		90	100		90	100		dB
POWER SUPPLY										
Rated Performance		±15			±15			±15		V
Operating	±5		±22	±5		±22	±5		±22	V
Power Supply Rejection Ratio										μV/V
Quiescent Current		1.7	2.8		1.7	2.8		2.0	2.8	mA
Power Consumption		50	85		50	85		50	85	mW
T_A = min								60	100	mW
T_A = max								75	115	mW
TEMPERATURE RANGE										
Operating Rated Performance	0		+70	0		+70	-55		+125	°C
Storage	-65		+150	-65		+150	-65		+150	°C

NOTES

[1] For supply voltages less than ±15V, the absolute maximum input voltage is equal to the supply voltage.

Specifications subject to change without notice.

Specifications shown in boldface are tested on all production units at final electrical test. Results from those tests are used to calculate outgoing quality levels. All min and max specifications are guaranteed, although only those shown in boldface are tested on all production units.

ORDERING GUIDE

Model	Temperature Range	Package[1]	Initial Off-Set Voltage
AD741CN	0 to +70°C	MINI-DIP (N8A)	6.0mV
AD741CH	0 to +70°C	TO-99	6.0mV
AD741JN	0 to +70°C	MINI-DIP (N8A)	3.0mV
AD741JH	0 to +70°C	TO-99	3.0mV
AD741KN	0 to +70°C	MINI-DIP (N8A)	2.0mV
AD741KH	0 to +70°C	TO-99	2.0mV
AD741LN	0 to +70°C	MINI-DIP (N8A)	0.5mV
AD741LH	0 to +70°C	TO-99	0.5mV
AD741H	-55°C to +125°C	TO-99	5.0mV
AD741SH	-55°C to +125°C	TO-99	2.0mV

NOTE

[1] See Section 19 for package outline information.

ABSOLUTE MAXIMUM RATINGS

Absolute Maximum Ratings	AD741, J, K, L, S	AD741C
Supply Voltage	±22V	±18V
Internal Power Dissipation	500mW[1]	500mW
Differential Input Voltage	±30V	±30V
Input Voltage	±15V	±15V
Storage Temperature Range	-65°C to +150°C	-65°C to +150°C
Lead Temperature (soldering, 60 seconds)	300°C	300°C
Output Short Circuit Duration	Indefinite[2]	Indefinite

NOTES

[1] Rating applies for case temperature to +125°C. Derate TO-99 linearity at 6.5mW/°C for ambient temperatures above +70°C.

[2] Rating applies for shorts to ground or either supply at case temperatures to +125°C or ambient temperatures to +75°C.

Typical Performance Curves

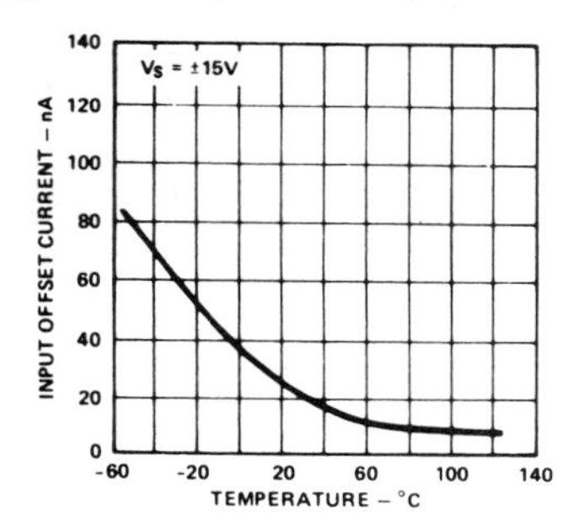

Figure 1. Offset Current vs. Temperature

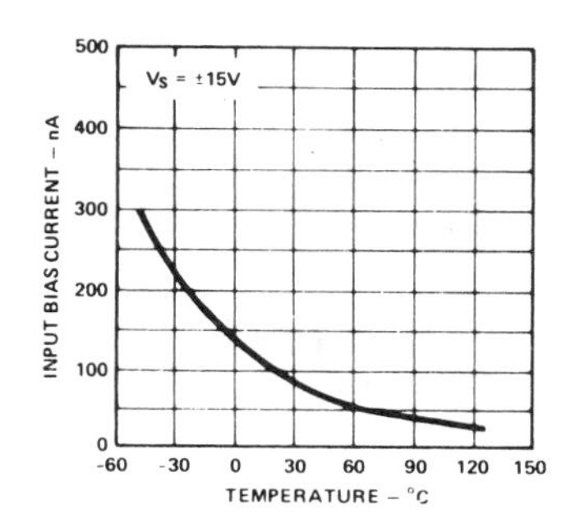

Figure 2. Bias Current vs. Temperature

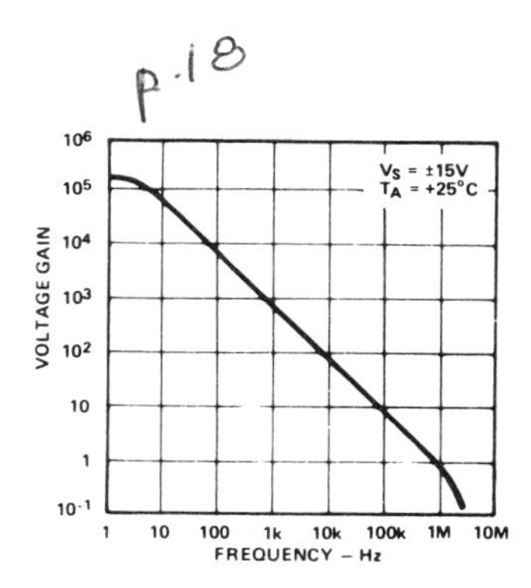

Figure 3. Open Loop Gain vs. Frequency

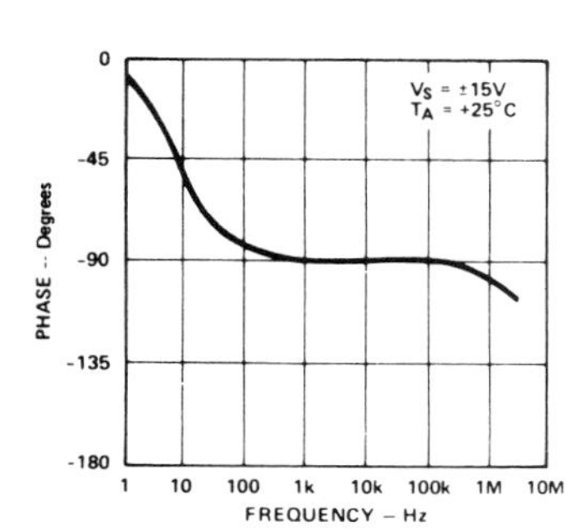

Figure 4. Open Loop Phase Response vs. Frequency

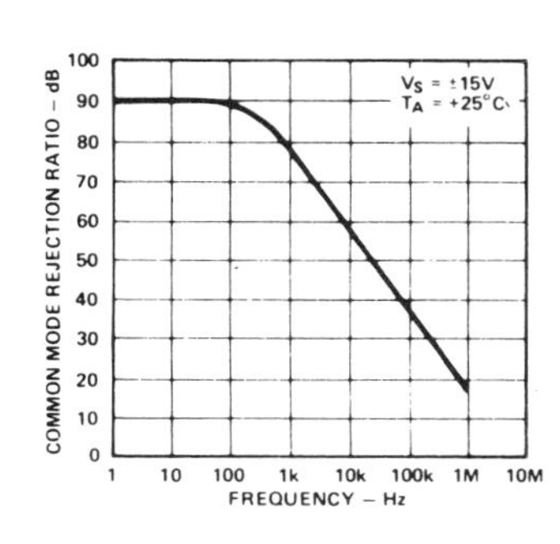

Figure 5. Common Mode Rejection vs. Frequency

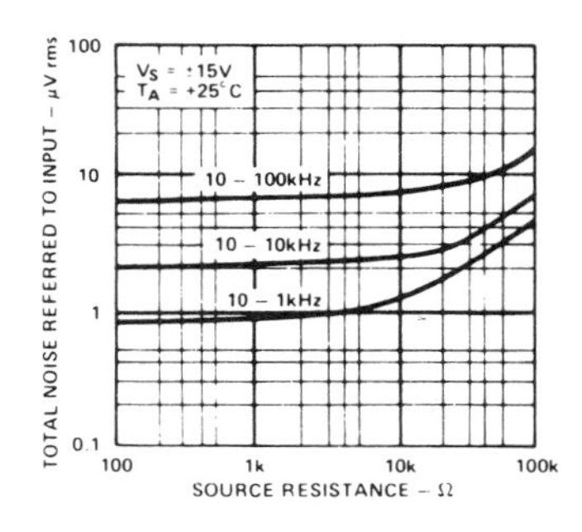

Figure 6. Broad Band Noise vs. Source Resistance

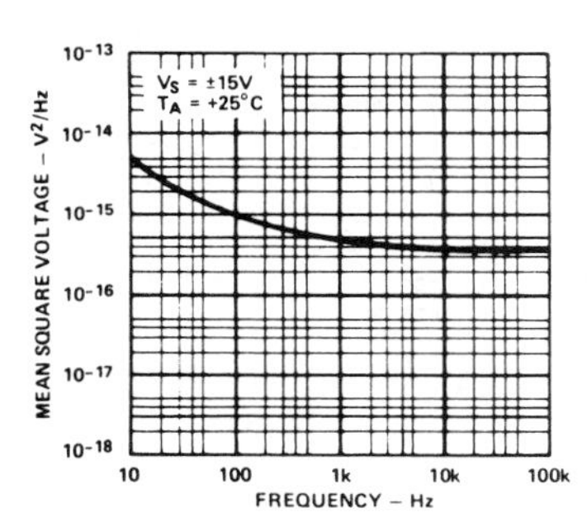

Figure 7. Input Noise Voltage vs. Frequency

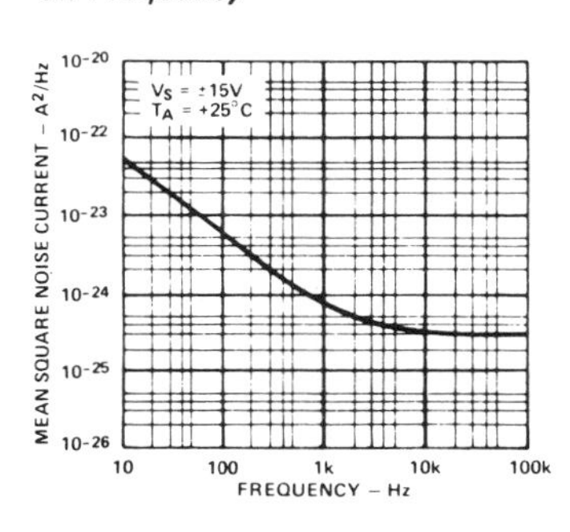

Figure 8. Input Noise Current vs. Frequency

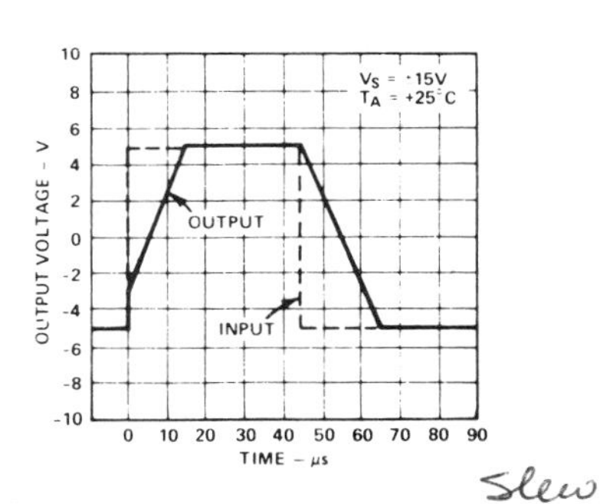

Figure 9. Voltage Follower Large Signal Pulse Response

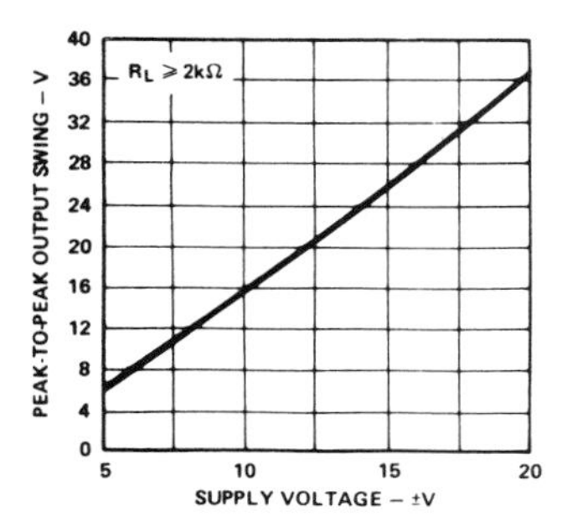

Figure 10. Output Voltage Swing vs. Supply Voltage

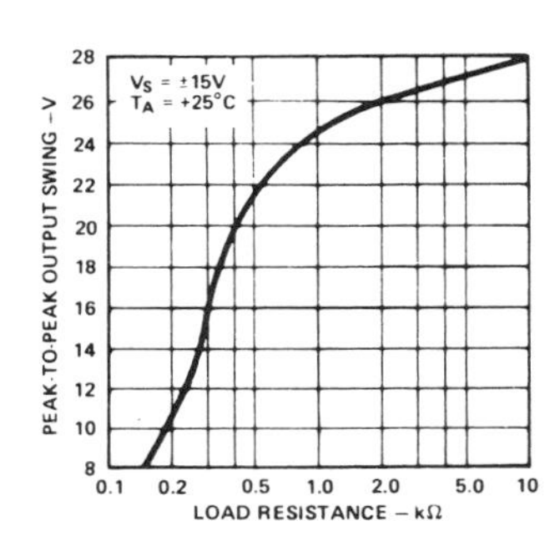

Figure 11. Output Voltage Swing vs. Load Resistance

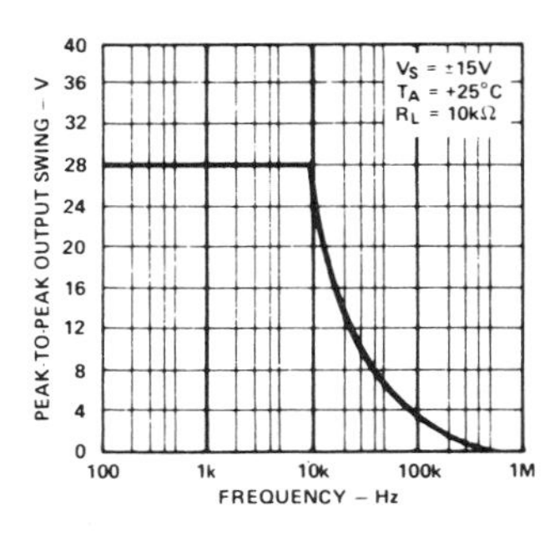

Figure 12. Output Voltage Swing vs. Frequency

General Purpose Low Cost IC Operational Amplifier

AD101A, AD201A, AD301A, AD301AL

FEATURES
Low Bias and Offset Current
Single Capacitor External Compensation for Operating Flexibility
Nullable Offset Voltage
No Latch-Up
Fully Short Circuit Protected
Wide Operating Voltage Range

AD101 SERIES FUNCTIONAL BLOCK DIAGRAMS

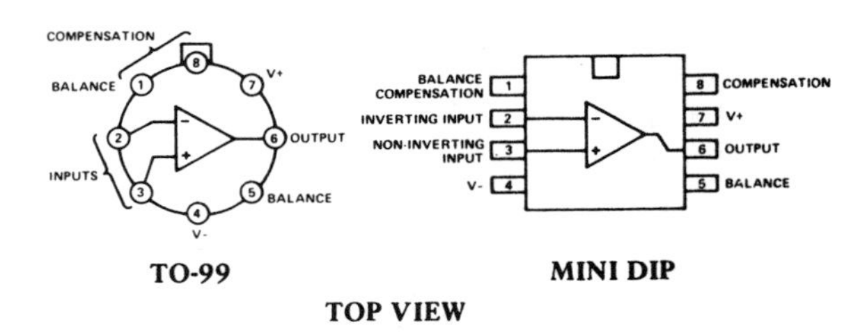

TO-99 MINI DIP
TOP VIEW

GENERAL DESCRIPTION

The Analog Devices AD101A, AD201A, AD301A and AD301AL are high performance monolithic operational amplifiers. All the circuits feature full short circuit protection, external offset voltage nulling, wide operating voltage range, and the total absence or "latch-up". Because frequency compensation is performed externally with a single capacitor (30pF maximum), the AD101A, AD201A, AD301A and AD301AL provide greater flexibility than internally compensated amplifiers since the degree of compensation can be fitted to the specific system application.

The AD101A and AD201A have identical specifications in the TO-99 package; the former guaranteed over the -55°C to +125°C temperature range, and the latter over -25°C to +85°C. The AD201A is also available in the mini-DIP package for high performance operation over the 0 to +70°C temperature range. The AD301A is specified for operation over the 0 to +70°C temperature range in both the TO-99 and mini-DIP packages. The AD301AL is the highest accuracy version of this series. Improved processing and additional electrical testing allow the user to achieve precision performance at low cost. The device provides substantially increased accuracy by reducing errors due to offset voltage (0.5mV max), offset voltage drift (5.0μV/°C max), bias current (30nA max), offset current (5nA max), voltage gain (80,000 min), PSRR (90dB min), and CMRR (90dB min). The AD301AL is also specified from 0 to +70°C and is available in the TO-99 can or 8-pin mini-DIP.

SCHEMATIC DIAGRAM

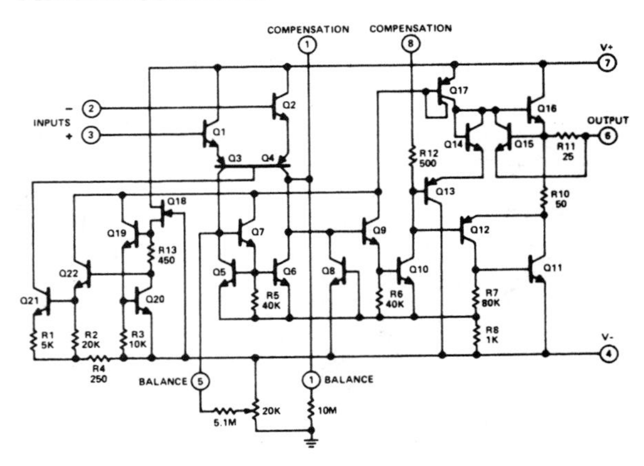

Courtesy of **Analog Devices, Inc.**

SPECIFICATIONS (typical @ +25°C and ±15V dc, unless otherwise specified)

ABSOLUTE MAXIMUM RATINGS	AD101A, AD201A, AD301A, AD301AL unless otherwise specified
Supply Voltage	
AD101A, AD201A	±22V
AD301A, AD301AL	±18V
Power Dissipation[1]	
TO-99 (Metal Can)	500mW
Dual In-Line (Mini-DIP)	500mW
Differential Input Voltage	±30V
Input Voltage[2]	±15V
Output Short Circuit Duration[3]	Indefinite
Operating Temperature Range	
AD101A	-55°C to +125°C
AD201A (TO-99)	-25°C to +85°C
AD201A (Mini-DIP)	0 to +70°C
AD301A, AD301AL	0 to +70°C
Storage Temperature Range	-65°C to +150°C
Lead Temperature (Soldering, 60sec)	300°C

ELECTRICAL CHARACTERISTICS (T_A = +25°C unless otherwise specified)[4]

Parameter	Conditions	AD101A/AD201A			AD301A			AD301AL			Units
		Min	Typ	Max	Min	Typ	Max	Min	Typ	Max	
Input Offset Voltage	$R_S \leq 50k\Omega$		0.7	2.0		2.0	7.5		0.3	0.5	mV
Input Offset Current			1.5	10		3	50		3	5	nA
Input Biz											
Input Bias Current			30	75		70	250		15	30	nA
Input Resistance		1.5	4		0.5	2		1.5	4		MΩ
Supply Current	$V_S = \pm 20V$		1.8	3.0							mA
	$V_S = \pm 15V$					1.8	3.0		1.8	3	mA
Large Signal Voltage Gain	$V_S = \pm 15V$, $V_{OUT} = \pm 10V$, $R_L \geq 2k\Omega$	50	160		25	160		80	300		V/mV
The Following Specifications Apply Over the Operating Temperature Ranges[4]											
Input Offset Voltage	$R_S \leq 10k\Omega$			3.0			10		0.5	1	mV
Input Offset Current				20			70		5	10	nA
Average Temp. Coefficient of Input Offset Voltage	$T_A(min) \leq T_A \leq T_A(max)$		3.0	15		6.0	30		2	5	μV/°C
Average Temp. Coefficient of Input Offset Current	$+25°C \leq T_A \leq T_A(max)$		0.01	0.1		0.01	0.3		0.01	0.1	nA/°C
	$T_A(min) \leq T_A \leq +25°C$		0.02	0.2		0.02	0.6		0.01	0.1	nA/°C
Input Bias Current				100			300		30	45	nA
Large Signal Voltage Gain	$V_S = \pm 15V$, $V_{OUT} = \pm 10V$, $R_L \geq 2k\Omega$	25			15			40	100		V/mV
Input Voltage Range	$V_S = \pm 20V$	±15									V
	$V_S = \pm 15V$				±12			±12			V
Common Mode Rejection Ratio	$R_S \leq 50k\Omega$	80	96		70	90		90	100		dB
Supply Voltage Rejection Ratio	$R_S \leq 50k\Omega$	80	96		70	96		90	100		dB
Output Voltage Swing	$V_S = \pm 15V$, $R_L = 10k\Omega$	±12	±14		±12	±14		±12	±14		V
	$V_S = \pm 15V$, $R_L = 2k\Omega$	±10	±13		±10	±13		±10	±13		V
Supply Current	$T_A = T_A(max)$, $V_S = \pm 20V$		1.2	2.5					1.8	3	mA

NOTES

[1] The maximum desirable junction temperature of the AD101A is +150°C; that of the AD201A, AD301A and AD301AL is +100°C. For operating at elevated temperatures, devices must be derated based upon a thermal resistance of +150°C/W, junction to ambient, or +45°C/W, junction to case. The thermal resistance of the Dual In-Line package is +160°C/W, junction to ambient.

[2] For supply voltages less than ±15V, the absolute maximum input voltage is equal to the supply voltage.

[3] For the AD301A and AD301AL continuous short circuit is allowed for case temperatures to +70°C and ambient temperatures to +55°C.

[4] Unless otherwise specified, these specifications apply for supply voltages and ambient temperatures of ±5V to ±20V and -55°C to +125°C for the AD101A, ±5V to ±20V and -25°C to +85°C for the AD201AH (0 to +70°C for the AD201AN), and ±5V to ±15V and 0 to +70°C for the AD301A and AD301AL.

Specifications subject to change without notice.

ORDERING GUIDE

MODEL	TEMP RANGE	ORDER NUMBER*	PACKAGE OPTION**
AD301AL	0 to +70°C	AD301AL	TO-99, N8A
AD201A	−25°C to +85°C	AD201A	TO-99, N8A
AD301A	0 to +70°C	AD301A	TO-99, N8A
AD101A	−55°C to +125°C	AD101AH	TO-99

*Add package type leter: H = TO-99, N = Mini DIP.
**See Section 19 for package outline information.

FREQUENCY COMPENSATION CIRCUITS

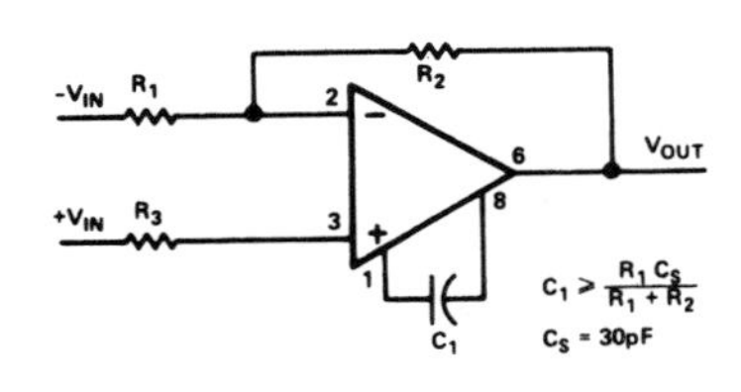

Figure 1. Single Pole Compensation

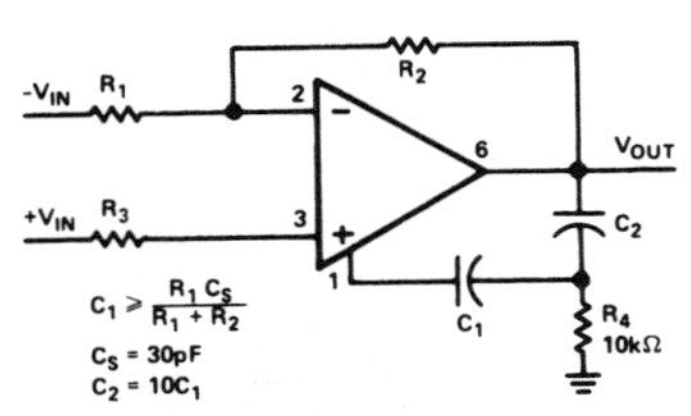

Figure 2. Two Pole Compensation

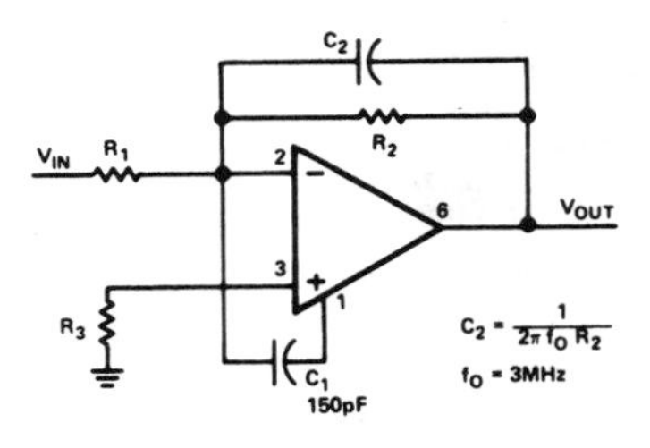

Figure 3. Feedforward Compensation

GUARANTEED PERFORMANCE CURVES (Curves apply over the Operating Temperature Ranges)

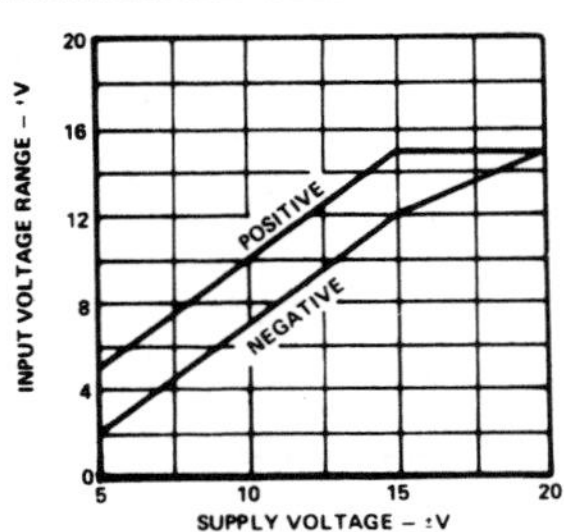

Input Voltage Range

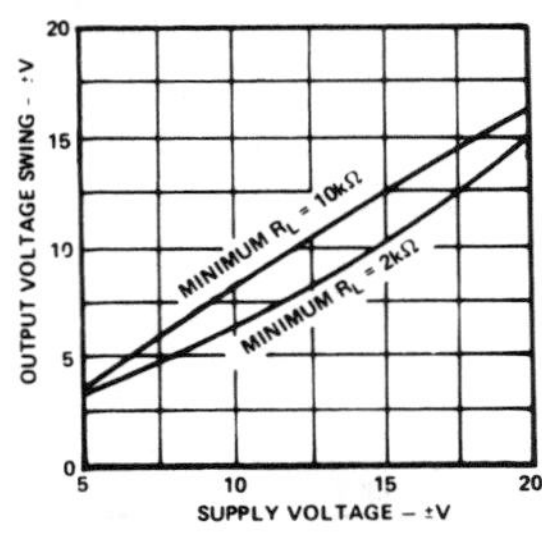

Output Swing

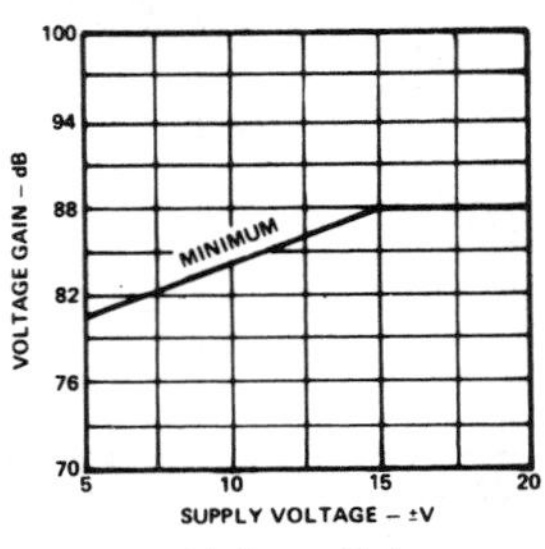

Voltage Gain

TYPICAL PERFORMANCE CURVES[4]

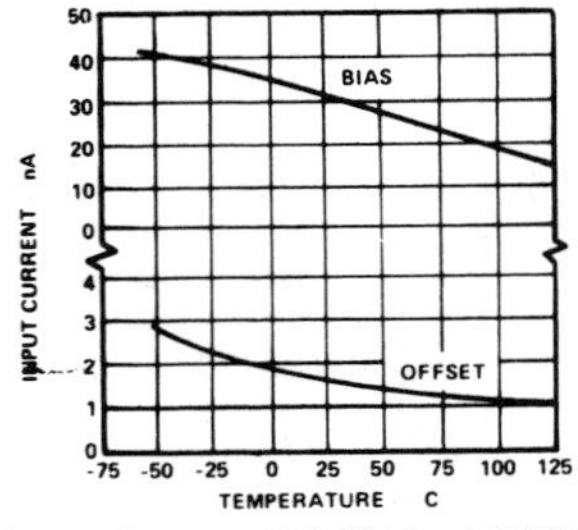

Input Current AD101A, AD201A

Input Current – AD301A

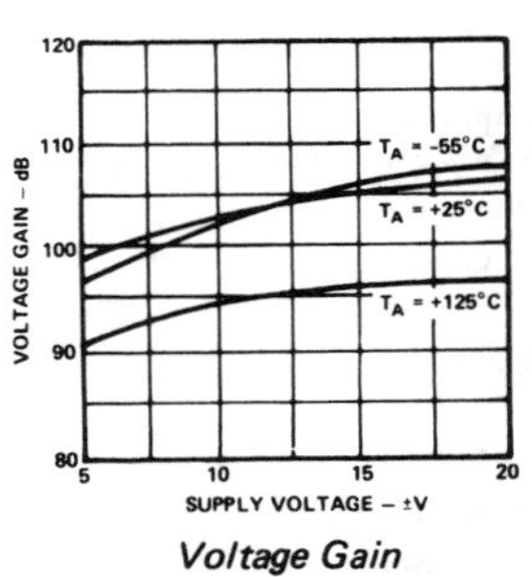

Voltage Gain

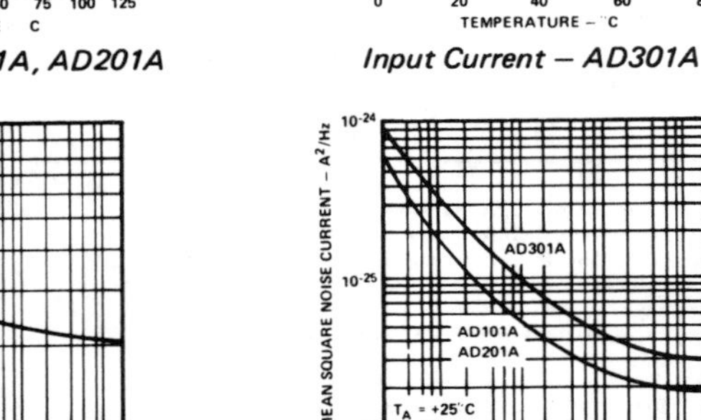

Input Noise Voltage

Input Noise Current

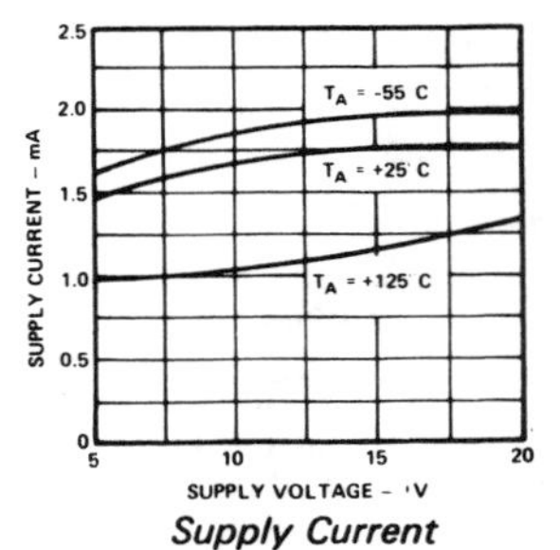

Supply Current

TYPICAL PERFORMANCE CURVES

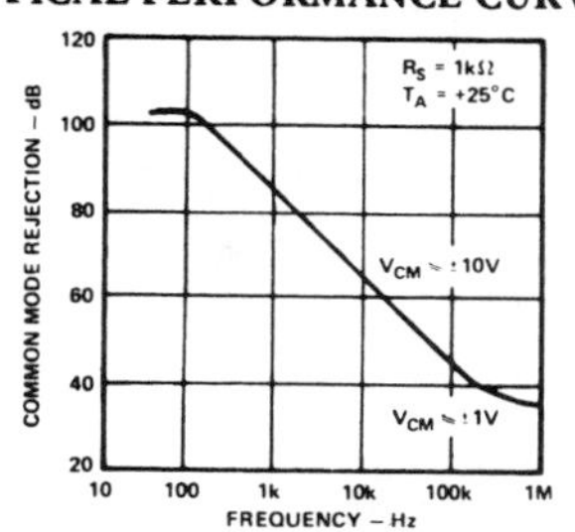

Common Mode Rejection

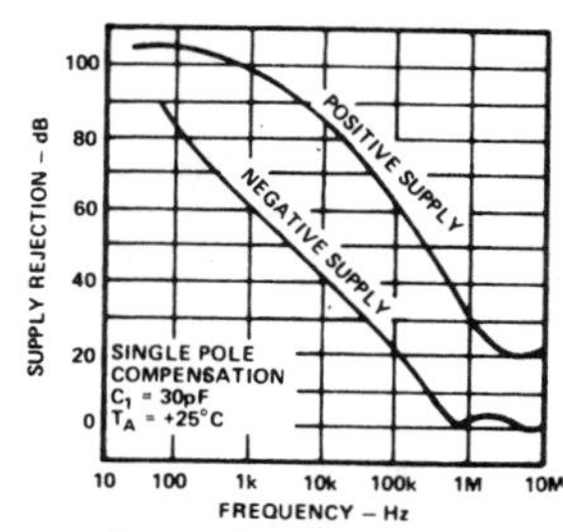

Power Supply Rejection

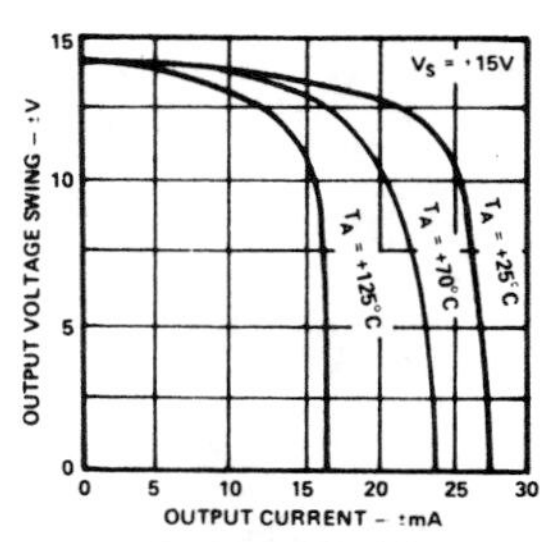

Current Limiting

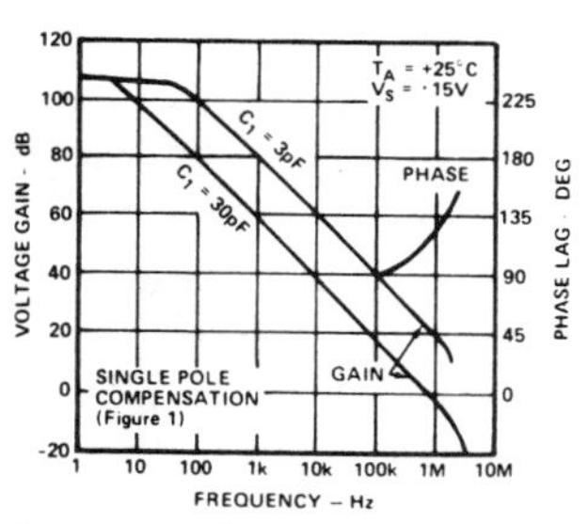

Open Loop Frequency Response

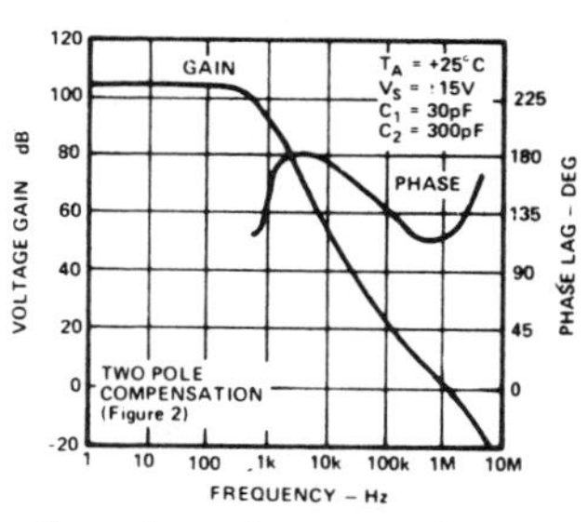

Open Loop Frequency Response

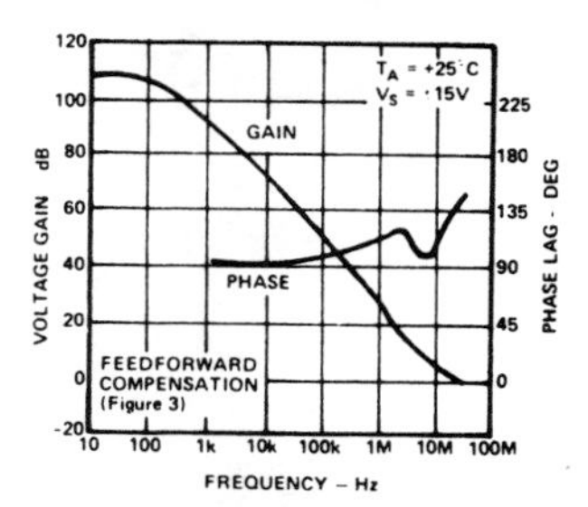

Open Loop Frequency Response

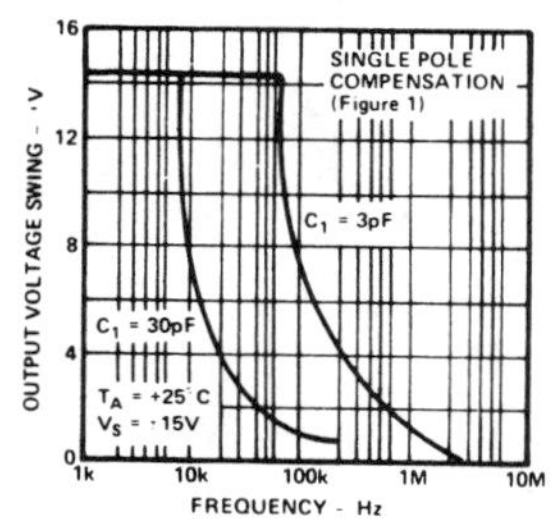

Large Signal Frequency Response

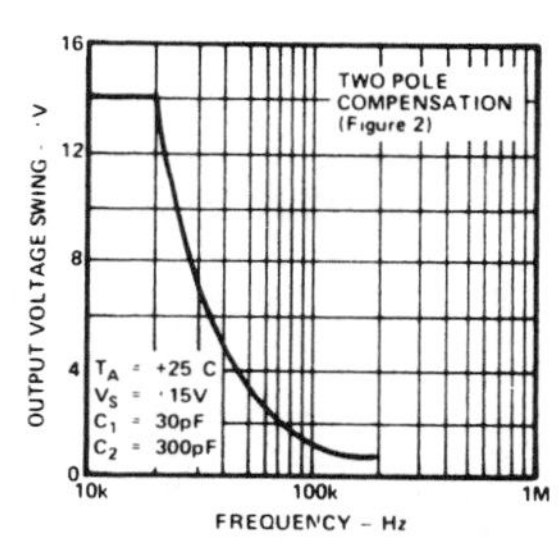

Large Signal Frequency Response

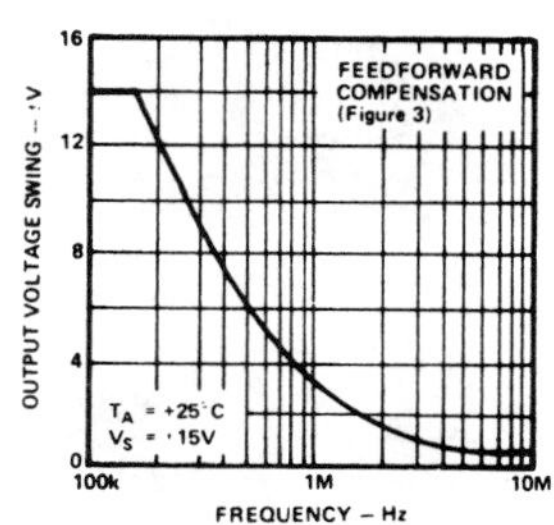

Large Signal Frequency Response

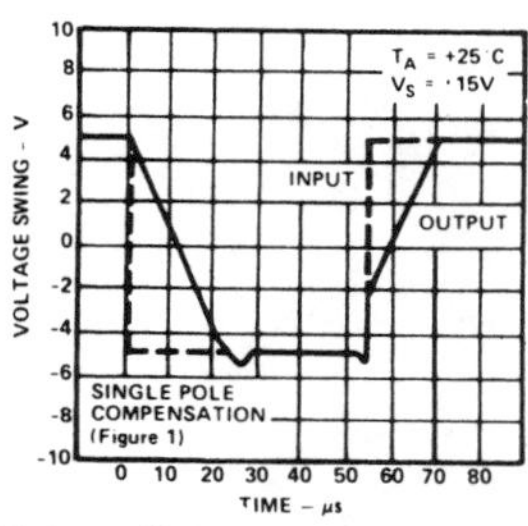

Voltage Follower Pulse Response

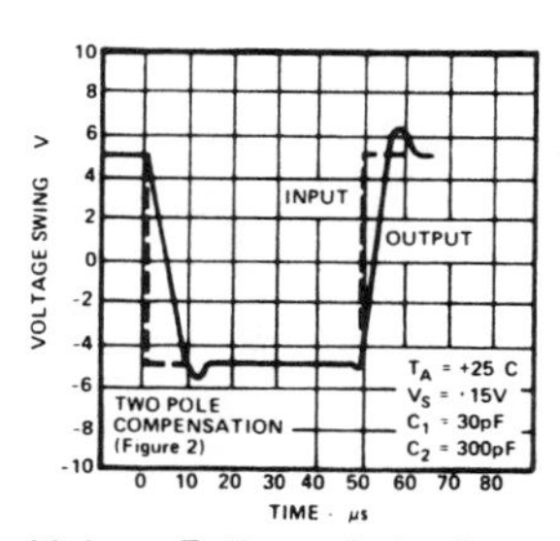

Voltage Follower Pulse Response

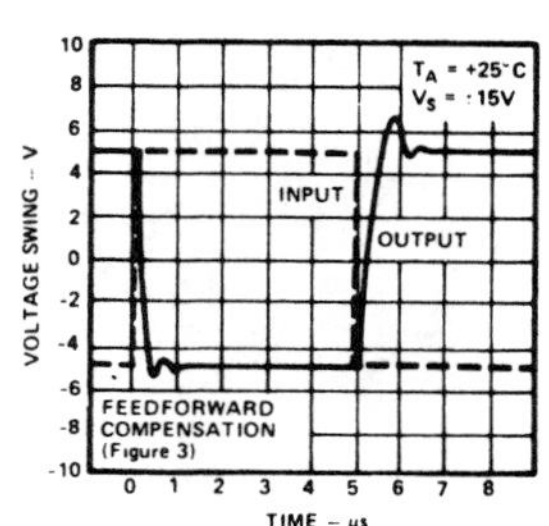

Inverter Pulse Response

Op-Amp Voltage-Level Detectors: Comparators

2

LEARNING OBJECTIVES

Upon completion of this chapter on the use of the op amp as a voltage-level detector, you will be able to:

- Properly classify inverting and noninverting zero-voltage-level detectors.
- Draw the output voltage wave form versus time and the output voltage versus input voltage transfer function for both the inverting and noninverting zero-crossing detectors.
- Build and use basic resistor divider networks for practical voltage references.
- Describe the characteristics of Analog Devices' AD584 pin-programmable voltage reference, and use it to create four different precise reference voltages.
- Connect light-emitting diodes (LEDs) to the output of an op amp and use them as visual indicators.
- Draw the output voltage waveform versus time for inverting and noninverting voltage-level detectors.
- Draw the output voltage versus input voltage transfer functions for inverting and noninverting voltage-level detectors.
- Describe the characteristics of an LM311 comparator.
- Use the LM311 comparator to interface with both TTL and CMOS logic.
- Predict the output state of the LM311 comparator when the strobe terminal is used.

2-0 INTRODUCTION

One of the simplest applications of an op amp is to use it as a voltage-level detector or comparator. To accomplish this, a reference voltage V_{ref} to be sensed is applied to one input of an op amp. V_{ref} may be 0 V, a positive voltage, or a negative voltage with respect to ground. The other input of the op amp has V_i, a time-varying voltage applied to it. When V_i crosses the reference voltage level, the output of the op amp will switch to either $+V_{sat}$ or $-V_{sat}$, depending on the type of comparator circuit.

In this way, output voltage V_o identifies two conditions as they occur at the inputs. The first is *when V_i equals V_{ref}*. This occurs at the instant the output switches from one saturation level to the other. The second condition is to detect if V_i is *above* or *below* V_{ref}. This can be determined by the polarity of V_o.

LEDs connected to the output terminal of an op amp provide a convenient way of displaying visually the input conditions of the comparator. For example, one LED, one op amp, and one diode will give a visual indication if a voltage V_1 is above or below reference voltage V_{ref}.

Comparators used to detect voltage levels other than zero volts require precise values for V_{ref}. Both positive and negative reference voltages can be developed with basic resistor divider circuits. However, when accuracy and stability are demanded, a precision reference voltage should be used. The AD584 pin-programmable reference voltage IC is one example of an excellent voltage reference. This device can output any one of four precise voltages by simple pin selection on the package.

2-1 CLASSIFICATION OF VOLTAGE COMPARATORS

Voltage comparators are divided into two *classifications*. They are the *noninverting* and *inverting comparators* shown in Fig. 2-1. This classification is specified by the input terminal to which V_i is applied. When V_i is attached to the (+) input of an op amp, a noninverting comparator is

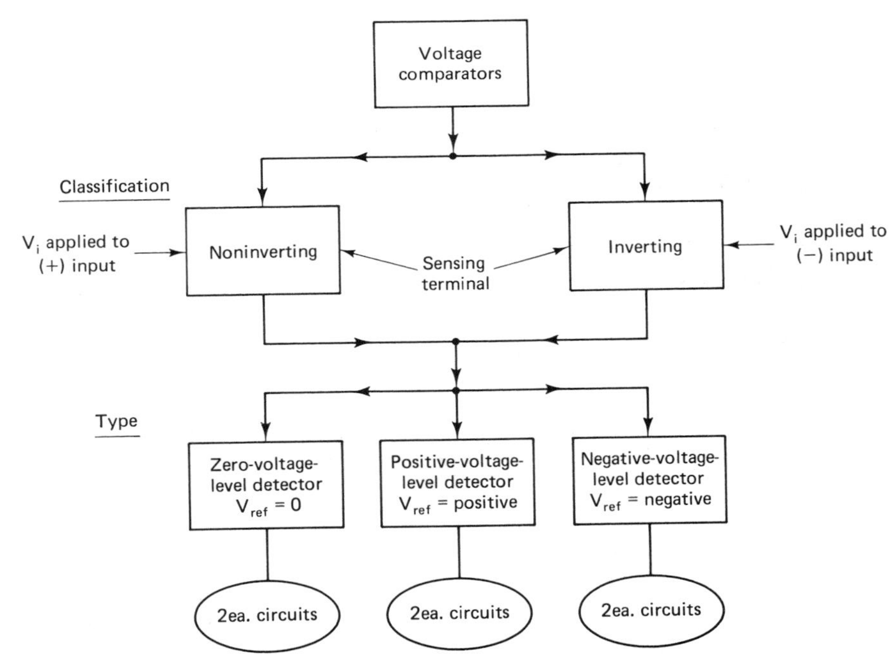

Figure 2-1 There are six possible voltage comparator circuits that can be built using op amps.

created. Move V_i to the (−) input and the op amp becomes an inverting comparator.

Inverting or noninverting comparators can be further subdivided into *types* depending on the polarity of reference voltage to be sensed. V_{ref} is applied to the other input terminal of the op amp. When V_{ref} is equal to 0 V, the comparator is a *zero-voltage-level detector* or zero-crossing detector. If V_{ref} is a positive voltage, the comparator type is a *positive-voltage-level detector.* When V_{ref} is negative, a *negative-voltage-level detector* is established.

2-2 SIMPLIFIED ANALYSIS OF VOLTAGE COMPARATORS

The magnitude and polarity of output voltage V_o in any comparator circuit is under the control of the input terminals. Equations (1-7) and (1-8) showed that V_o depended on the magnitude and polarity of the differential input voltage E_d. The equations are repeated here for convenience.

$$E_d = V_{(+)} - V_{(-)} \quad (2\text{-}1a)$$

$$V_o = A_{OL} E_d \quad (2\text{-}1b)$$

In a noninverting comparator, V_i is connected as $V_{(+)}$, and V_{ref} is connected as $V_{(-)}$. Figure 2-2a shows graphically the following:

1. $V_o = +V_{sat}$ when $V_i = V_{(+)}$ is above (more positive than) $V_{ref} = V_{(-)}$.
2. $V_o = -V_{sat}$ when $V_i = V_{(+)}$ is below (less positive than) $V_{ref} = V_{(-)}$.

The following example illustrates the operation of noninverting comparators.

EXAMPLE 2-1

Determine the magnitude and polarity of V_o if the input conditions shown in Fig. 2-3 are applied to the noninverting comparator of Fig. 2-2a.

SOLUTION: (a) $V_o = +V_{sat}$ since $V_i = +3$ V is *above* $V_{ref} = 0$ V; (b) $V_o = -V_{sat}$ since $V_i = -3$ V is *below* $V_{ref} = +2$ V; (c) $V_o = +V_{sat}$ since $V_i = -2$ V is *above* $V_{ref} = -5$ V.

For an *inverting* comparator, V_i is connected to $V_{(-)}$ and V_{ref} to $V_{(+)}$. Figure 2-2b shows graphically the following:

1. $V_o = -V_{sat}$ when $V_i = V_{(-)}$ is above (more positive than) $V_{ref} = V_{(+)}$.
2. $V_o = V_{sat}$ when $V_i = V_{(-)}$ is below (less positive than) $V_{ref} = V_{(+)}$.

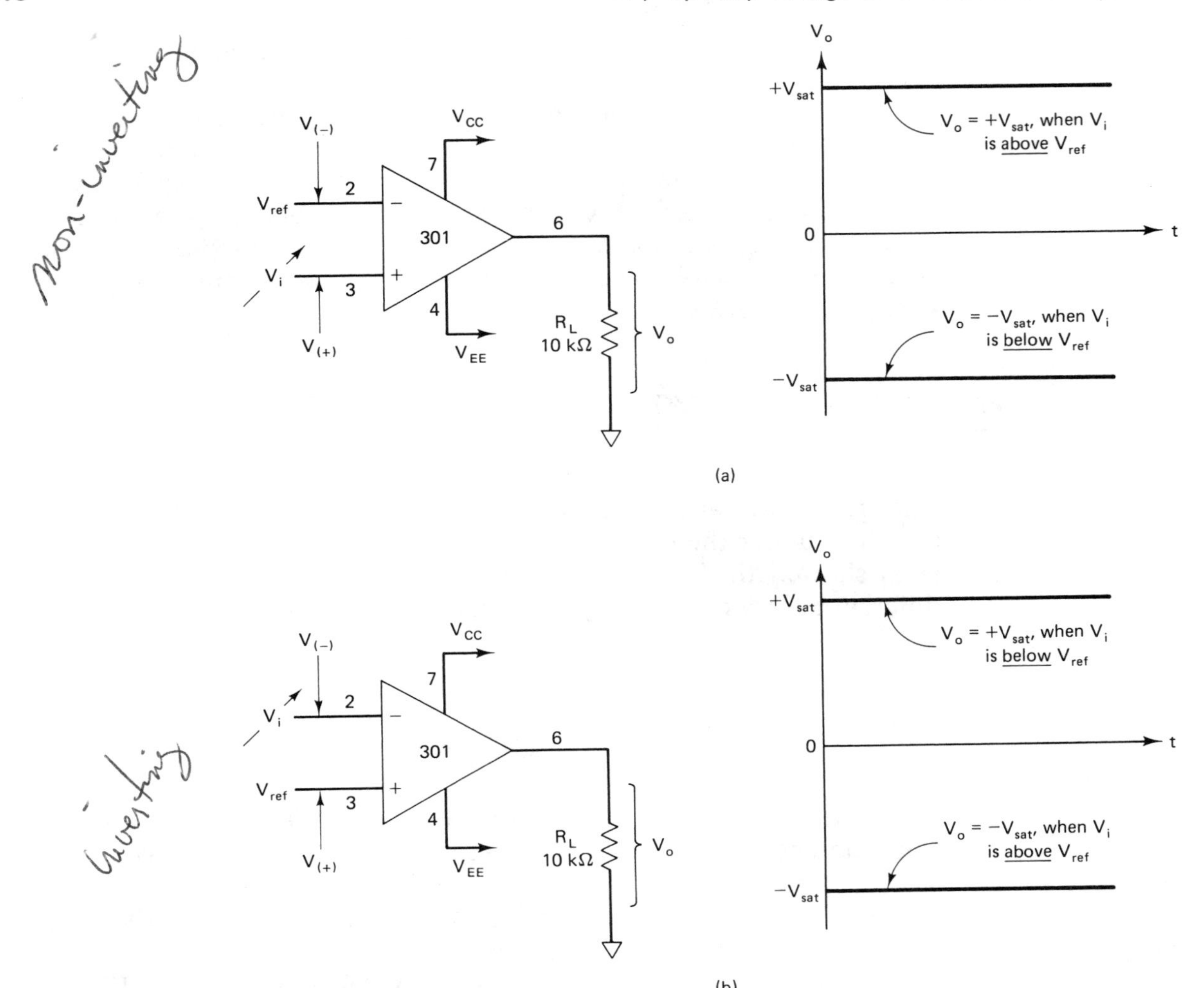

Figure 2-2 Basic operation of both noninverting and inverting comparators. (a) Since E_i is applied to pin 3 the circuit is a noninverting comparator; (b) since V_i is applied to pin 2 the circuit is an inverting comparator.

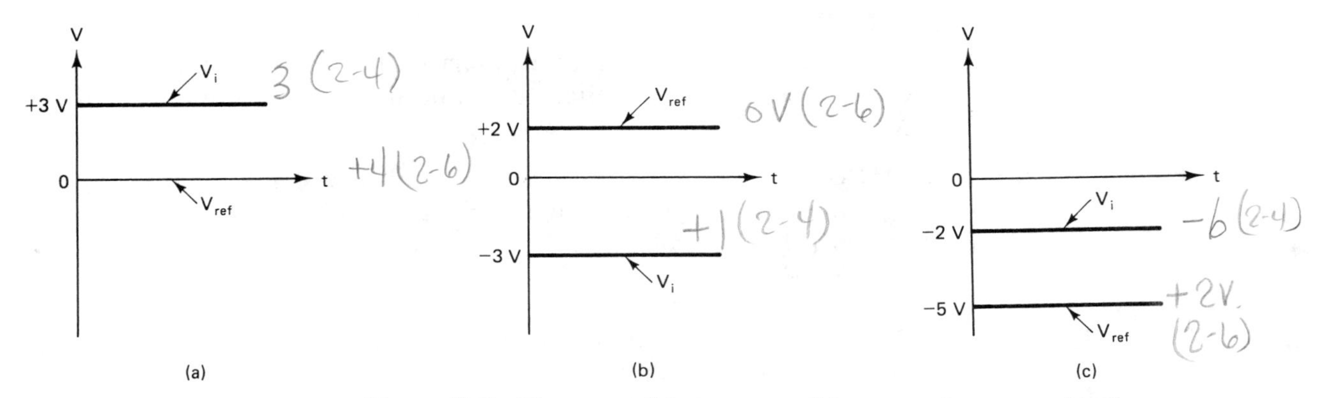

Figure 2-3 Three possible input conditions are shown graphically for the comparators shown in Fig. 2-2. The sign of E_d and V_o depends on how V_i and V_{ref} are connected to the op amp. (a) The voltage E_i = *+3 V is above* V_{ref} = 0 V; (b) the voltage $E_i = -3$ V is below. $V_{ref} = +2$ V.

EXAMPLE 2-2

Determine the magnitude and polarity of V_o if the input voltages of Fig. 2-3 are applied to the inverting comparator of Fig. 2-2b.

SOLUTION: (a) $V_o = -V_{sat}$ since V_i is *above* $V_{ref} = 0$ V; (b) $V_o = +V_{sat}$ since $V_i = -3$ V is *below* $V_{ref} = +2$ V; (c) $V_o = -V_{sat}$ since $V_i = -2$ V is *above* $V_{ref} = -5$ V.

Example 2-1 shows that V_o is *above* 0 V at $+V_{sat}$ if V_i is *above* V_{ref} (noninverting action). Example 2-2 shows that V_o is *below* 0 V at $-V_{sat}$ if V_i is *above* V_{ref} (inverting action).

2-3 ZERO-VOLTAGE-LEVEL DETECTORS

2-3.1 Noninverting Zero-Crossing Detector

A practical noninverting zero-voltage-level detector is shown in Fig. 2-4a. The 301 op amp has been selected for its moderately fast slew rate specifications of 10 V/μs rather than the 741 with a slew rate of only 0.5V/μs. This means that the output voltage of the 301 op amp can change 20 times faster than the output voltage of the 741.

V_i is applied to the noninverting sense terminal of the comparator as a $\pm$5-V peak sine wave at a test frequency of 100 Hz. The reference terminal is connected to ground. Thus $V_{ref} = 0$ V. Figure 2-4b shows the plot of V_i versus time, V_{ref} versus time, and V_o versus time.

The time plots of Fig. 2-4b should be interpreted as follows:

1. When V_o is at $+V_{sat}$, V_i is above the reference.
2. When V_o switches from $-V_{sat}$ to $+V_{sat}$, V_i is crossing 0 V and going positive (point *a*).
3. When V_o is at $-V_{sat}$, V_i is below the reference.
4. When V_o switches from $+V_{sat}$ to $-V_{sat}$, V_i is crossing 0 V and going negative (point *b*).

These observations show noninverting activity.

A more convenient form for representing characteristics of a comparator is the transfer function in Fig. 2-4c. The vertical output axis is a plot of V_o, while V_i is plotted along the horizontal axis. V_{ref} is shown at 0 V on the horizontal axis.

When V_i crosses 0 V and goes positive, V_o switches to $+V_{sat}$ and remains at $+V_{sat}$, while V_i is positive. When V_i crosses 0 V going negative, V_o switches to $-V_{sat}$ and remains at $-V_{sat}$ while V_i is negative. Noninverting comparators have a transfer function that is characterized by a plot that is *up and to the right.*

2-3.2 Inverting Zero-Crossing Detector

Figure 2-5a shows a practical inverting zero-voltage-level detector using a 301 op amp as the comparator. V_i is applied to the inverting *sense* terminal and set to $\pm$5 V peak at 100 Hz. The noninverting terminal is

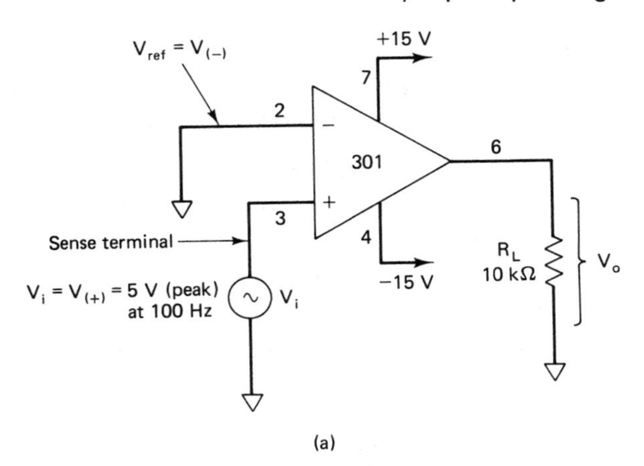

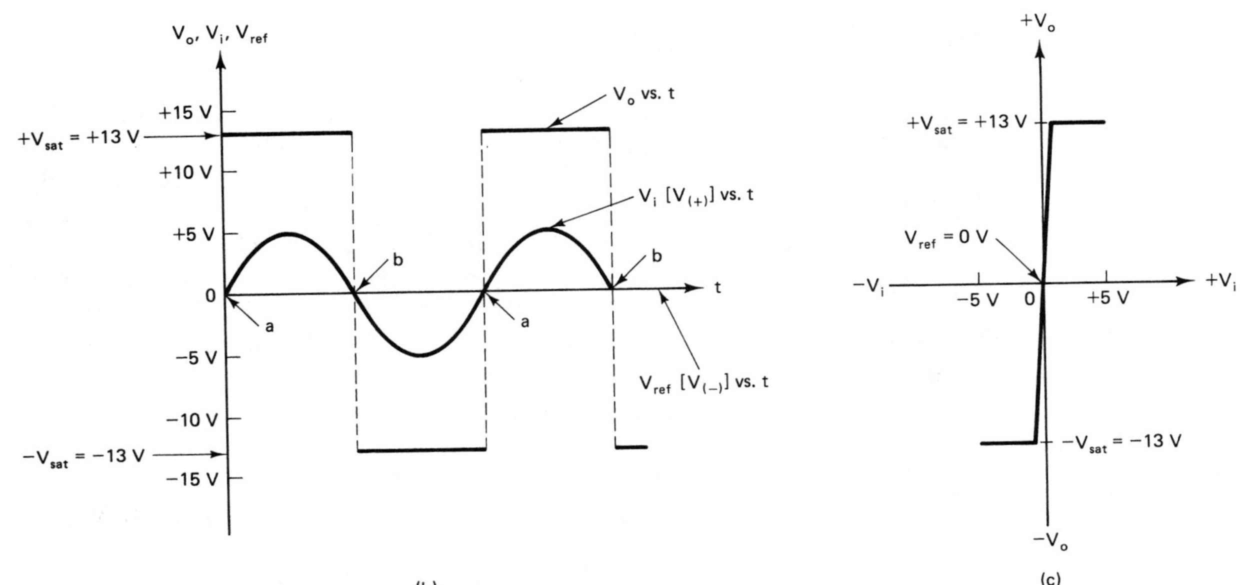

Figure 2-4 A noninverting comparator may be used as a zero-voltage level detector. (a) The 301 op amp used as a noninverting zero-voltage level detector; (b) $V_o = +V_{sat}$ *when* V_i is above V_{ref}, $V_o = -V_{sat}$ when V_i is below V_{ref}; (c) V_o versus V_i transfer function.

connected to a zero voltage reference by a ground. Figure 2-5b shows V_i versus time, V_{ref} versus time, and V_o versus time. When V_i crosses V_{ref} and goes above it (point *a*), V_o switches to $-V_{sat}$. When V_i crosses V_{ref} and goes below it (point *b*), V_o switches to $+V_{sat}$.

The output voltage wave shape for an inverting zero-crossing detector should be interpreted as follows:

1. When V_o switches from $+V_{sat}$ to $-V_{sat}$ (or from $-V_{sat}$ to $+V_{sat}$), V_i is crossing 0 V (points *a* and *b*).
2. When V_o is at $+V_{sat}$, V_i is below the reference. When V_o is at $-V_{sat}$, V_i is above the reference.

Figure 2-5c shows the transfer function of the inverting zero-crossing detector.

When V_i crosses 0 V going positive, V_o switches to $-V_{sat}$. When V_i crosses 0 V going negative, V_o switches to $+V_{sat}$. Inverting comparators

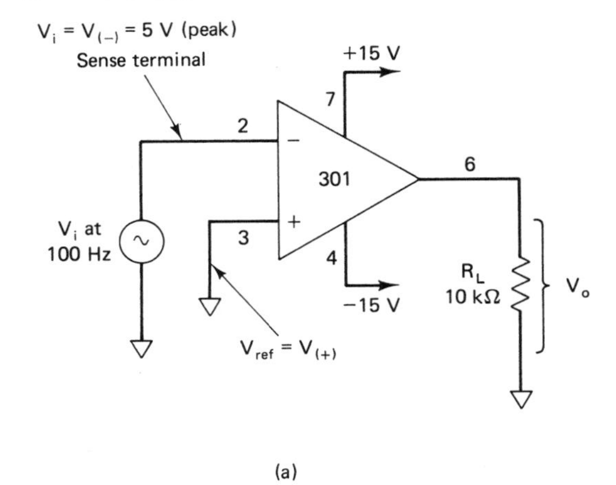

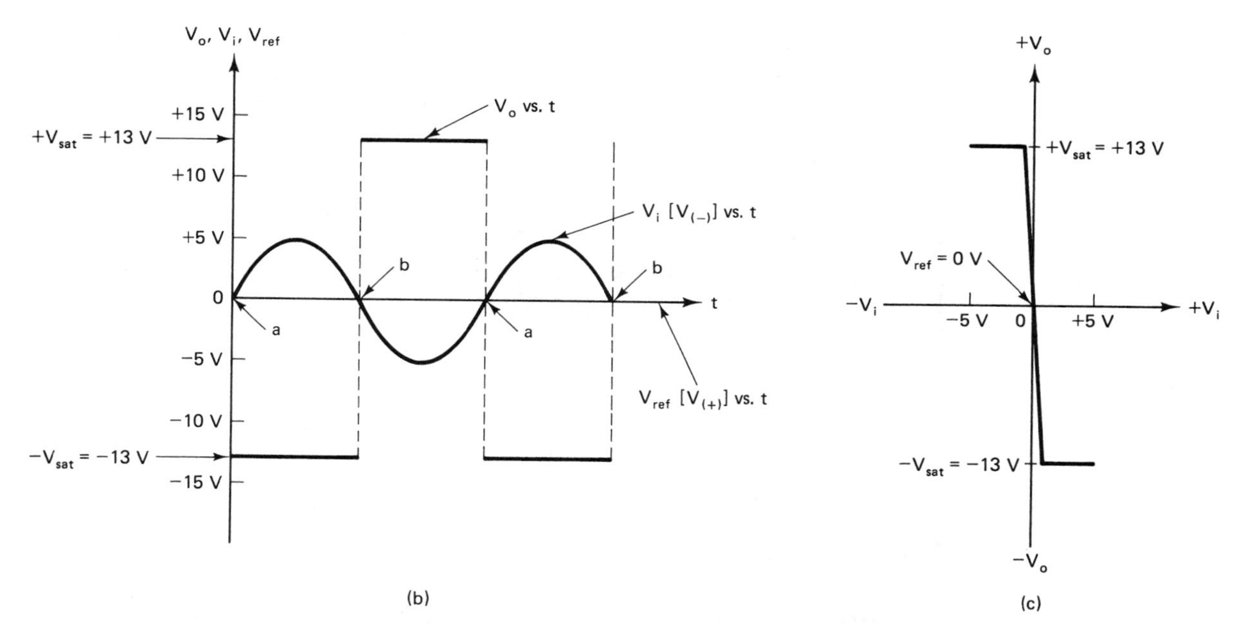

Figure 2-5 An inverting comparator may be used as a zero-voltage level detector. (a) The 301 op amp used as an inverting zero-voltage level detector; (b) $V_o = +V_{sat}$ when V_i is below V_{ref}, $V_o = -V_{sat}$ *when* V_i is above V_{ref}; (c) V_o versus V_i transfer function.

have a transfer function that is characterized by a plot that is *up and to the left.*

2-4 POSITIVE AND NEGATIVE VOLTAGE REFERENCES

2-4.1 Basic Resistor Divider for a Single-Ended Reference Voltage

It is often necessary to supply a precise voltage reference other than zero volts to one input terminal of a comparator. When V_{ref} must be a positive voltage (measured with respect to ground), use the resistor-divider circuit in Fig. 2-6. We assume that the op-amp input draws no current. Then R_1 and R_2 are in series and conducting the same current I.

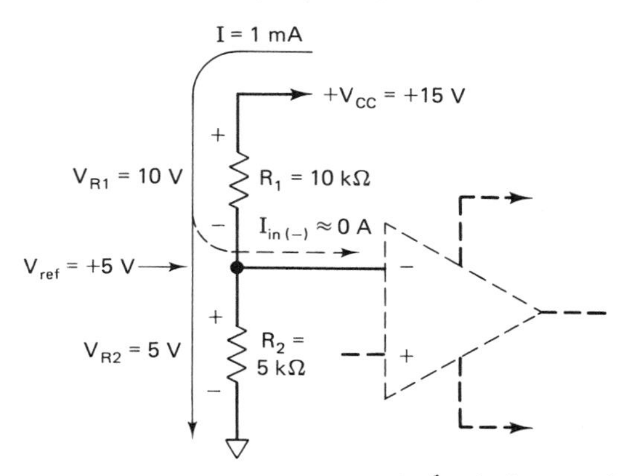

Figure 2-6 A fixed voltage reference is made from a simple resistor-divider circuit.

V_{ref} can be set to any desired voltage by first picking a convenient value for I such as 1 mA. Then calculate the required resistor values. Example 2-3 shows how to design a +5 V reference.

EXAMPLE 2-3

Obtain a reference of +5 V using the op-amp power supply.

SOLUTION: Since only a positive reference voltage is specified, select the V_{CC} = 15 V supply. Select a convenient value for I of 1 mA. Then, since $V_{ref} = V_{R2}$, the value of R_2 is

$$R_2 = \frac{V_{ref}}{I} = \frac{+5\text{ V}}{1\text{mA}} = 5\text{ k}\Omega$$

The voltage across R_1 is $V_{R1} = V_{CC} - V_{ref}$, and $V_{R1} = (15\text{ V}) - (5\text{ V}) =$ 10 V, and $R_1 = \dfrac{10\text{ V}}{1\text{mA}} = 10\text{k}\Omega$. Use 1% resistors for both R_1 and R_2 to achieve the required precision.

We can conclude from Example 2-3 that if the current I is chosen to be 1 mA, *every volt required to be developed across a resistor corresponds to 1 kΩ of resistance.*

2-4.2 Basic Resistor Divider for Positive and Negative References

When the reference voltage must be adjustable, R_2 is replaced by a potentiometer. If reference voltage must be adjustable between a positive and a negative limit, use the solution shown in Fig. 2-7. Three series resistors (R_1, R_2, and R_3) are connected between the V_{CC} and V_{EE} terminals of the op amp power supply. V_{ref} can be set to any value between positive limit V_A and negative limit V_B. The following example illustrates the design of an adjustable reference.

EXAMPLE 2-4

Build an adjustable reference so that V_{ref} varies between ±2.5 V.

SOLUTION: Using the ±15-V bipolar op-amp power supply, first determine the voltage across each resistor as follows:

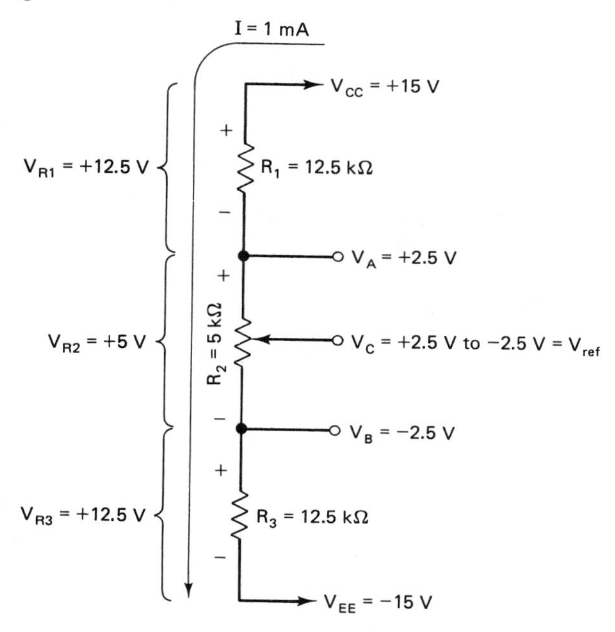

Figure 2-7 Adjustable voltage reference capable of setting V_{ref} to any value between −2.5 and 2.5 V.

$$V_{R1} = V_{CC} - V_A = (+15\text{ V}) - (+2.5\text{ V}) = 12.5\text{ V}$$
$$V_{R2} = V_A - V_B = (+2.5\text{ V}) - (-2.5\text{ V}) = 5\text{ V}$$
$$V_{R3} = V_B - V_{EE} = (-2.5\text{ V}) - (-15\text{ V}) = 12.5\text{ V}$$

Choose $I = 1$ mA so that each volt will require 1 kΩ. Then both R_1 and R_3 must be 12.5 kΩ resistors and R_2 a 5-kΩ potentiometer.

Practical applications of these resistor divider-type voltage references will be shown throughout this book since they are an easy method of establishing dc reference voltages. When the requirements become more restrictive and a precision reference source is required, the solution is slightly more complex. In the next section, a precision voltage reference, the AD584 integrated circuit from Analog Devices, is introduced to solve just such problems.

2-4.3 AD584 Pin-Programmable Precision Voltage Reference

The AD584 is a precision voltage reference. It is an integrated circuit capable of maintaining an accurate and stable output voltage and is packaged in a round 8-pin TO-99 case (Figure. 2-8a). The AD584 achieves high output accuracy by laser trimming of its internal resistors. The four output voltages available by pin programming are +2.500, +5.000, +7.500, and +10.000 V.

Figure 2-8b shows dc power applied to pin 8 and ground to pin 4. This supply voltage can range from 4.5 to 30 V. However, with a typical supply of +15 V, the supply current required by this device is only 0.75 mA. To program the output voltage for, say, 5.000 V, connect a jumper between pin 2 and the output, pin 1, as in Fig. 2-8b.

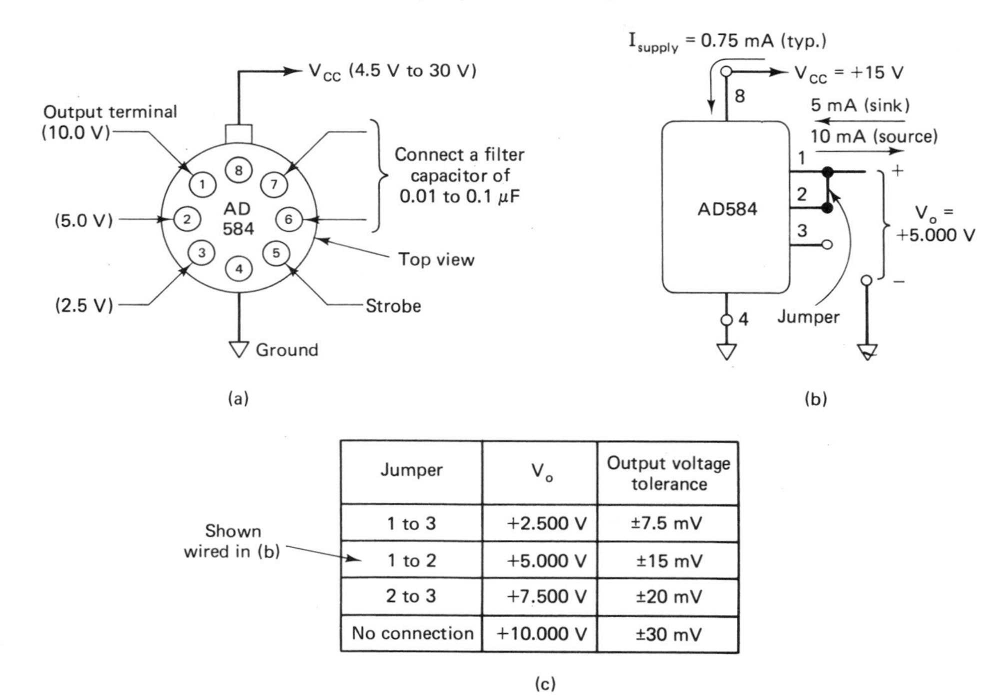

Jumper	V_o	Output voltage tolerance
1 to 3	+2.500 V	±7.5 mV
1 to 2	+5.000 V	±15 mV
2 to 3	+7.500 V	±20 mV
No connection	+10.000 V	±30 mV

(c)

Figure 2-8 AD584 pin-programmable voltage reference: (a) top view of a TO-99 package used to house the AD584 precision voltage reference; (b) AD584 shown symbolically as a simple rectangle with pin 1 always the output; (c) pin connections and output voltage tolerances for the four possible reference levels.

Figure 2-8c shows how to program the pin connections to obtain any one of the four output voltages. Figure 2-8c also shows the output voltage tolerance specified by the manufacturer for each of the programmed output voltages. Note that even at the maximum output of 10.000 V, the output is expected to be within ±0.03 V (±30 mV) of that value.

The output terminal of the AD584 can source a maximum load current of 10 mA or sink a maximum output current of 5 mA. It can accomplish this without change in output voltage.

An additional feature of the AD584 is its strobe terminal at pin 5. Left open, the device operates normally. Connect pin 5 to approximately 0 V (<200 mV) and the output voltage at pin 1 goes to 0 V. Current drain on the power supply is reduced to 0.1 mA to conserve power. This strobe feature allows for direct interface with the output port of a personal computer. In this way the computer can control the presence (or absence) of a reference level.

EXAMPLE 2-5

Show the wiring required to establish a precise +7.500-V reference at the inverting input of an op amp using the AD584 IC.

SOLUTION: Refer to Fig. 2-9. Note that the output, pin 1, is connected to the op amp's inverting input terminal. To set $V_{ref} = +7.500$ V, simply connect a jumper from pin 2 to pin 3.

The AD584 precision reference IC is used in this book when an exact voltage level is to be established.

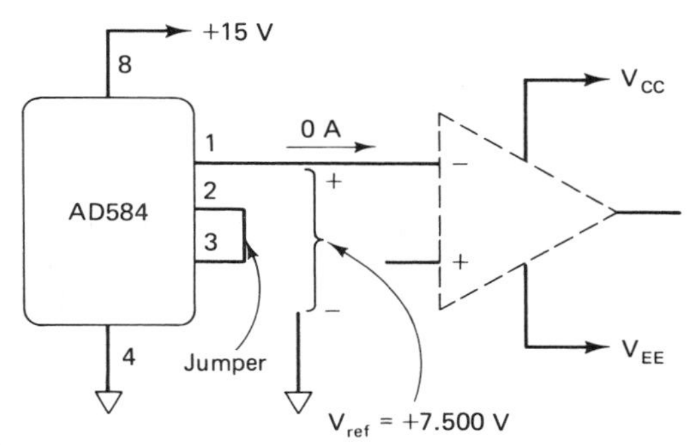

Figure 2-9 The AD584 voltage reference provides precisely +7.500 V as V_{ref} to the inverting input of an op amp.

2-5 NONINVERTING VOLTAGE-LEVEL DETECTORS

2-5.1 Positive Reference-Voltage-Level Detector

A noninverting positive reference-voltage-level detector is shown in Fig. 2-10a. V_i is a 5-V (peak) triangle wave at 100 Hz and is connected directly to the noninverting *sense* terminal of the comparator. The inverting *reference* terminal is connected to the output of an AD584, which is pin-programmed for $V_{ref} = +2.500$ V.

Figure 2-10b shows the plots of V_i versus time, V_{ref} versus time, and V_o versus time for this noninverting positive-voltage-level detector. Figure 2-10b shows that when V_i is above $V_{ref} = +2.500$ V, V_o is at $+V_{sat}$. When V_i is below V_{ref}, V_o is at $-V_{sat}$. The output voltage waveshape for any noninverting voltage-level detector can be interpreted as follows:

1. When V_o switches from $-V_{sat}$ to $+V_{sat}$, V_i is crossing V_{ref} and going above it.
2. When V_o switches from $+V_{sat}$ to $-V_{sat}$, V_i is crossing V_{ref} and going below it.

The transfer function of Fig. 2-10a is shown in Fig. 2-10c. The moment V_i crosses $V_{ref} = +2.500$ V and goes above (more positive) it, V_o switches to $+V_{sat}$. When V_i falls below $V_{ref} = +2.500$ V, V_o switches to $-V_{sat}$.

2-5.2 Negative Reference Voltage-Level Detector

Figure 2-11a illustrates a noninverting negative-voltage-level detector. V_i is applied to the noninverting *sense* terminal. The inverting *reference* terminal is connected to a resistor-divider network set to $V_{ref} = -2.5$ V.

Plots of V_i versus time, V_{ref} versus time, and V_o versus time are shown in Fig. 2-11b. The transfer function is shown in Fig. 2-11c.

1. $V_o = +V_{sat}$ when V_i is above V_{ref}.
2. $V_o = -V_{sat}$ when V_i is below V_{ref}. This noninverting action occurs even though V_{ref} is a negative voltage.

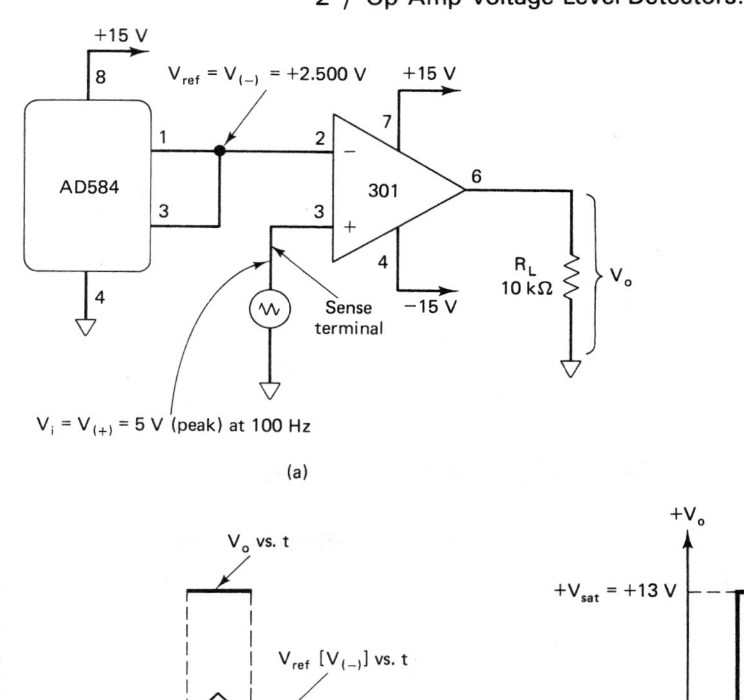

Figure 2-10 A noninverting comparator detects a positive voltage level: (a) noninverting, positive voltage level detector; (b) $V_o = V_{sat}$ when V_i is above V_{ref}; $V_o = -V_{sat}$ when V_i is below V_{ref}; (c) V_o versus V_i transfer function with a positive V_{ref}.

2-6 LIGHT-EMITTING DIODES AS VISUAL INDICATORS

2-6.1 Forward Biasing a LED

In all voltage-level-detecting applications thus far, V_o has been either at $+V_{sat}$ or $-V_{sat}$. It is sometimes useful to make a visual display for the condition of V_o without using a voltmeter or oscilloscope.

One type of visual display is the light-emitting diode (LED). Applications include go/no go, true/false, and pass/fail indicators. The LED is a solid-state diode that emits light. Color depends on the material used for its construction. The LED symbol is shown in Fig. 2-12a.

As with a conventional semiconductor diode, the anode terminal (A) must be more positive than the cathode (K) in order for the LED to emit light. For most diodes, a forward current $I_{LED(F)}$ of about 10 to 30 mA is required for proper illumination. Figure 2-12b illustrates the expected

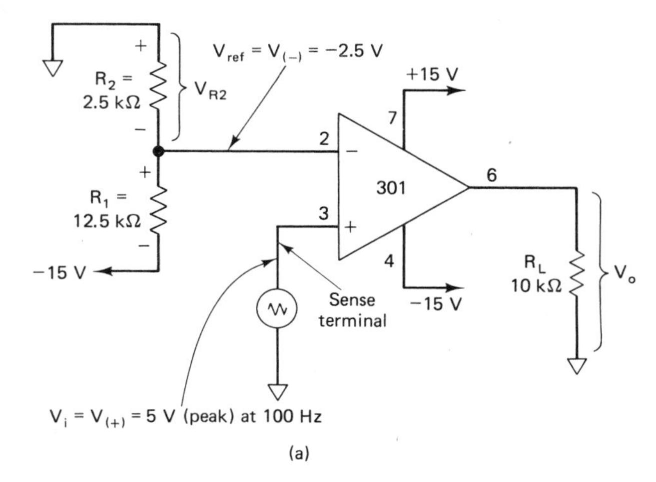

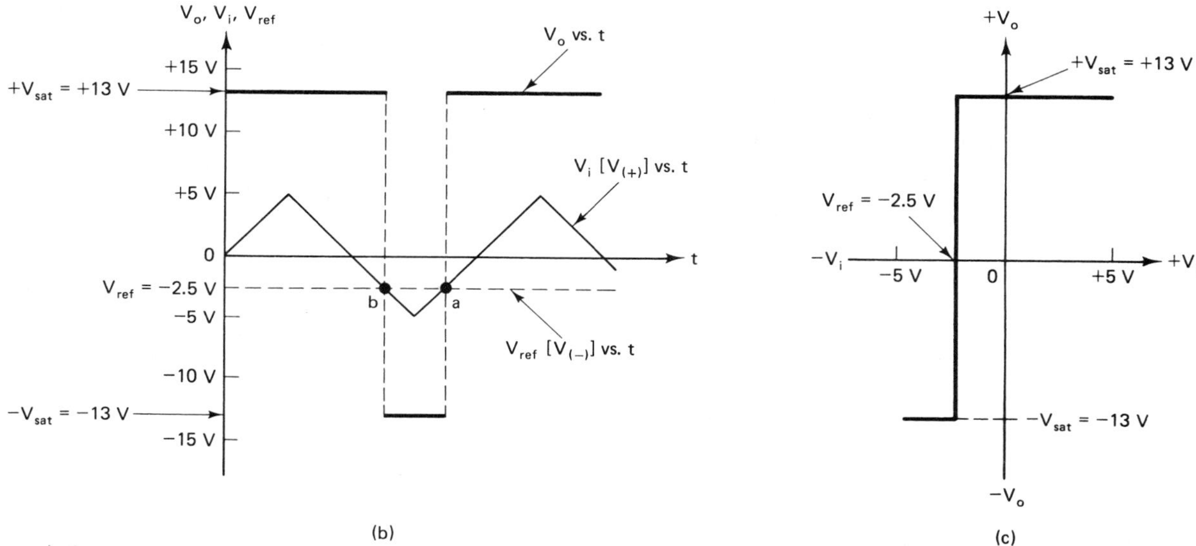

Figure 2-11 A noninverting comparator detects a negative voltage level: (a) 301 as a noninverting negative reference-voltage level detector; (b) $V_o = V_{sat}$ when V_i is above V_{ref}; $V_o = -V_{sat}$ when V_i is below V_{ref}; (c) V_o versus V_i transfer function with a negative V_{ref}.

forward turn-on voltage $V_{LED(F)}$ for the three most popular LED colors when $I_{LED(F)}$ is flowing.

Figure 2-12c shows the op-amp connections that light a LED when V_o is at $+V_{sat}$. If the short-circuit current I_{SC} of the op amp is between the 10 to 30 mA rating of the LED (I_{SC} = 25 mA for the 741), no series current-limiting resistor is required. For op amps without short-circuit protection, a 680-Ω current limiting resistor R_o is needed to prevent damage to the LED.

2-6.2 Reverse Biasing a LED

When the output terminal of an op amp goes negative to $-V_{sat} = -13$ V, the LED is reverse biased and destroyed (Fig. 2-13a). Most LEDs can withstand a reverse voltage of about 3 V; therefore, a conventional silicon diode must be connected across the LED in an inverse-parallel configuration to protect the LED. If V_o goes negative as in Fig. 2-13a, the op-amp

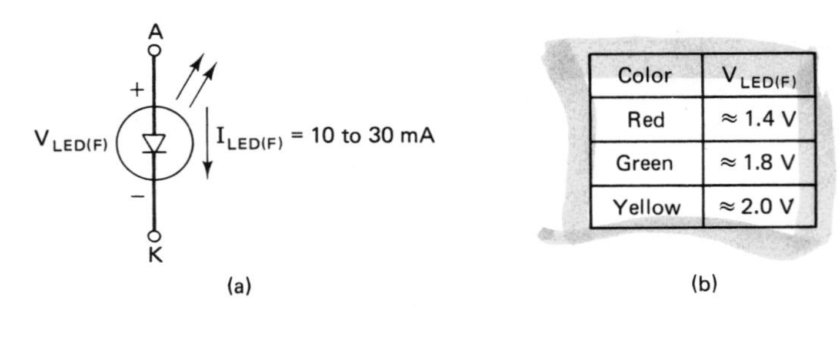

Color	$V_{LED(F)}$
Red	≈ 1.4 V
Green	≈ 1.8 V
Yellow	≈ 2.0 V

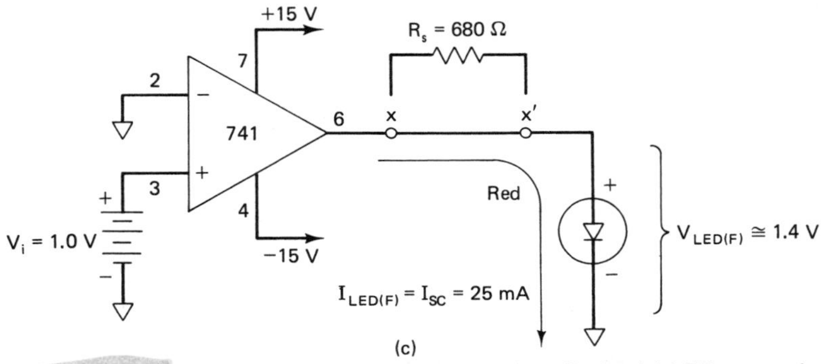

Figure 2-12 Forward biasing a light-emitting diode: (a) LED properly forward biased and emitting light; (b) approximate values of $V_{LED(F)}$ for various-colored LEDs; (c) add a series-limiting resistor between points x and x' for all op amps without internal short-circuit protection.

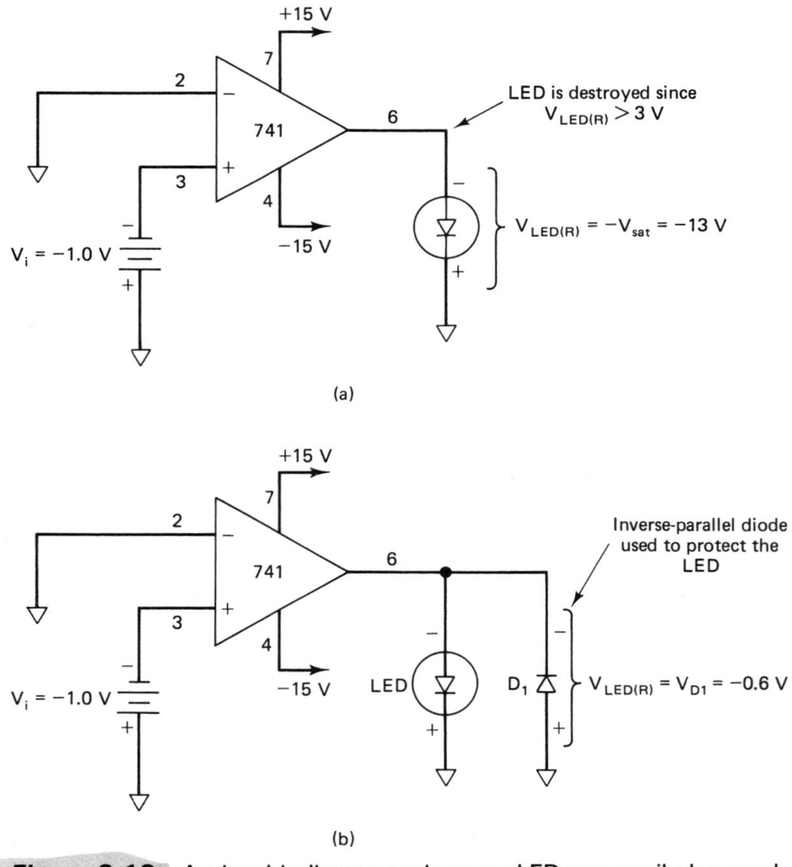

Figure 2-13 A visual indicator such as an LED can easily be used with an op amp. (a) If a reverse voltage of greater than about 3 V is developed across the LED, it will be destroyed; (b) adding an inverse parallel diode as shown will prevent $-V_{sat}$ from appearing across the LED and destroying the device.

output is clamped to −0.6 V, the turn-on voltage of D_1. Therefore, the reverse voltage on the LED can never exceed 0.6 V and is protected from damage.

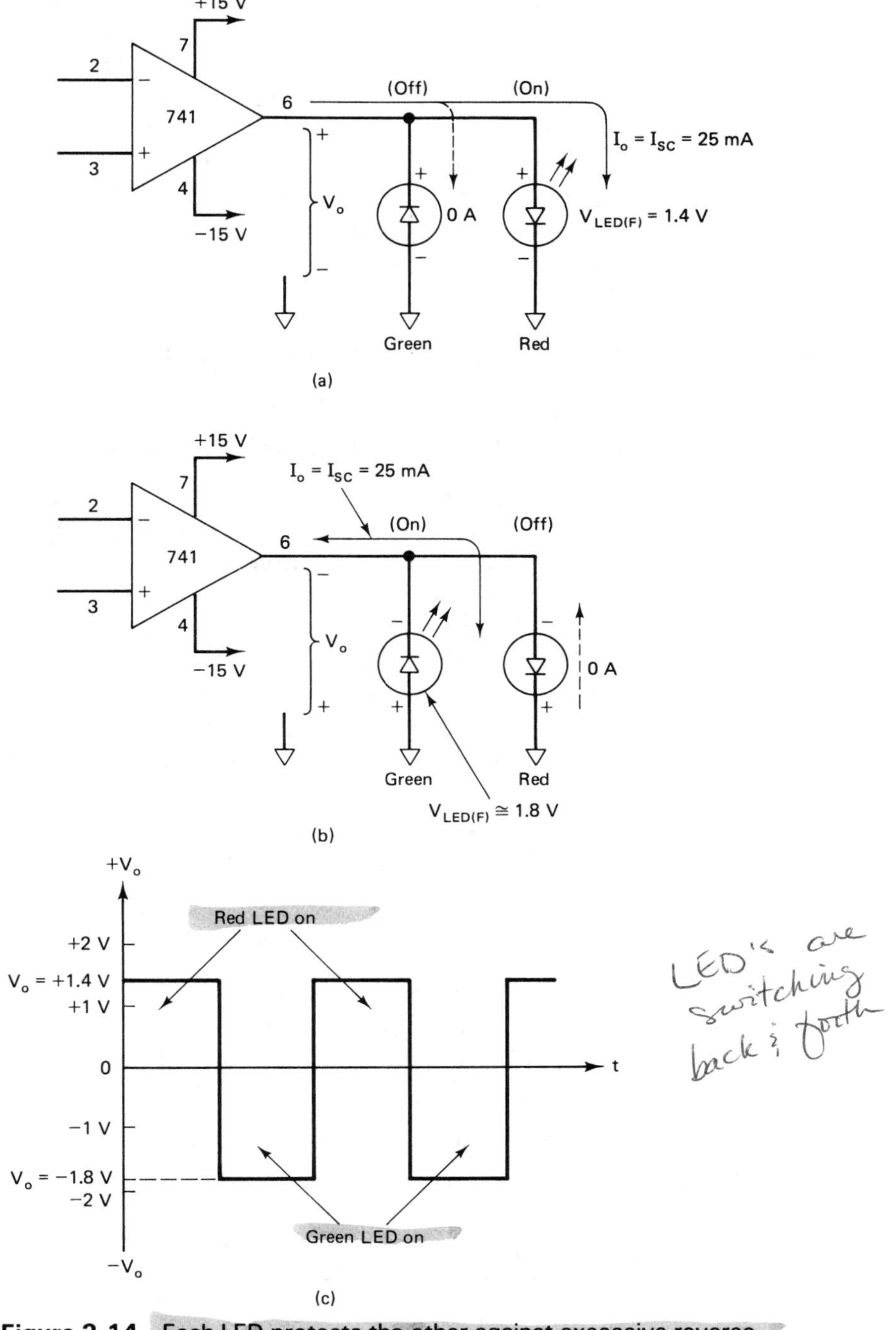

Figure 2-14 Each LED protects the other against excessive reverse bias. (a) The red LED is on with $V_{LED(F)}$ equal to 1.4 V; the green LED is protected with only 1.4 V of reverse bias. (b) The green LED is on with $V_{LED(F)}$ equal to 1.8 V; the red LED is protected by the green LED; (c) Voltage waveshapes when LEDs are used at the op-amp output terminal.

EXAMPLE 2-6

Use two LEDs at the output of an op amp. The red LED should light when the output goes *positive.* A green LED should light when the output goes *negative.*

SOLUTION: Use a 741 with $I_{SC} = 25$ mA, as in Fig. 2-14, to eliminate the need for a current-limiting resistor R_S. Two LEDs are connected to pin 6 in an *inverse-parallel* configuration. When V_o goes positive (Fig. 2-14a), the red LED, which is forward-biased, goes on and clamps pin 6 at 1.4 V. The green LED is reverse-biased with only 1.4 V and is protected.

When V_o goes negative (Fig. 2-14b), the green LED goes on and clamps pin 6 at -1.8 V. The red LED is reverse-biased with only 1.8 V and is protected by the green LED. Figure 2-14c shows the output voltage waveshapes for the conditions of this problem.

2-7 INVERTING VOLTAGE-LEVEL DETECTORS

2-7.1 Postitive Reference-Voltage-Level Detector with LED Visual Indicators

An inverting positive reference-voltage-level detector is shown in Fig. 2-15a. V_i is connected directly to the inverting *sense* terminal of the comparator. The noninverting *reference* terminal is connected to a +2.5-V positive reference voltage.

Figure 2-15b shows the plots of V_i versus time, V_{ref} versus time, and V_o versus time. When V_i crosses the 2.5-V reference and goes above it (point *a*), V_o aims for $-V_{sat}$. However, the green LED turns on and V_o is clamped to -1.8 V. When V_i crosses the reference and goes below it (point *b*), V_o switches positive—aiming for $+V_{sat}$ and turning on the red LED. V_o is clamped to +1.4 V.

The output voltage waveshape for any inverting voltage-level detector can be interpreted as follows:

1. V_o is negative when V_i is above V_{ref}.
2. V_o is positive when V_i is below V_{ref}.

This interpretation is seen in the transfer function of Fig. 2-15c. When V_i goes above $V_{ref} = +2.5$ V, V_o switches negative to turn on the green LED and is clamped to -1.8 V. When V_i falls below V_{ref}, V_o switches positive to turn on the red LED. V_o is clamped to 1.4 V.

2-7.2 Negative Reference-Voltage-Level Detector with LED Visual Indicators

Figure 2-16a illustrates an inverting negative reference-voltage-level detector. V_i is applied to the inverting *sense* terminal, and a negative V_{ref} is applied to pin 3, the *reference* terminal.

Plots of V_i versus time, V_{ref} versus time, and V_o versus time are shown in Fig. 2-16b. Note that V_o goes negative when V_i is above V_{ref} to light the green LED. V_o goes positive when V_i is below V_{ref} to light the red LED. Transition from positive to negative, and vice versa, occurs at $V_{ref} = -2.5$ V.

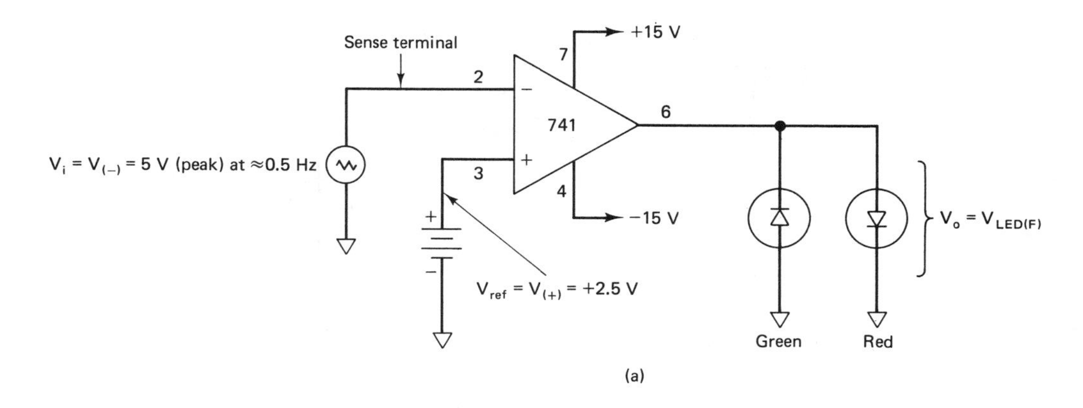

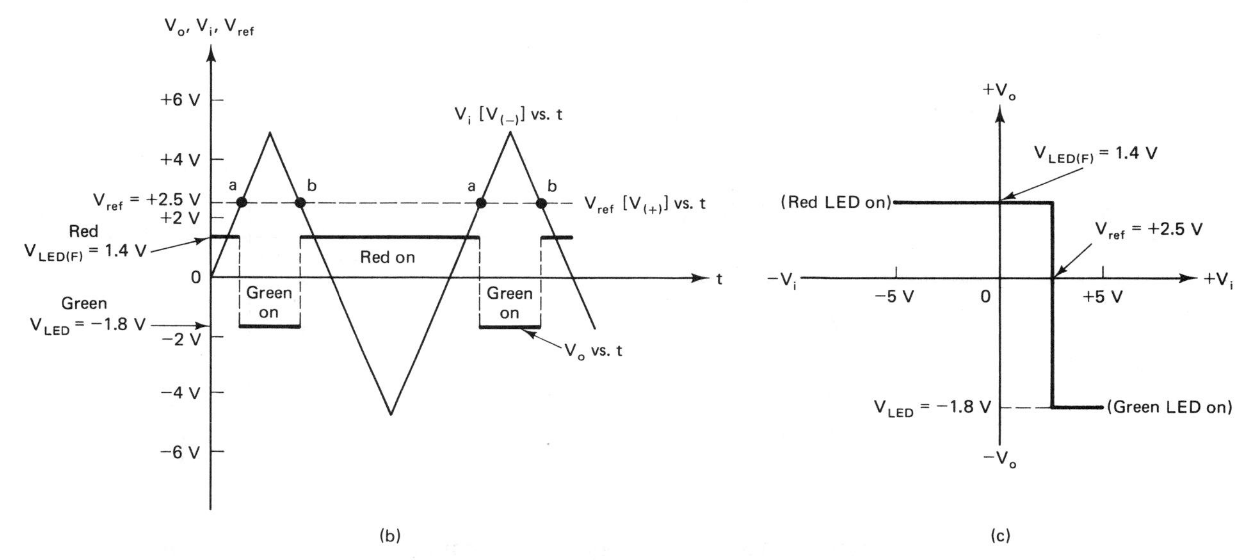

Figure 2-15 An inverting comparator may be used to detect positive voltage LED outputs. (a) Inverting positive reference-voltage-level detector with LED outputs; (b) V_o goes negative (green LED is on) when V_i is above V_{ref}, V_o goes positive (red LED is on) when V_i is below V_{ref}; (c) V_o versus V_i transfer function with positive reference and LED outputs.

2-8 LM311 PRECISION VOLTAGE COMPARATOR

2-8.1 General Performance Specifications

National Semiconductors LM311 is a commercial-grade voltage comparator designed for superior performance over an operating temperature range of 0 to 70°C. Its military version, the LM111, operates over a wider temperature range, −55 to 125°C. Both have become industrial standards. Figure 2-17a illustrates the pinout for an 8-pin mini-DIP package. Several other versions are also available, such as the LH2311 dual comparator, the LF311 with FET input transistors, and the LP311 low-power comparator. Standard LM311 comparators can operate from a single +5-V supply voltage at pin 8. Pin 4 is grounded. They can also be operated with dual supplies as large as ±18 V. When powered from a ±15-V supply, the LM311 draws a positive supply current of 5.1 mA and a negative supply

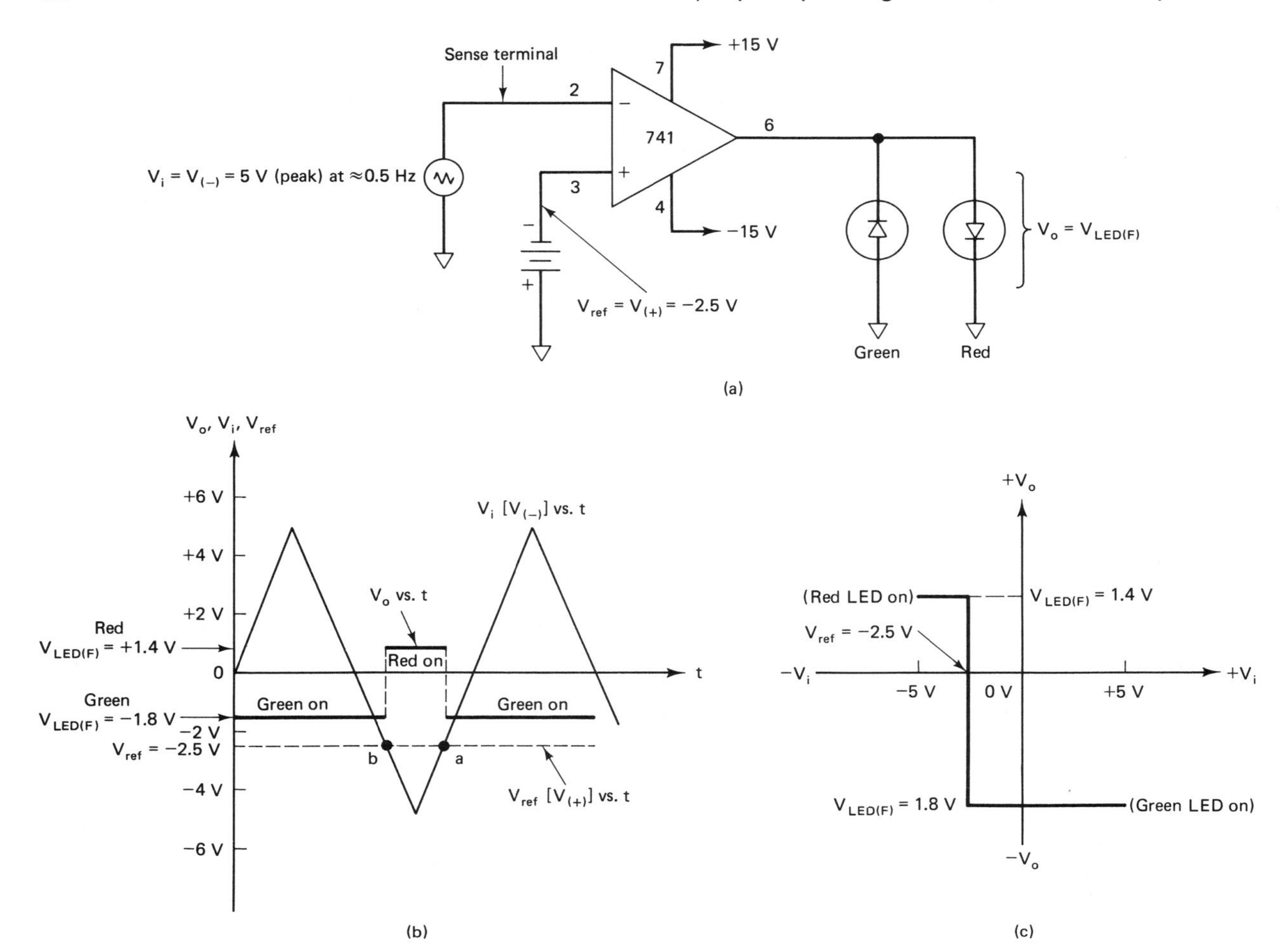

Figure 2-16 An inverting comparator may be used to detect negative voltages. (a) Inverting negative reference-voltage-level detector with LED outputs; (b) V_o goes negative (green LED is on) when V_i is above V_{ref}; V_o goes positive (red LED is on) when V_i is below V_{ref}; (c) V_o versus V_i transfer function with negative reference and LED outputs.

current of typically 4.1 mA. For this reason, the LP311 low-power device is recommended for portable applications requiring batteries.

2-8.2 Output Terminal

Figure 2-17b shows the open-collector configuration provided at the output terminal pin 7 of the LM311. If pin 1 is grounded, the output can be modeled as a switch that is either *open* (output transistor in cutoff) or *shorted* (output transistor in saturation).

An open-collector output allows the LM311 to be compatible with both TTL and CMOS logic levels. In Fig. 2-18 one end of a pull-up resistor is connected to the output terminal at pin 7 and the other end to the appropriate digital power supply V^{++}. For TTL logic V^{++} equals +5 V and for CMOS logic V^{++} is normally +15V. The output voltage V_o at pin 7 is "high" or at V^{++} when the output switch is open. No current flows through the pull-up resistor R and therefore pin 7 is *pulled up* to +5 V (+15 V for CMOS logic).

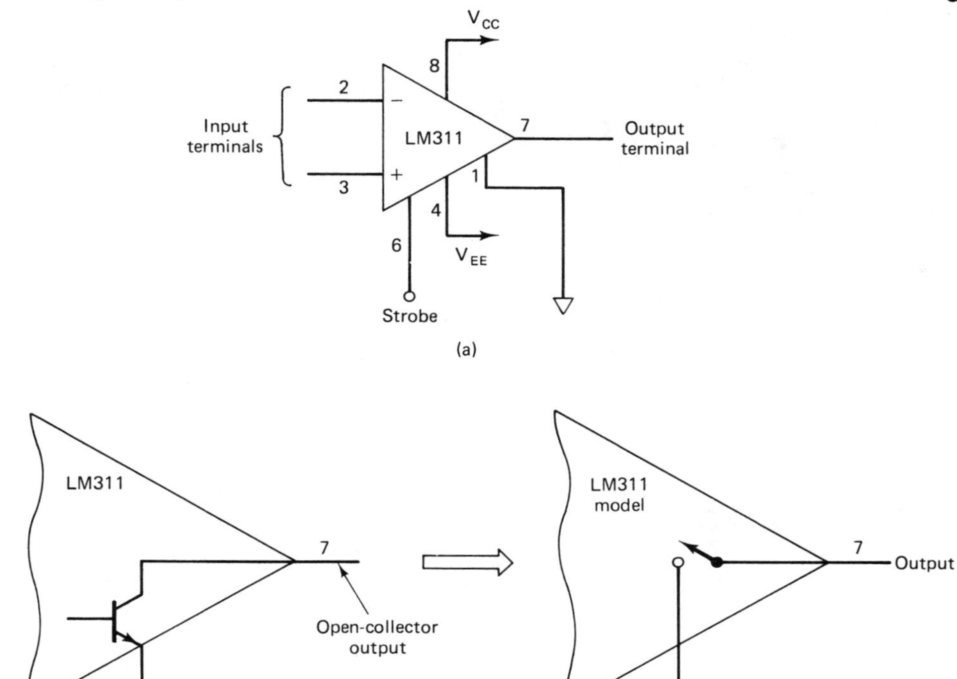

Figure 2-17 LM311 comparator: (a) pin configuration of the comparator housed in a standard 8-pin mini-DIP package; (b) output of the comparator is either an open switch or a switch connecting the output to ground.

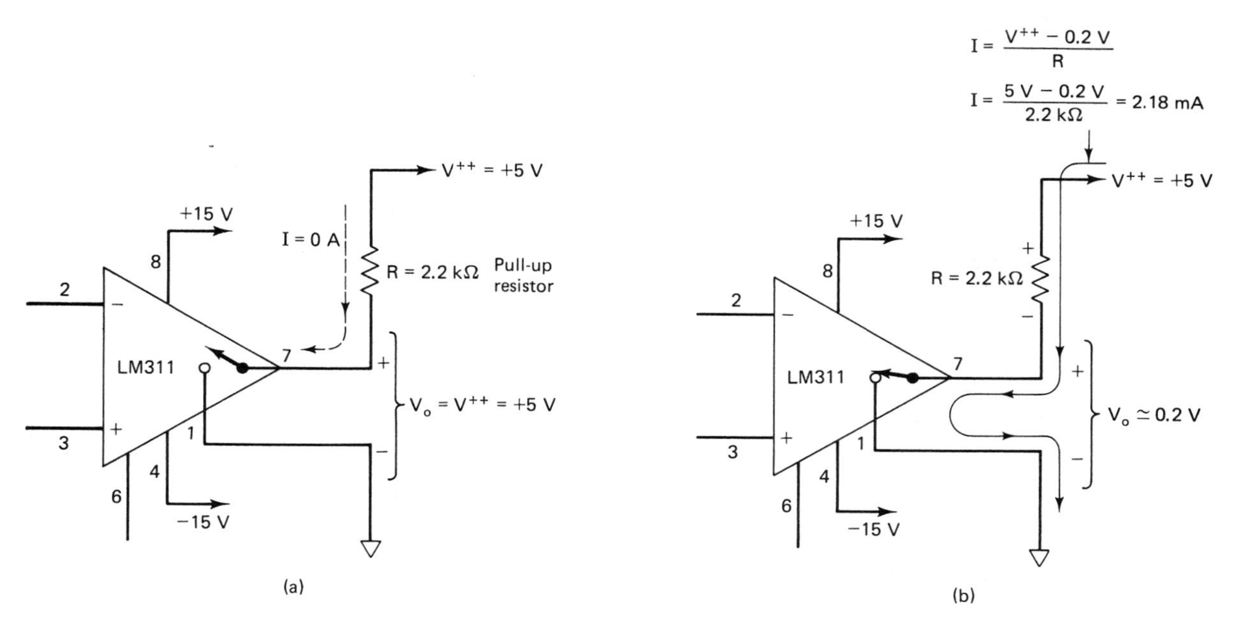

Figure 2-18 The output of an LM311 comparator can be either at V^{++} or approximately 0 V (ground). (a) With the output switch open, V_o *at* pin 7 is pulled up to the V^{++} supply; (b) when the output switch is closed, V_o at pin 7 drops to the saturation voltage of the output transistor, about 0.2 V.

If the output switch is shorted, then, as in Fig. 2-18b, V_o is "low" or at approximately 0.2 V. Current I_{sink} flows through the pull-up resistor R and into the saturated output transistor. The current is usually limited by R to 40 mA or less.

2-8.3 Input Terminals

Output voltage V_o, or the state of the output switch, is dependent on the magnitude and polarity of differential input voltage E_d. In Fig. 2-19a, the inverting input is grounded and +1 V is applied to the noninverting input. E_d is +1 V. For these input conditions, the output switch is open.

When $V_{(+)}$ is above $V_{(-)}$, V_o is high at a level of V^{++} (output switch open).

In Fig. 2-19b, the inverting input is grounded and −1 V is applied to the noninverting input. E_d is equal to −1 V. For these input conditions, the output switch is shorted.

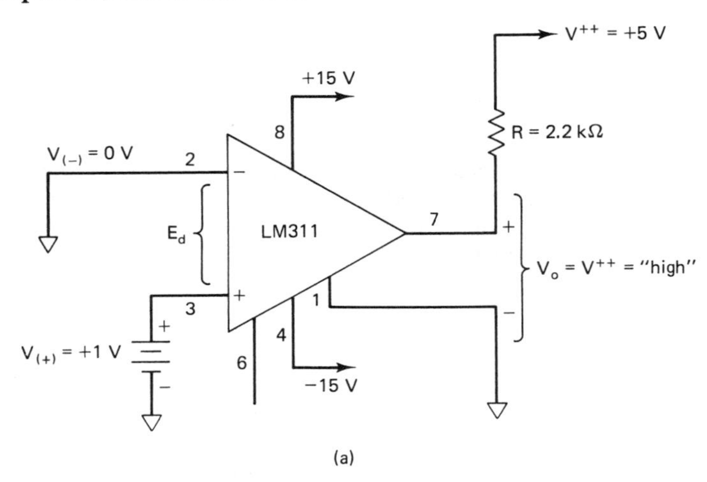

(a)

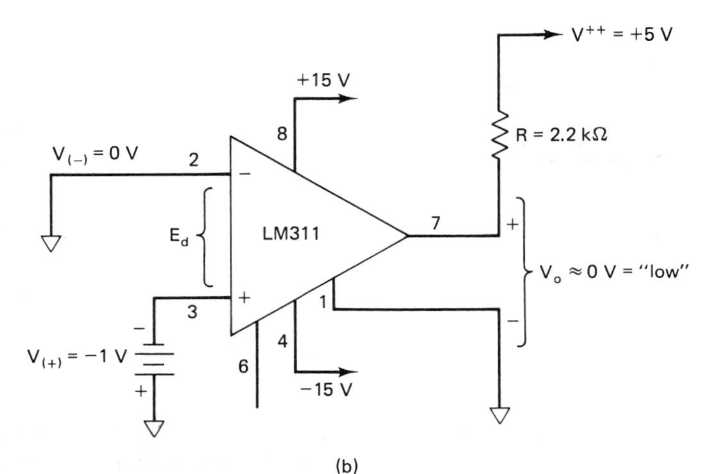

(b)

Figure 2-19 The output of an LM311 comparator; like that of any other comparator, is controlled by its input terminals. (a) When $V_{(+)}$ is above $V_{(-)}$, $V_o = V^{++}$ = "high"; (b) when $V_{(+)}$ is $V_{(-)}$, $V \simeq 0$ V = "low."

When $V_{(+)}$ is below $V_{(-)}$, V_o is low at a level of 0.2 V ($\approx$ 0 V) (output switch is closed).

2-8.4 Strobe Terminal

Strobe terminal pin 6 of the LM311 is usually left open to allow normal control of V_o via the input terminals. When the strobe terminal is grounded through a 10-k Ω resistor, the output switch is open. V_o is forced to V^{++} and will remain at V^{++} as long as the strobe is grounded. This strobe feature allows the LM311 to be controlled by the output port of a computer. The strobe current should be limited to about 3 mA. The LM311 will be used in Chapter 3, where practical applications of voltage comparators are considered.

End of Chapter Exercises

Name: ____________________

Date: __________ **Grade:** ___

A. FILL-IN THE BLANKS

Fill in the blanks with the best answer.

1. There are two main classifications for voltage level detectors __________ and __________ .
2. Voltage level detectors can detect zero, __________ or __________ reference voltages.
3. If a voltage level detector has an inverting characteristic, V_i is connected to the __________ input.
4. If $V_o = +V_{sat}$ in an inverting zero crossing detector, V_i is __________ 0 V.
5. Two resistors, 6 kΩ and 9 kΩ can be connected to a 15V supply to give V_{refs} of __________V or __________V.
6. Strap pins 1 to 3 of an AD584 to obtain V_{ref} = __________ .
7. A forward biased red LED requires a current of approximately __________mA and a forward voltage of __________V.
8. LEDS must not be reverse-biased by more than __________V.
9. If strobe is open, and $E_d = +1$V for an LM311, the output switch is __________ .

B. TRUE/FALSE

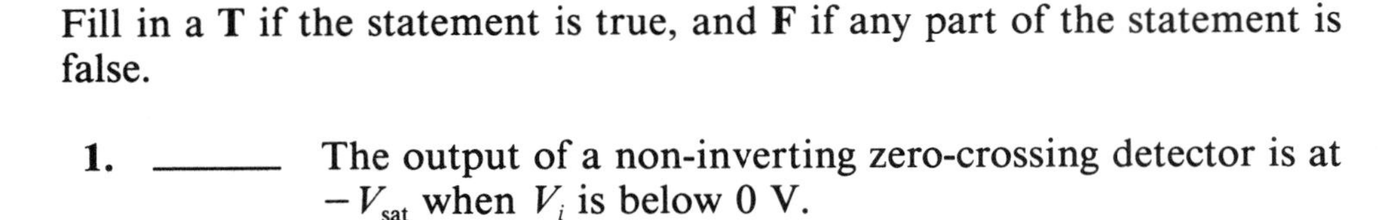

Fill in a **T** if the statement is true, and **F** if any part of the statement is false.

1. ______ The output of a non-inverting zero-crossing detector is at $-V_{sat}$ when V_i is below 0 V.

2. ______ V_o should be at $+V_{sat}$ when V_i is below 5 V. Wire V_i to the non-inverting input and -5 V to the inverting input.

3. ______ Connect V_{ref} to the inverting input to make a non-inverting voltage level detector.

4. ______ To wire LEDs in inverse parallel, connect anode 1 to cathode 2, and cathode 1 to anode 2.

5. ______ A red LED is connected between output of a 741 and ground. When forward biased it will clamp the output to 1.4V and draw all of the op amp's short-circuit current.

6. ______ The LM311 has a push-pull output just like the 741.

7. ______ Disconnect the LM311's strobe terminal to give E_d control over V_o.

8. ______ A pull-up resistor is required to bring high the output of an LM 311.

C. CIRCLE THE CORRECT ANSWER

Circle the correct answer for each statement.

1. When V_i is sensed by the non-inverting input, the circuit is classified as a (non-inverting, inverting) voltage level detector.
2. V_i is applied to pin 3 and 5V to pin 2 of a 301. This circuit is classified as a(n) (inverting, non-inverting) level detector. It detects (-5V, $+5$V).
3. V_o is at $+V_{sat}$ or $-V_{sat}$ when V_i is below and above ground respectively. This circuit is called a zero-crossing (non-inverting, inverting) detector.
4. An inverting positive voltage level detector gives (positive, negative) transitions in V_o when V_i crosses V_{ref} in the positive direction.
5. Pin (1, 2, 3) is the output terminal for an AD584 wired for an output of 10.000V.
6. A non-inverting 741 zero-crossing detector is to light a green LED when V_i is above 0 V, and a red LED when V_i is below 0 V. Connect

the green LED's (cathode, anode) to pin 6 and the red LED's (cathode, anode) to pin 6.

7. When E_d of an LM311 is positive, its output switch is (open, closed) and V_o is at (0 V, V^{++}). Assume strobe is open.

D. MATCHING

Match the name or symbol in column **A** with the statement that matches best in column **B**.

	Column A		Column B
1. ______	AD584	**a.**	V_i wired to pin 2 of 741
2. ______	LM311	**b.**	V_i wired to pin 3 of 741
3. ______	$V_{ref} = 0$ V	**c.**	precision voltage comparator
4. ______	inverting	**d.**	zero crossing detector
5. ______	non-inverting	**e.**	precision voltage reference

PROBLEMS

2-1. Name two comparator classifications and explain how each is identified.

2-2. Name the three possible level detectors.

2-3. Give the comparator classification and type for Fig. 2-2a if the conditions of
(a) Fig. 2-3a,
(b) Fig. 2-3b, and
(c) Fig. 2-3c are applied.

2-4. Repeat Example 2-1 if in
(a) Fig. 2-3a, $V_i = 3$ V;
(b) Fig. 2-3b, $V_i = +1$ V;
(c) Fig. 2-3c, $V_i = -6$ V.

2-5. Give the comparator classification and type for Fig. 2-2b if the conditions of
(a) Fig. 2-3a,
(b) Fig. 2-3b, and
(c) Fig. 2-3c are applied.

2-6. Repeat Example 2-2 if in
(a) Fig. 2-3a, $V_{ref} = +4$ V;
(b) Fig. 2-3b, $V_{ref} = 0$ V;
(c) Fig. 2-3c, $V_{ref} = +2$ V.

2-7. In Fig. 2-4a, make $V_{CC} = +18$ V and $V_{EE} = -18$ V.
(a) What is V_{ref}?
(b) What is V_o when V_i is above V_{ref}?
(c) What is V_o when V_i is below V_{ref}?
(d) Sketch the transfer function.

2-8. In Fig. 2-5a, change V_{CC} and V_{EE} to +12 V and −12 V, respectively.
(a) What is V_{ref}?
(b) What is V_o when V_i is above V_{ref}?
(c) What is V_o when V_i is below V_{ref}?
(d) Sketch the transfer function.

2-9. Repeat Example 2-3 for V_{ref} of
(a) +2 V;
(b) +4 V;
(c) +8 V.

2-10. Show the complete schematic (with pin numbers) for the AD584 precision voltage reference to obtain V_o of
(a) +2.500 V;
(b) +5.000 V;
(c) +7.500 V;
(d) +10.000 V.

2-11. What is the output voltage V_o for each of the conditions of Problem 2-10 if pin 5 of the AD584 is grounded?

2-12. Redraw Fig. 2-10a if $V_{ref} = +5.00$ V and replot Fig. 2-10b and Fig. 2-10c if V_i is changed to a 10-V (peak) triangle wave.

2-13. Redesign the resistor-divider network of Fig. 2-11a for a $V_{ref} = -5.000$ V. Redraw the circuit. Replot Fig. 2-11b and Fig. 2-11c if V_i is changed to a 10-V (peak) triangle wave.

2-14. What is $V_{LED(F)}$ for the following colors:
(a) red;
(b) green;
(c) yellow.

2-15. What is the maximum value of V_{LEDR} that can reverse bias a LED?

2-16. Repeat Example 2-6 using two LEDs at the output of an op amp. A green LED should light when the output goes positive and a yellow LED should light when the output goes negative.

2-17. Replace the green LED in Fig. 2-15a with red and the red LED with yellow. Replot Fig. 2-15b and c.

2-18. Replace the green LED in Fig. 2-16a with a yellow and the red LED with a green. Replot Fig. 2-15b and c.

2-19. Draw the complete schematic of a noninverting, +2.500-V level detector using the LM311 comparator and AD584 precision voltage reference. Indicate all pin numbers. Make the output compatible with TTL logic.

2-20. The strobe terminal of an LM311 comparator is at 0 V. What is V_o?

Laboratory Exercise 2

Name: ____________________

Date: __________ **Grade:** ___

VOLTAGE COMPARATORS

OBJECTIVES: Upon completion of this laboratory exercise on voltage comparators, you will be able to (1) display the V_o versus time and transfer functional of a noninverting zero-crossing detector; (2) design and test a voltage comparator; (3) design a bipolar voltage reference; (4) test noninverting voltage-level detectors; and (5) correctly add LEDs to the op amp.

REFERENCE: Chapter 2

PARTS LIST

1	301 op amp	3	10-kΩ resistors
1	741 op amp	1	red LED
1	10-kΩ trim pot	1	green LED
1	220-Ω $\frac{1}{2}$-w resistor		

Procedure A: Display V_o versus Time and V_o versus V_i Curves for a Non-inverting Zero-Crossing Detector

1. Build the noninverting zero-crossing detector shown in Fig. L2-1. Connect V_{CC} and V_{EE} to pins 7 and 4, respectively. Apply dc power to

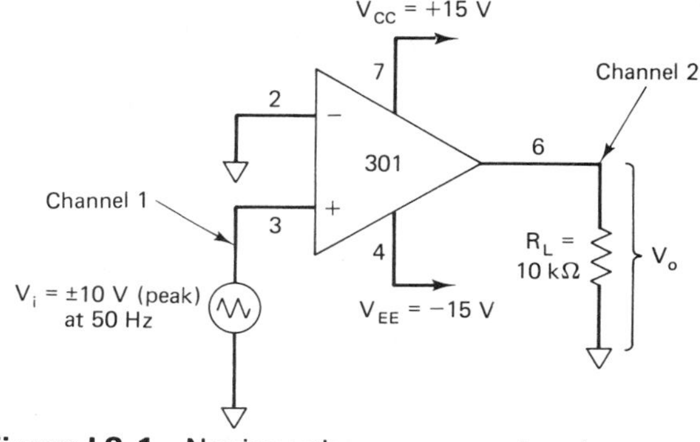

Figure L2-1 Noninverting zero-crossing detector.

the op amp *before* applying the input voltage V_i. Set V_i to a ±10-V (peak) triangle wave at a test frequency of 50 Hz.

2. Connect channel 1 of the CRO to monitor V_i and channel 2 to monitor V_o. Switch the CRO to *dc coupling.*
3. On the V_i versus time curve shown in Fig. L2-2, plot V_o versus time. Include, as part of this sketch, V_{CC} and V_{EE} shown as *dashed* lines. From your sketch of V_o versus time, find the following values: $+V_{sat}$

   = __________ ; $-V_{sat}$ = __________ .
4. What is the value of V_{ref} in the circuit shown in Fig. L2-1? V_{ref}

   = __________ . Label V_{ref} on Fig. L2-2. State what happens to V_o

   as V_i crosses V_{ref} and goes above it? ______________________

   ____

   State what happens to V_o as V_i crosses V_{ref} and goes below it. ____

   ____
5. To display the transfer function for the comparator shown, switch the CRO's time base to (XY). Ground both channels and zero the spot in the center of the scope face (see Fig. L2-3).
6. Switch to *dc coupling* and sketch the transfer function on Fig. L2-3. Include as part of this sketch both V_{CC} and V_{EE} shown as *dashed* lines. From your sketch of V_o versus V_i, again measure $\pm V_{sat}$. $+V_{sat}$

   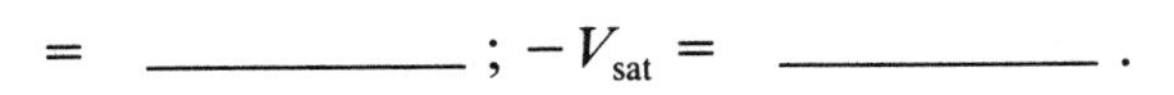

   = __________ ; $-V_{sat}$ = __________ .

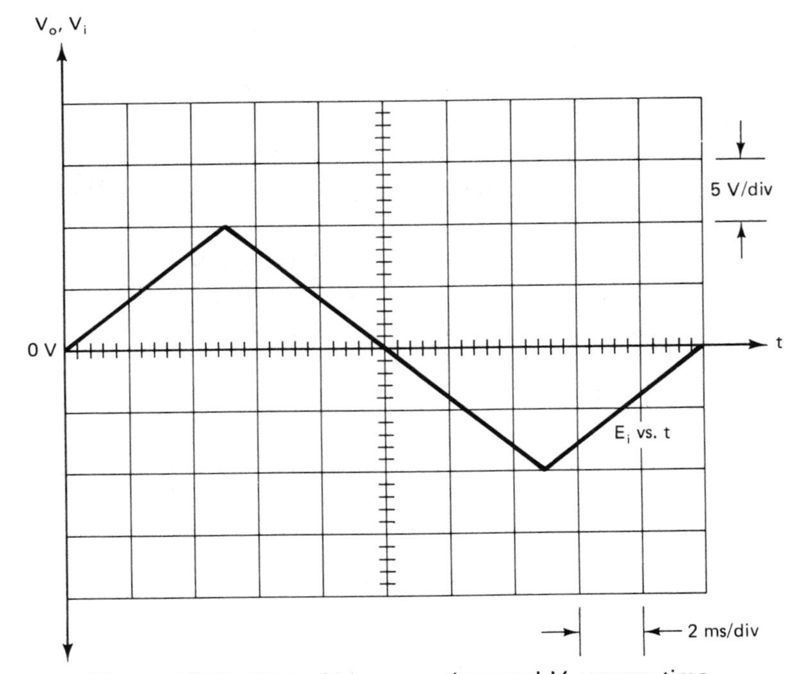

Figure L2-2 Plot of V_i versus time and V_o versus time.

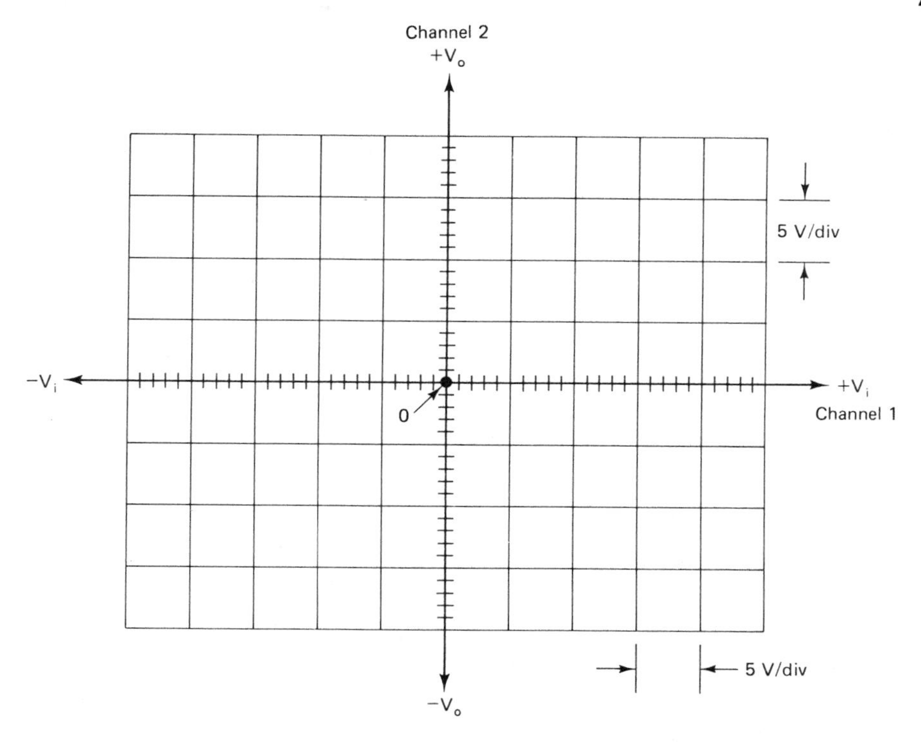

7. What characteristic of Fig. L2-3 indicates that it is the transfer function of a noninverting comparator? ____________________

____

Procedure B: Design and Testing of a Voltage Comparator

8. The transfer function of a voltage comparator is shown in Fig. L2-4. Extract from this curve the following information:

$+V_{sat}$ = __________; $-V_{sat}$ = __________; V_{CC} = __________; V_{EE} = __________; $+V_i$ (peak) = __________; $-V_i$ (peak) = __________; V_{ref} = __________; classification and type of comparator ____________________

9. Use the transfer function of Fig. L2-4 and the information in step 8 to draw the comparator circuit required to produce Fig. L2-4. Label all parts and identify all wiring on the circuit labeled as Fig. L2-5.
10. Build your design as shown in Fig. L2-5. Apply dc power (V_{CC} and V_{EE}) to the circuit. Set V_i to the peak value found in step 8. Adjust V_i to a 50-Hz triangle wave.
11. Connect channel 1 of the CRO to monitor V_i and channel 2 to monitor V_o. Switch the scope to dc coupling and on a time base of 2 mS/div to observe both V_i and V_o simultaneously.

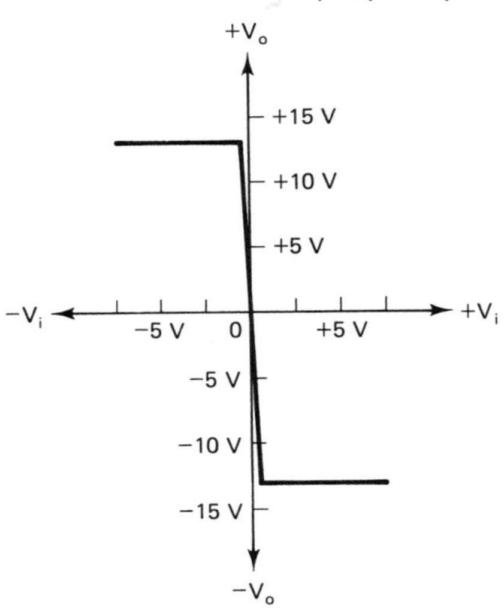

Figure L2-4 Transfer function.

12. Sketch both V_i versus time and V_o versus time on Fig. L2-6. Include on this sketch both V_{CC} and V_{EE} as dashed lines. What is the value of V_{ref} for Fig. L2-5? V_{ref} = ________. Label V_{ref} on Fig. L2-6.

13. What happens to V_o when V_i crosses V_{ref} and goes above it? ____

____

What happens to V_o when V_i crosses V_{ref} and goes below it? ____

____

14. Switch the CRO's time base to (XY) in order to display the V_o versus V_i transfer function. Ground both CRO channels and zero the spot in the center of the scope face (see Fig. L2-7).

15. Switch to dc coupling and sketch the transfer function on Fig. L2-7. Include, as part of this sketch, both V_{CC} and V_{EE} as dashed lines.

16. What must be done to Fig. L2-5 to reduce the output voltage V_o to approximately ±5 V and not $\pm V_{sat}$? Explain. ____________

____

(*Hint:* $\pm I_{SC} = \pm 25$ mA for the 741 op amp. Make the necessary change to Fig. L2-5 and measure V_o.) $\pm V_o = \pm$ ________.

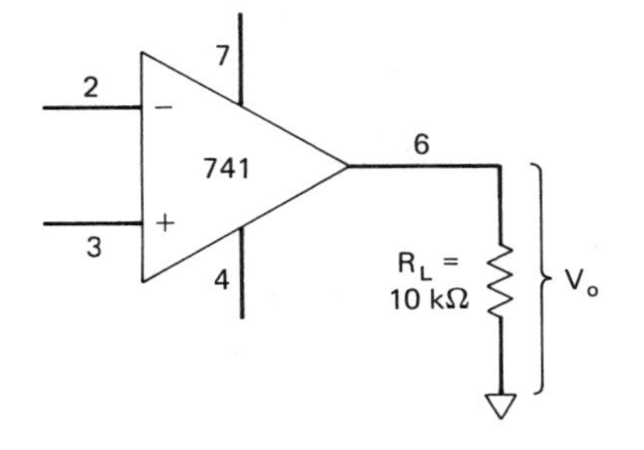

Figure L2-5 Voltage comparator.

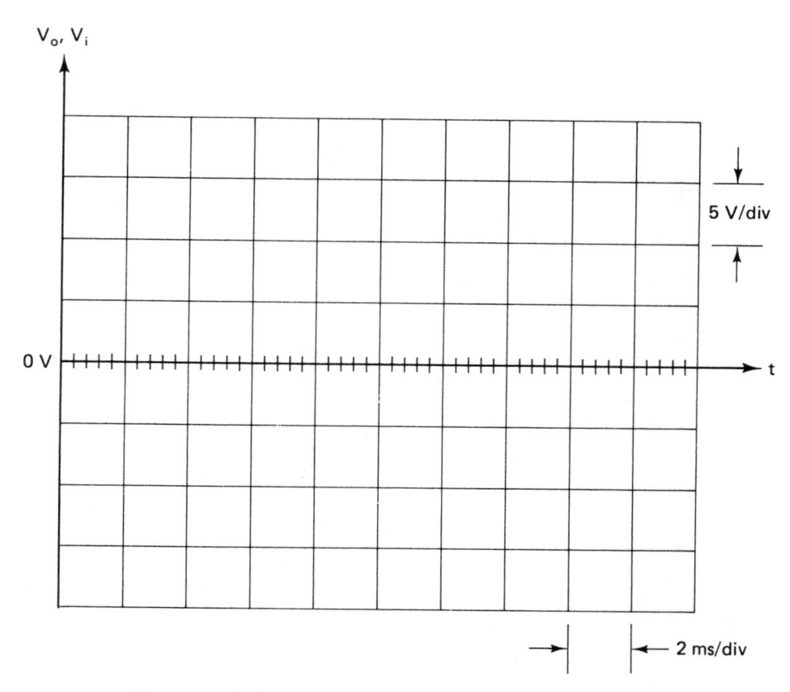

Figure L2-6 Plot of V_i versus t and V_o versus t.

Procedure C: Practical Voltage Reference Design

17. Figure L2-8 shows the partial solution to a practical voltage reference circuit. Complete the design by solving for both R_1 and R_2. V_A is to be +5 V, $V_B = -5$ V, and V_C is adjustable between ±5 V. (a) Solve

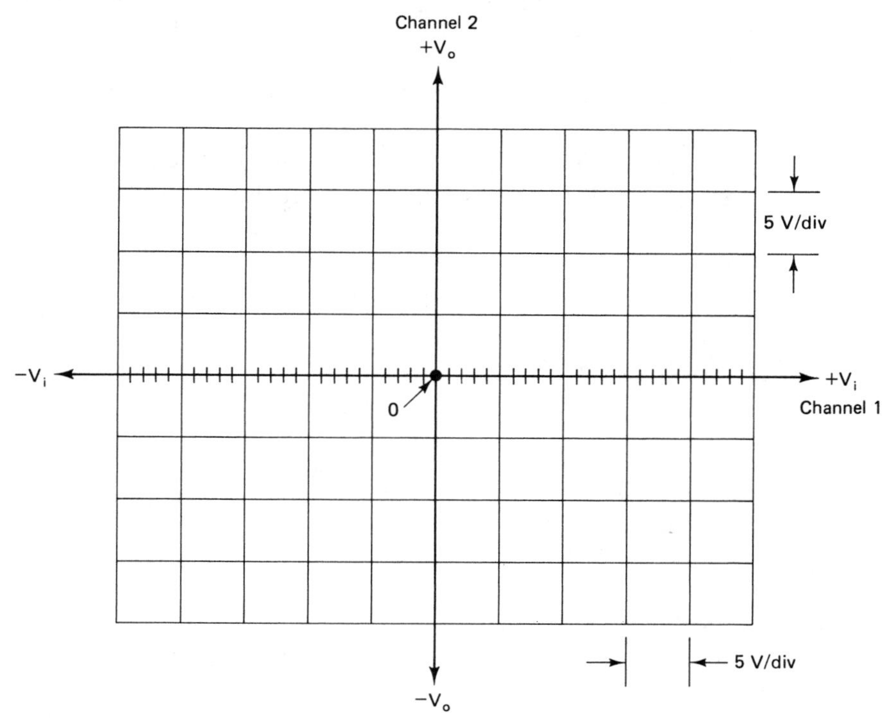

Figure L2-7 Plot of V_o versus V_i transfer function.

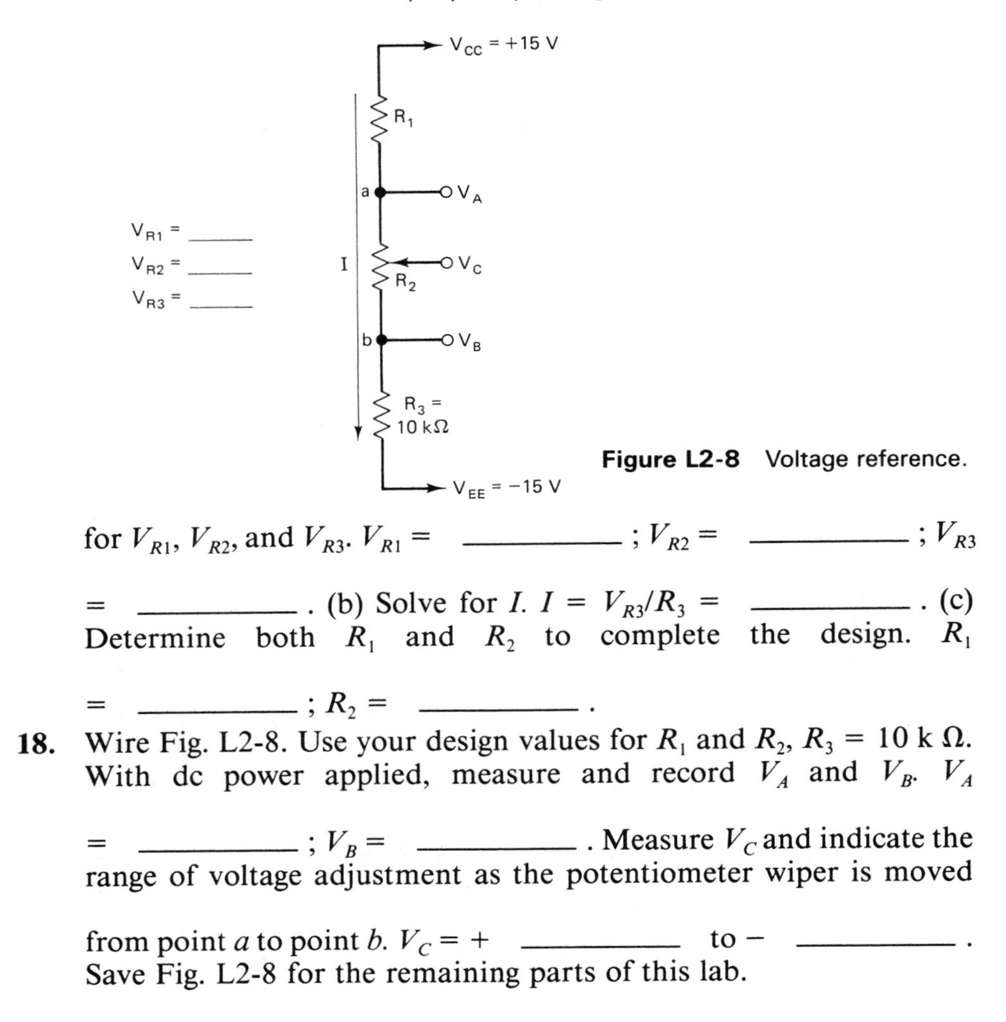

Figure L2-8 Voltage reference.

for V_{R1}, V_{R2}, and V_{R3}. V_{R1} = ______________; V_{R2} = ______________; V_{R3} = ______________. (b) Solve for I. $I = V_{R3}/R_3$ = ______________. (c) Determine both R_1 and R_2 to complete the design. R_1 = ______________; R_2 = ______________.

18. Wire Fig. L2-8. Use your design values for R_1 and R_2, $R_3 = 10\text{ k}\,\Omega$. With dc power applied, measure and record V_A and V_B. V_A = ______________; V_B = ______________. Measure V_C and indicate the range of voltage adjustment as the potentiometer wiper is moved from point a to point b. V_C = + ______________ to − ______________. Save Fig. L2-8 for the remaining parts of this lab.

Procedure D: Noninverting Voltage-Level Detector

19. Wire Fig. L2-9 using the practical voltage reference circuit of Fig. L2-8 to establish V_{ref} on the (−) input.

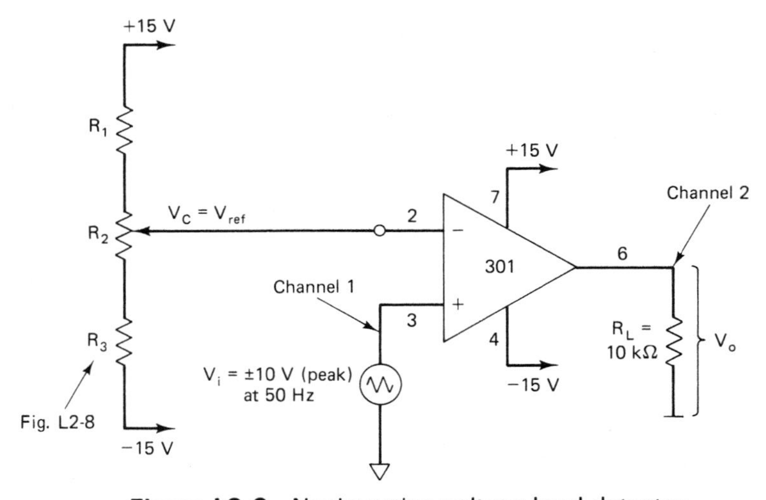

Figure L2-9 Noninverting voltage-level detector.

20. Connect a digital multimeter (DMM) on dc volts to measure V_{ref}, and adjust R_2 to set V_{ref} at +4 V. Adjust V_i to ±10 V peak at 50 Hz and display V_i with channel 1, as shown in Fig. L2-10. Sketch V_o on Fig. L2-10a. At what value of V_i does V_o switch from $-V_{sat}$ to $+V_{sat}$?

V_i = ____________ . At what value of V_i does V_o switch from $+V_{sat}$ to

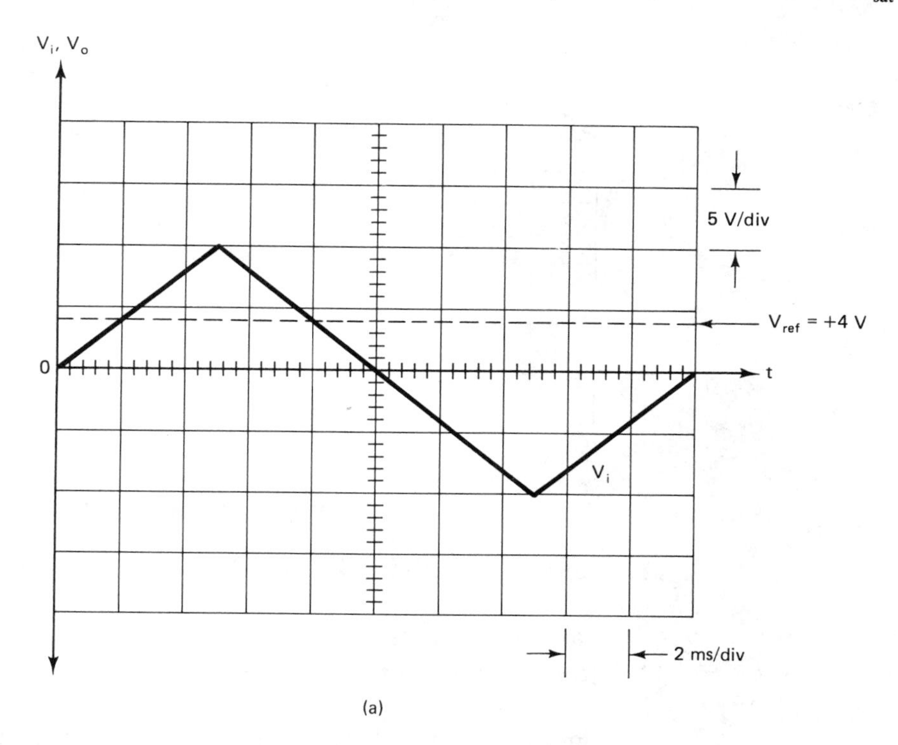

(a)

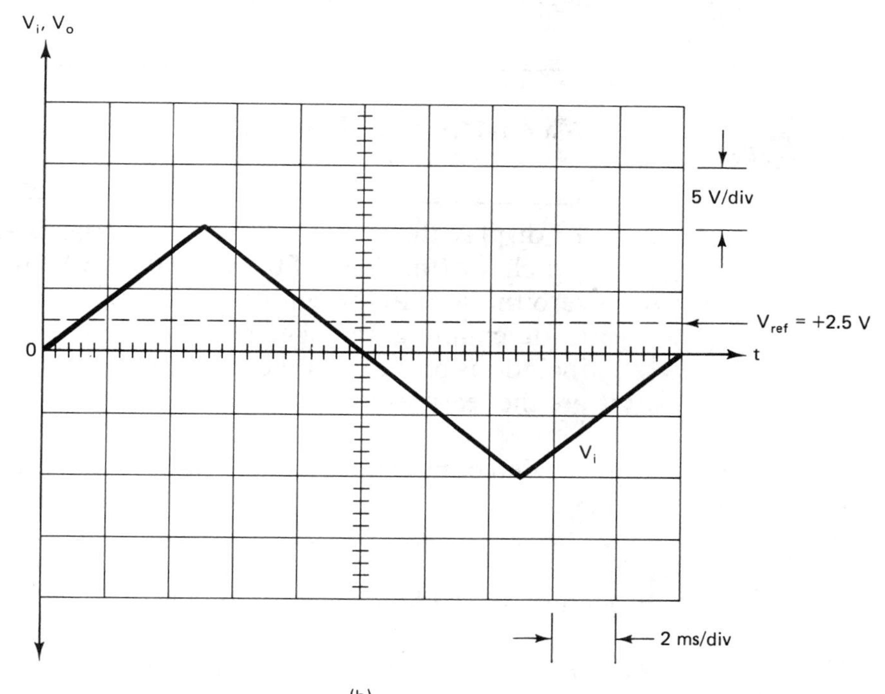

(b)

Figure L2-10 Plots of V_i versus t.

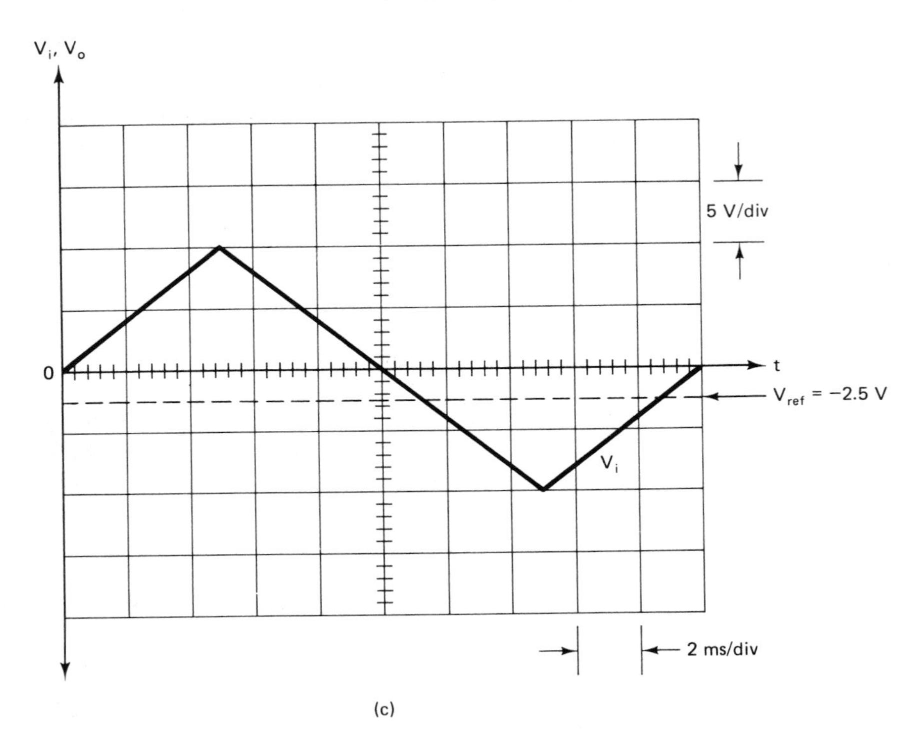

(c)

Figure L2-10 (cont'd)

21. Repeat step 20 for $V_{ref} = +2.5$ V. Sketch V_o on Fig. L2-10b. At what value of V_i does V_o switch from $-V_{sat}$ to $+V_{sat}$? $V_i =$ __________ .

22. Repeat step 20 for $V_{ref} = -2.5$ V. Sketch V_o on Fig. L2-10c. What happens to V_o when V_i crosses V_{ref} and goes above it? __________

____

What happens to V_o when V_i crosses V_{ref} and goes below it? ____

____

23. To display the transfer function of Fig. L2-9 when $V_{ref} = -2.5$ V, switch the time base of the CRO to (XY). Ground both channels and zero the spot in the center of the scope face (see Fig. L2-11). Switch to dc coupling and sketch the curve on Fig. L2-11. Label V_{ref} and include as part of this sketch both V_{CC} and V_{EE} as dashed lines. What are the features of this sketch that indicate a noninverting comparator detecting a nonzero voltage? __________________

____

Procedure E: Adding LEDs to an Op Amp

24. Wire the comparator test circuit shown in Fig. L2-12, taking care to install the LEDs properly.

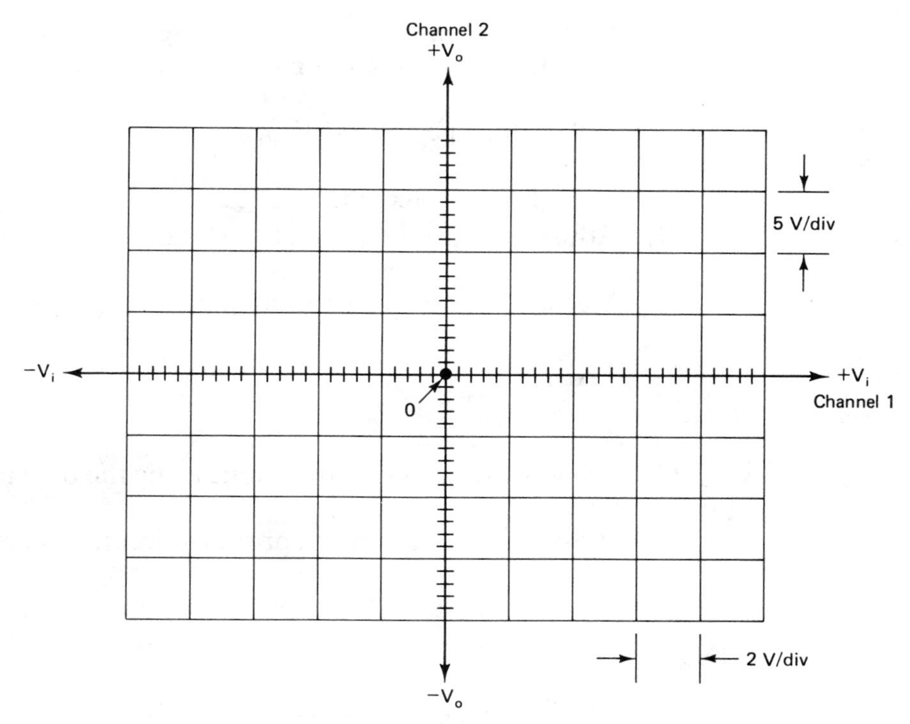

Figure L2-11 V_o versus V_i transfer function.

25. Adjust R_2 until the green LED is "on" and the red LED is "off." Measure and record the following dc voltages with a DMM. V_o = ________; $V_{LED(F)}$ = ________ (green); $V_{LED(R)}$ = ________ (red). Measure the output current of the op amp at points c–c'. I_o = ________.

26. Answer the following questions using the data taken in step 25. (a) Is the op amp in saturation or short-circuit protection? ________.

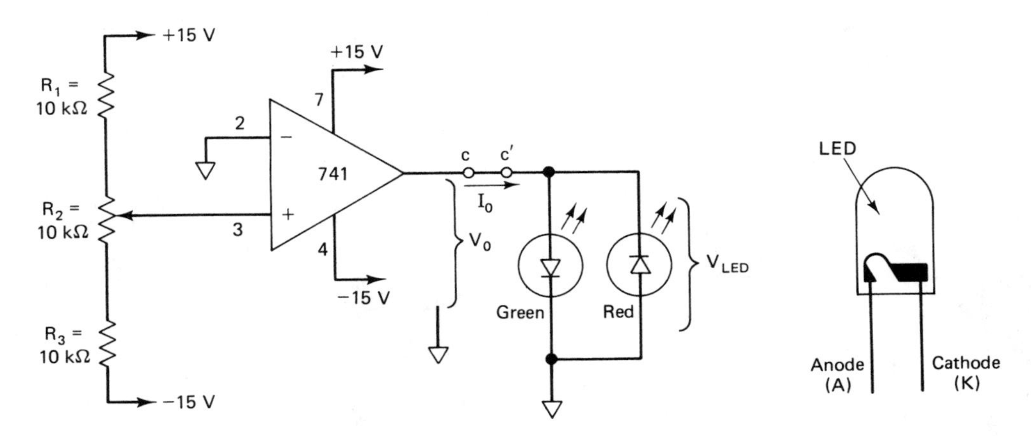

Figure L2-12 Test circuit.

(b) Is the op amp "sinking" or "sourcing" current? ________ .
(c) What is the maximum reverse voltage that can exist across a LED before damage occurs? $V_{LED(R)}$ ________ (maximum). (d) Is the "off" LED protected? ________ .

27. Adjust R_2 until the red LED goes "on" and the green LED goes "off." Measure and record the following: $V_{LED(F)}$ = ________ (red); $V_{LED(R)}$ = ________ (green); V_o = ________ ; I_o = ________ .

28. Answer the following questions using the data taken in step 27. (a) Is the op amp in saturation or short-circuit protection? ________ .
(b) Is the op amp "sinking" or "sourcing" current? ________ .
(c) What is the recommended forward current that an "on" LED can handle? $I_{LED(F)}$ = ________ . (d) Is $I_{LEF(F)}$ for the "on" LED limited to a reasonable value? ________ . (e) What is limiting I_o to the values you measured in steps 25 and 27? ________

____

CONCLUSION:

Applications for Voltage Comparators

3

LEARNING OBJECTIVES

Upon completion of this chapter on the use of voltage comparators, you will be able to:

- Interface 10-element LED bar-graph arrays with comparator ICs to construct useful instruments such as light-column voltmeters.
- Adjust the light-column voltmeter to read any desired voltage level in linear increments.
- Describe the LM358 dual op amp that operates from a single power supply voltage.
- Use the LM358 op amp as the monitor in an overvoltage (crowbar) protection circuit for a dc power supply.
- Interconnect a voltage reference, SCR, and comparator to produce a practical crowbar protection circuit.
- Describe the characteristics of a basic window detector using two LM311 comparators.
- Connect LEDs to the output terminals of a window detector circuit to monitor the status of a dc power supply.
- Graph the typical voltage levels of standard TTL logic.
- Use the standard TTL voltage levels to construct a voltage-divider circuit for a TTL logic probe.
- Construct a TTL logic probe, using the LM358 op amp and a resistor-divider network.

3-0 INTRODUCTION

We will learn how easy it is to use an op amp as a voltage comparator by presenting several useful circuits. Our first circuit combines a 10-element bar-graph LED array with National Semiconductors LM3914 dot/bar display driver integrated circuit. The result is a high-quality light-column voltmeter. The 10 individual comparator and voltage references required for this type of application are all contained within the LM3914 IC. This assures accuracy and lowers the final parts count.

Our next example is providing overvoltage protection (crowbar protection) for a dc power supply. Required is a voltage comparator op amp that can operate from a single supply voltage. The LM358 dual op amp is just such a device. It will be used as the voltage detection device along with an SCR, voltage reference IC, and a few support components. The result is an effective overvoltage protection circuit for any dc power supply.

Protecting a power supply from overvoltage is accomplished with a single op-amp comparator. But monitoring the status of that same supply will require two additional op amps connected as a window detector. A window detector can determine when a voltage is above, below, or within prescribed limits. The output of the window detector is attached to LEDs to provide a visual indication of status.

Our last illustration will show how to make a window-detector circuit that is a practical and portable logic probe for TTL logic. A green LED will indicate the presence of a logic 1, while a red LED will indicate logic 0.

3-1 LIGHT-COLUMN VOLTMETER

3-1.1 Bar-Graph Indicator Circuits

Historically, voltmeters and voltage indicating devices were constructed using analog-type panel meters. For visibility, the resultant systems had to be large. The mechanical movements would become unreliable in portable applications after being dropped. Panel meters can be improved by replacing the meter movement with the more aesthetically pleasing solid-state bar-graph indicator. The result is a smaller display with virtually no reliability problems.

Bar graphs are typically designed around a series of 10 LEDs assembled in a convenient DIP-style package. Functional circuit design is quite simple using one of the many dot/bar driver ICs as the interface between the voltage to be measured and the LED bar-graph indicator. Because of its high reliability and circuit simplicity, this type of display has found wide acceptance in automotive dashboards, audio or medical electronics, and in many industrial applications. Bar-graph indicators can be used as a go/no go indicator, VU meter, or a basic voltmeter, to name just a few applications.

3-1.2 HDSP-4830 Ten-Element Bar-Graph Array

The HDSP-4830 (see data sheets, pp 106-111) 10-element bar-graph array is manufactured by Hewlett-Packard and housed in a standard 20-pin DIP-style package as shown in Fig. 3-1a. This high-efficiency red LED

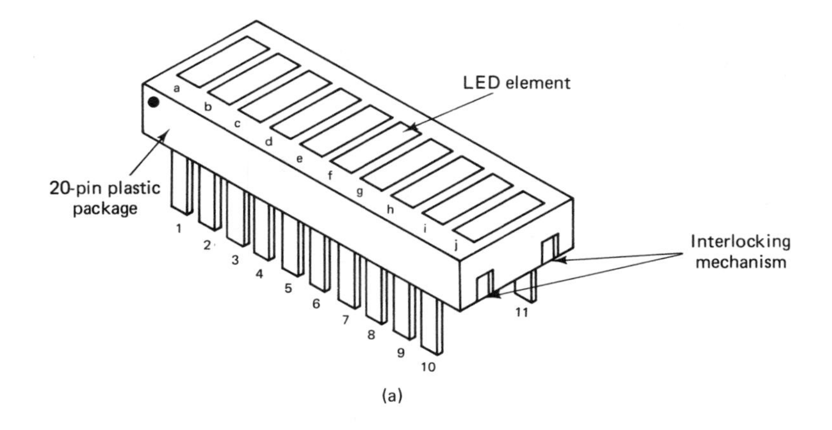

(a)

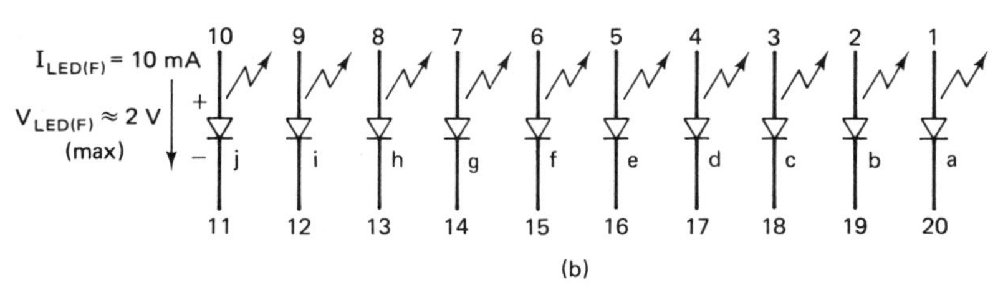

(b)

Figure 3-1 Bar-graph displays are packaged in DIP-style packages with leads provided for individual LEDs. (a) Package design for the HDSP-4830 10-element bar-graph display; (b) schematic representation of the HDSP-4830 10-element bar-graph array.

array is just one in a series of dot/bar-graph arrays manufactured by HP. The series includes a standard red (HDSP-4820), yellow (HDSP-4840), or green (HDSP-4850) array as well as several multicolored arrays.

Each LED segment of the HDSP-4830 display in Fig. 3-lb is wired separately and requires only 10 mA of forward current to provide the proper luminous intensity. With $I_{LED(F)}$ = 10 mA, the forward voltage $V_{LED(F)}$ across the device will not exceed a maximum voltage of about 2.0 V. As with any light-emitting diode, the LEDs within the HDSP-4830 case will be destroyed if reverse voltages, $V_{LED(R)}$, in excess of 3.0 V are applied.

3-1.3 LM3914 Dot/Bar Display Driver IC

Figure 3-2 shows a simplified functional block diagram of the LM3914 dot/bar display driver IC. This chip contains almost all the electronics required to develop an accurate, rugged light-column type of voltmeter. All that will be required, in addition to the LM3914, is a 10-element LED display and a few passive components.

The LM3914 operates from a single supply ranging from 3 to 25 V. Pin 3 is connected to V_{CC} and pin 2 to ground. Internally, there is an accurate 10-resistor voltage-divider network, precision internal voltage source, and an array of 10 inverting-type comparators. They each compare the level of input signal at pin 5 with the voltage levels established on the resistor-divider network.

Figure 3-2 shows how the 1.25-V internal reference voltage source is

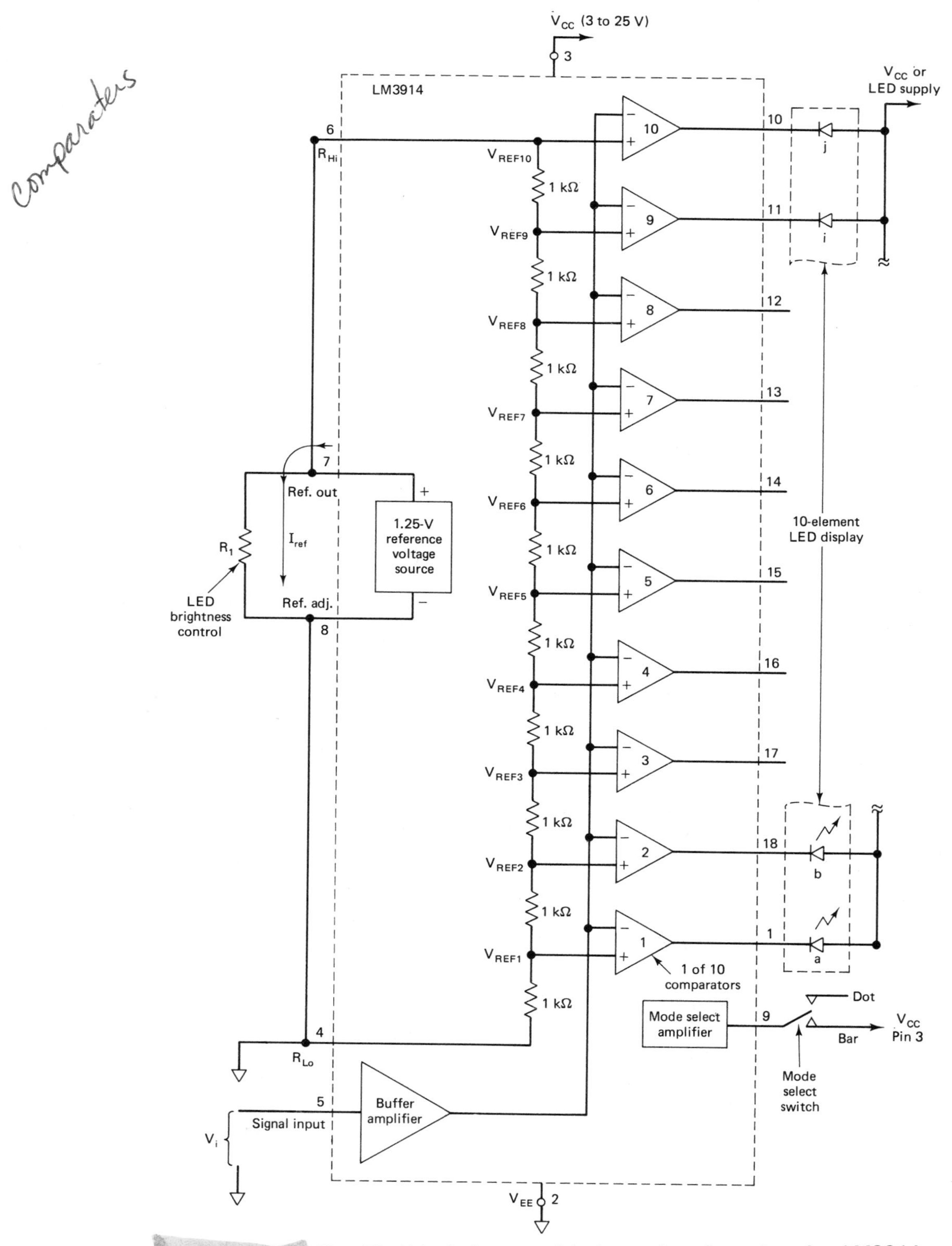

Figure 3-2 Simplified block diagram of the internal configuration of an LM3914 dot/bar display driver IC.

connected directly across the internal resistor string if pin 6 is connected to pin 7, pin 8 to pin 4, and pin 4 to the ground. The precision 1.25-V reference is then divided evenly by each 1-kΩ resistor to make the voltage across each resistor exactly 125 mV. When, for example, the input signal V_i is greater than $V_{REF1} = 125$ mV, the output of comparator 1 is enabled and LED *a* lights. As V_i exceeds $V_{REF2} = 250$ mV, comparator 2's output is enabled and LED *b* also goes on. This continues until V_i exceeds $V_{REF10} =$ 1.25 V enabling comparator 10 and lighting LED *j*.

An additional feature of the LM3914 is the ability to adjust the LED's brightness ($I_{LED(F)}$) with a single-current programming resistor R_1 connected across pins 7 and 8. Current I_{ref} is set by R_1. Each LED current $I_{LED(F)}$ will be approximately 10 times $I_{ref.}$ The desired LED current $I_{LED(F)}$ is set by calculating R_1 from

$$R_1 = \frac{12.5 \text{ V}}{I_{LED(F)}} \tag{3-1}$$

EXAMPLE 3-1

(a) Determine the current programming resistor R_1 for an $I_{LED(F)}$ of 20 mA. (b) Solve for I_{ref} flowing through R_1.

SOLUTION (a) Use Eq. (3-1) to find R_1

$$R_1 = \frac{12.5 \text{ V}}{20 \text{ mA}} = 625\ \Omega$$

(b) Since the voltage across R_1 is the internal voltage of 1.25 V,

$$I_{ref} = \frac{V_{ref}}{R_1} = \frac{1.25 \text{ V}}{625\ \Omega} = 2 \text{ mA}$$

Two modes of display are possible with the LM3914 IC. The first, called *dot display,* is enabled by leaving pin 9, the *mode select* pin, unconnected. In the dot mode, a single LED is on at any one time. A second mode, the *bar display,* is made possible by connecting pin 9 to pin 3 (V_{CC}). Bar mode operation is characterized by all LEDS being on if their driver's reference voltage is below V_i. In the next section we will connect the bar-graph display of Fig. 3-1 to the LM3914 of Fig. 3-2 and construct a practical voltmeter.

3-1.4 0- to 5-V Light-Column Voltmeter

Figure 3-3 shows the complete schematic for a light-column voltmeter capable of displaying, in the bar mode, input voltages V_i from 0 to 5 V.

The HDSP-4830 high-efficiency red bar-graph display will conduct 10 mA per LED. The LM3914's current programming resistor, R_1, is found from Eq. (3-1):

$$R_1 = \frac{12.5 \text{ V}}{I_{LED(F)}} = \frac{12.5 \text{ V}}{10 \text{ mA}} = 1.25 \text{ k}\Omega$$

Program voltage V_{out} must be set to the *full-scale value of* V_i or 5.0 V. V_{out} will be set by installing R_2 between pins 8 and 4. In Fig. 3-3, V_{out} is determined by the 1.25-V reference voltage V_{ref}, R_1, and R_2 from

$$V_{out} = \frac{V_{ref}(R_1 + R_2)}{R_1} \tag{3-2}$$

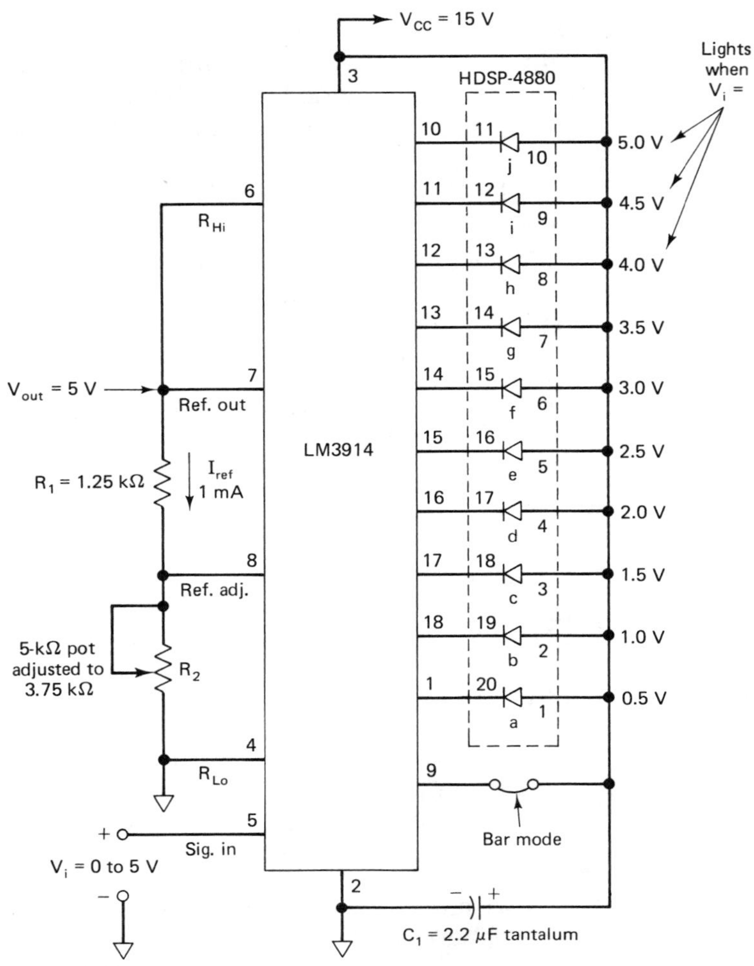

Figure 3-3 A 0-to 5-V dc light-column voltmeter using the LM3914 display driver chip in the bar mode.

If the desired output voltage is known, Eq. (3-2) can be rearranged to find R_2:

$$R_2 = \left(\frac{V_{out} - V_{ref}}{V_{ref}}\right)R_1 \tag{3-3}$$

EXAMPLE 3-2

Determine the value of R_2 to obtain an output voltage range of 5 V. From Eq. (3-1), R_1 equals 1.25 kΩ to set $I_{LED(F)}$ = 10 mA.

SOLUTION Find R_2 from Eq. (3-3):

$$R_2 = \left(\frac{5\text{ V} - 1.25\text{ V}}{1.25\text{ V}}\right) 1.25\text{ k}\Omega = 3.75\text{ k}\Omega$$

R_2 is usually an adjustable resistor as shown in Fig. 3-3. V_{out} can then be trimmed to 5.00 V precisely. A 10-turn pot is desirable for this type of application for a finer adjustment.

As a final practical note, a 2.2-μF tantalum (or 10-μF aluminum capacitor should be installed between the LED power supply terminal and ground if the leads to the LED supply exceed 6 in.

EXAMPLE 3-3

Determine (a) the input voltage range and (b) the current flow through an on LED if the resistors in Fig. 3-3 are $R_1 = 833\ \Omega$ and $R_2 = 5.833\ \text{k}\Omega$.

SOLUTION (a) Using Eq. (3-1) gives us $R_1 = 12.5\text{V}/I_{\text{LED(F)}}$. Therefore,

$$I_{\text{LED(F)}} = \frac{12.5\text{ V}}{R_1} = \frac{12.5\text{ V}}{833\ \Omega} = 15\text{ mA}$$

(b) From Eq. (3-2),

$$V_{\text{out}} = V_{\text{REF}}\left(\frac{R_1 + R_2}{R_1}\right)$$

$$= 1.25\text{ V}\left(\frac{833\ \Omega + 5833\ \Omega}{833\ \Omega}\right) = 1.25\text{ V}(8.00) = 10\text{ V}$$

Thus $V_i = 0$ to 10 V.

3-2 OVERVOLTAGE PROTECTION FOR DC POWER SUPPLIES

3-2.1 Theory of Operation for Crowbar Protection Circuits

A common application of comparators is in the overvoltage protection circuitry for a digital power supply. Conventional transistor-transistor-logic (TTL) normally is operated from a 5-V regulated power supply. If the regulator's output voltage exceeds a maximum of about 5.5 V, the logic it powers can be destroyed. Figure 3-4 shows a typical TTL power supply with an added monitor block and crowbar protection switch.

In Fig. 3-4, a transformer bridge rectifier and capacitor converts a 115-V ac line voltage to an unregulated 12 to 17 V dc across the 1000μF filter capacitor. Next, a 7805 three-terminal IC regulator converts this unregulated voltage to a regulated 5 V.

Fuse F is placed in series with the output of the unregulated supply and the input of the 7805 IC. A noninverting voltage-level detector

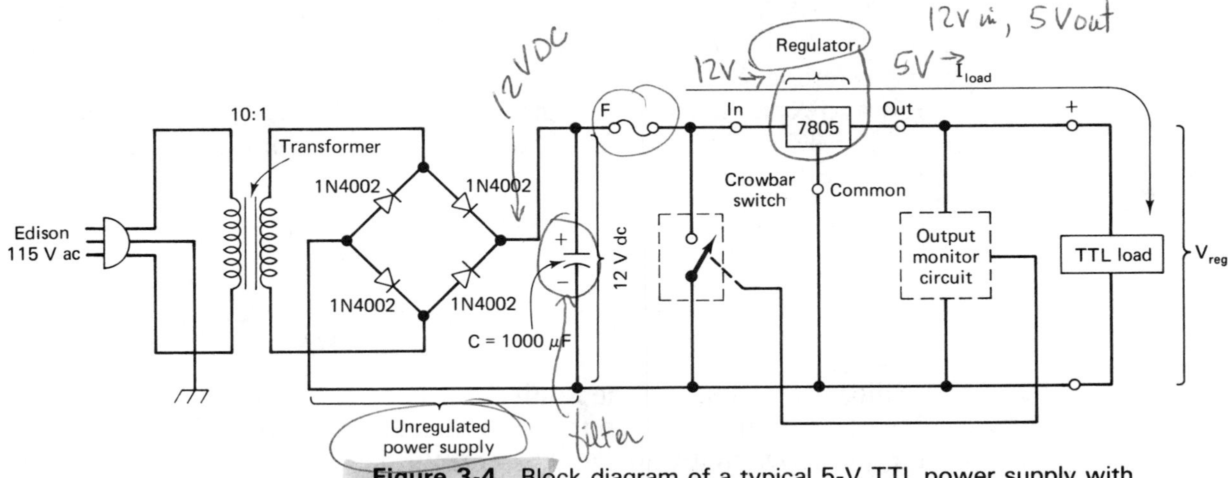

Figure 3-4 Block diagram of a typical 5-V TTL power supply with crowbar (overvoltage) protection.

constantly monitors the regulator's output voltage. Suppose that a fault occurs in the regulator, causing its output to rise to 5.25 V. This is 0.25 V below the damage point. The rise will be detected by the monitor and it will gate on (close) the normally open crowbar switch. When the crowbar switch [actually, a silicon-controlled rectifier (SCR)] is gated on, its low-resistance path to ground will cause the fuse to blow. The blown fuse disconnects the defective regulator chip from the unregulated supply and protects the digital logic from an overvoltage. This type of circuitry is often called *crowbar* protection. The low "on" resistance of the SCR acts like a crowbar, shorting out the unregulated power supply.

3-2.2 Characteristics of the LM358 Dual Op Amp

The crowbar protection circuitry of the preceding section requires a comparator that will operate from a single supply. An excellent choice for this application is National Semiconductor's LM358 dual operational amplifier. Two low-power op amps are packaged in a single 8-pin mini-DIP (Fig. 3-5a). Power is applied simultaneously to both op amps when pin 8 is connected to V_{CC} and pin 4 to ground. V_{CC} can range from +3 to +30 V and the power supply current drain at idle is only 500μA.

Figure 3-5b shows some of the characteristics of each op amp in the package. Note first that the input bias currents are only 45 nA (45×10^{-9}A) and are therefore assumed to draw negligible current.

The output limits are $\pm V_{sat}$. When the output is at $+V_{sat}$, V_o will go to $V_{CC} - 1.5$ V. For example, if V_{CC} is +5 V, $+V_{sat} = 5\text{ V} - 1.5\text{ V} = 3.5\text{ V}$. When the output is at $-V_{sat}$, however, V_o will actually go to 0 V or ground, (unlike the 741).

The LM358 can source an ouput current of 40 mA, or sink 20 mA. This single supply op amp with a few other components will form the monitor circuitry detailed in the next section.

3-2.3 Crowbar Protection for a +5-V Digital Power Supply

One of the LM358 op amps in Fig. 3-6 compares the regulator's output voltage via R_1 and R_2 with a reference voltage. The AD584 establishes a precise reference voltage of 2.50 V on the inverting input, pin 2.

R_1 and R_2 monitor the regulator's output V_{reg}. Under normal operation, $V_{reg} = +5$V. The voltage at the noninverting input pin 3 is calculated from the voltage-divider rule.

$$V_{R2} = V_{(+)} = V_{reg}\left(\frac{R_2}{R_1 + R_2}\right)$$

$$= +5\text{ V}\left(\frac{2.5\text{ k}\Omega}{2.75\text{ k}\Omega + 2.5\text{ k}\Omega}\right) = 2.38\text{ V} \qquad (3\text{-}4)$$

Since the voltage on the noninverting input is below the voltage on the inverting input V_o will go to $-V_{sat} = 0$ V. No current will flow through R_3 and diode D_1. With no gate current, the SCR crowbar switch will be "off."

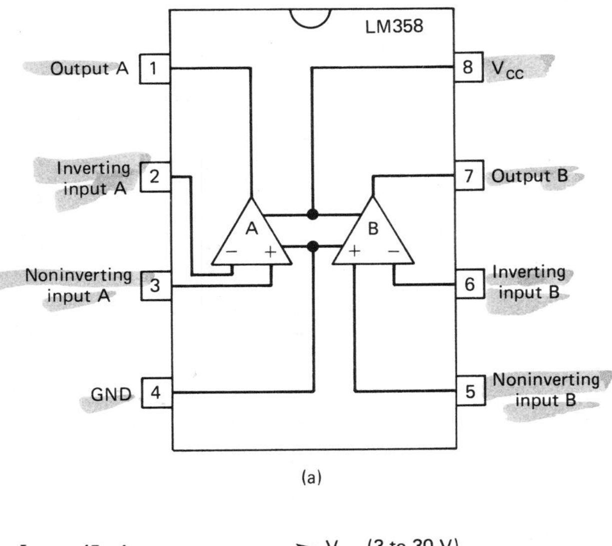

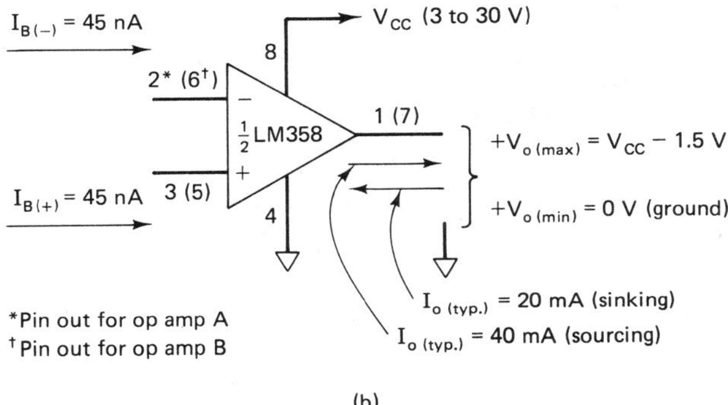

Figure 3-5 The LM358 is a low-powered dual op amp capable of single-supply operation. (a) The LM358 dual op amp in an 8-pin mini-DIP; (b) some of the basic characteristics of an LM358 op amp.

When a fault occurs causing V_{reg} to exceed the "trip" point of 5.25 V, the voltage on the noninverting input will exceed

$$V_{(+)} = 5.25\ \text{V}\left(\frac{2.5\ \text{k}\Omega}{2.75\ \text{k}\Omega + 2.5\ \text{k}\Omega}\right) = +2.5\ \text{V}$$

A fault condition thus causes $V_{(+)}$ to be slightly above $V_{(-)} = 2.50$ V. This input switches V_o of the LM358 to $+V_{sat} = 3.5$ V. V_o forward biases D_1 to force gate current into the SCR and turn it "on." Fuse F blows the moment the SCR anode current exceeds the fuse's rated value. The series limiting resistor R_3 is calculated in the following example.

EXAMPLE 3-4

Calculate the value of R_3 required to turn on the SCR. Assume that I_{GT} must be a minimum gate current of 5 mA with a minimum gate voltage of V_{GT} of 0.5 V.

SOLUTION When $V_o = +V_{sat,}$ the output of the op amp feeds three

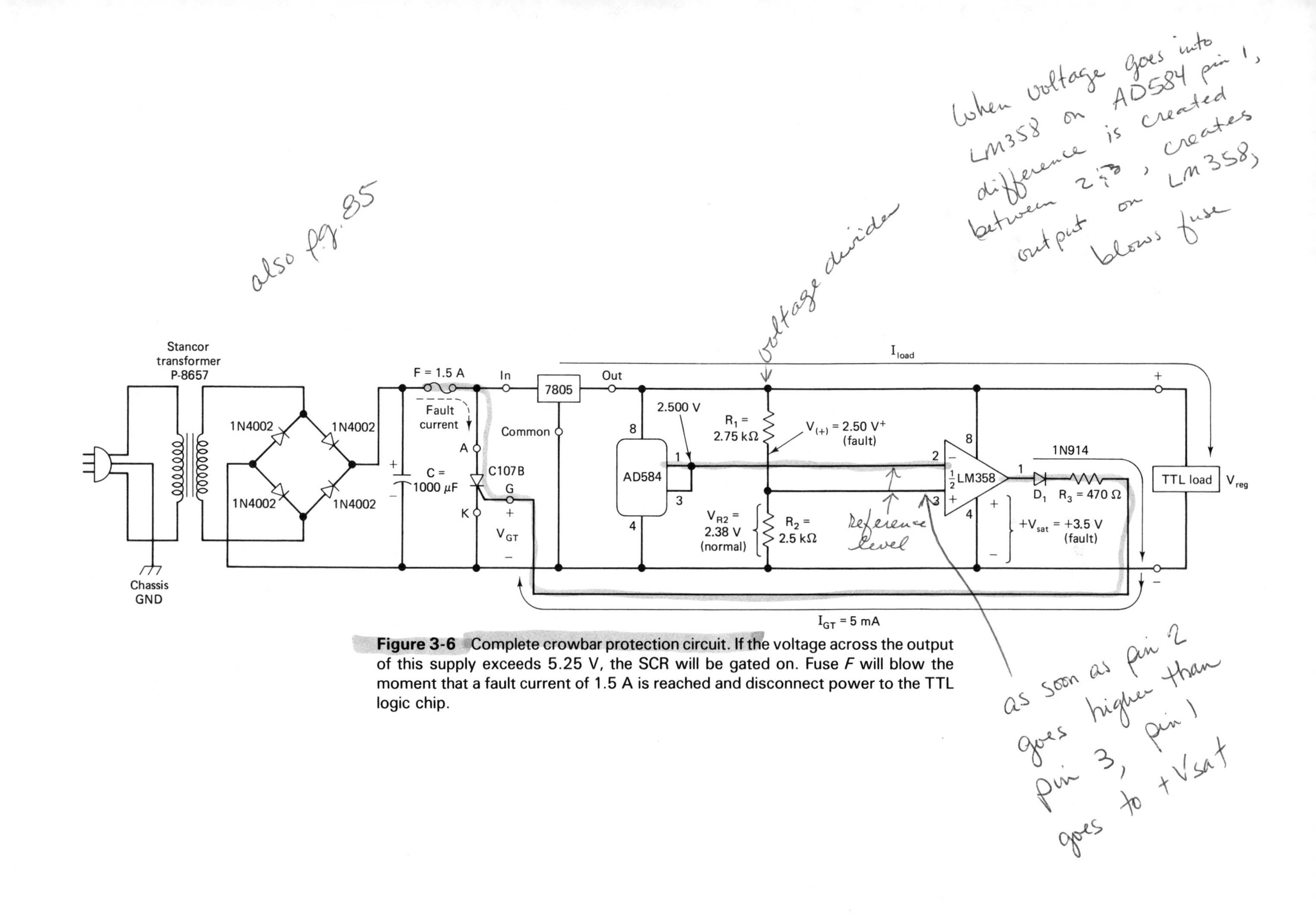

Figure 3-6 Complete crowbar protection circuit. If the voltage across the output of this supply exceeds 5.25 V, the SCR will be gated on. Fuse *F* will blow the moment that a fault current of 1.5 A is reached and disconnect power to the TTL logic chip.

voltage drops. These are $V_{R3} = I_{GT}R_3$, V_{D1} and V_{GT}. Thus $+V_{sat} = I_{GT}R_3 + V_{D1} + V_{GT}$. Therefore, R_3 is calculated from

$$R_3 = \frac{+V_{sat} - V_{D1} - V_{GT}}{I_{GT}} = \frac{3.5\text{ V} - 0.6\text{ V} - 0.5\text{ V}}{5\text{ mA}}$$

$$= \frac{2.4\text{ V}}{5\text{ mA}} = 480\Omega$$

Select a standard value 470-Ω resistor for R_3.

3-3 WINDOW COMPARATORS

3-3.1 Basic Window Detector

The circuit of Fig. 3-7 is called a window detector. It monitors V_i and indicates when this voltage is either above, below, or within two prescribed voltage limits. The noninverting input of comparator A is attached to both the inverting input of comparator B and V_i. An upper reference voltage V_{UR} of +2.0 V is applied to the inverting input of comparator A and a lower reference voltage V_{LR} of −2.0 V is applied to the noninverting input of comparator B.

Circuit operation of the basic window comparator in Fig. 3-7 can be summarized as follows:

1. When V_i is between V_{UR} and V_{LR} (i.e., $-2.0\text{V} < V_i < 2.0\text{V}$), the outputs of both comparators V_{O1} and V_{O2} are = 0 V.
2. When V_i is above $V_{UR} = 2.0$ V, $V_{O1} = V_{CC} = +15$ V and $V_{O2} =$ 0 V.

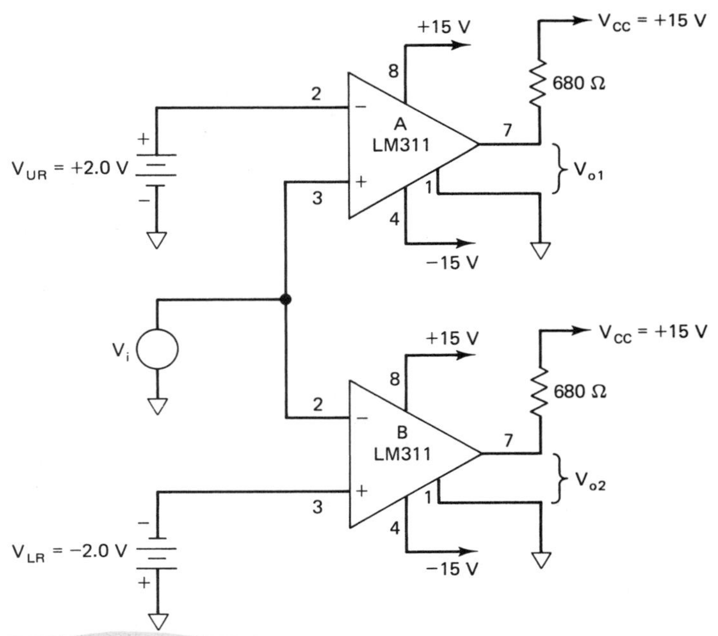

Figure 3-7 Basic window detector using two LM311 op-amp comparators.

3. When V_i is below $V_{LR} = -2$ V, $V_{O1} = 0$ V and $V_{O2} = V_{CC} = +15$ V.

The utility of this type of circuit depends on how the output terminals of both comparators are connected to visual indicators such as LEDs. In the next sections we illustrate several output configurations.

3-3.2 Window Detectors with LED Indicators

A dc power supply for TTL logic must be regulated to 5.0 V. If this supply voltage should exceed 5.5 V, the logic can be destroyed. If the supply drops below 4.5 V, the logic may exhibit marginal operation. For these reasons, the status of a TTL power supply is monitored by the window detector of Fig. 3-8.

The power supply's output is connected to the input terminal (X) of the window detector. R_1 is adjusted until V_{UR} is 5.5 V and R_2 adjusted until V_{LR} is 4.5 V. A 680Ω resistor is added to the output of each comparator to limit LED current to about 20 mA when the LM311's output switch is open (E_d = +). Protection diodes are not required since neither V_{O1} or V_{O2} can go negative.

When the TTL power supply is at its normal level of 5.0 V, the output of each comparators A and B is 0 V and both LEDS are off, indicating normal operation. If, however, the TTL power supply exceeds $V_{UR} = +5.5$ V, V_{O1} goes positive to light the red LED and indicate a high-voltage fault. The yellow LED is off because $V_{O2} = 0$ V.

If the TTL supply falls below $V_{LR} = +4.5$ V, V_{O2} will go positive and

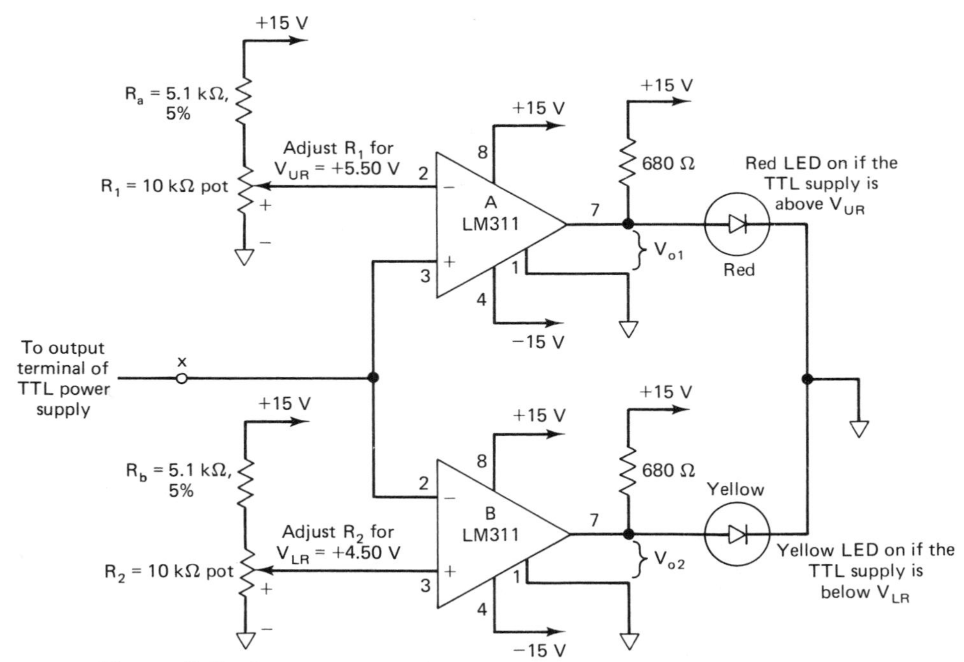

Figure 3-8 Status monitor for a TTL power supply. Normal operation is indicated by both LEDs being off. The red LED indicates an overvoltage fault and the yellow LED indicates a low-voltage fault.

the yellow LED will go on to indicate a low-voltage fault. V_{O1} is now 0 V and the red LED is off.

It may be desirable to provide an indication for normal operation. For example, a green LED should light when the TTL supply is above 4.5 V or below 5.5 V, to signify normal operation. Figure 3-8 can be modified as shown in Fig. 3-9. The two 1N914 diodes combine the outputs of both comparators into a *wired-or* configuration. When the input voltage is within the window of compared voltages ($V_{LR} < V_i < V_{UR}$), V_{O1} and V_{O2} are both 0 V. V_y is also 0 V and the 2N2222 transistor is cut off. Therefore, a current can flow from the V_{CC} supply via 680Ω into the green LED. This circuit provides an indication of normal operation when the green LED is on.

When the TTL power supply input voltage is outside the window, either voltage V_{O1} or V_{O2} goes positive. Therefore, V_y is at a positive voltage. The transistor is driven into saturation with V_{CE} at about 0.2 V. The LED is off under these conditions.

3-4 PORTABLE LOGIC PROBE FOR TTL LOGIC

3-4.1 Design Objectives for the Logic Probe

Our last application of voltage comparators will be to design a portable logic tester. Two LEDs are used as the outputs. When a logic 1 is being measured, the green LED is to be *on* and a red LED will *light* for a logic 0.

Make the tester portable by selecting a standard 9-V transistor battery as the probe's power supply. Since each logic level tested will require a separate comparator, use the single-package LM358 dual op amp. Before the design can be completed, both logic levels have to be defined.

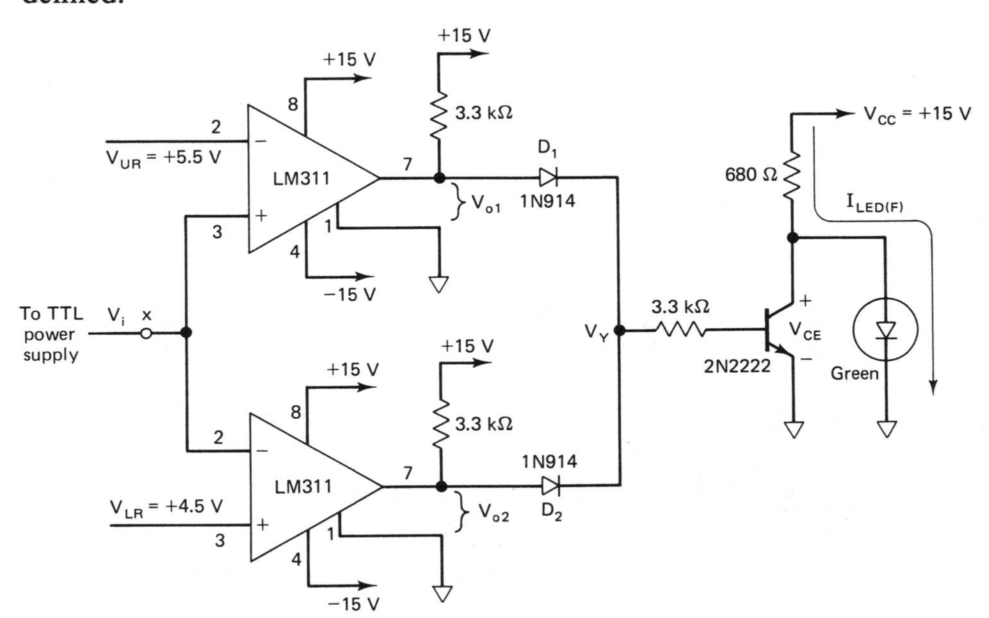

Figure 3-9 In this window detector normal operation is indicated by the green LED being on.

3-4.2 Graphical Representation of TTL Logic Levels

Figure 3-10a shows the logic levels for standard TTL logic. A logic 1 is defined as any voltage between the minimum value of +2.4 V up to the maximum value of V_{CC} = +5 V. If we label the +2.4 V as V_{REF1}, our probe must light the green LED for measured levels above V_{REF1}.

A logic 0 is defined as any voltage between the maximum value of +0.8 V down to the minimum value of 0 V. Label +0.8 V as V_{REF0}. Our probe must then light the red LED for measured voltages below V_{REF0}.

The area between V_{REF1} and V_{REF0} is a guard or undefined area, that is, neither a logic 1 or a logic 0. Thus neither LED should light.

3-4.3 Establishing V_{REF1} and V_{REF0}

V_{REF1} and V_{REF0} will be the two voltage reference levels at the inputs of the probe's comparators. A simple resistor-divider circuit, to establish these reference voltages, is designed in the following example.

EXAMPLE 3-5

Pick values for the resistor-divider circuit of Fig. 3-10b to set V_{REF1} = +2.4 V and V_{REF0} = 0.8 V.

SOLUTION Select a current I of 1 mA and solve for V_{R1}, V_{R2}, and V_{R3}, as follows:

$$V_{R1} = V_{CC} - V_{REF1} = (9\text{ V}) - (2.4\text{ V}) = 6.6\text{ V}$$

$$V_{R2} = V_{REF1} - V_{REF0} = (2.4\text{ V}) - (+0.8\text{ V}) = 1.6\text{ V}$$

$$V_{R3} = V_{REF0} = 0.8\text{ V}$$

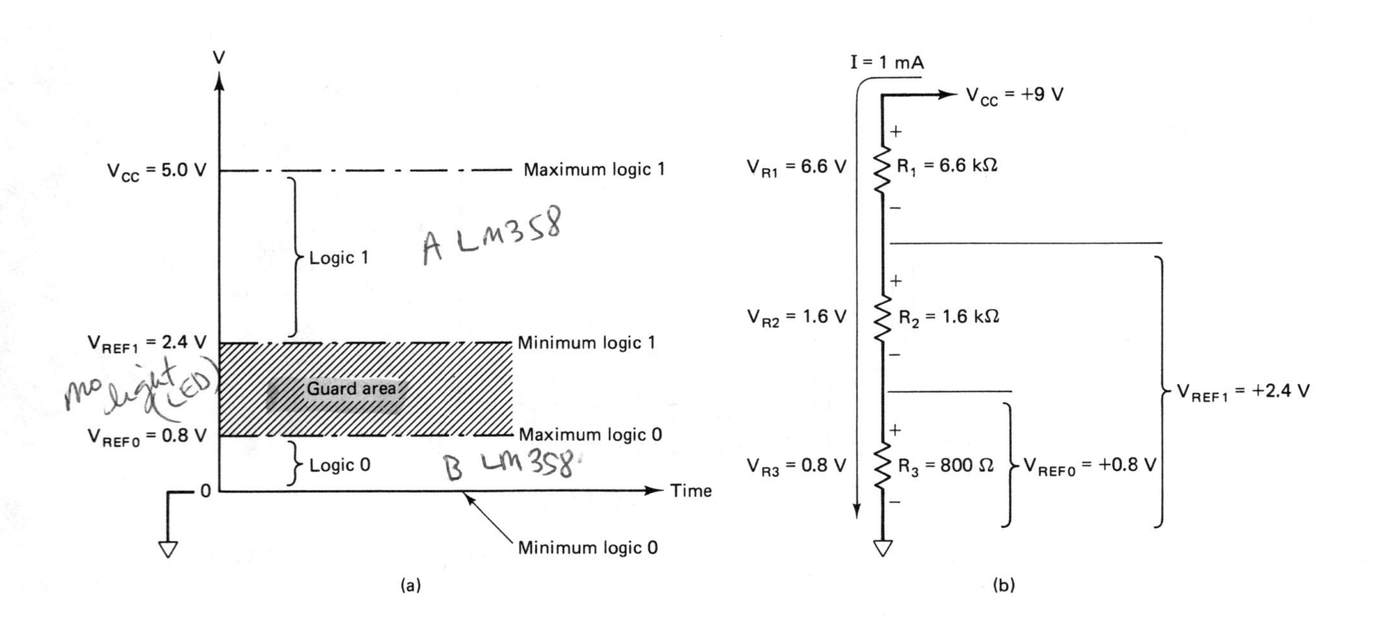

If I = 1 mA, 1 V requires 1 kΩ. Then, R_1 = 6.6 kΩ, R_2 = 1.6 kΩ, R_3 = 800 Ω.

3-4.4 The Completed Logic Probe

A practical probe design is shown in Fig. 3-11. A logic 1 is detected by noninverting comparator A that has a reference voltage of +2.4 V. When the probe contacts a logic 1 voltage level, the output of op amp A goes positive to $+V_{sat}$ = (+9 V) − (1.5 V) = +7.5 V, lighting the green LED. A 270Ω resistor limits the current to approximately 20 mA.

A logic 0 is detected by inverting comparator B with a reference voltage set at 0.8 V. When the probe contacts a logic 0 voltage level, the output of op amp B goes positive to $+V_{sat}$ = +7.5 V, lighting the red LED.

Since the probe is powered from a single 9-V battery, the output of the op amps can never go negative; therefore, no LED protection diodes are necessary.

A final note on grounds. The probe is built around a 9-V battery with its negative terminal as the circuit's reference or *analog ground.* The *digital ground* of the circuit to be tested must be connected to this analog ground.

In Chapter 4, we turn our attention away from applications of comparators as voltage monitors. We complete our study of comparators by showing how they can provide control.

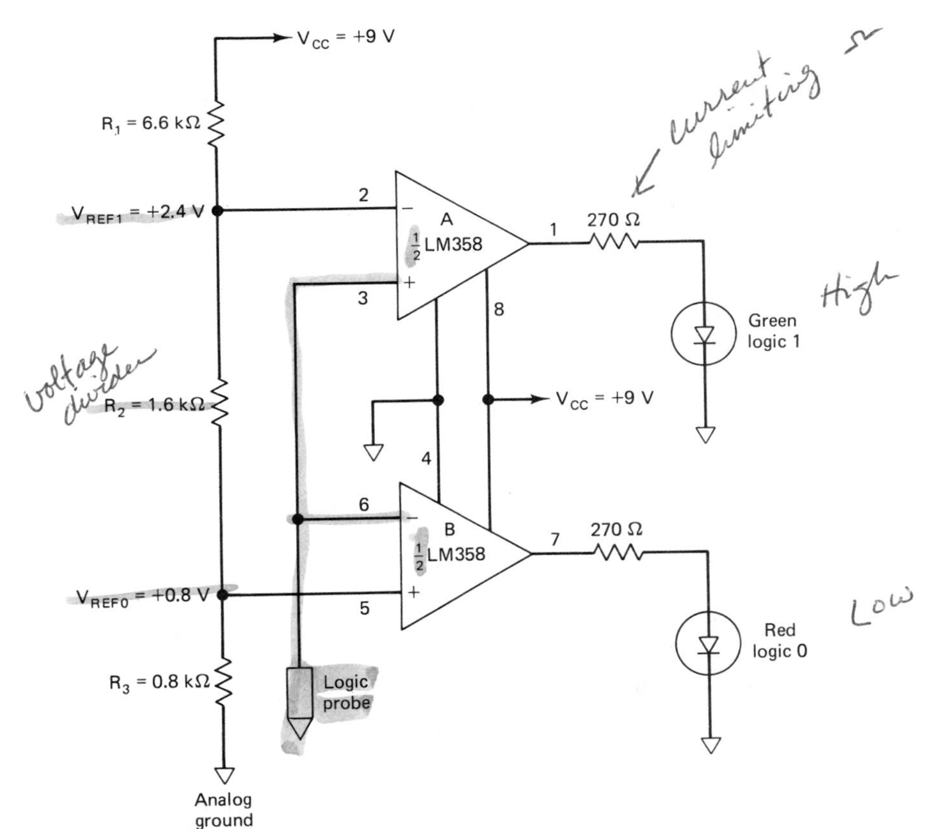

Figure 3-11 Complete logic probe circuit for TTL logic using a single LM358 package.

End of Chapter Exercises

Name: ____________________

Date: __________ **Grade:** ___

A. FILL-IN THE BLANKS

Fill in the blanks with the best answer.

1. When 10 LEDs are fabricated into a single 20-pin dual in-line package, the product is called a __bar-graph__ array.
2. The __HDSP-4830__ is a dot/bar display driver. It contains ten comparators whose outputs drive ten LEDs with a current determined by a single external __________.
3. A crowbar overvoltage protection circuit __________ a regulator's supply voltage, and __________ a fuse to disconnect the __________ supply from the regulated supply.
4. The LM358 is a __dual__ op amp that can be powered from a single supply voltage.
5. The semiconductor device that crowbars a fuse with a short circuit is called a(n) __________.
6. Window detectors compare an input voltage with __________ reference voltage(s) and needs the same number of op amps.
7. The __LM358__ is a special purpose precision comparator whose output is usually connected to V^{++} via a __reference level__ resistor.
8. When the strobe terminal of an LM311 is grounded via a 10kΩ

resistor, and differential input E_d is negative, the output switch is __________.

9. The logic zero voltage for TTL is below __________.

B. TRUE/FALSE

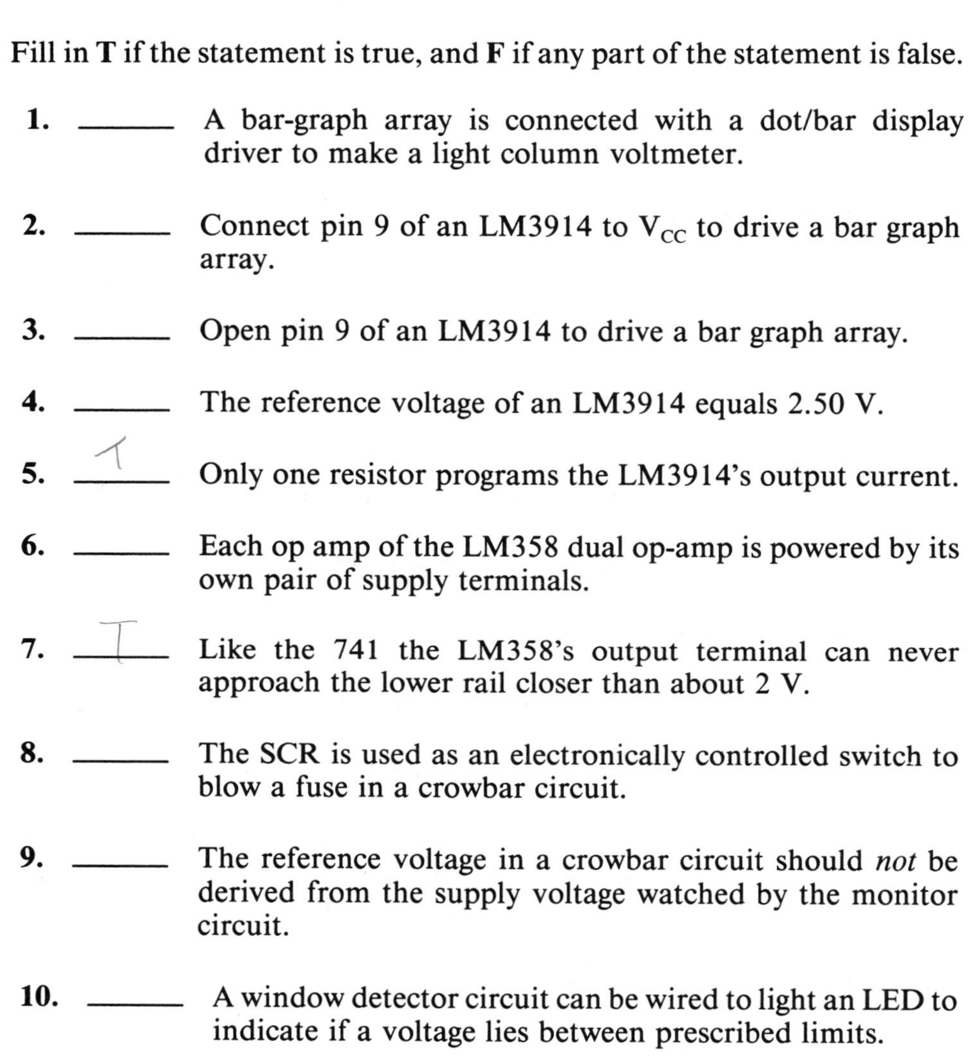

Fill in **T** if the statement is true, and **F** if any part of the statement is false.

1. ______ A bar-graph array is connected with a dot/bar display driver to make a light column voltmeter.
2. ______ Connect pin 9 of an LM3914 to V_{CC} to drive a bar graph array.
3. ______ Open pin 9 of an LM3914 to drive a bar graph array.
4. ______ The reference voltage of an LM3914 equals 2.50 V.
5. ______ Only one resistor programs the LM3914's output current.
6. ______ Each op amp of the LM358 dual op-amp is powered by its own pair of supply terminals.
7. ______ Like the 741 the LM358's output terminal can never approach the lower rail closer than about 2 V.
8. ______ The SCR is used as an electronically controlled switch to blow a fuse in a crowbar circuit.
9. ______ The reference voltage in a crowbar circuit should *not* be derived from the supply voltage watched by the monitor circuit.
10. ______ A window detector circuit can be wired to light an LED to indicate if a voltage lies between prescribed limits.

C. CIRCLE THE CORRECT ANSWER

Circle the correct answer for each statement.

1. The (LM311, LM3914) is a dot/bar display driver IC.
2. Each LED of a bar graph array requires (10 to 30 mA, 50 to 100 mA).
3. A crowbar protection circuit uses a (comparator, SCR) to monitor the output of a regulated power supply.
4. A crowbar circuit uses a (switch, SCR) to blow a fuse.

5. The reference voltage of the output monitor of a crowbar circuit (should, should not) be derived from a voltage reference source.
6. Input bias currents of the LM358 dual op amp are typically (50nA, 500 nA).
7. The LM311 precision comparator has an (open-collector, push-pull) output circuit.
8. Window comparators require (one, two) reference voltages.
9. TTL has a logic 1 level between 5.0 and (2.4, 0.8) volts.

D. MATCHING

Match the name or symbol in column **A** with the statement that matches best in column **B**.

	COLUMN A		COLUMN B
1. ______	AD584	**a.**	dot/bar display driver
2. ______	LM358	**b.**	precision comparator
3. ______	LM311	**c.**	dual op-amp
4. ______	LM3914	**d.**	voltage reference
5. ______	C107B	**e.**	SCR

PROBLEMS

3-1. **(a)** What is the expected value of voltage developed across one element of the HDSP-4830 display if $I_{LED(F)} = 10$ mA?
(b) What is the maximum reverse voltage that can be applied across an element of the HDSP-4830 display?

3-2. Refer to Fig. 3-2.
(a) What is the range of dc supply voltages that can be applied to pin 3 of the LM3914?
(b) What is the comparator classification for the devices inside the LM3914?
(c) What is the total resistance between pins 4 and 6?
(d) A precision 1.25 V reference voltage exists between which two pins of the LM3914?

3-3. Determine the value of current programming resistor R_1 in Fig. 3-2 to set $I_{LED(F)}$ to
(a) 15 mA;
(b) 25 mA.

3-4. Solve for the reference current I_{ref} for each of the conditions of Problem 3-3.

3-5. What must be done to pin 9 of the LM3914 to obtain

(a) a bar mode display;

(b) a dot mode display?

3-6. Redesign Fig. 3-3 to make a 0- to 10-V voltmeter. I_{LEDF} must be set to 30 mA and the voltmeter must operate in the dot mode.

3-7. Repeat Example 3-3 for $R_1 = 2.5\ \text{k}\Omega$ and $R_2 = 22.5\ \text{k}\Omega$.

3-8. Refer to Fig. 3-5 and set V_{CC} to +15 V. Determine

(a) $V_{o(max)}$;

(b) $V_{o(min)}$?

3-9. Repeat Example 3-4 for $I_{GT} = 1$ mA and $V_{GT} = 0.8$ V.

3-10. **(a)** Determine the current flow through the 680Ω resistor of Fig. 3-7 when V_{O1} goes to 0 V.

(b) What is the current flow when $V_{O1} = V_{CC}$?

3-11. Redraw Fig. 3-8 to monitor the status of a 6-V power supply. $V_{UR} = 7.5$ V and $V_{LR} = 5.0$ V. Recalculate all resistors and limit the current through both LEDs to ≃ 25 mA.

3-12. Repeat Example 3-5 for a $V_{CC} = +15$ V. Assume that I remains constant at 1 mA.

3-13. Recalculate the LED current-limiting resistor in Fig. 3-11 if $V_{CC} = +15$ V and I_{LEDF} must be held below 20 mA.

Laboratory Exercise 3

Name: ______________________

Date: __________ **Grade:** ___

LIGHT-COLUMN VOLTMETER

OBJECTIVES: Upon completion of this laboratory exercise on the design and test of a light-column voltmeter, you will be able to (1) measure $V_{LED(F)}$ at rated $I_{LED(F)}$; (2) measure some of the more important characteristics of the LM3914 dot/bar display driver and program the LM3914's output current for your LEDs; (3) design a 0- to 10-V voltmeter; and (4) produce displays in either the dot or bar mode.

REFERENCE: Chapter 3, HDSP-4830 data sheet.

PARTS LIST

1	HDSP-4830 bar-graph array	1	8.2-kΩ resistor
1	LM3914 dot/bar display driver	1	1-kΩ trim pot
1	330-Ω resistor	2	10-kΩ trim pots
1	1.2-kΩ resistor	1	1N914 diode

Procedure A: Characteristics of the HDSP-4830

1. Hewlett-Packard manufactures the HDSP-4830 (pp 106-111) 10-element bar-graph array in a 20-pin standard dual-in-line package, as shown in Fig. L3-1. This high-efficiency red display outputs the proper light intensity when a forward current, $I_{LED(F)}$, of 10 mA is passed through each diode. Under these current conditions, a forward voltage of about 2.0 V (maximum) is expected across each diode. Fig. L3-2 shows that each diode is separately mounted between adjacent pins.
2. Test the individual elements of this display by first building the test circuit of Fig. L3-3. Use the breadboard's 5-V supply and a 330-Ω resistor to limit the current to approximately 10 mA. Note that the 1N914 diode is included here to protect the LED segments from an accidental reverse voltage. Connect LED *a* as shown. Measure both $I_{LED(F)}$ and $V_{LED(F)}$ and record your results in Fig. L3-4.

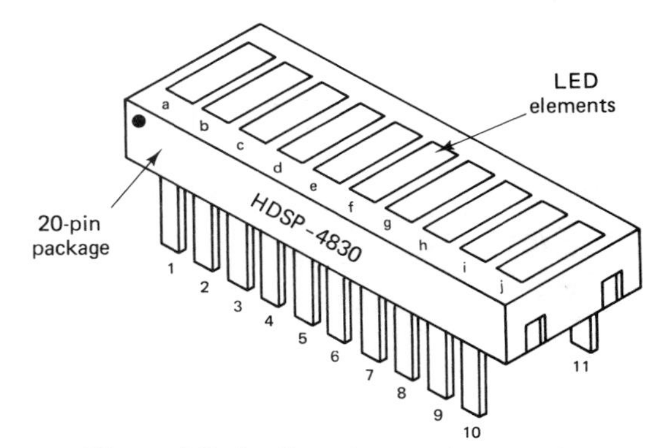

Figure L3-1 Ten-element bar graph.

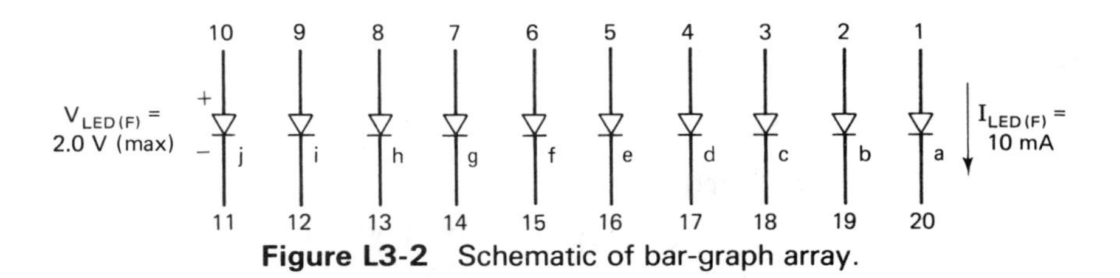

Figure L3-2 Schematic of bar-graph array.

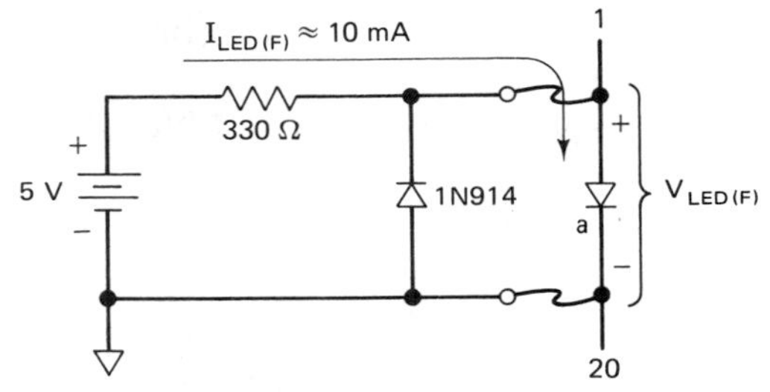

Figure L3-3 Test circuit.

3. Repeat step 2 for both LED *e* and LED *j*. (See Fig. L3-2 for the correct pin numbers.) Record your results in Fig. L3-4.
4. Take the average of both $V_{LED(F)}$ and $I_{LED(F)}$ and record these results in Fig. L3-4. Are the average values of $V_{LED(F)}$ and $I_{LED(F)}$ below the manufacturer's specification of 2.0 V (maximum) and 10 mA (typical)? __________

LED	$V_{LED(F)}$	$I_{LED(F)}$
a		
e		
j		
Average		

Figure L3-4

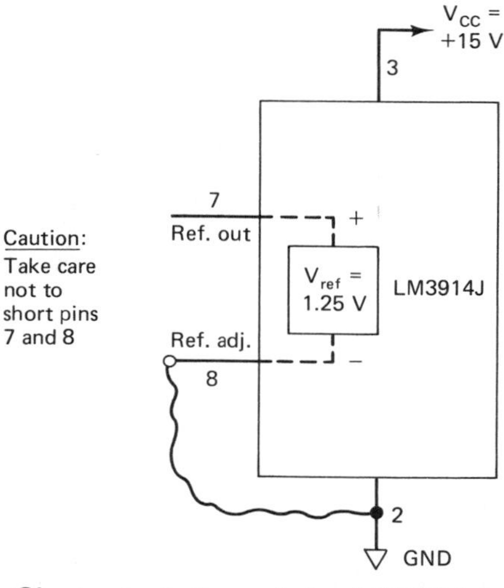

Figure L3-5 Display driver.

Procedure B: Characteristics of the LM3914

5. Build Fig. L3-5 to test the characteristics of an LM3914 dot/bar display driver IC from National Semiconductor. Apply +15 V to pin 3 and ground pins 2 and 8. Measure the internal reference voltage between pins 7(+) and 8(−). V_{ref} = __________.
6. To connect V_{ref} to the internal resistor divider of ten l-kΩ resistors, connect pin 7 to pin 6 (R_{Hi}) and pin 8 to both pin 4 (R_{LO}) and pin 2 ground. These connections are shown in Fig. L3-6.
7. A single *current programming resistor* R_1 across pins 7 and 8 can be used to set LED brightness. Select R_1 using the following equation:

$$R_1 = \frac{10(V_{ref})}{I_{LED}} = \frac{10(1.25\text{ V})}{I_{LED}} = \frac{12.5\text{ V}}{I_{LED}}$$

(*Note:* To interface the HDSP-4830 bar graph to the LM3914 will require an I_{LED} of 10 mA.) R_1 = __________. Using the closest standard value resistor, connect R_1 as shown in Fig. L3-6.

8. Remeasure V_{ref}. V_{ref} = __________. Has this value changed substantially from the measurement in step 5? __________.
9. The supply current required by pin 3 of the LM3914 is approximately 6 mA for the conditions above. Measure and record I_{supply}. I_{supply} = __________.
10. Wire the bar-graph array as shown in Fig. L3-7. Also, add a 8.2-kΩ resistor and a 1-kΩ pot as shown for test. Adjust the 1-kΩ pot until all LEDs are off. This test circuit can be used to show the accuracy of the internal 10-resistor-divider network of the LM3914. With the wiring shown in Fig. L3-7, the precise V_{ref} = 1.25 V is connected directly across the 10-resistor internal divider, which will segment V_{ref} into 10 precise threshold voltages. Each threshold voltage will be 125 mV above the last.

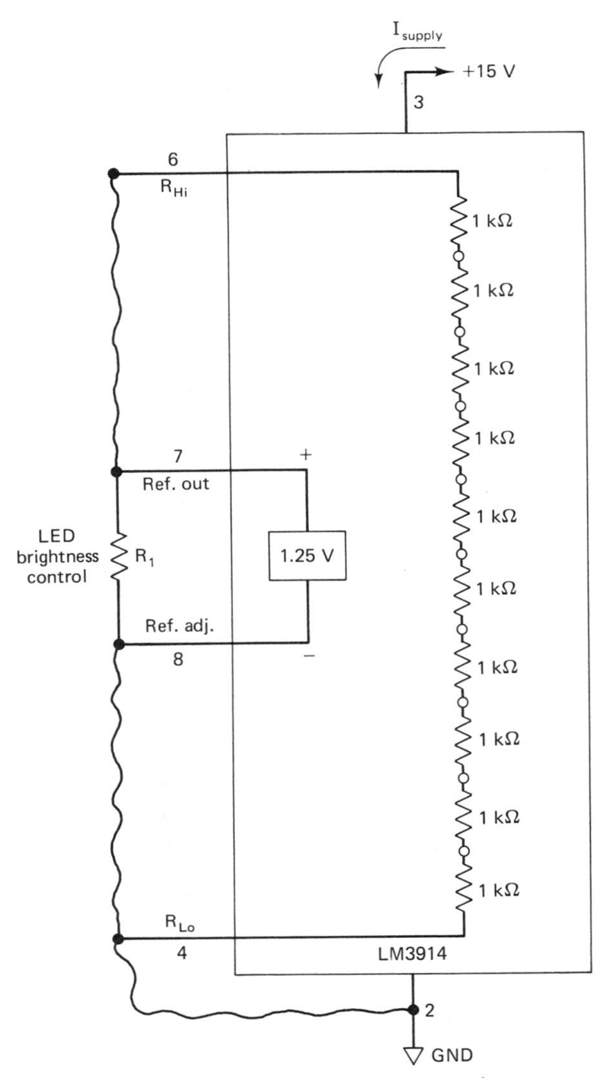

Figure L3-6 Brightness control.

11. Monitor V_i with a DMM. *Slowly* adjust the 1-kΩ pot until LED lights. Record this threshold voltage (value of V_i to make LED *a* light) in Fig. L3-8.
12. Repeat step 11 for each additional LED. Do these measured values compare favorably with the theoretical values shown in Fig. L3-7? __________.

Procedure C: Design of a 0- to 10-V Light-Column Voltmeter

13. To expand Fig. L3-7 to a 0- to 10-V voltmeter, the reference voltage V_{ref} across the internal resistor-divider network must be expanded to a maximum of 10 V. This is accomplished by adding the *voltage programming resistor* R_2. The value of R_2 is determined by using the following equation:

$$R_2 = \left(\frac{V_{o(max)} - V_{ref}}{V_{ref}}\right)R_1$$

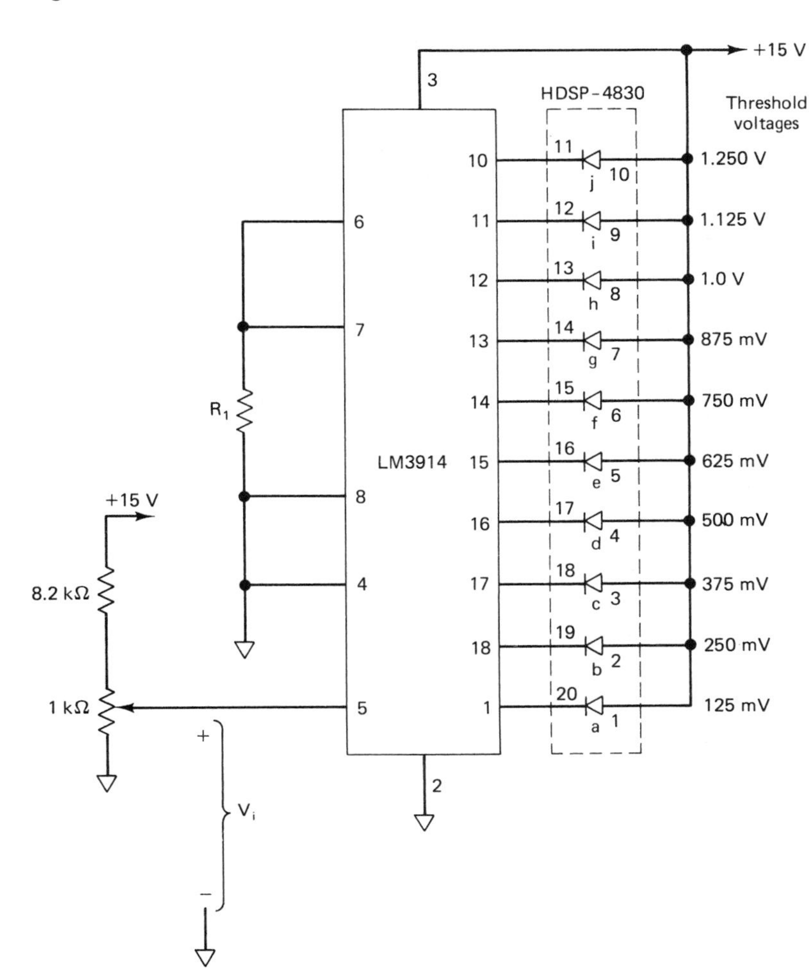

Figure L3-7 Test circuit.

LED	Threshold voltage, V_i
a	
b	
c	
d	
e	
f	
g	
h	
i	
j	

Figure L3-8
Threshold voltages.

where $V_{o(\max)}$ is the maximum value of the voltmeter range. Solve for R_2 to build a 0- to 10-V meter. R_2 = __________. Add R_2 as shown in Fig. L3-9.

14. Set R_2 by first connecting the DMM between pins 6 and 4. Adjust R_2 until $V_o = V_{o(\max)} = +10.0$ V.
15. To test this voltmeter design, add a 3.9-kΩ resistor and a 10-kΩ trim pot to your design (see Fig. L3-10). Adjust the 10-kΩ trim pot until all LEDs are "off." Monitor V_i with a DMM. *Slowly* adjust the 10-kΩ pot until LED lights. Record this threshold voltage in Fig. L3-11.
16. Repeat step 15 for each additional LED. Do the measured values compare favorably with the theoretical values shown on Fig. L3-9?

Procedure D: Dot and Bar Mode Pin

17. The designs presented thus far have been with pin 9 left open. The result has been displays represented by a single LED light at any one time.
18. A second mode, the *bar* display, is made possible by connecting pin 9

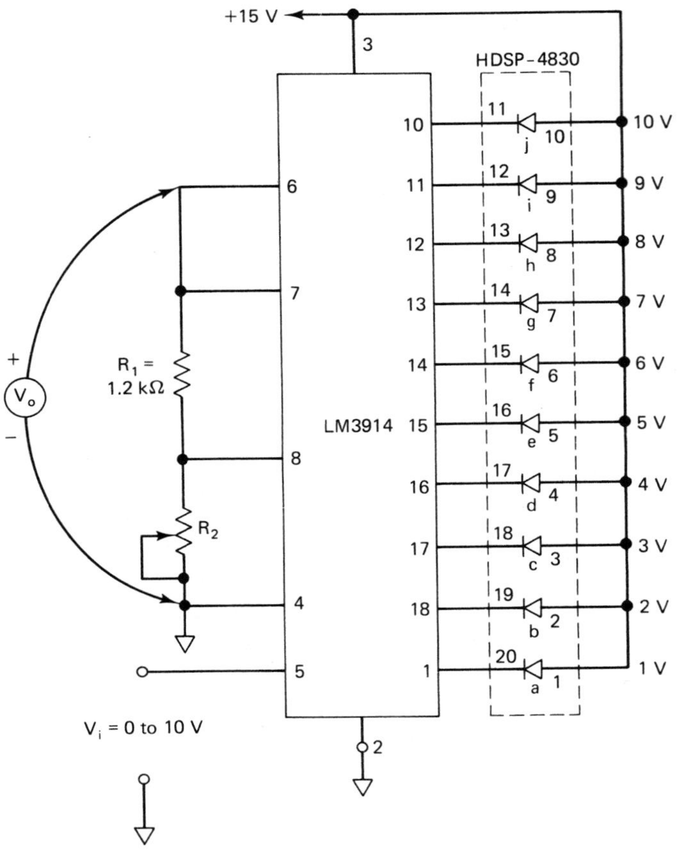

Figure L3-9 0- to 10-V voltmeter.

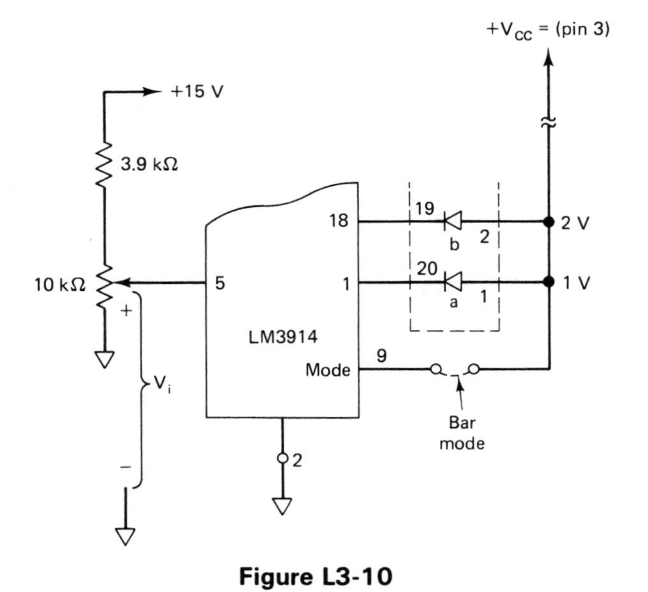

Figure L3-10

LED	Threshold voltage, V_i
a	
b	
c	
d	
e	
f	
g	
h	
i	
j	

Figure L3-11 Threshold voltages for a 0- to 10-V meter.

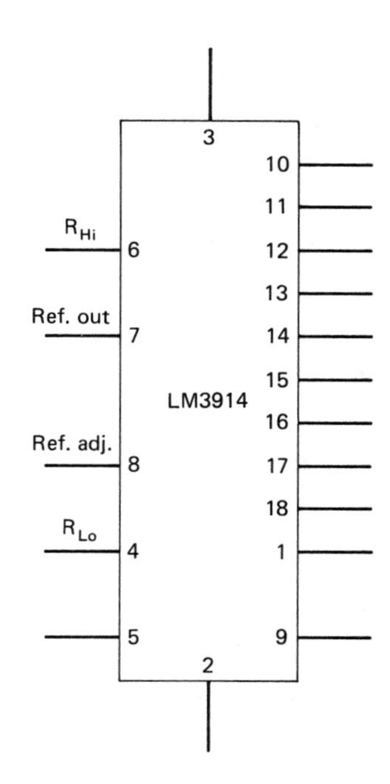

Figure L3-12

to V_{CC} = +15 V (pin 3 of the LM3914.) Bar mode operation is characterized by all LEDs being on if their threshold voltages are below V_i.

19. Jump pin 9 to $+V_{CC}$ = 15 V, as shown in Fig. L3-10. Adjust the 10-kΩ pot to vary V_i from 0 to 10 V and observe the results.

CONCLUSION: Design a 0- to 5-V light-column voltmeter to produce a bar-type display using LEDs that require a forward current of 15 mA. Show the complete circuit schematic below.

10-ELEMENT BAR GRAPH ARRAY

RED HDSP-4820
HIGH-EFFICIENCY RED HDSP-4830
YELLOW HDSP-4840
HIGH PERFORMANCE GREEN HDSP-4850
MULTICOLOR HDSP-4832
MULTICOLOR HDSP-4836

TECHNICAL DATA NOVEMBER 1984

Features

- CUSTOM MULTICOLOR ARRAY CAPABILITY
- MATCHED LEDs FOR UNIFORM APPEARANCE
- END STACKABLE
- PACKAGE INTERLOCK ENSURES CORRECT ALIGNMENT
- LOW PROFILE PACKAGE
- RUGGED CONSTRUCTION—RELIABILITY DATA SHEETS AVAILABLE
- LARGE, EASILY RECOGNIZABLE SEGMENTS
- HIGH ON-OFF CONTRAST, SEGMENT TO SEGMENT
- WIDE VIEWING ANGLE
- CATEGORIZED FOR LUMINOUS INTENSITY
- HDSP-4832/-4836/-4840/-4850 CATEGORIZED FOR DOMINANT WAVELENGTH

Applications

- INDUSTRIAL CONTROLS
- INSTRUMENTATION
- OFFICE EQUIPMENT
- COMPUTER PERIPHERALS
- CONSUMER PRODUCTS

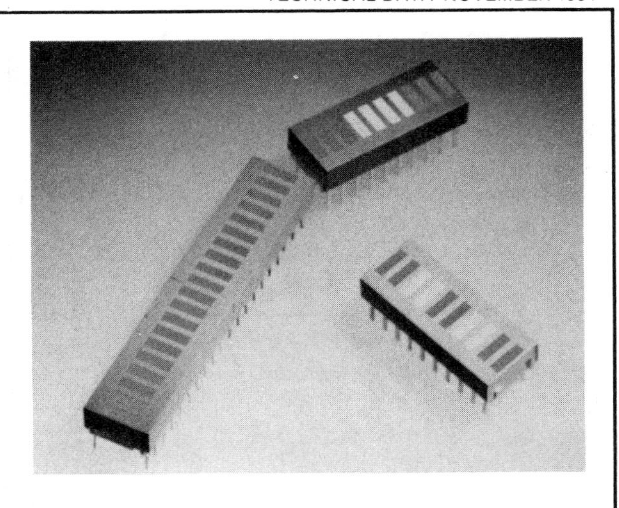

Description

These 10-element LED arrays are designed to display information in easily recognizable bar graph form. The packages are end stackable and therefore capable of displaying long strings of information. Use of these bar graph arrays eliminates the alignment, intensity, and color matching problems associated with discrete LEDs. The HDSP-4820/-4830/-4840/-4850 each contain LEDs of just one color. The HDSP-4832/-4836 are multicolor arrays with High-Efficiency Red, Yellow, and Green LEDs in a single package. CUSTOM MULTICOLOR ARRAYS ARE AVAILABLE WITH MINIMUM DELIVERY REQUIREMENTS. CONTACT YOUR LOCAL DISTRIBUTOR OR HP SALES OFFICE FOR DETAILS.

Package Dimensions

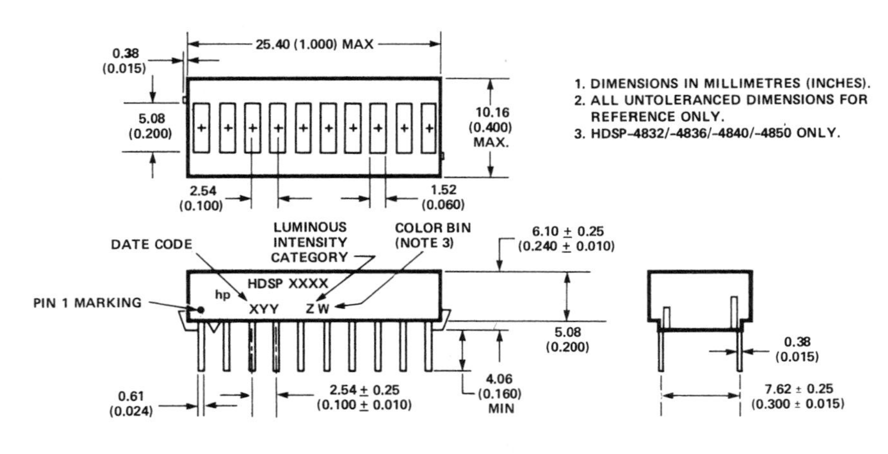

Absolute Maximum Ratings[9]

Parameter	HDSP-4820	HDSP-4830	HDSP-4840	HDSP-4850
Average Power Dissipation per LED ($T_A = 25°C$)[1]	125 mW	125 mW	125 mW	125 mW
Peak Forward Current per LED	150 mA[2]	90 mA[3]	60 mA[3]	90 mA[3]
DC Forward Current per LED	30 mA[4]	30 mA[5]	20 mA[6]	30 mA[7]
Operating Temperature Range	-40° C to +85° C			−20° C to +85° C
Storage Temperature Range	-40° C to +85° C			
Reverse Voltage per LED	3.0 V			
Lead Soldering Temperature (1.59 mm (1/16 inch) below seating plane)[8]	260° C for 3 sec			

NOTES:
1. Derate maximum average power above $T_A = 25°C$ at 1.67 mW/° C. This derating assumes worst case $R_{\Theta J\text{-}A} = 600°$ C/W/LED.
2. See Figure 1 to establish pulsed operating conditions.
3. See Figure 6 to establish pulsed operating conditions.
4. Derate maximum DC current above $T_A = 63°C$ at 0.81 mA/° C per LED. This derating assumes worst case $R_{\Theta J\text{-}A} = 600°$ C/W/LED. With an improved thermal design, operation at higher temperatures without derating is possible. See Figure 2.
5. Derate maximum DC current above $T_A = 50°C$ at 0.6 mA/° C per LED. This derating assumes worst case $R_{\Theta J\text{-}A} = 600°$ C/W/LED. With an improved thermal design, operation at higher temperatures without derating is possible. See Figure 7.
6. Derate maximum DC current above $T_A = 70°C$ at 0.67 mA/° C per LED. This derating assumes worst case $R_{\Theta J\text{-}A} = 600°$ C/W/LED. With an improved thermal design, operation at higher temperatures without derating is possible. See Figure 8.
7. Derate maximum DC current above $T_A = 37°C$ at 0.48 mA/° C per LED. This derating assumes worst case $R_{\Theta J\text{-}A} = 600°$ C/W/LED. With an improved thermal design, operation at higher temperatures without derating is possible. See Figure 9.
8. Clean only in water, Isopropanol, Ethanol, Freon TF or TE (or equivalent) and Genesolve DI-15 (or equivalent).
9. Absolute maximum ratings for the HER, Yellow, and Green elements of the multicolor arrays are identical to the HDSP-4830/-4840/-4850 maximum ratings.

Internal Circuit Diagram

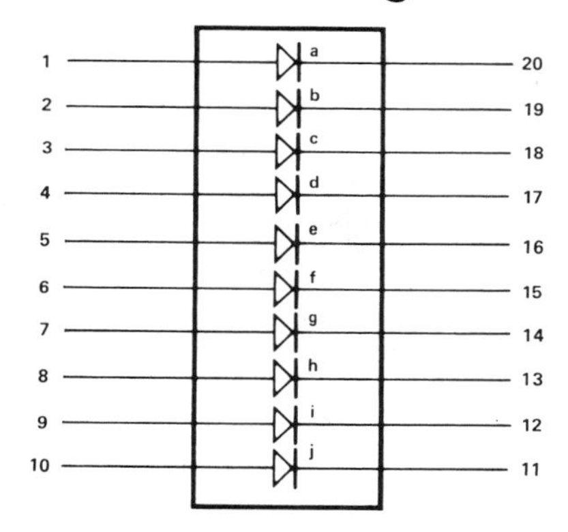

Pin	Function	Pin	Function
1	Anode a	11	Cathode j
2	Anode b	12	Cathode i
3	Anode c	13	Cathode h
4	Anode d	14	Cathode g
5	Anode e	15	Cathode f
6	Anode f	16	Cathode e
7	Anode g	17	Cathode d
8	Anode h	18	Cathode c
9	Anode i	19	Cathode b
10	Anode j	20	Cathode a

Multicolor Array Segment Colors

Segment	HDSP-4832 Segment Color	HDSP-4836 Segment Color
a	HER	HER
b	HER	HER
c	HER	Yellow
d	Yellow	Yellow
e	Yellow	Green
f	Yellow	Green
g	Yellow	Yellow
h	Green	Yellow
i	Green	HER
j	Green	HER

Electrical/Optical Characteristics at $T_A = 25°$ C[4]

RED HDSP-4820

Parameter	Symbol	Test Conditions	Min.	Typ.	Max.	Units
Luminous Intensity per LED (Unit Average)[1]	I_F	I_F = 20 mA	610	1250		μcd
Peak Wavelength	λ_{PEAK}			655		nm
Dominant Wavelength[2]	λ_d			645		nm
Forward Voltage per LED	V_F	I_F = 20 mA		1.6	2.0	V
Reverse Voltage per LED	V_R	I_R = 100 μA	3	12[5]		V
Temperature Coefficient V_F per LED	$\Delta V_F/°C$			-2.0		mV/°C
Thermal Resistance LED Junction-to-Pin	$R_{\Theta J\text{-}PIN}$			300		°C/W/ LED

HIGH-EFFICIENCY RED HDSP-4830

Parameter	Symbol	Test Conditions	Min.	Typ.	Max.	Units
Luminous Intensity per LED (Unit Average)[1]	I_V	I_F = 10 mA	900	3500		μcd
Peak Wavelength	λ_{PEAK}			635		nm
Dominant Wavelength[2]	λ_d			626		nm
Forward Voltage per LED	V_F	I_F = 20 mA		2.1	2.5	V
Reverse Voltage per LED	V_R	I_R = 100 μA	3	30[5]		V
Temperature Coefficient V_F per LED	$\Delta V_F/°C$			-2.0		mV/°C
Thermal Resistance LED Junction-to-Pin	$R_{\Theta J\text{-}PIN}$			300		°C/W/ LED

YELLOW HDSP-4840

Parameter	Symbol	Test Conditions	Min.	Typ.	Max.	Units
Luminous Intensity per LED (Unit Average)[1]	I_V	I_F = 10 mA	600	1900		μcd
Peak Wavelength	λ_{PEAK}			583		nm
Dominant Wavelength[2,3]	λ_d		581	585	592	nm
Forward Voltage per LED	V_F	I_F = 20 mA		2.2	2.5	V
Reverse Voltage per LED	V_R	I_R = 100 μA	3	40[5]		V
Temperature Coefficient V_F per LED	$\Delta V_F/°C$			-2.0		mV/°C
Thermal Resistance LED Junction-to-Pin	$R_{\Theta J\text{-}PIN}$			300		°C/W/ LED

GREEN HDSP-4850

Parameter	Symbol	Test Conditions	Min.	Typ.	Max.	Units
Luminous Intensity per LED (Unit Average)[1]	I_V	I_F = 10 mA	600	1900		μcd
Peak Wavelength	λ_{PEAK}			566		nm
Dominant Wavelength[2,3]	λ_d			571	577	nm
Forward Voltage per LED	V_F	I_F = 10 mA		2.1	2.5	V
Reverse Voltage per LED	V_R	I_R = 100 μA	3	50[5]		V
Temperature Coefficient V_F per LED	$\Delta V_F/°C$			-2.0		mV/°C
Thermal Resistance LED Junction-to-Pin	$R_{\Theta J\text{-}PIN}$			300		°C/W/ LED

NOTES:

1. The bar graph arrays are categorized for luminous intensity. The category is designated by a letter located on the side of the package.
2. The dominant wavelength, λ_d, is derived from the CIE chromaticity diagram and is that single wavelength which defines the color of the device.
3. The HDSP-4832/-4836/-4840/-4850 bar graph arrays are categorized by dominant wavelength with the category designated by a number adjacent to the intensity category letter. Only the yellow elements of the HDSP-4832/-4836 are categorized for color.
4. Electrical/optical characteristics of the High-Efficiency Red elements of the HDSP-4832/-4836 are identical to the HDSP-4830 characteristics. Characteristics of Yellow elements of the HDSP-4832/-4836 are identical to the HDSP-4840. Characteristics of Green elements of the HDSP-4832/-4836 are identical to the HDSP-4850.
5. Reverse voltage per LED should be limited to 3.0 V Max.

HDSP-4820

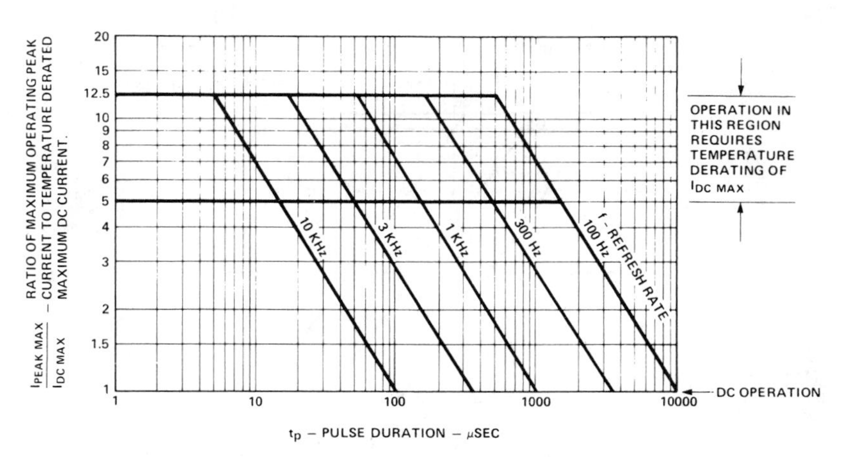

Figure 1. Maximum Tolerable Peak Current vs. Pulse Duration

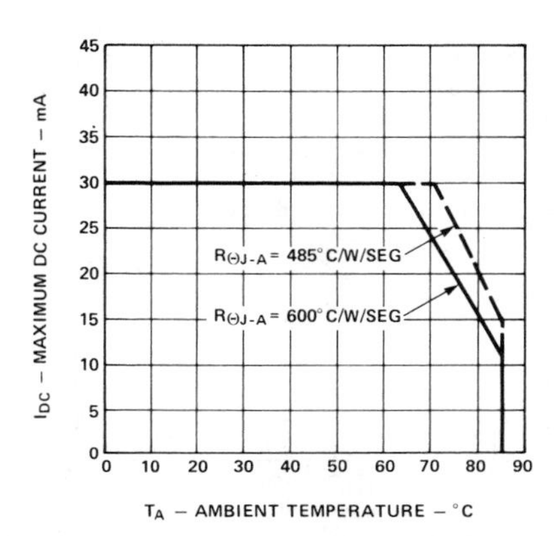

Figure 2. Maximum Allowable D.C. Current per LED vs. Ambient Temperature. Deratings based on Maximum Allowable Thermal Resistance, LED Junction-to-Ambient on a per LED basis. T_{JMAX} = 100° C

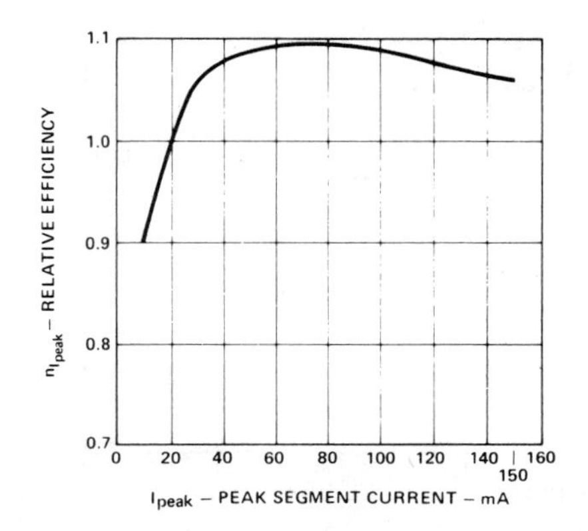

Figure 3. Relative Efficiency (Luminous Intensity per Unit Current) vs. Peak Segment Current

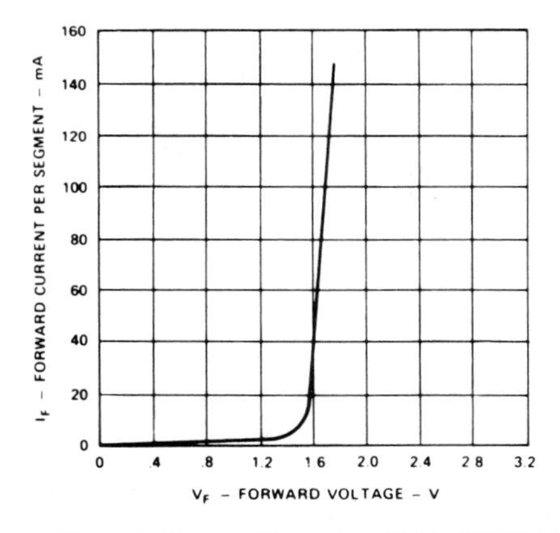

Figure 4. Forward Current vs. Forward Voltage

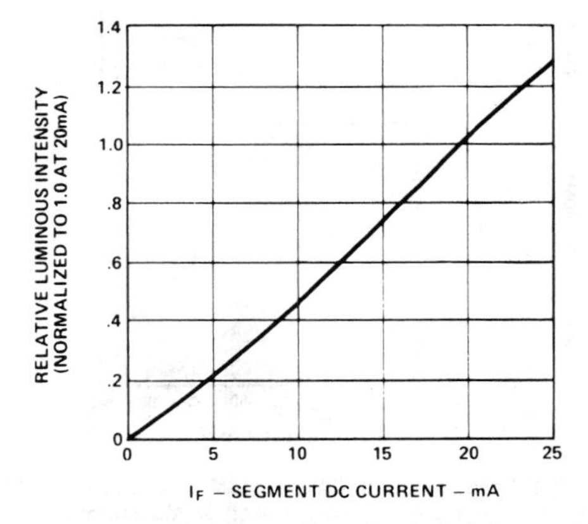

Figure 5. Relative Luminous Intensity vs. D.C. Forward Current

For a Detailed Explanation on the Use of Data Sheet Information and Recommended Soldering Procedures, See Application Note 1005.

HDSP-4830/-4840/-4850

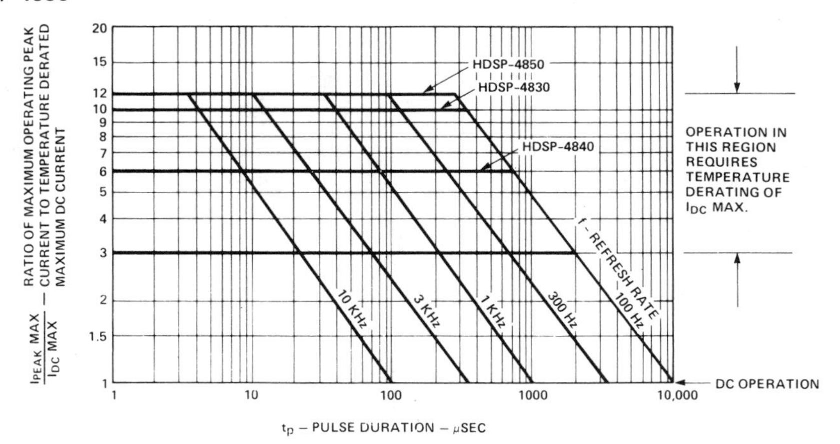

Figure 6. HDSP-4830/-4840/-4850 Maximum Tolerable Peak Current vs. Pulse Duration

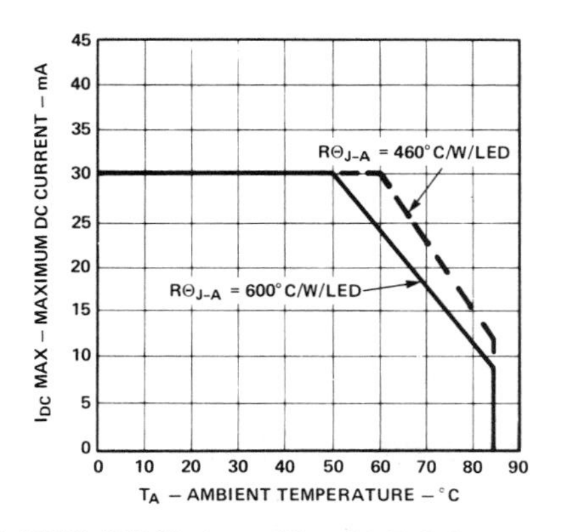

Figure 7. HDSP-4830 Maximum Allowable D.C. Current per LED vs. Ambient Temperature. Deratings Based on Maximum Allowable Thermal Resistance Values, LED Junction-to-Ambient on a per LED basis. $T_{J\ MAX} = 100°$ C.

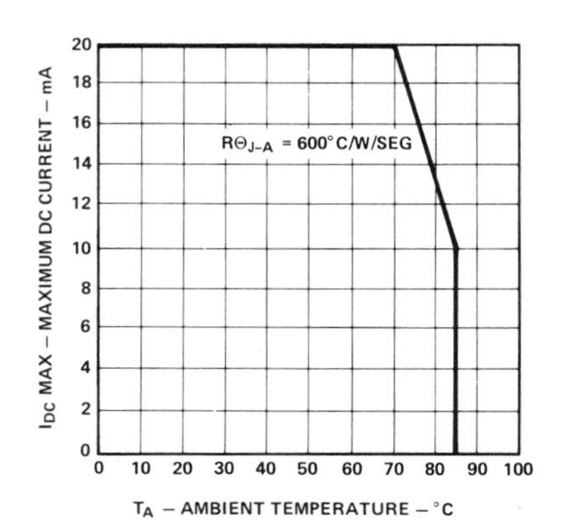

Figure 8. HDSP-4840 Maximum Allowable D.C. Current per LED vs. Ambient Temperature. Deratings Based on Maximum Allowable Thermal Resistance Values, LED Junction-to-Ambient on a per LED basis. $T_{J\ MAX} = 100°$ C.

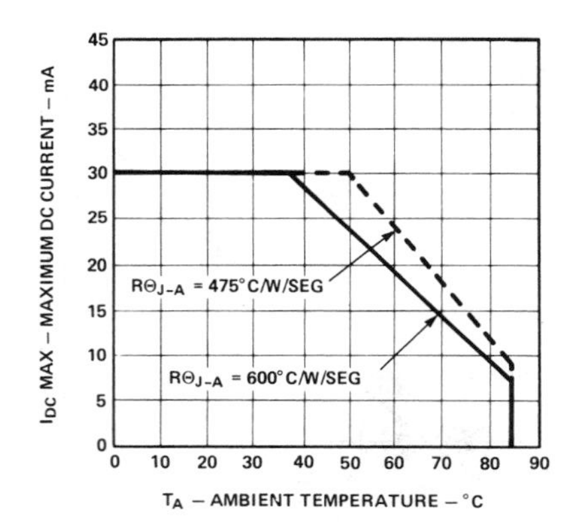

Figure 9. HDSP-4850 Maximum Allowable D.C. Current per LED vs. Ambient Temperature. Deratings Based on Maximum Allowable Thermal Resistance Values, LED Junction-to-Ambient on a per LED basis. $T_{J\ MAX} = 100°$ C.

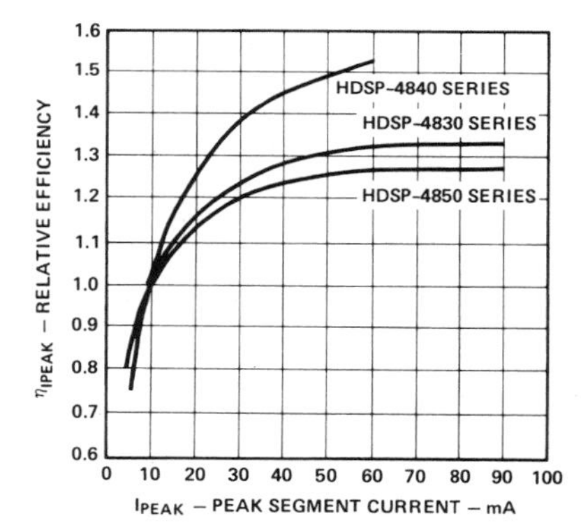

Figure 10. Relative Efficiency (Luminous Intensity per Unit Current) vs. Peak Segment Current

For a Detailed Explanation on the Use of Data Sheet Information and Recommended Soldering Procedures, See Application Note 1005.

HDSP-4830/-4840/-4850

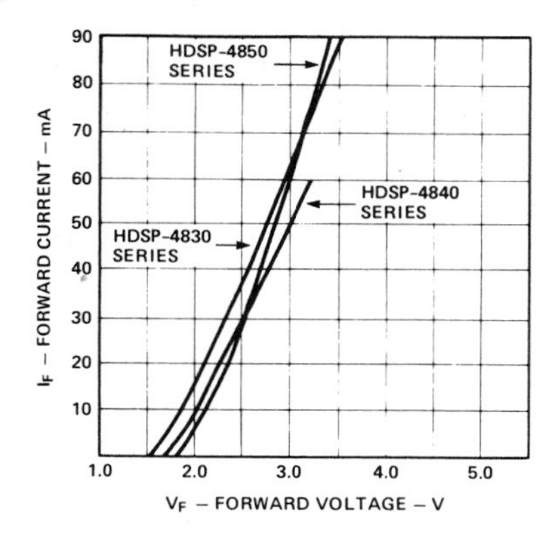

Figure 11. Forward Current vs. Forward Voltage

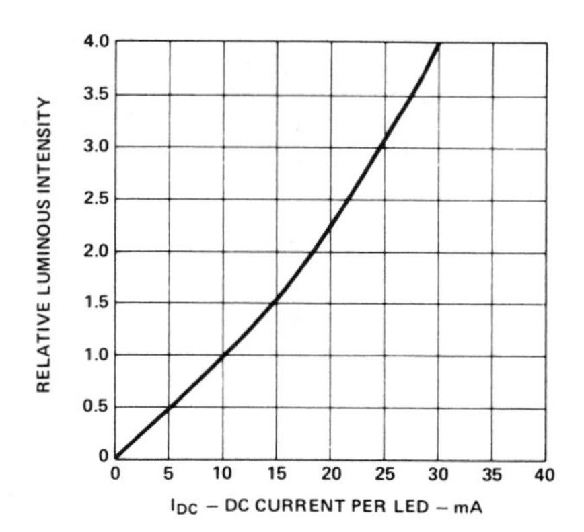

Figure 12. HDSP-4830/-4840/-4850 Relative Luminous Intensity vs. D.C. Forward Current

Electrical

These versatile bar graph arrays are composed of ten light emitting diodes. The light from each LED is optically stretched to form individual elements. The diodes in the HDSP-4820 bar graph utilize a Gallium Arsenide Phosphide (GaAsP) epitaxial layer on a Gallium Arsenide (GaAs) Substrate. The HDSP-4830/-4840 bar graphs utilize a GaAsP epitaxial layer on a GaP substrate to produce the brighter high-efficiency red and yellow displays. The HDSP-4850 bar graph array utilizes a GaP epitaxial layer on a GaP substrate. The HDSP-4832/-4836 multicolor arrays have high efficiency red, yellow, and green LEDs in one package.

These display devices are designed to allow strobed operation. The typical forward voltage values, scaled from Figure 4 or 11, should be used for calculating the current limiting resistor value and typical power dissipation. Expected maximum V_F values, for the purpose of driver circuit design and maximum power dissipation, may be calculated using the following $V_{F\ MAX}$ models.

HDSP-4820 (Red)

$V_{F\ MAX} = 1.75\ V + I_{PEAK}\ (12.5\Omega)$
For: $I_{PEAK} \geq 5$ mA

HDSP-4830/-4840 (High Efficiency Red/Yellow)

$V_{F\ MAX} = 1.75V + I_{PEAK}\ (38\Omega)$
For $I_{PEAK} \geq 20$ mA

$V_{F\ MAX} = 1.6V + I_{DC}\ (45\Omega)$
For: 5 mA $\leq I_{DC} \leq$ 20 mA

HDSP-4850 (Green)

$V_{F\ MAX} = 2.0V + I_{PEAK}\ (50\Omega)$
For: $I_{PEAK} > 5$ mA

Refresh rates of 1 KHz or faster provide the most efficient operation resulting in the maximum possible time averaged luminous intensity.

The time averaged luminous intensity may be calculated using the relative efficiency characteristic shown in Figures 3 and 10. The time averaged luminous intensity at T_A = 25° C is calculated as follows:

$$I_{V\ TIME\ AVG} = \left[\frac{I_{F\ AVG}}{I_{F\ SPEC\ AVG}}\right](\eta I_{PEAK})\ (I_{V\ SPEC})$$

Example: For HDSP-4830 operating at I_{PEAK} = 50 mA, 1 of 4 Duty Factor

$$\eta I_{PEAK} = 1.35\ (\text{at } I_{PEAK} = 50\text{ mA})$$

$$I_{V\ TIME\ AVG} = \left[\frac{12.5\text{ mA}}{10\text{ mA}}\right](1.35)\ 2280\ \mu cd = 3847\ \mu cd$$

For Further Information Concerning Bar Graph Arrays and Suggested Drive Circuits, Consult HP Application Note 1007 Entitled "Bar Graph Array Applications".

For more information, call your local HP sales office listed in the telephone directory white pages. Ask for the Components Department. Or write to Hewlett-Packard: **U.S.A.** — P.O. Box 10301, Palo Alto, CA 94303-0890. **Europe** — P.O. Box 999 1180 AZ Amstelveen, The Netherlands. **Canada** — 6877 Goreway Drive, Mississauga, L4V 1M8, Ontario. **Japan** — Yokogawa-Hewlett-Packard Ltd., 3-29-21, Takaido-Higashi, Suginami-ku, Tokyo 168. **Elsewhere** in the world, write to Hewlett-Packard Intercontinental, 3495 Deer Creek Road, Palo Alto, CA 94304.

Printed in U.S.A. Data Subject to Change — Obsoletes 5954-0859 (4/84) 5954-0869 (11/84)

Positive Feedback and On/Off Controls

4

LEARNING OBJECTIVES

Upon completion of this chapter on positive feedback and on/off control, you will be able to:

- Describe the operation of a thermostat and tell why it must have the property of hysteresis.
- Draw an analogy between set-point temperature and set-point voltage and between hysteresis temperature and hysteresis voltage.
- Show how hysteresis provides noise immunity for comparator circuits.
- Describe hysteresis action in an inverting or noninverting comparator.
- Make a line voltage synchronizer to synchronize devices or circuits with 60-Hz line voltages.
- Describe the operating principles for on/off control circuits.
- Show how to interface a low-voltage control circuit with a high-voltage load.
- Show the similarities between the output–input characteristics of a heat or air-conditioning control system and a comparator circuit with hysteresis.
- Tell why hysteresis implies that a circuit or system has the property of "memory."
- Make an electronic equivalent of a mechanical temperature control.

4-0 INTRODUCTION

In this chapter we employ applications involving on/off control to introduce the action of positive feedback in electronic circuits. Any circuit with positive feedback exhibits properties of *hysteresis* or *memory.* This curious property is always necessary to make useful circuits such as on/off controllers or oscillators that generate signals. We introduce this idea of circuit memory by examining the action of a controller that is familiar to all.

4-1 ON/OFF CONTROL BASICS

4-1.1 Thermostat

Suppose that the thermostat of an air conditioner is set at T_{set} = 65°F. Suppose also that the control circuitry was designed incorrectly to turn on above 65°F, and off below 65°F. If temperature slowly approached 65°F, the air conditioner would chatter on and off. This same condition can happen in the comparator circuits of Chapter 2 when V_i (temperature analogy) slowly approaches V_{ref} (set-point analogy). To solve this problem, all control circuits must have the property of hysteresis or memory.

4-1.2 Hysteresis (Memory)

Thermostat controls of air conditioners actually respond to two temperatures, as shown in Fig. 4-1a. If temperature is below low reference temperature T_{LR} = 63°F, the controller issues an unconditional command to turn the air conditioner off. If temperature is above the high reference temperature T_{UR} = 67°F, the controller commands the air conditioner to turn on.

Suppose that the temperature is 70°F and the air conditioner is on. The room cools down past T_{UR} and T_{set} to T_{LR}. When temperature drops below T_{LR}, the air conditioner turns off (points C to D in Fig. 4-1a). The room warms past T_{LR} and T_{set} to T_{UR} before the air conditioner turns on again (points A to B).

Note that the air conditioner can be *either* on *or* off in the temperature range T_{LR} = 63°F to T_{UR} = 67°F. This region is called the *hysteresis* or *memory* region. The interpretation is as follows:

1. If the air conditioner *is* off, it remains off in the hysteresis region (A to B). That is, the control "remembers" the last "off" state.

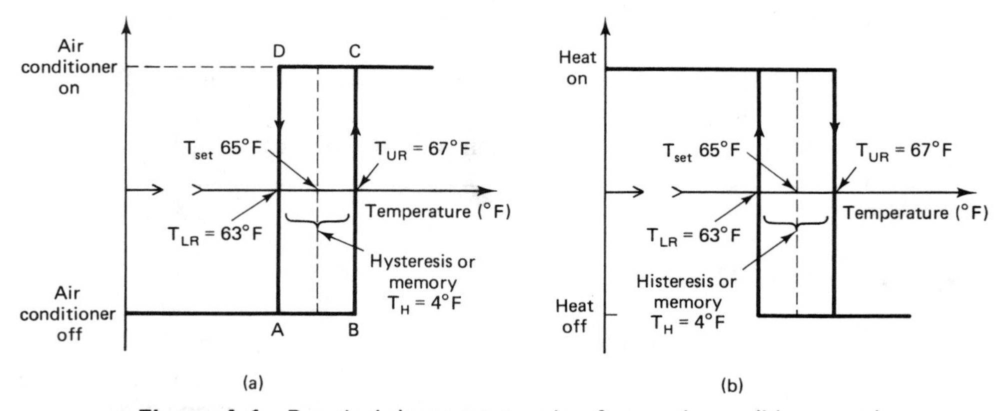

Figure 4-1 Practical thermostat action for an air conditioner and a heater. (a) An air conditioner set at 65°F turns on at 67°F and turns off at 63°F; (b) a heater set at 65°F turns off at 67°F and turns on at 63°F.

2. If the air conditioner *is* on, it remains on in the hysteresis region (C to D). That is, the control "remembers" the last "on" state.

Now the air conditioner will *not* chatter when temperature hovers at 65°F. The uncertainty has been removed by hysteresis.

4-1.3 Hysteresis in a Heater Control Circuit

Hysteresis action is reviewed by an example.

EXAMPLE 4-1

A heater's wall thermostat is set by the occupant to $T_{set} = 65°F$. The occupant believes that temperature will be held at this *set-point* temperature. In reality, the thermostat has the control characteristics of Fig. 4-1b. (a) Is the heater on or off at 65°F? (b) When is the heater unconditionally on *or* off? (c) What is the hysteresis or memory range of the thermostat?

SOLUTION (a) If the heater was on at 65°F, it remains on until the room warms to 67°F. If the heater was off at 65°F, it remains off until the room cools to 63°F. In short, the heater control remembers the last unconditional command.

(b) The heater will be unconditionally *on* if temperature is below $T_{LR} = 63°F$ and unconditionally *off* if temperature is above $T_{UR} = 67°F$.

(c) The hysteresis temperature is $T_H = T_{UR} - T_{LR}$ or $67°F - 63°F = 4°F$. This is the memory range.

The need for the hysteresis has been shown with respect to the familiar mechanical wall thermostat. In the next sections we show how hysteresis is added to electrical control circuits. These circuits are essentially comparators modifed by adding positive feedback.

4-2 POSITIVE FEEDBACK GIVES NOISE IMMUNITY

4-2.1 Positive Feedback Creates Two Stable States

Positive feedback occurs in an op-amp circuit when there is a connection between the output terminal and the noninverting input terminal as in Fig. 4-2. Positive feedback also causes a circuit to have two stable "memory" states.

In *both* Fig. 4-2a and b, input V_i equals 0 V. As will be shown, V_i lies within the memory or hysteresis range (Section 4-1). To analyze a circuit with positive feedback, one must first make an assumption. Then use cause and effect to prove or disprove that assumption. This technique is illustrated by asking the question: What is the value of V_o in Fig. 4-2 if $V_i = 0$ V?

One person might assume that V_o is stable at $+V_{sat}$ as in Fig. 4-2a. If this is true, an *upper* reference voltage V_{UR} is fed back to the (+) input of the op amp. V_{UR} depends on the resistors and value of $+V_{sat}$. It is equal to

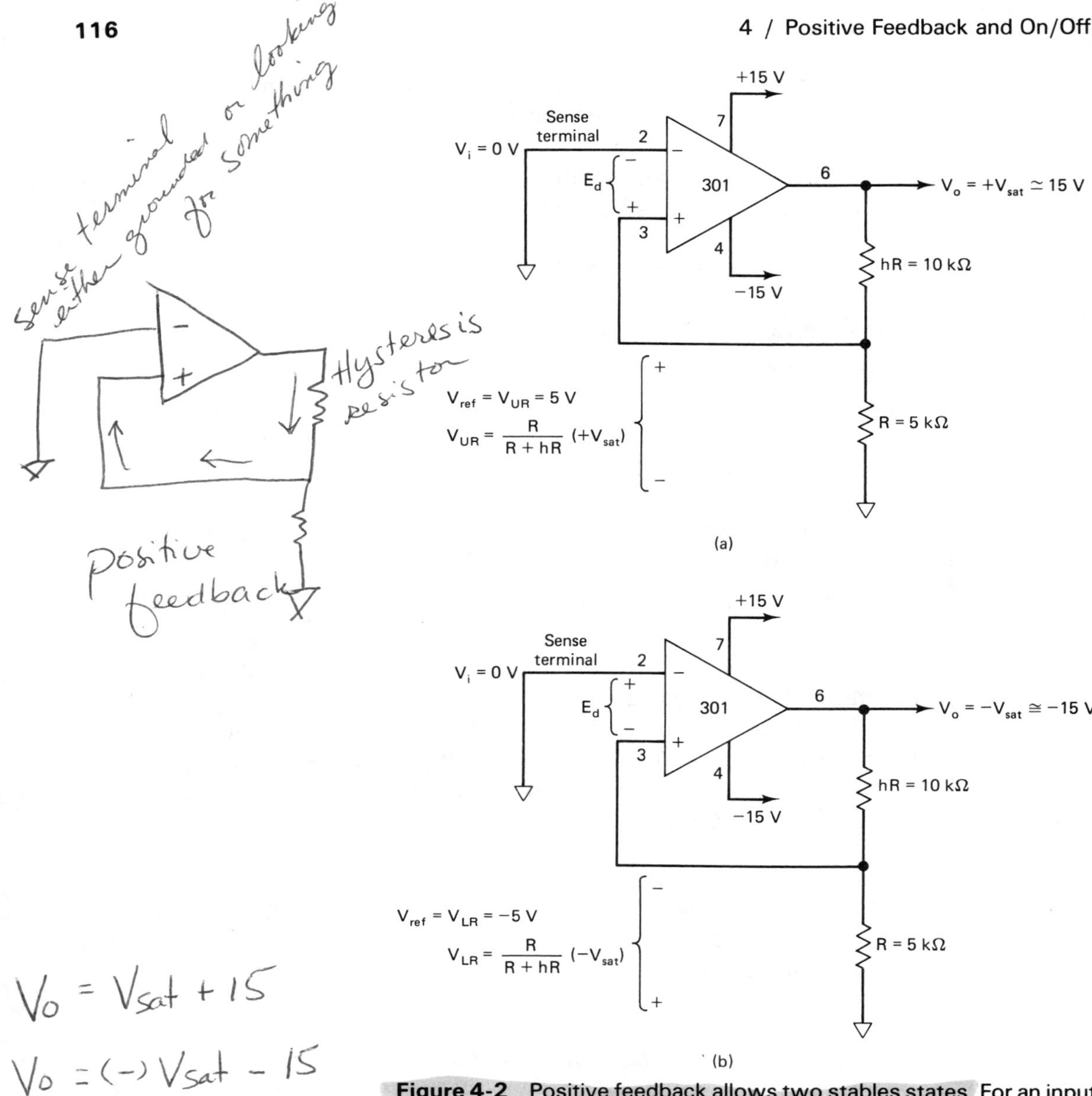

Figure 4-2 Positive feedback allows two stables states. For an input $V_i = 0$, in the hysteresis or memory range, output V_O can be stable at either $+V_{sat}$ or $-V_{sat}$. (a) If input E_i equals zero, V_O can be stable at V_{sat}, (b) if input V_i equals zero, V_O can also be stable at $-V_{sat}$.

$$V_{UR} = \frac{R}{R + hR}(+V_{sat}) = \frac{1}{1 + h}(+V_{sat}) \tag{4-1a}$$

where

$$h = \frac{hR}{R} \qquad (hR \text{ means "hysteresis resistor"}) \tag{4-1b}$$

In Fig. 4-2a, $h = 10\text{ k}\Omega/5\text{ k}\Omega = 2$ and $1/(1 + h) = 1/3$. Assume for simplicity that $+V_{sat} = +15$ V; then $V_{UR} = +5$ V. From Chapter 2, $E_d = +5$ V and V_o must be at $+V_{sat}$. The first assumption is correct.

A second person might assume that V_o is at $-V_{sat}$ as in Fig. 4-2b. If so, a new lower reference voltage V_{LR} will be established at

$$V_{LR} = \frac{R}{R + hR}(-V_{sat}) = \frac{1}{1 + h}(-V_{sat}) \tag{4-2}$$

V_{LR} will equal $(\frac{1}{3})(-15\text{ V}) = -5$ V. Since E_d is clearly negative, V_o will be stable at $-V_{sat}$. The second assumption is *also* correct.

These observations prove that positive feedback can cause two stable memory states. Figure 4-2 is an inverting comparator since the input voltage to be sensed, V_i, is wired to the $(-)$ input. However, V_{ref}, at the $(+)$ input, now has *two* possible values: (1) *upper reference* voltage, V_{UR}, and (2) *lower reference* voltage, V_{LR}. V_{ref} depends only on the output $\pm V_{sat}$ and the resistor divider R and hR.

In the next section we show how to change from one state to another and how hysteresis is created by positive feedback.

4-2.2 Hysteresis Action in an Inverting Comparator

In the positive feedback circuit of Fig. 4-3a, assume that V_i is at $+10$ V. See point A on the graphs of Fig. 4-3b and c. If V_o is at $+V_{sat} = +15$ V or $-V_{sat} = -15$ V, V_{ref} will be either $V_{UR} = +5$ V or $V_{LR} = -5$ V. E_d will then equal, using Eq. (1-7), $V_{UR} - V_i = +5\text{ V} - 10\text{ V} = -5$ V or $V_{LR} - V_i = 5\text{ V} - 10\text{ V} = -15$ V. For either possibility, E_d is negative. Therefore, V_o is unconditionally sent to $A_{OL}E_d = -V_{sat}$.

Since V_o is at $-V_{sat}$, V_{ref} is set at $V_{LR} = -5$ V from Eq. (4-2). Let V_i go negative beginning from $+10$ V at point A in Fig. 4-3b. When V_i crosses below $V_{LR} = -5$ V at point B, pin 3 is above pin 2 in potential and E_d goes positive. This forces V_o to snap positive to $+V_{sat}$. V_o now resets V_{ref} to $V_{UR} = +5$ V.

Refer to time interval B to C in Fig. 4-3b. V_i drops to -10 V and then rises toward $+10$ V. When V_i crosses above the *new* reference voltage $V_{UR} = +5$ V at point C'. E_d goes negative. V_o snaps to $-V_{sat}$ and sets V_{ref} back to $V_{LR} = -5$ V.

Hysteresis action is more visible from the plot of V_o versus V_i in Fig. 4-3c. Beginning at point A, $V_o = -V_{sat}$ and sets $V_{ref} = V_{LR}$. When V_i goes from $+10$ V to -5 V, V_o remains at $-V_{sat}$ (see A to B). When V_i drops below $V_{LR} = -5$ V, V_o snaps to $+V_{sat}$ at point B and changes V_{ref} to $V_{UR} = +5$ V. V_o remains at $+V_{sat}$ until V_i crosses V_{UR} at point C'. Here V_o snaps negative and changes V_{ref} back to V_{LR}.

In Fig. 4-3c, the memory range occupies the region between V_{LR} and V_{UR}. This range defines the circuit's hysteresis voltage V_H from

$$V_H = V_{UR} - V_{LR} \tag{4-3}$$

Thus, for the circuit of Fig. 4-3, $V_H = 5\text{ V} - (-5\text{ V}) = +10$ V.

4-2.3 Hysteresis Gives Noise Immunity

The circuit of Fig. 4-4a is intended to act as an inverting zero-crossing detector. That is, V_o snaps positive to indicate that V_i crossed 0 V going negative. Assume that a very large noise voltage appears in series with input signal V_i. As shown in Fig. 4-4, V_o responds to both the $+10$-V peak signal and the 5-V noise voltages.

The 10- and 5-kΩ resistors in Fig. 4-4b add hysteresis to the circuit. Assume for simplicity that $\pm V_{sat} = \pm 15$ V. From Eqs. (4-1) and (4-2), V_{UR}

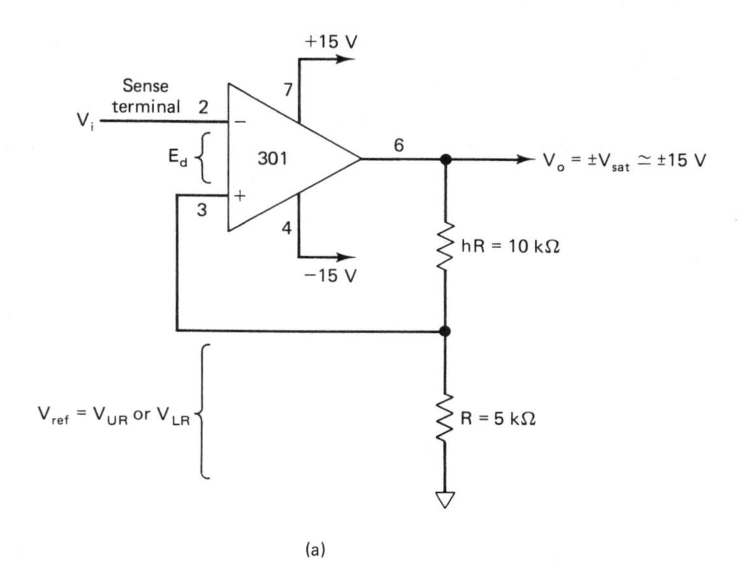

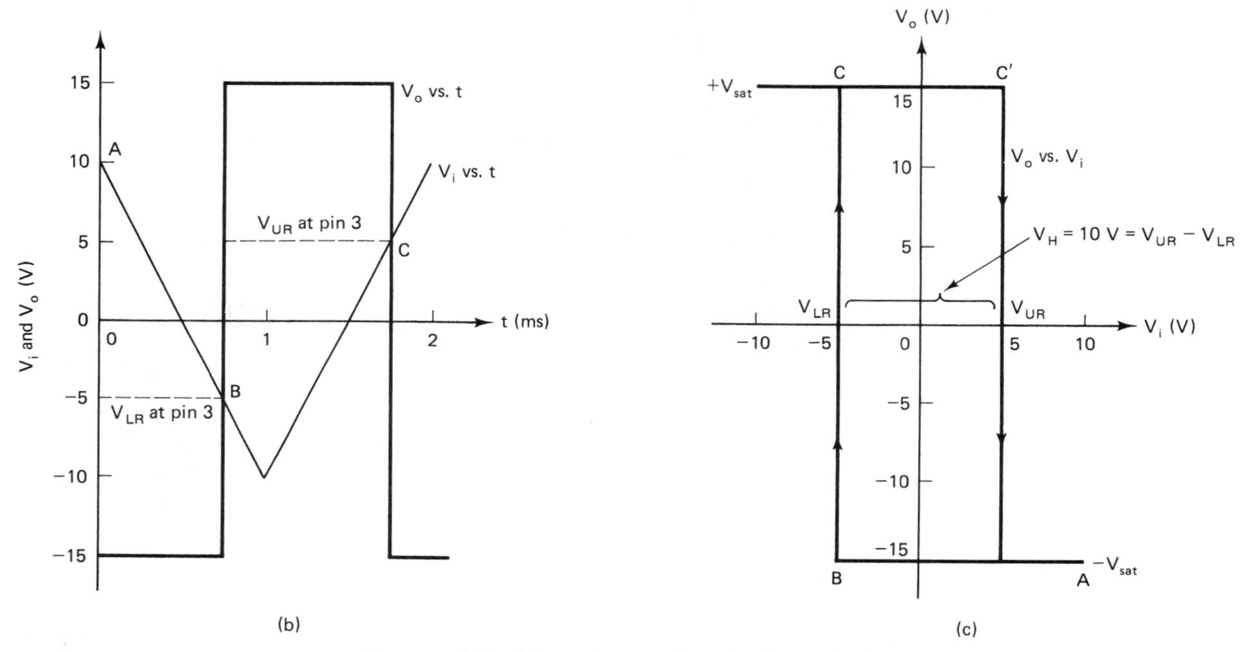

Figure 4-3 Waveshapes for the inverting comparator in (a) are shown in (b) and (c). For simplicity, assume that $\pm V_{sat} = \pm 15$ **V. (a) Positive feedback adds hysteresis to an inverting comparator; (b) input** V_i versus t and V_O versus t; (c) V_O versus V_i.

$= +5$ V and $V_{LR} = -5$ V. At time zero V_i is about 12.5 V, causing V_o to equal -15 V. V_{ref} must equal $V_{LR} = -5$ V. V_i and the noise voltage must drop below -5 V at point A before V_o snaps to $+15$ V. The new reference voltage is now $V_{UR} = +5$ V. When V_i, plus noise, cross -5 V at point B, V_{LR} voltage is no longer the reference. Therefore, the false crossings have been eliminated. True, V_o does not change at precisely the instant where V_o crosses 0 V. However, we have no false crossings in the presence of a prohibitively large noise signal.

A close inspection of the waveshapes in Fig. 4-4b leads to the conclusion that hysteresis voltage V_H will guarantee immunity against

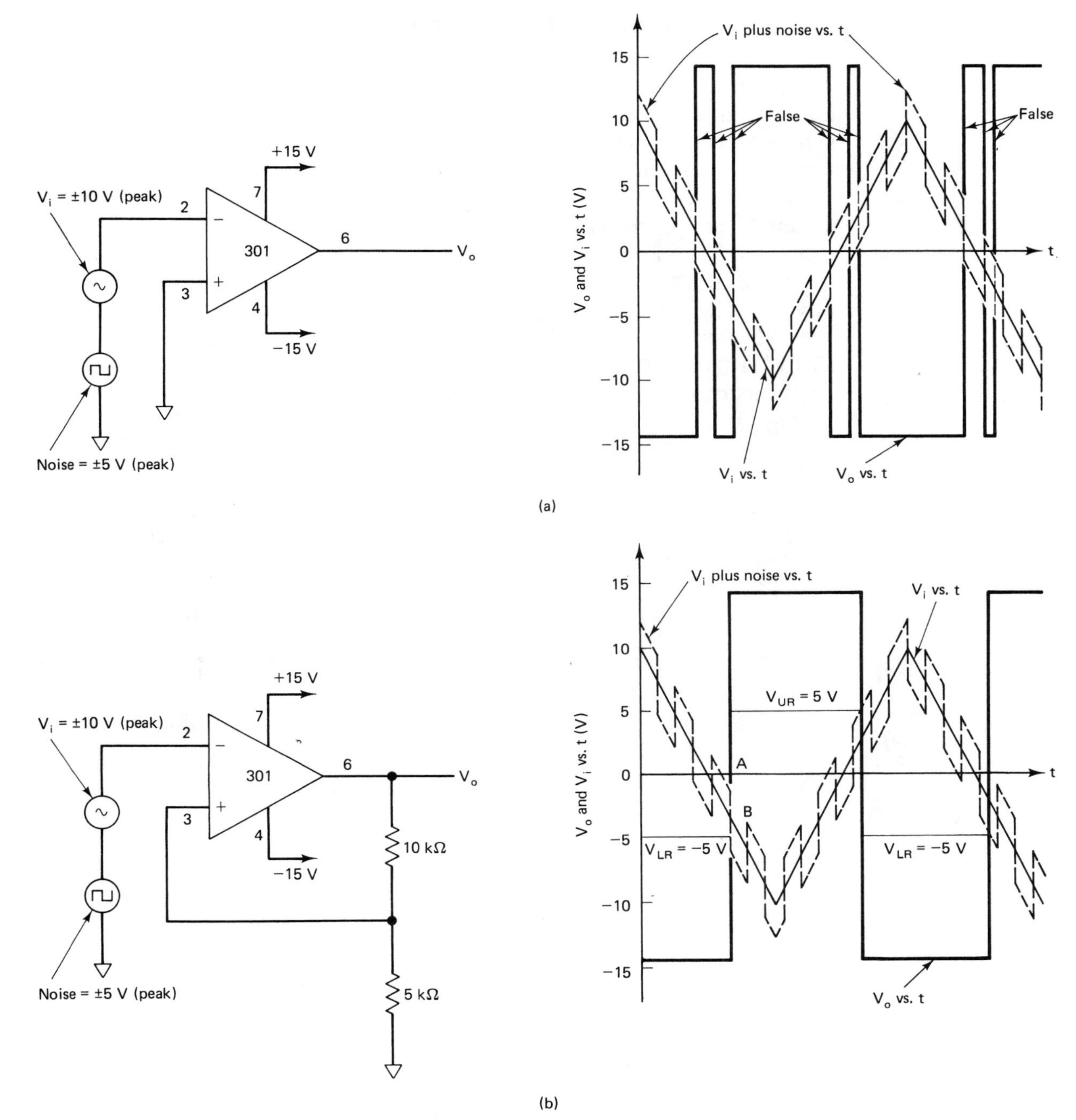

Figure 4-4 (a) V_O responds to both signal and noise voltage; (b) hysteresis eliminates false changes of V_O. In (a), V_i crosses zero volts three times, V_O crosses zero volts nine times due to noise voltage; in (b), V_i crosses zero volts three times, V_O crosses zero volts three times. V_O has no false crossings.

false output crossings for peak-to-peak noise immunity voltage less than V_H. That is,

$$\text{p-p noise voltage immunity} = V_H = V_{UR} - V_{LR} \tag{4-4}$$

We employ the principles above to solve a practical problem by an example in the next section.

4-2.4 Line Voltage Synchronizer

Temperature or motor control circuits, clock circuits, ac-to-ac converter circuits, and other applications require that a device or system be synchronized to the local line voltage. More specifically, the application requires a signal (such as a change in V_o) when the power company's line voltage crosses 0 V. The circuit of Fig. 4-5a is a line voltage synchronizer. When V_o snaps positive as in Fig. 4-5b, it means that the line voltage crossed 0 V and is going positive.

The transformer performs three jobs: It isolates the power line from any direct electrical connection to the low-voltage electronics. It reduces the 115-V rms line voltage to a value compatible with the op amp. V_i is a (3.15 V rms) × (1.41) = 4.4 V peak sine wave shown in Fig. 4-5b. Finally, the transformer is wired to give a 180° phase shift between primary line voltage and secondary voltage V_i.

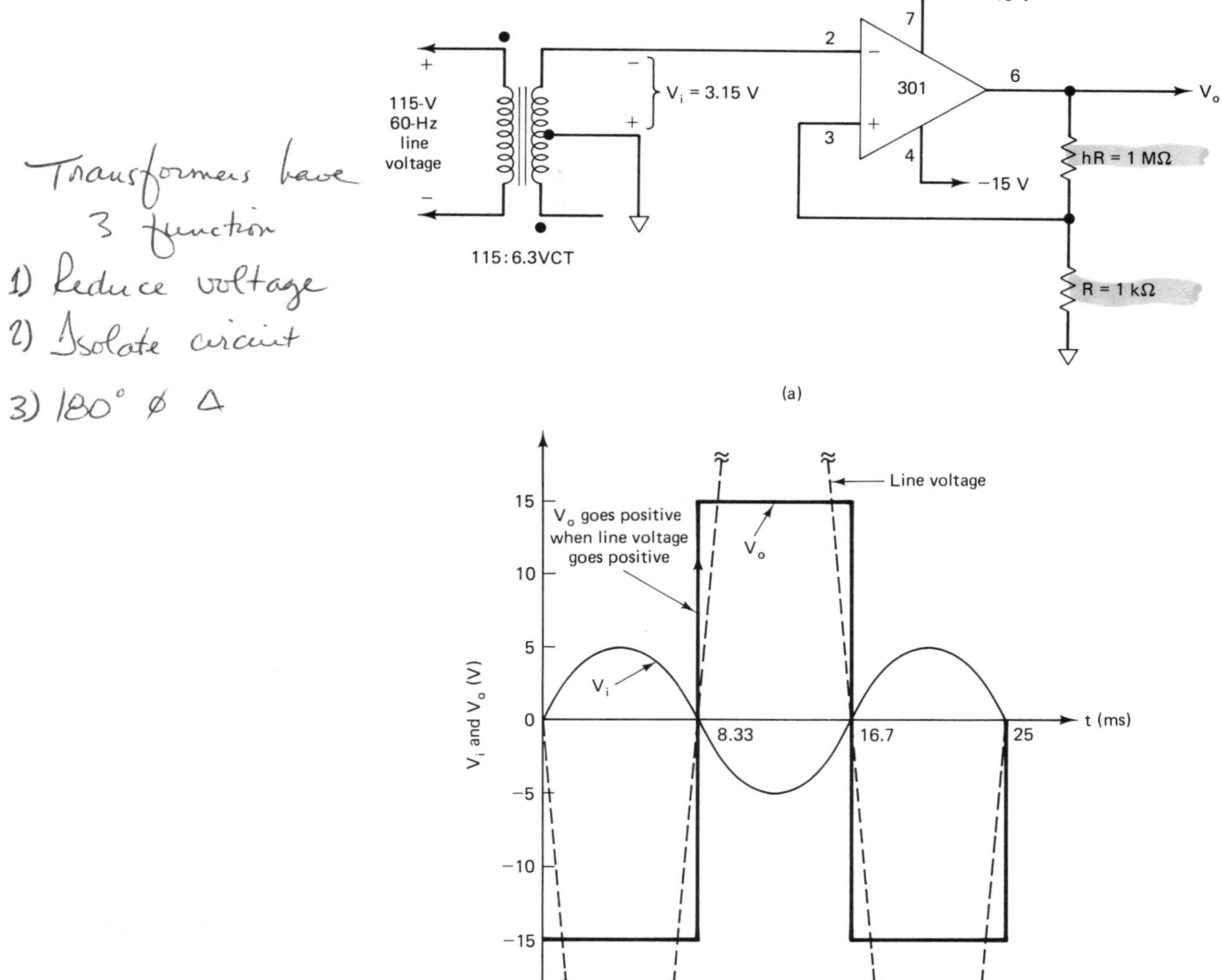

Figure 4-5 Line voltage synchronizer. *hR* and *R* add about 30 MV of hysteresis to provide noise immunity. (a) Changes in V_o are synchronized with line voltage zero crossings; (b) waveshapes for the circuit in (a).

In Fig. 4-5b, V_i is a down-scaled, inverted version of line voltage. The op amp is wired as an *inverting* comparator. The transformer inversion (Fig. 4-5a) cancels the op amp's inversion, so V_o is in phase with line voltage. When line voltage crosses 0 V going positive, V_o snaps positive. This transition of V_o is used to synchronize other circuits with line voltage.

EXAMPLE 4-2

Calculate (a) V_{UR}, (b) V_{LR}, and (c) noise immunity for the line voltage synchronizer in Fig. 4-5.

SOLUTION (a) Assume that $\pm V_{sat} = \pm 13$ V. From Eq. (4-1a),

$$V_{UR} = \frac{R}{R + hR}(+V_{sat}) = \frac{1\ \text{k}\Omega}{(1 + 1000)\ \text{k}\Omega}(+13\ \text{V}) = \frac{13\ \text{V}}{1001\ \text{k}}$$

$$\simeq +13\ \text{mV}$$

(b) From Eq. (4-2),

$$V_{LR} = \frac{R}{R + hR}(-V_{sat}) = \frac{1\ \text{k}\Omega}{(1 + 1000)\ \text{k}\Omega}(-13\ \text{V}) \approx -13\ \text{mV}$$

(c) Peak-to-peak noise voltage immunity is equal to hysteresis voltage V_H. From Eq. (4-4),

$$V_H = V_{UR} - V_{LR} = +13\ \text{mV} - (-13\ \text{mV}) = +26\ \text{mV}$$

4-2.5 Hysteresis Guidelines

It is sound engineering practice to add some hysteresis to any comparator circuit. When in doubt, arbitrarily add about 20 mV of hysteresis to impart some noise immunity. Hysteresis also increases the speed at which V_o changes from one state to the other.

One convenient design guideline is to choose a 1 MΩ for resistor *hR*. For each kilohm in resistor *R*, V_{UR}, and V_{LR} will equal 1/1000 of $+V_{sat}$ and $-V_{sat}$, respectively, or approximately ±13 mV.

4-3 ON/OFF CONTROL CIRCUITS

4-3.1 Circuit Operation

The basic on/off control circuit in Fig. 4-6a has positive feedback. *Hysteresis adjust resistor hR* is connected between output and (+) input of the op amp. Input V_i is connected via *set-point adjust resistor sR* to the (+) input of the op amp. Therefore, the circuit is noninverting in nature.

Assume that V_{ref} is a negative voltage in Fig. 4-6a and that V_i is 0 V. E_d would be negative forcing V_o to $-V_{sat}$. See the "0" points in Fig. 4-6b and c. V_i must rise to some positive voltage where E_d would just go positive and switch to V_o to $+V_{sat}$. This value of V_i defines upper references voltage V_{UR}, shown as point *A* in Fig. 4-6b and c.

When $V_o = +V_{sat}$, it opposes the effect of V_{ref} and holds pin 3 positive. V_i must drop to a value equal to lower reference voltage V_{LR}. Here E_d goes negative to switch V_o back to $-V_{sat}$ (see point *B* in Fig. 4-6b and c).

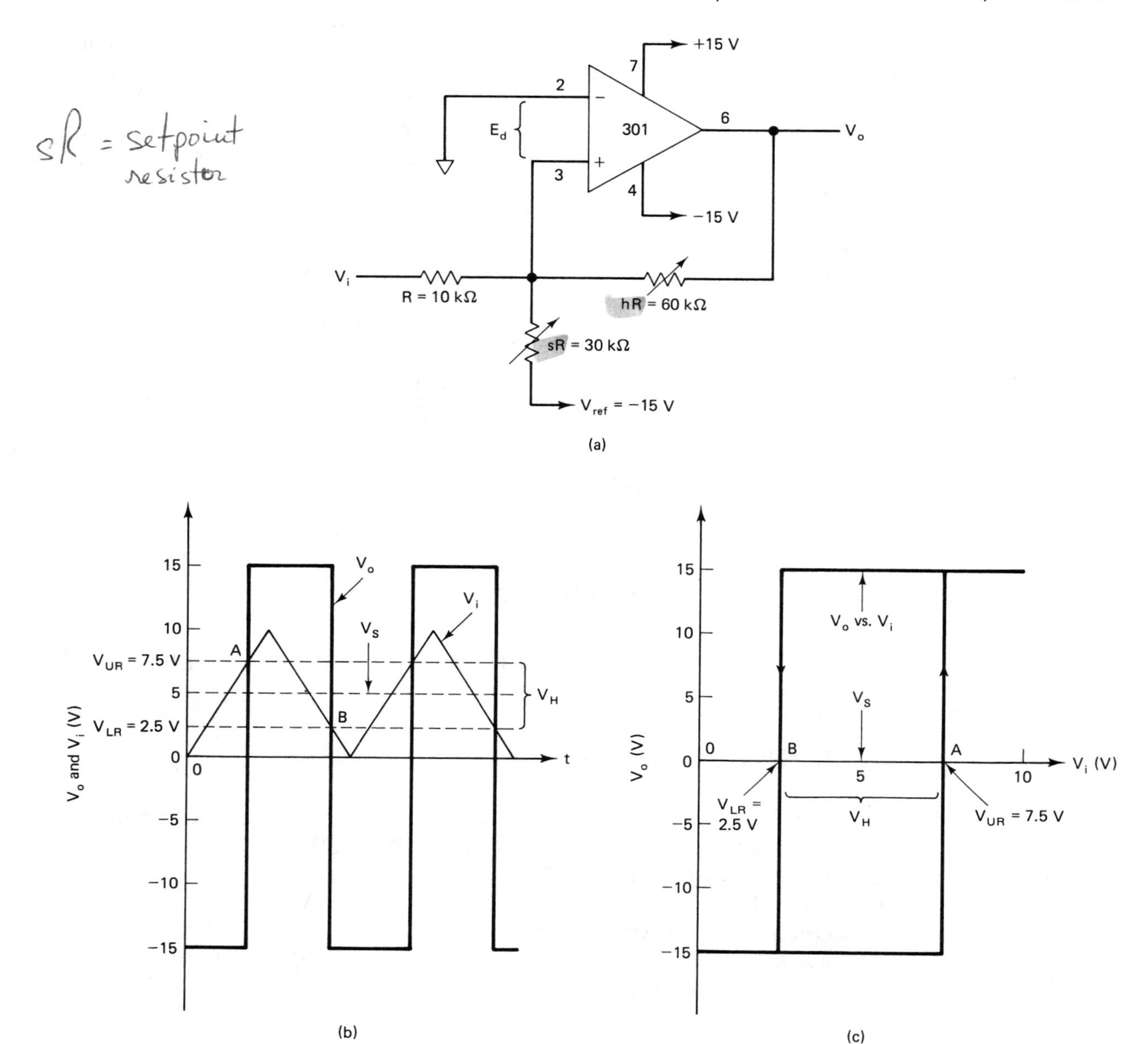

Figure 4-6 Operation of the basic on/off control circuit is illustrated by the time-varying wave shapes in (b) and the output–input characteristic in (c). (a) Basic on–off control circuit where resistor *sR* determines the set-point voltage and resistor *hR* determines the hysteresis voltage; (b) waveshapes for V_o and V_i; (c) V_o versus V_i.

4-3.2 Set-Point Voltage

Recall the discussion about operation of a wall thermostat for heat control. The operator dials 65°F as the *set-point* temperature. This is analogous to the *set-point* voltage V_S in Fig. 4-6b and c. Resistor *sR* in Fig. 4-6a allows adjustment of the *set-point* voltage just as a dial on the wall thermometer allows adjustment of the *set-point* temperature.

Hysteresis is set in the wall thermometer by calibrating technicians. They use a magnet, small heater, and a bimetallic strip. In Fig. 4-6a, electrical hysteresis is adjusted by *hysteresis resistor hR*.

4-3.3 Performance Equations

In the on/off control circuit of Fig. 4-6, set-point voltage V_S is given by

$$V_S = \frac{-V_{ref}}{s} \tag{4-5a}$$

where

$$s = \frac{sR}{R} \tag{4-5b}$$

Hysteresis voltage V_H is given by

$$V_H = \frac{(+V_{sat}) - (-V_{sat})}{h} \tag{4-6a}$$

where

$$h = \frac{hR}{R} \tag{4-6b}$$

All of the equations above are valid if the magnitudes of $-V_{sat}$ and $-V_{sat}$ are reasonably equal.

If you know both V_S and V_H, then V_{UR} and V_{LR} are found from

$$V_{UR} = V_S + \frac{V_H}{2} \tag{4-7a}$$

and

$$V_{LR} = V_S - \frac{V_H}{2} \tag{4-7b}$$

4-3.4 Circuit Analysis

Circuit analysis is reviewed by an example.

EXAMPLE 4-3

Assume that $\pm V_{sat} = \pm 15$ V in Fig. 4-6a. Find (a) set-point voltage V_S; (b) hysteresis voltage V_H; (c) upper reference voltage V_{UR}; (d) lower voltage reference V_{LR}.

SOLUTION (a) From Eqs. (4-5a) and (4-5b),

$$s = \frac{sR}{R} = \frac{30\ \text{k}\Omega}{10\ \text{k}\Omega} = 3 \qquad V_S = \frac{-V_{ref}}{s} = -\frac{-15\ \text{V}}{3} = +5\ \text{V}$$

Note: V_{ref} must be a negative value for V_S to be positive.

(b) From Eqs. (4-6a) and (4-6b),

$$h = \frac{hR}{R} = \frac{60\ \text{k}\Omega}{10\ \text{k}\Omega} = 6 \qquad V_H = \frac{(+15\ \text{V}) - (-15\ \text{V})}{6} = \frac{30\ \text{V}}{6} = 5\ \text{V}$$

(c) From Eq. (4-7a),

$$V_{UR} = V_S + \frac{V_H}{2} = 5\ \text{V} + \frac{5\ \text{V}}{2} = +7.5\ \text{V}$$

(d) From Eq. (4-7b),

$$V_{LR} = V_S - \frac{V_H}{2} = 5\text{ V} - \frac{5\text{ V}}{2} = +2.5\text{ V}$$

4-4 HEATING SYSTEM ELECTRONIC CONTROL

4-4.1 Basic Heat Control System

A wall thermostat converts room temperature to a motion that closes or opens low-voltage contacts. These contacts control a relay that turns an oil burner motor or electric heater on or off.

A block diagram for the electronic equivalent control of a home heating system is presented in Fig. 4-7a. The first block represents a temperature-to-voltage converter. An input temperature range of 0 to 100°F gives an output voltage range of 0 to 10 V. This voltage now becomes input V_i to the control circuit in Fig. 4-7b (or also Fig. 4-6).

Suppose that the set-point temperature is to equal T_{set} = 65°F with a hysteresis temperature of T_H = 4°F. That is, heat should be *on* when temperature is *equal to or below* T_{LR} = 63°F. Heat should be *off* when

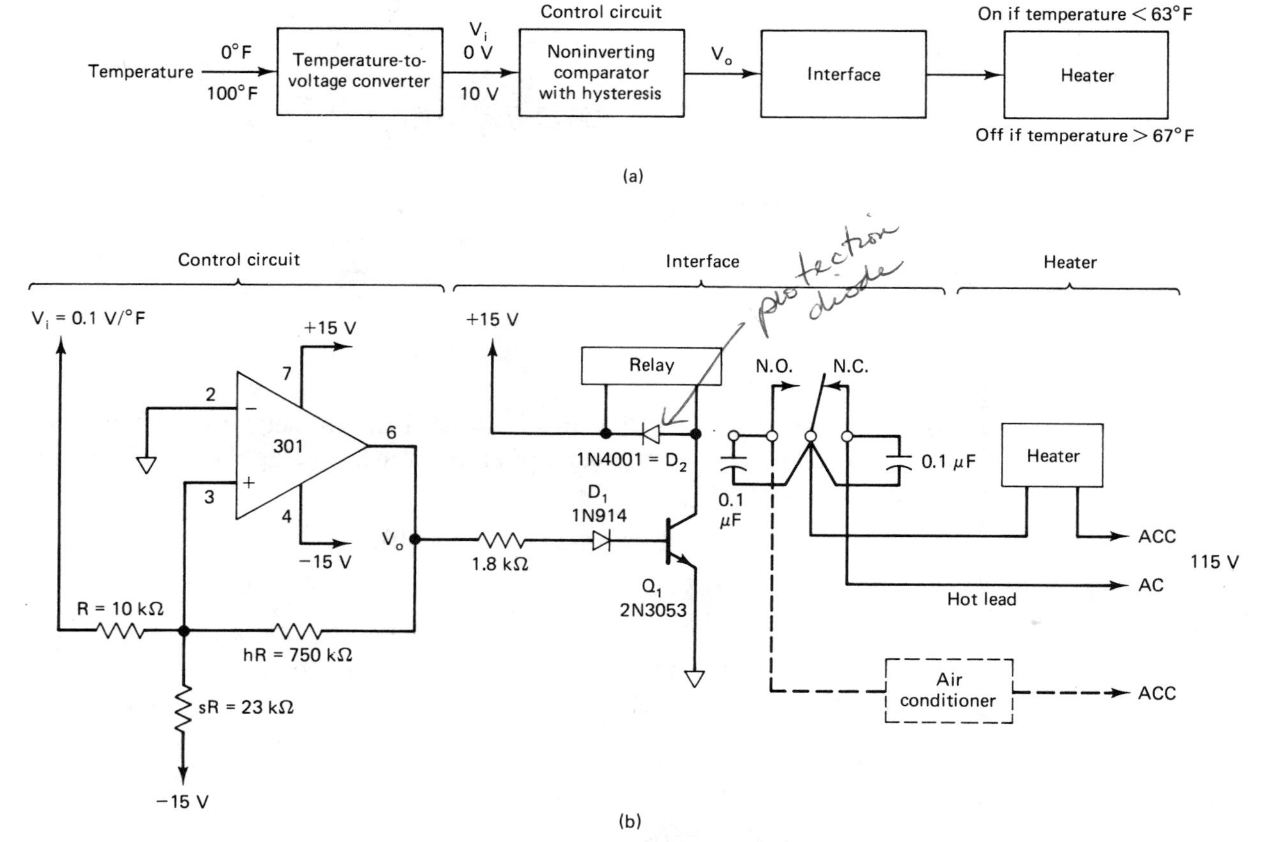

Figure 4-7 (a) Block diagram of a heating system; (b) electronic control circuit and interface.

temperature *equals or exceeds* $T_{UR} = 67°F$. The corresponding values for temperature-dependent voltage V_i would be

$$V_{LR} = 6.3 \text{ V} \qquad V_S = 6.5 \text{ V} \qquad V_{UR} = 6.7 \qquad V_H = 0.4 \text{ V}$$

The low-voltage control circuit of Fig. 4-7b needs an interface to the high-voltage heater, as does its mechanical counterpart. The electronic interface circuit is discussed in Section 4-4.3, but first we analyze the electronic control circuit.

4-4.2 Electronic Counterpart of a Mechanical Temperature Control

Let us examine, by an example, how the on/off control voltages of Section 4-4.1 are realized in Fig. 4-7b.

EXAMPLE 4-4

For the control circuit in Fig. 4-7b, find (a) V_S; (b) V_H; (c) V_{UR}; V_{LR}.

SOLUTION Assume for simplicity that $\pm V_{sat} = \pm 15$ V.

(a) From Eqs. (4-5a) and (4-5b),

$$s = \frac{sR}{R} = \frac{23 \text{ k}\Omega}{10 \text{ k}\Omega} = 2.3 \qquad V_S = -\frac{V_{ref}}{S} = -\frac{-15 \text{ V}}{2.3} \simeq 6.5 \text{ V} \simeq (65°\text{F})$$

(b) From Eqs. (4-6a) and (4-6b),

$$h = \frac{hR}{R} = \frac{750 \text{ k}\Omega}{10 \text{ k}\Omega} = 75 \qquad V_H = \frac{(15 \text{ V}) - (-15 \text{ V})}{75} = \frac{30 \text{ V}}{75} \simeq 0.4 \text{ V} \simeq (4°\text{F})$$

(c) and (d) From Eqs. (4-7a) and (4-7b),

$$V_{UR} = V_S + \frac{V_H}{2} = 6.5 \text{ V} + \frac{0.4 \text{ V}}{2} = 6.7 \text{ V} \simeq (67°\text{F})$$

$$V_{LR} = V_S - \frac{V_H}{2} = 6.5 \text{ V} = \frac{0.4 \text{ V}}{2} = 6.3 \text{ V} \simeq (63°\text{F})$$

Example 4-4 shows one specific analogy between a thermal mechanical control system and its electronic counterpart. The interface circuit is discussed next.

4-4.3 Operation of the Interface Circuit

The interface circuit of Fig. 4-7b is made from a resistor, two diodes, a transistor, and a relay. Transistor Q_l amplifies the op amp's low output current ($\approx$6 mA) enough to activate the relay. Diode D_1 protects the base of Q_1 against excessive reverse bias when V_o of the op amp is at $-V_{sat}$. Diode D_2 absorbs the relay's inductive "kick" to protect the collector of Q_1 when Q_1 turns off. The relay's contacts isolate the 115-V line voltage from the low-voltage electronics.

To analyze circuit operation, assume that temperature is below 63°F and V_i is below $V_{LR} = 6.3$ V. Output V_o of the op amp will be at $-V_{sat}$. D_1 is reverse biased so that Q_1 is cut off. The relay is deenergized. Its wiper

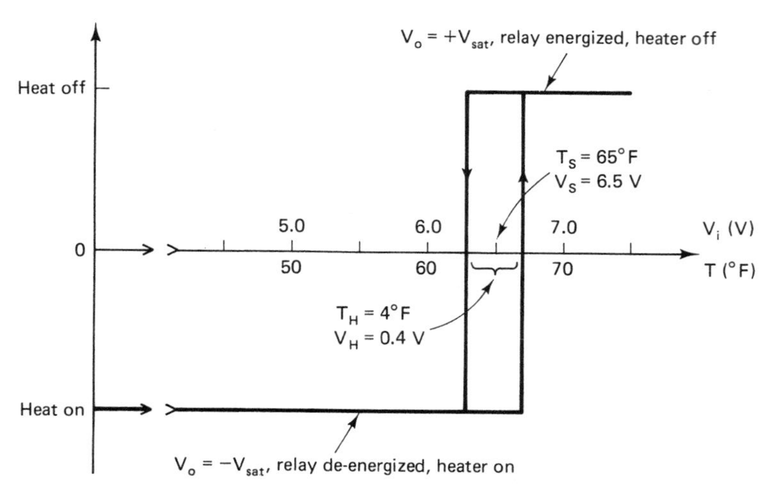

Figure 4-8 Operational summary for the heater control circuit of Figure 4-6.

touches the normally closed (N.C.) contact, completing the line voltage circuit to turn the heater on.

Assume that room temperature rises until V_i exceeds 6.7 V (≈67°F). V_o snaps positive to $+V_{sat}$, turning Q_1 on and energizing the relay. The relay's wiper transfers to the normally open (N.O.) contact, breaking the heater circuit to shut the heater off. The control sequence is reviewed and summarized.

4-4.4 Summary

Operation of the electronic-controlled heater system of Fig. 4-7b is summarized by the output-input characteristic in Fig. 4-8. V_S = 6.5 V, and consequently, T_S = 65° is adjusted by set-point resistor sR. Hysteresis is adjusted by hysteresis resistor *hR*.

If temperature is below 63°F, or above 67°F, the heater is unconditionally on or off, respectively. In the "memory" range from 63 to 67°F, the heater stays on if it was on, or stays off if it was off. Hysteresis gives the circuit the capability of "remembering" the last unconditional command.

EXAMPLE 4-5

Convert the heater control of Fig. 4-7b into an air conditioner or refrigerator control. The cooler should be *off* when the temperature is below T_{LR} = 63°F and *on* when the temperature is above T_{UR} = 67°F.

SOLUTION The example is offered to show the exceptional versatility of the ordinary relay with transfer contacts. That is, a wiper can transfer a "HOT" ac lead from N.C. to N.O. contacts, or vice versa.

In Fig. 4-7b the relay is deenergized when temperature is below 63°F. Since the air conditioner should be off, simply wire its motor to the N.O. contact in Fig. 4-7b. If room temperature is above 67°F, the relay energizes to turn the air conditioner on.

End of Chapter Exercises

Name: ____________________

Date: __________ **Grade:** ___

A. FILL-IN THE BLANKS

Fill in the blanks with the best answer.

1. A circuit with positive feedback displays the property of __________.

2. Positive feedback causes an op amp circuit to have __________ stable memory states.

3. The stable states of an op amp with positive feedback are __________ and __________.

4. Positive feedback occurs in an op amp circuit when a connection is made between the output terminal and the __________ input terminal.

5. The two stable output voltage states of an op amp with positive feedback are __________ and __________.

6. Hysteresis voltage V_H occupies the range of voltages between __________ and __________.

7. The set-point voltage of Fig. 4-6 can be changed by adjusting the __________ resistor.

8. In order to change the hysteresis voltage in Fig. 4-6, adjust the __________ resistor.

9. It is required that the set-point voltage in Fig. 4-6 be a negative value. V_{ref} must therefore be chosen to be a __________ voltage.

10. Diode __________ protects the collector of Q_1 in Fig. 4-7 from the relay's inductive "kick".

B. TRUE/FALSE

Fill in **T** if the statement is true, and **F** if any part of the statement is false.

1. ______ V_{UR} is always more positive than V_{LR}.
2. ______ The set-point voltage, V_S, is always greater than V_{LR} and less than V_{UR}.
3. ______ Adjustments of the hR resistor in Fig. 4-6 changes both V_H and V_S.
4. ______ D_1 and D_2 of Fig. 4-7 are included as protection for Q_1.
5. ______ Sound engineering practice requires that 20 volts of hysteresis voltage be added to impart some noise immunity to any comparator circuit.

C. CIRCLE THE CORRECT ANSWER

Circle the correct answer for each statement.

1. The output of Fig. 4-3 switches from $+V_{sat}$ to $-V_{sat}$ at (V_{UR}, V_{LR}).
2. The output of Fig. 4-3 switches from $-V_{sat}$ to $+V_{sat}$ at (V_{UR}, V_{LR}).
3. Sound engineering practice requires that (some, no) hysteresis be added to a comparator to guard against noise.
4. Hysteresis (increases, decreases) the speed at which V_0 switches from one stable state to the other.
5. Which voltage lies symmetrically between V_{UR} and V_{LR}? ($+V_{sat}$, V_{set}).

D. MATCHING

Match the name or symbol in column **A** with the statement that matches best in column **B**.

	COLUMN A	COLUMN B
1. ______	s	a. $V_{UR} - V_{LR}$
2. ______	hR	b. adjusts V_s
3. ______	V_{ref}	c. Hysteresis Voltage
4. ______	V_H	d. sR/R
5. ______	sR	e. adjusts V_H

PROBLEMS

4-1. A water heat control is programmed for a set-point temperature of 160°F and a hysteresis temperature of 20°F. At what temperature does the water heater turn off and on?

4-2. Why does an on/off temperature control need hysteresis?

4-3. In the circuit of Fig. 4-2, hR = 13 kΩ and R = 2 kΩ. Assume that $\pm V_{sat} = \pm 15$ V. Find

(a) V_{UR};
(b) V_{LR};
(c) V_H.

4-4. In Problem 4-3, assume that V_i is a +5 V (peak) triangle wave. Sketch the resulting V_o versus t and V_o versus V_i. Show V_{UR}, V_{LR}, and V_H.

4-5. What is the noise voltage immunity for the circuit of Problems 4-3 and 4-4?

4-6. In the on/off control circuit of Fig. 4-6a, resistor sR is changed to 50 kΩ, and hR to 100 kΩ. Find

(a) V_S;
(b) V_H;
(c) V_{UR} and V_{LR}.

4-7. In the heater control circuit of Fig. 4-7b, which resistor would you increase or decrease to
(a) decrease set-point temperature to 62°F;
(b) increase hysteresis temperature to 6°F?

4-8. What does the transistor and relay do in the interface circuit of Fig. 4-7b?

4-9. Sketch the output versus input characteristics of a heating system with a set temperature of 62°F and a hysteresis temperature of 6°F. The output is to be plotted on the vertical axis to show if the heater is on or off. Input temperature is plotted on the horizontal axis.

Laboratory Exercise 4

Name: ____________________

Date: __________ **Grade:** ___

POSITIVE FEEDBACK AND ON/OFF CONTROLS

OBJECTIVES: Upon completion of this laboratory exercise on comparators with positive feedback, you will be able to (1) measure both V_{UR} and V_{LR}; (2) display V_i versus t and V_o versus t; (c) plot the transfer function of a comparator with positive feedback; (4) analyze an on/off control circuit, and (5) test the control circuit.

REFERENCE: Chapter 4

PARTS LIST

2	301 op amps	1	33-kΩ resistor
2	2.2-kΩ resistors	1	10-kΩ trim pot
1	8.2-kΩ resistor	1	50-kΩ trim pot
2	10-kΩ resistors	1	100-kΩ trim pot

Procedure A: Measuring V_{UR} and V_{LR}

1. The circuit shown in Fig. L4-1 is an inverting comparator with positive feedback. Measure and record the actual values of both hR and R and use these values in all calculations relating to Fig. L4-1.

   $hR =$ __________; $R =$ __________. Using a DMM, measure and record the power supply voltages, $V_{CC} =$ __________; $V_{EE} =$ __________. Also, uses these values for calculations.

2. Assume that $+V_{sat}$ is 2 V below V_{CC} and $-V_{sat}$ is 2 V above V_{EE}. Calculate the upper and lower reference voltages (V_{UR} and V_{LR}) expected in Fig. L4-1.

$$V_{UR} = \frac{+V_{sat}}{1 + h} \quad \text{and} \quad V_{LR} = \frac{-V_{sat}}{1 + h}$$

   where $h = hR/R$. $V_{UR} = +$ __________; $V_{LR} = -$ __________.

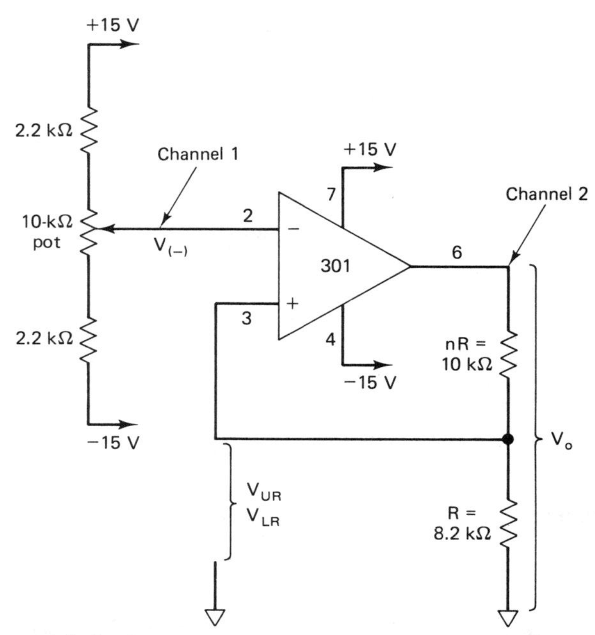

Figure L4-1 Inverting comparator with positive feedback.

3. Wire Fig. L4-1. Connect Channel 1 to measure $V_{(-)}$ and Channel 2 to monitor V_o. Set the sensitivity of both CRO channels to 5 V/div. With both channels switched to GND, zero both traces in the center of the scope face. Switch the CRO to dc coupling.
4. Initially set the 10-kΩ pot of Fig. L4-1 to $V_{(-)} = +10$ V. Record the magnitude and polarity of V_o. V_o = ____________. Is V_o at $-V_{sat}$? ____________.
5. Slowly adjust the 10-kΩ pot until V_o switches from $-V_{sat}$ to $+V_{sat}$. Record the voltage $V_{(-)} = V_{LR}$ at this first transition point. V_{LR} = ____________. Does this measured value compare favorably with the prediction in step 2? ____________.
6. Reset the 10-kΩ pot until $V_{(-)}$ is at -10 V. Record the magnitude and polarity of V_o. V_o = ____________. Is V_o at $+V_{sat}$? ____________.
7. Slowly adjust the 10-kΩ pot until V_o just switches from $+V_{sat}$ to $-V_{sat}$. Record the voltage $V_{(-)} = V_{UR}$ at this second transition point. V_{UR} = ____________. Does this measured value compare favorably with the prediction in step 2? ____________.

Procedure B: Display V_i versus t and V_o versus t

8. Modify Fig. L4-1 to include the triangle wave function generator shown in Fig. L4-2. Set V_i to ± 10 V (peak) at a frequency of 100 Hz. Connect the scope as shown.
9. Obtain the waveshape of V_i shown on Fig. L4-3. Include on Fig.

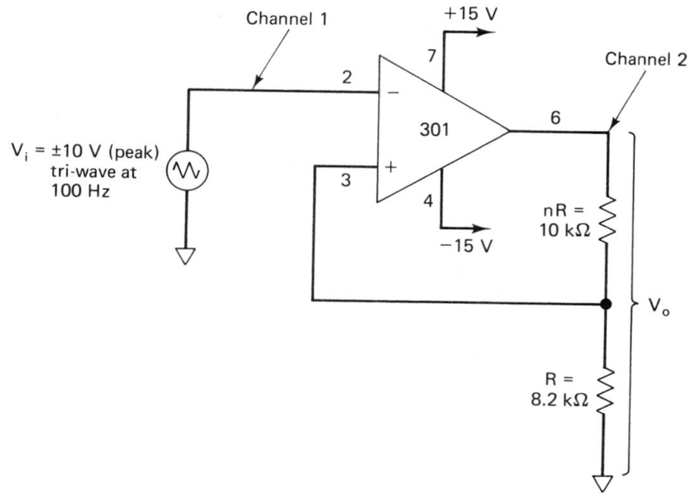

Figure L4-2 Test circuit to plot V_i versus t and V_o versus t.

L4-3, as dashed lines, the measured value of V_{UR} (step 5) and V_{LR} (step 7). Label these lines as V_{UR} and V_{LR}.

10. Carefully plot V_o versus t on Fig. L4-3. Label the transition points from $-V_{sat}$ to $+V_{sat}$ as point x on Fig. L4-3. Label the transition points from $+V_{sat}$ to $-V_{sat}$ as point y. Point x occurs at __________ (V_{UR} or V_{LR}). Point y occurs at __________ (V_{UR} or V_{LR}).

Procedure C: Transfer Function of a Comparator with Positive Feedback

11. To obtain the transfer function of the comparator shown in Fig. L4-2, switch the CRO's time base to (XY). Ground both channels

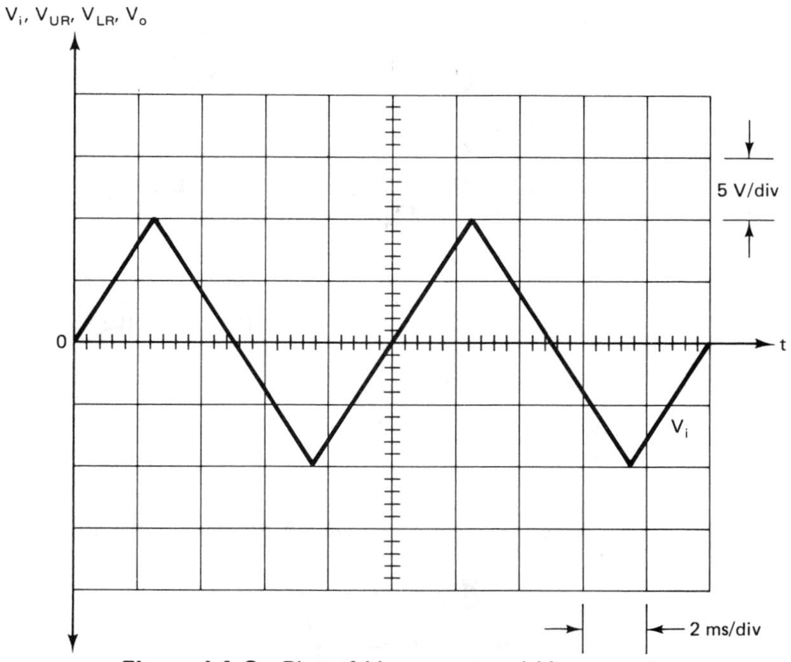

Figure L4-3 Plot of V_i versus t and V_o versus t.

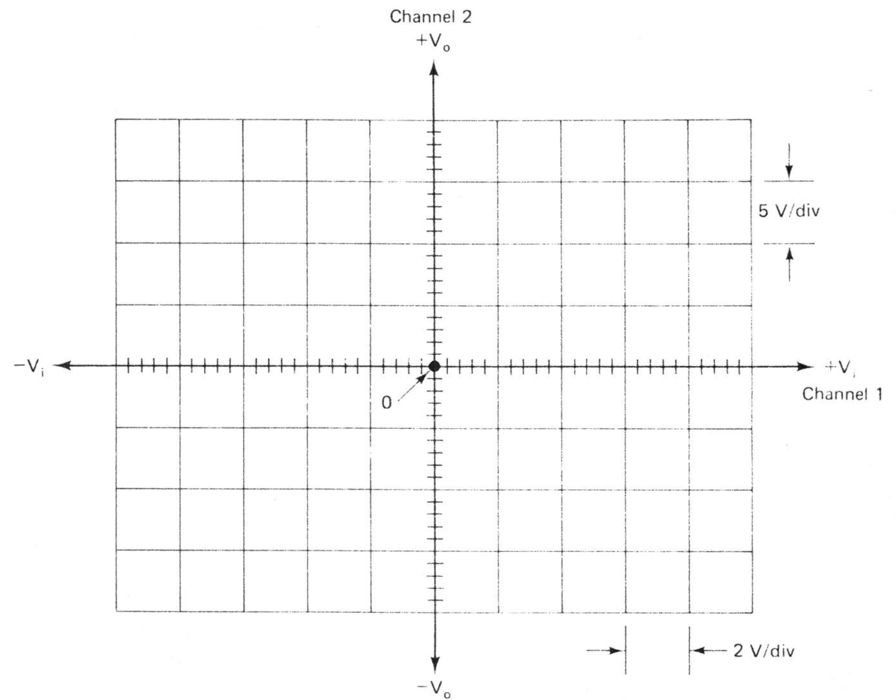

Figure L4-4 Plot of V_O versus V_i.

and zero the spot in the center of the scope face (see Fig. L4-4). Channel 1 is set to 2 V/div and channel 2 to 5 V/div. Switch to dc coupling.

12. Accurately sketch the V_o versus V_i transfer function in Fig. L4-4. Label the following items: V_{UR}, V_{LR}, V_S, V_H, $+V_{sat}$, and $-V_{sat}$. Using Fig. L4-4, measure and record the values for the items above. V_{UR} = __________; V_{LR} = __________; V_s = __________; V_H = __________; $+V_{sat}$ = __________; $-V_{sat}$ __________.

Procedure D: On/Off Control Analysis

13. The on/off control circuit of Fig. L4-5 is to be analyzed for both the set-point voltage V_s and the hysteresis voltage V_H. Assume that $\pm V_{sat} = \pm 13$ V for this problem.
14. Calculate the value of s from the equation s = sR/R. s = __________. The set-point voltage is then determined by $V_s = -(V_{ref})/s$. V_S = __________. (*Note:* V_{ref} is a negative number that results in a positive value of set-point voltage V_S).
15. Calculate the value of h from the equation h = hR/R. h = __________. The value of hysteresis voltage V_H is then obtained by $V_H = [(+V_{sat}) - (-V_{sat})]/h$. V_H = __________.
16. From the results above, solve for both the upper and lower reference

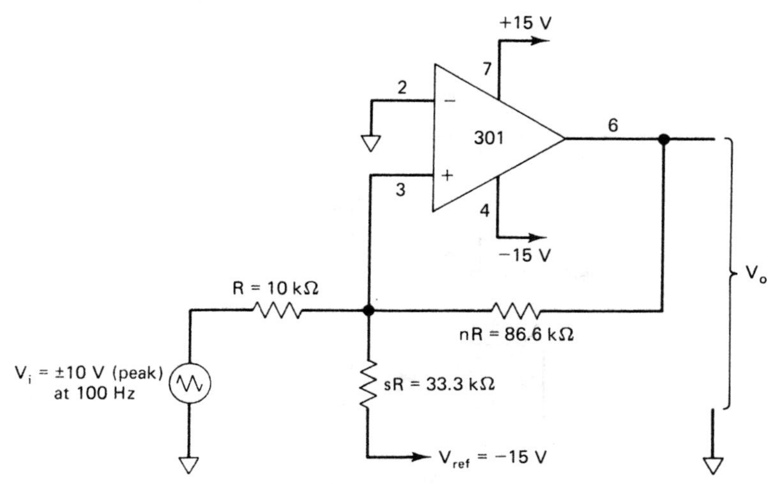

Figure L4-5 On/off controller.

voltages V_{UR} and V_{LR} using the equations provided: $V_{UR} = V_S + V_H/2$ and $V_{LR} = V_S - V_H/2$. V_{UR} = __________; V_{LR} = __________.

17. Use Fig. L4-6 to sketch the expected transfer function of Fig. L4-5. Include on this sketch the following information: V_H (hysteresis voltage), V_S (set-point voltage), V_{UR} (upper reference), V_{LR} (lower reference). Also show, as dashed lines, $\pm V_{sat}$.

Procedure E: Testing the Control Circuit

18. Wire the circuit shown in Fig. L4-7. Use a 33-kΩ fixed resistor and 100-kΩ pot in series to set hR = 86.6 kΩ and install this combina-

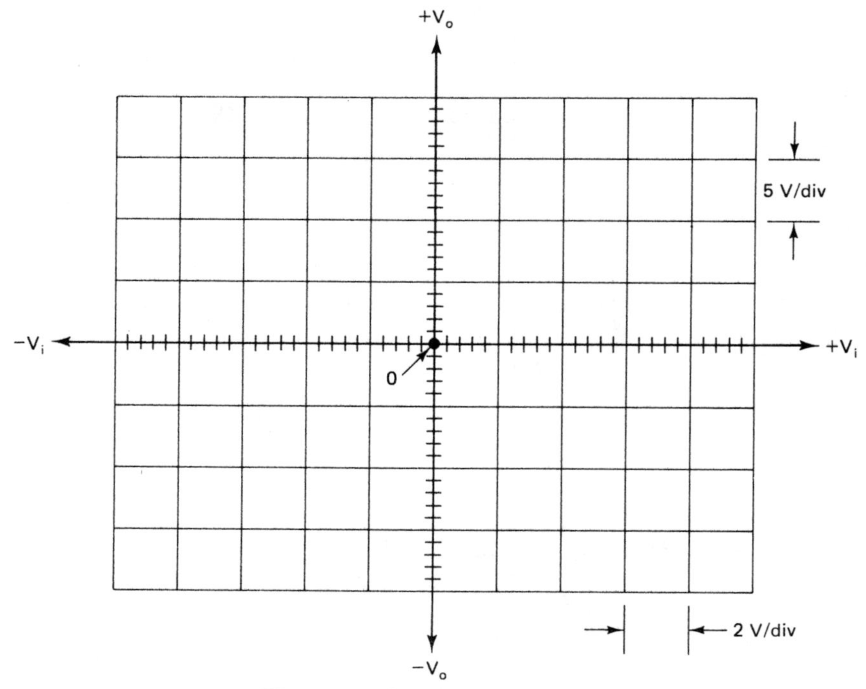

Figure L4-6 Transfer curve.

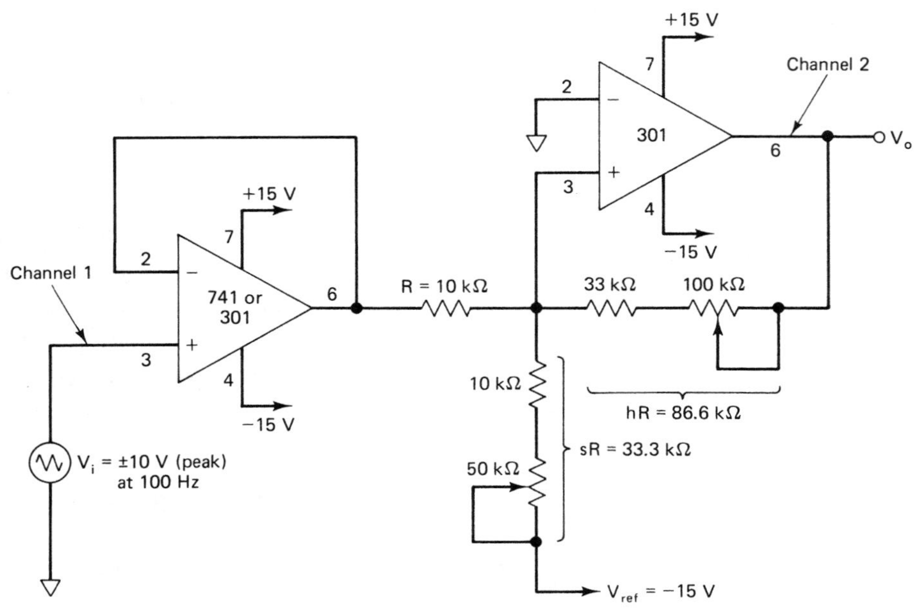

Figure L4-7 On/off control circuit.

tion in Fig. L4-7. Build the sR resistor from a 10-kΩ fixed resistor and 50-kΩ pot adjusted to sR = 33.3 kΩ. Set V_i to a ±10-V (peak) triangle wave at a test frequency of 100 Hz. Connect the CRO as shown with channel 1 at 2 V/div and channel 2 at 5 V/div.

19. Switch the scope to (XY) and with both channels grounded, zero the spot as shown in Fig. L4-8. Then switch to dc coupling to obtain the transfer curve of the on/off controller.

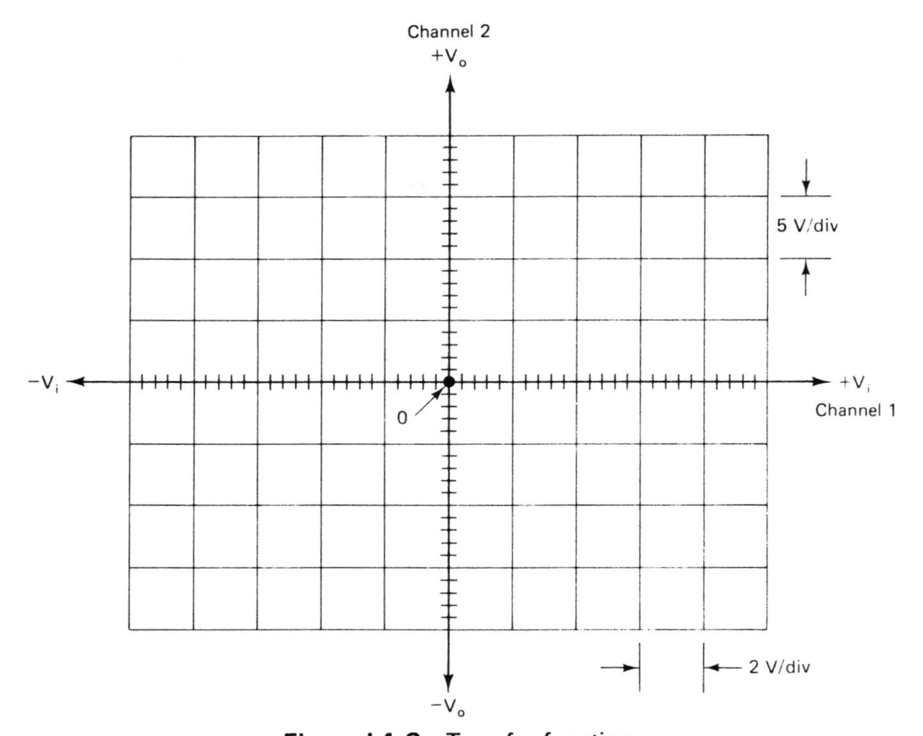

Figure L4-8 Transfer function.

20. Sketch the waveshape on Fig. L4-8. Measure: V_{UR} = ________; V_{LR} = ________; V_S = ________; V_H = ________; $+V_{sat}$ = ________; $-V_{sat}$ = ________. Label each of these values on Fig. L4-8.
21. Do the measured values shown in Fig. L4-8 compare favorably with the predictions of Fig. L4-6? ________.
22. While watching the CRO adjust the 100-kΩ pot. Does V_H change and V_S remain constant? ________. Now adjust the 50-kΩ pot. Does V_S change and V_H remain constant? ________.
23. Change V_{ref} from −15 V (i.e., V_{EE}) to +15 V (i.e., V_{CC}). What happens to V_S? ____

____

CONCLUSION:

Inverting and Noninverting Amplifiers

5

LEARNING OBJECTIVES

Upon completion of this chapter on operational amplifiers, you will be able to:

- Describe how negative feedback works to maintain the op amp in the linear region and out of saturation.
- Analyze op-amp circuits with negative feedback and calculate all currents and voltages.
- Determine the closed-loop voltage gain A_{CL} by inspection of the amplifier circuit.
- Set the input resistance of an inverting amplifier by simply selecting the proper value of input resistor R_i.
- Design a TEE network inverting amplifier to get large values of both input resistance and gain.
- Approximate the input and output resistances of a noninverting amplifier.
- Control amplification of a programmable-gain amplifier (PGA) with a computer.
- Determine the closed-loop gain for a voltage follower.
- Approximate the input and output resistances of a voltage-follower amplifier.
- Use a voltage-follower amplifier to improve the input resistance characteristics of a typical analog multimeter.

5-0 INTRODUCTION

In the last few chapters, discussion has centered on how the op amp is used as a switch. In Chapters 2 and 3, with no feedback applied, the op-amp output voltage V_o was in either positive or negative saturation.

Positive feedback was added to the op amp in Chapter 4, but this only altered the point at which the output would change from one saturation level to the other. The op amp was still being operated as a switch in the *nonlinear* mode.

We begin with an introduction to the concepts of negative feedback. The use of negative feedback around an op amp has the effect of forcing the op amp to operate as an amplifier in the *linear* mode between the limits of $\pm V_{sat}$. Negative feedback provides the basis for the development of two types of amplifier. They are the *inverting* and *noninverting* configurations. With negative feedback around an op amp, the resulting amplifier will have a *closed-loop gain* A_{CL} that is set by external resistors. Closed-loop gain A_{CL} will be independent of *open-loop gain* A_{OL} of the op amp. In this way, input signals applied to the amplifier will be amplified and faithfully reproduced at its output.

Each amplifier configuration presented, begins with a circuit analysis. For each amplifier type, applications follow that demonstrate their usefulness.

5-1 THEORY OF NEGATIVE FEEDBACK

Negative feedback can be established around an op amp by providing a path for dc current to flow from the output terminal back to the inverting input terminal. Two forms of negative feedback are shown in the test circuits of Fig. 5-1. In both circuits the negative *feedback loop* is closed. For Fig. 5-1a, the entire output voltage is fed directly back to the inverting input. In Fig. 5-1b, only a fraction of V_o, the voltage across R_i, is fed back

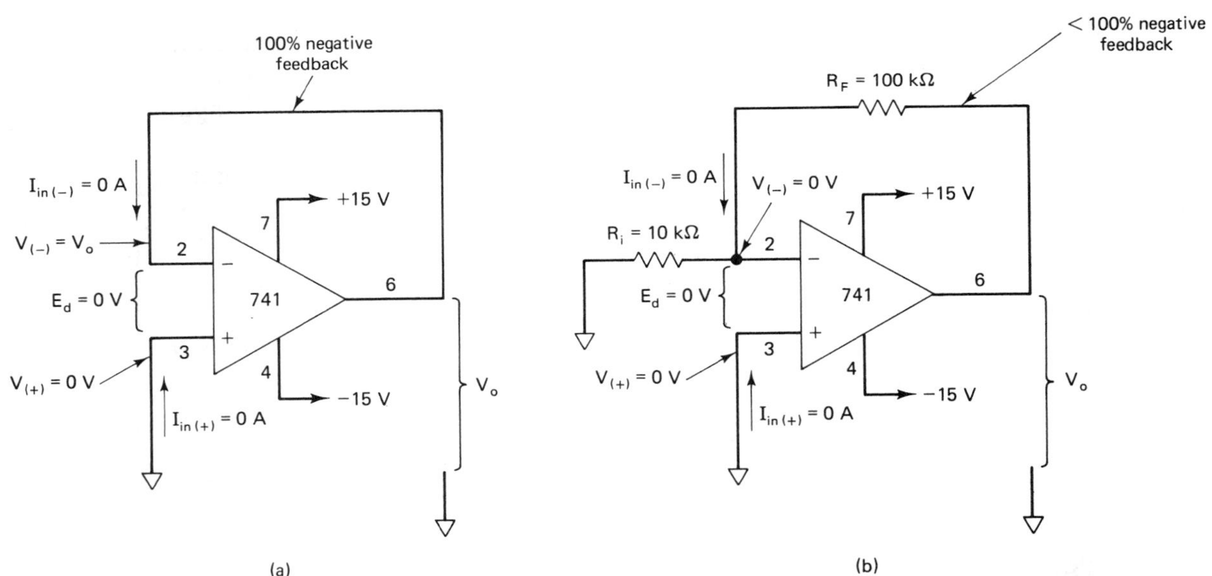

Figure 5-1 Negative feedback forces the voltage on the inverting input to equal the voltage on the noninverting input. (a) The entire output voltage V_o is fed back to the inverting input terminal; (b) the amount of voltage fed back in this circuit is set by the voltage divider rule.

to the inverting input terminal. Both test circuits have their noninverting input terminal connected to a zero-voltage reference or ground.

Negative feedback action, in both circuits, forces the voltage at the inverting input $V_{(-)}$ to be driven to the voltage on the noninverting input $V_{(+)}$. The first assumption to be made for analyzing op amps with negative feedback is then as follows:

1. The voltage on the inverting input must equal the voltage on the noninverting input terminal:

$$V_{(-)} = V_{(+)} \tag{5-1}$$

Since the voltage $V_{(+)}$ on the circuits of Fig. 5-1 is equal to 0 V, then the voltage $V_{(-)}$ must also equal 0 V. The second assumption for analysis then becomes:

2. The differential input voltage E_d must equal zero volts:

$$E_d = 0 \text{ V} \tag{5-2}$$

Our last assumption applies to all op amps with very large input resistances.

3. The current into either input terminal of an op amp is zero amperes:

$$I_{in(-)} = I_{in(+)} = 0 \text{ A} \tag{5-3}$$

Feedback action can be verified by building both test circuits in Fig. 5-1 and measuring V_o. V_o will be forced to approximately 0 V dc if the negative feedback mechanism is working properly. V_o will not be in plus or minus saturation as in previous chapters. We use the assumptions of this section to analyze all the circuits in this chapter, beginning with the inverting amplifier.

5-2 INVERTING AMPLIFIER

5-2.1 Circuit Analysis

A positive input voltage V_i is applied through an input resistor R_i of 10 K Ω to the (−) input terminal of the inverting amplifier (Fig. 5-2). The (+) terminal of the op amp is connected to ground and negative feedback is supplied through a 100-k Ω feedback resistor labeled R_F. With negative feedback present, the voltage E_d=0 V (assumption 2). Note that the (−) input of the op amp is at a *virtual ground potential.*

V_i is applied to one side of R_i and the other side of R_i is at virtual ground (Fig. 5-2); therefore, the input current I_i is set by

$$I_i = \frac{V_i}{R_i} \tag{5-4}$$

The polarity of V_i sets the direction of I_i. All of I_i must flow through

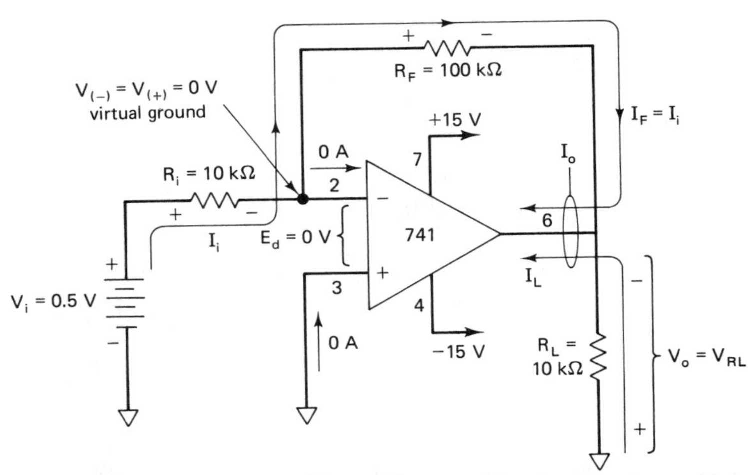

Figure 5-2 Inverting amplifier with a positive input voltage V_i applied. If the polarity of V_i is reversed, all current directions and voltage polarities would be reversed.

feedback resistor R_F (assumption 3) and into the op amp's output terminal.

The current through R_F is called *feedback current* I_F. Since both I_F and I_i are the same;

$$I_F = I_i \tag{5-5}$$

Because I_F flows through R_F, the voltage across R_F or V_{RF} is given by

$$V_{\text{RF}} = I_F R_F = I_i R_F \tag{5-6}$$

One side of R_F is at virtual ground and the other side is connected to the output terminal. The output voltage V_o must then be equal in magnitude to V_{RL}, or

$$V_o = \text{V}_{\text{RL}} = V_{\text{RF}} = I_i R_F \tag{5-7}$$

Note that in Fig. 5-2, the output terminal pin 6 and consequently V_o goes negative when V_i is positive. Thus the circuit is inverting in nature. The phase angle, θ, between V_o and V_i is then

$$\theta = 180° \tag{5-8}$$

In Fig. 5-2, the voltage across R_L is V_o. Load current I_L is set by V_o and R_L, or

$$I_L = \frac{V_o}{R_L} \tag{5-9}$$

The output current of the op amp, I_o, furnishes *both* feedback and load current.

$$I_o = I_F + I_L \tag{5-10}$$

The maximum recommended value of I_o is 5 mA for the 741.

EXAMPLE 5-1

Using the values shown in Fig. 5-2, solve for all the currents and voltages.

SOLUTION From Eq. (5-4), the current I_i is

$$I_i = \frac{V_i}{R_i} = \frac{0.5\text{ V}}{10\text{ k}\Omega} = 0.05\text{ mA } (50\ \mu\text{A})$$

Using Eq. (5-5), solve for I_F.

$$I_F = I_i = 0.05\text{ mA } (50\ \mu\text{A})$$

The voltage across the feedback resistor R_F is given from Eq. (5-6).

$$V_{\text{RF}} = I_F R_F = 0.05\text{ mA} \times 100\text{ k}\Omega = 5\text{ V}$$

From Eq. (5-7) the values of V_o and V_{RL} are

$$V_o = V_{\text{RL}} = V_{\text{RF}} = 5\text{ V}$$

The load current I_L is found from Eq. (5-9):

$$I_L = \frac{V_o}{R_L} = \frac{5\text{ V}}{10\text{ k}\Omega} = 0.5\text{ mA}$$

and using Eq. (5-10), the op-amp output current I_o is

$$I_o = I_L + I_F = 0.5\text{ mA} + 0.05\text{ mA} = 0.55\text{ mA}$$

5-2.2 Closed-Loop Voltage Gain

Closed-loop voltage gain A_{CL} for the inverting amplifier of Fig. 5-2 is defined as the ratio of output voltage V_o to input voltage V_i:

$$A_{\text{CL}} = \frac{V_o}{V_i} \tag{5-11}$$

Substitution of Eqs. (5-4) and (5-6) into Eq. (5-11) yields the following expression for the magnitude (not sign) of A_{CL}.

$$A_{\text{CL}} = \frac{V_o}{V_i} = \frac{I_F R_F}{I_i R_i} = \frac{I_i R_F}{I_i R_i}$$

$$= \frac{R_F}{R_i} \tag{5-12}$$

Equation (5-12) reveals that the *magnitude* of closed loop gain for the inverting amplifier is simply the ratio of R_F and R_i, independent of the op-amp open-loop gain A_{OL}. To account for the fact that there is a 180° phase shift between V_o and V_i, a minus sign has to be added to Eq. (5-12). Equation (5-12) then becomes

$$A_{\text{CL}} = -\frac{R_F}{R_i} \tag{5-13}$$

our gain equation for an inverting amplifier. That is, when V_i goes positive, V_o goes negative.

EXAMPLE 5-2

(a) Solve for both A_{CL} and V_o for the circuit shown in Fig. 5-2. (b) Find V_o if V_i in Fig. 5-2 is changed to 0.75 V.

SOLUTION (a) Using Eq. (5-13), the closed-loop gain is

$$A_{CL} = -\frac{R_F}{R_i} = -\frac{100\ \text{k}\Omega}{10\ \text{k}\Omega} = -10$$

From Eq. (5-13) the output voltage equals $V_o = -A_{CL}\ V_i = (-)(10)(0.5)\text{V} = -5\text{V}$. The minus sign indicates that V_o is a negative voltage measured with respect to ground.

(b) The gain remains unchanged at $A_{CL} = -10$. With $V_i = -0.75$ V, the output voltage equals $V_o = (-10)(-0.75\ \text{V}) = +7.5$ V. A plus sign indicates that V_o is a positive voltage measured with respect to ground. Note in both (a) and (b) V_o is opposite in polarity (180°) from V_i.

5-2.3 Input and Output Resistance

Remember from basic circuit theory that when a resistor is connected across a voltage source, it draws a current from the supply. The smaller the load resistor, the larger the current drawn. If the current drawn exceeds the capacity of the source, its terminal voltage will drop. This is called *source loading* and should be avoided. In general, to minimize source loading, the resistor connected to the voltage source should be as large as possible.

Figure 5-3 shows that the input of an inverting amplifier is connected directly to V_i and draws from it a current I_i. To minimize source loading, input resistance R_{in} should be as large as possible. The input resistance seen by V_i is $R_{in} = V_i/I_i$.

From Eq. (5-4), $R_i = V_i/I_i$; therefore, the input resistance of an inverting amplifier equals R_i:

$$R_{in} = R_i \tag{5-14}$$

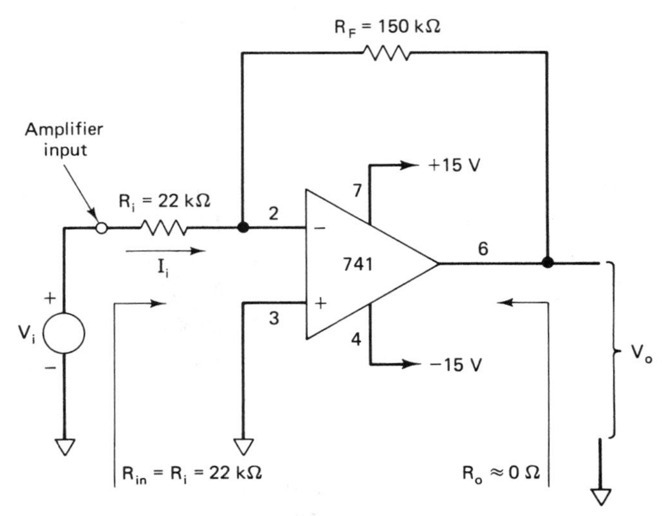

Figure 5-3 The input resistance R_{in} of an inverting amplifier is simply R_i. R_o is approximately o Ω for all amplifiers.

By inspection of Fig. 5-3, $R_{in} = 22\ k\Omega$. Any value of R_{in} can be set simply by selecting R_i. Generally, the value of R_i for an inverting amplifier should exceed 2 kΩ.

Output resistance R_o of an op amp (with negative feedback) can be assumed to be approximately 0 Ω. R_o is defined as the resistance seen looking into the output terminal of the op amp to ground. Since $R_o = 0\ \Omega$, the op amp is almost a perfect voltage source of value V_o, with an internal resistance R_o of 0 Ω, until I_o exceeds the manufacturer's guarantee of usually 5 mA (practically 10 mA).

5-2.4 V_o versus V_i Transfer Function

The transfer function for an inverting amplifier is similar to the plots for comparators introduced in Chapter 2. A time-varying signal V_i is applied as an input to the amplifier as in Fig. 5-4a. This causes a time varying output V_o. A plot of V_o versus V_i is a transfer function like the one illustrated in Fig. 5-4b.

From Eq. (5-13), the gain of Fig. 5-4a is $A_{CL} = -R_F/R_i = -470\ k\Omega/47\ k\Omega = -10$. Therefore, when $V_i = +1$ V (peak), $V_o = -10$ V (peak). Both V_i versus t and V_o versus t are plotted in Fig. 5-4a to highlight this circuit's ability to amplify V_i and produce an inherent phase shift of 180°.

The solid line shown in Fig. 5-4b indicates that V_o varies from +10 V through 0 to −10 V as V_i varies from −1 V through 0 to +1 V. The slope of this solid line then represents the gain of the amplifier and is given by

$$A_{CL} = \frac{\text{rise}}{\text{run}} \tag{5-15}$$

EXAMPLE 5-3

Calculate the gain of Fig. 5-4a from its transfer function in Fig. 5-4b.

SOLUTION From Eq. (5-15),

$$A_{CL} = \frac{\text{rise}}{\text{run}} = \frac{+10\text{ V}}{-1\text{ V}} = -10$$

The presence of a minus sign indicates a 180° phase shift, as expected. As long as V_i is below ±1.3 V, V_o will be in its linear region, with a value of less than $\pm V_{sat} = \pm 13$ V. Increase V_i, for example, to +2 V (peak) and V_o will be clipped at $-V_{sat} = -13$ V. See the dashed lines in Fig. 5-4b.

5-3 HIGH-GAIN, HIGH-INPUT RESISTANCE INVERTING AMPLIFIER

5-3.1 Limitations of the Basic Inverting Amplifier

A major limitation of the inverting amplifier configuration is that its input resistance is set by the size of R_i. R_i may have to be a relatively small value in order to obtain large values of voltage gain. When a design requires high input resistances coupled with large values of closed-loop gain, the inverting amplifier is unsuitable.

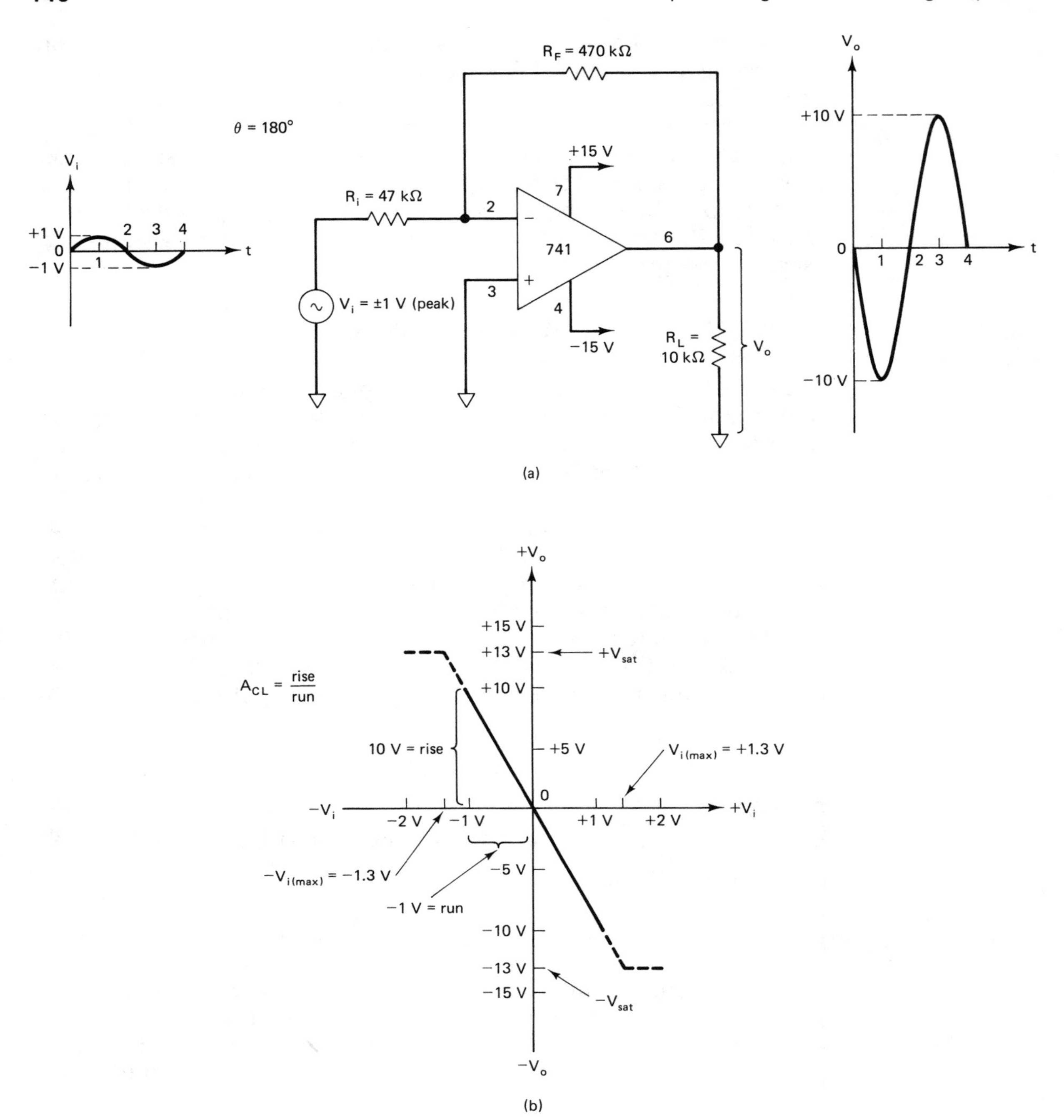

Figure 5-4 V_i versus t, V_O versus t, and V_O versus V_i plots of an inverting amplifier. (a) An ac signal applied to an inverting amplifier exhibits amplification on phase shift; (b) the transfer function of an inverting amplifier has a slope that is down and to the right.

Consider the real problem of building an inverting amplifier having an input resistance R_{in} of 1 MΩ and a gain A_{CL} of 500. The solution at first seems quite simple. Select an R_i resistor of 1 MΩ to satisfy the input specification [Eq. (5-14)] and determine the feedback resistor R_F from Eq. (5-12), where $R_F = (500)(1\ \text{M}\Omega) = 500\ \text{M}\Omega$. But a 500-MΩ resistor is a special item. Large-ohmic-value resistors are very costly, physically large, and prone to being inaccurate. For these reasons the use of high-ohmic-value resistors is undesirable.

A more practical solution to the foregoing problem is to create an "equivalent" large-value feedback resistor using a three-resistor combination of standard value component, configured in a "tee" pattern. This approach is presented in the next section.

5-3.2 Tee Network Feedback Resistor

Figure 5-5a shows an equivalent large-value feedback resistance constructed by using three individual resistors in a "tee" pattern. To achieve the input resistance specification, R_i is chosen to be 1 MΩ. Feedback resistors R_1 and R_2 are selected to be each 1 MΩ and the *current injector* resistor R is initially not connected at points A–A'.

To illustrate this circuit's ability to create large values of voltage gain, assume that V_i is a positive 10-mV signal as shown. Since the voltage $V_{(-)} = 0$ V, the input current is set by $I_i = V_i/R_i = 10$ mV/1 MΩ $= 0.01$ μA. With negative feedback $I_{in(-)} = 0$ A and all of the current I_i must flow through both R_1 and R_2. The voltage developed across each resistor is $V_i =$ (0.01 μA)(1 MΩ) = 10 mV, with the polarity as shown in Fig. 5-5a. V_o is equal to the sum of the voltages developed across both R_1 and R_2 or $V_o = -2V_i = -20$ mV. For a 10-mV input, the output is −20 mV or a gain $A_{CL} = -2$. Next, set current injector R to equal 1 MΩ and connect it to the circuit at points A−A' as in Fig. 5-5b. Resistors R and R_1 are attached directly at point A, with the other end of R connected to ground and R_1 connected to virtual ground. R is in parallel with R_1 and the voltage across the current injector resistor is set by the constant V_i voltage across R_1. The polarity is as shown. Adding R to create the tee network causes an additional current of value $I_i = V_i/R = 10$ mV/1 MΩ = 0.01 μA to be injected from R into R_2. This additional I_i component of current through R_2 develops another V_i drop across R_2. V_o is again the sum of the voltages developed across both R_1 and R_2. With R added, $V_o = -(2V_i + V_i) = -(30$ mV), or a gain of $A_{CL} = -3$.

If the value of R shown in Fig. 5-5c is reduced to, say, 100 kΩ, a current of $I = V_i = V_i/R = 10$ mV/100 kΩ = 0.1 mA = $10I_i$ is developed in resistor R and then injected into R_2, creating an additional voltage across it of $10V_i$. V_o then equals $-(2V_i + 10V_i) = -(120$ mV), a gain of $A_{CL} = -12$.

Note that the example above shows that as R is lowered, more current is injected into feedback resistor R_2 to increase the circuit's closed-loop gain A_{CL} without using huge feedback resistors.

5-3.3 Inverting Amplifier Design with Tee Network Feedback

Using the information presented in the preceding section, the tee network amplifier is redrawn and relabeled in Fig. 5-6. Since the series input resistor and both feedback resistors are equal in value, relabel these components as $R_i = R_1 = R_2 = nR$. The ratio of how much larger the feedback resistor nR is to R (the current injector resistor) can then be stated as

$$n = \frac{nR}{R} \tag{5-16}$$

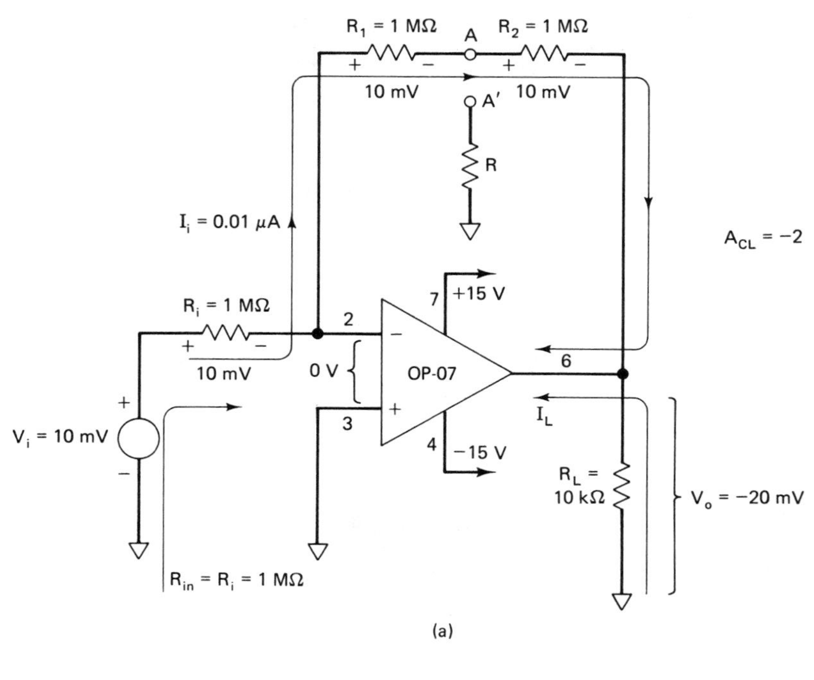

(a)

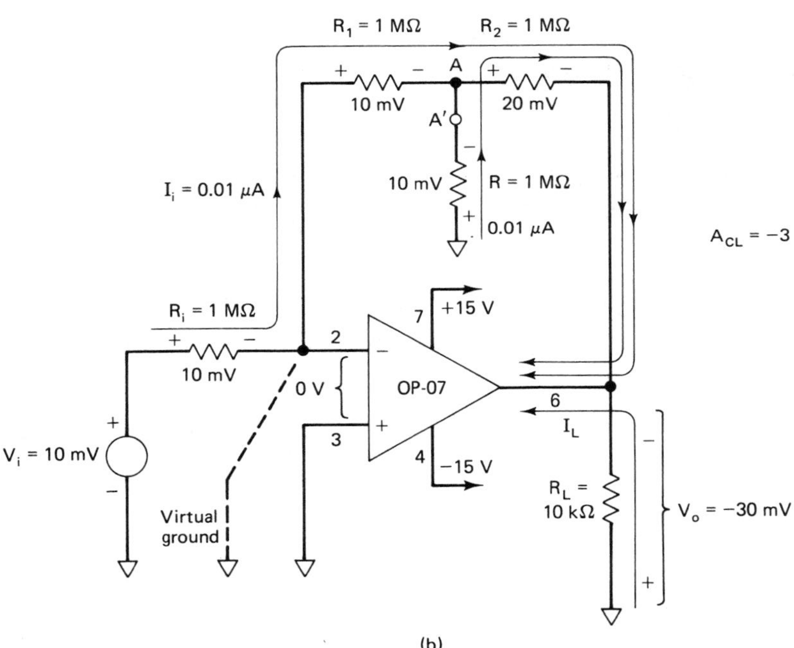

(b)

Figure 5-5 As the value of R in the tee network is made smaller, the circuit gain increases. (a) With R disconnected, the circuit gain is -2; (b) with $R = R_i$, the circuit gain is -3; (c) if $R = \frac{1}{10} R_i$, the circuit gain is -12.

This circuit's closed-loop gain is then determined from the equation

$$A_{CL} = -(2 + n) \tag{5-17}$$

Input resistance R_{in} is set by selecting the desired value of $R_i = nR$. This value should be limited to a maximum of approximately 10 MΩ.

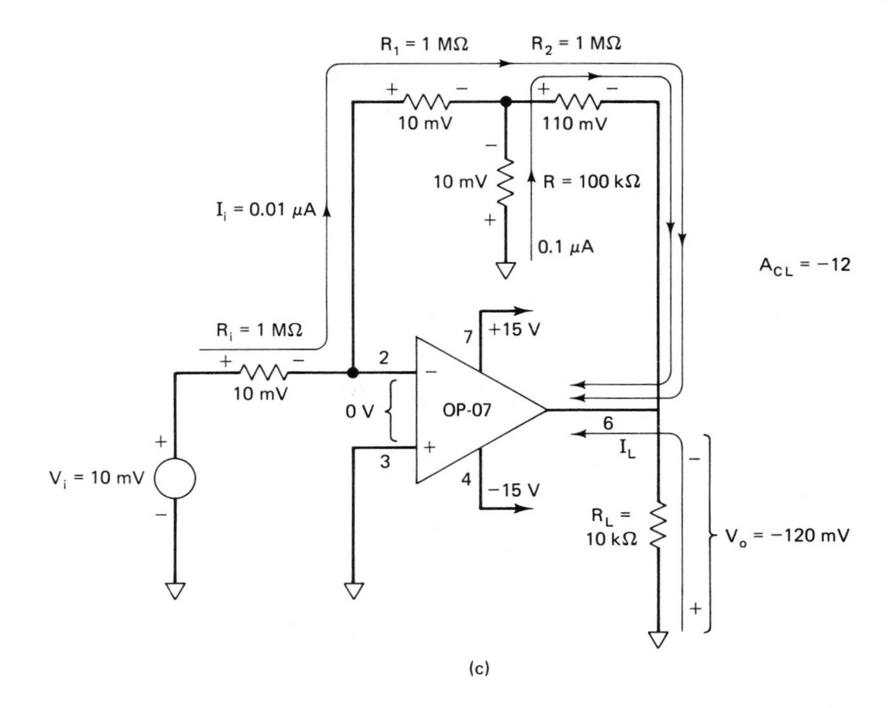

Then to set the gain, an appropriate value of current injector resistor R is chosen from Eq. (5-16).

EXAMPLE 5-4

Make a tee network inverting amplifier with an input resistance of 1 MΩ and a gain of −500.

SOLUTION Select $nR = 1\ \text{M}\Omega$ to get the required input resistance. Find n from Eq. (5-17).

$$A_{CL} = -500 = -(2 + n)$$

$$n = 498$$

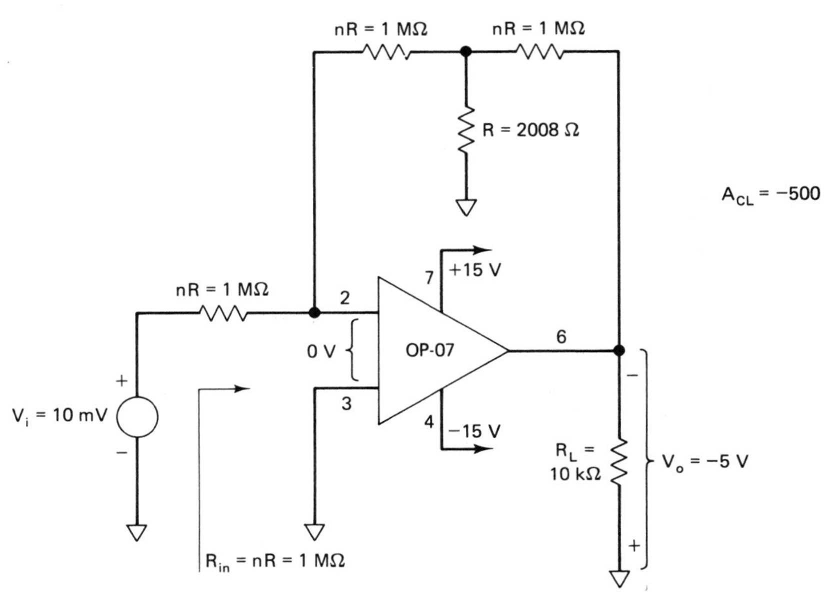

Figure 5-6 In a tee network, inverting amplifier nR sets the input resistance, and $A_{CL} = -(2 + n)$.

Calculate R from Eq. (5-16).

$$n = 498 = \frac{nR}{R} = \frac{1\ \text{M}\Omega}{R}$$

$$R = \frac{nR}{R} = \frac{1\ \text{M}\Omega}{498} = 2008\ \Omega$$

5-4 NONINVERTING AMPLIFIER

5-4.1 Circuit Analysis

A noninverting amplifier is shown in Fig. 5-7. In this configuration a positive V_i is applied directly to the (+) input terminal and negative feedback is provided to the (−) input terminal via R_F and R_i. The differential voltage E_d must equal 0 V (assumption 1) and the voltage $V_{(-)}$ must equal $V_{(+)}$ (assumption 2).

For the noninverting amplifier, input voltage is connected directly to the (+) input terminal $V_i = V_{(+)} = V_{(-)}$. V_i appears across R_i, causing the current direction shown with a value of

$$I_i = \frac{V_i}{R_i} \qquad (5\text{-}18)^*$$

Note that if an asterisk (*) is added to a noninverting amplifier equation, it signifies that an identical equation exists for the inverting amplifier.

Assume again that no current flows into the inverting input terminal of the op amp (assumption 3). The entire current I_i must then flow through feedback resistor R_F and

$$I_F = I_i \qquad (5\text{-}19)^*$$

The direction of current I_F sets the polarity of voltage across R_F as

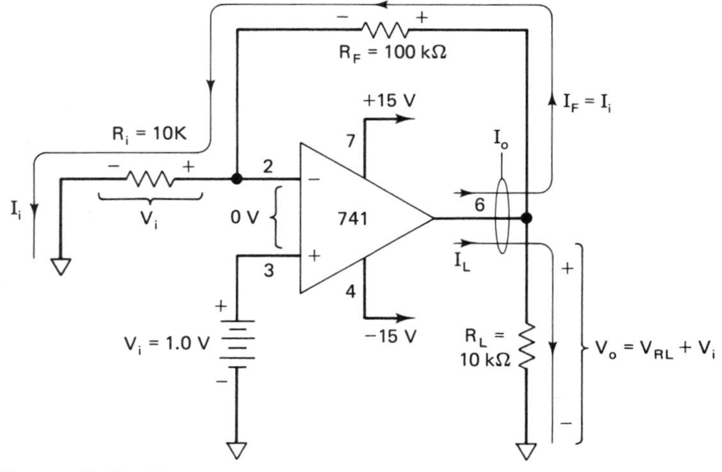

Figure 5-7 Noninverting amplifier with a positive input voltage applied. If the polarity of V_i is reversed, all current directions and voltage polarities would be reversed.

indicated in Fig. 5-7. The I_F current is supplied from the output terminal of the op amp. V_{RF} can be calculated from

$$V_{RF} = I_F R_F = I_i R_F \tag{5-20)*$$

The output voltage V_o is the potential developed across load R_L. R_L is connected between the output terminal of the op amp and ground. One side of the feedback resistor R_F is connected from the output terminal in series with R_i to ground. R_L is in parallel with the series combination of $R_F + R_i$. V_o can be expressed as

$$V_o = V_{RF} + V_{Ri} \qquad (5\text{-}21a)$$

or

$$V_o = V_{RF} + V_i \qquad (5\text{-}21b)$$

Since the positive side of R_F is connected to the output terminal of the op amp, the output voltage V_o across R_L is going positive with respect to ground. This is consistent with the thought that *a positive-going input voltage V_i applied to the noninverting input terminal will result in a positive-going output voltage.* There is no phase shift between V_i and V_o for a noninverting amplifier.

$$\theta = 0° \qquad (5\text{-}22)$$

I_L must flow from the output terminal of the op amp and through R_L as shown in Fig. 5-7. I_L is determined from

$$I_L = \frac{V_o}{R_L} \qquad (5\text{-}23)^*$$

The op amp's output current I_o is the sum of both load current I_L and feedback current I_F.

$$I_o = I_L + I_F \qquad (5\text{-}24)^*$$

EXAMPLE 5-5

Using the values shown in Fig. 5-7, solve for all the currents and voltages.

SOLUTION From Eq. (5-18) the value of I_i is

$$I_i = \frac{V_i}{R_i} = \frac{1.0 \text{ V}}{10 \text{ k}\Omega} = 0.1 \text{ mA}$$

Use Eq. (5-19) to solve for I_F.

$$I_F = I_i = 0.1 \text{ mA}$$

The voltage developed across the feedback resistor R_F is given from Eq. (5-20).

$$V_{RF} = I_F R_F = (0.1 \text{ mA})(100 \text{ k}\Omega) = 10\text{V}$$

Output voltage V_o is found using Eq. (5-21b).

$$V_o = V_{RF} + V_i = 10V + 1V = 11V$$

Load current I_L then becomes

$$I_L = \frac{V_o}{R_L} = \frac{11\text{ V}}{10\text{ k}\Omega} = 1.1\text{ mA}$$

5-4.2 Closed-Loop Voltage Gain

Closed-loop voltage gain for the noninverting amplifier configuration of Fig. 5-7 is defined as

$$A_{CL} = \frac{V_o}{V_i} \tag{5-25}*$$

Substitution of Eqs. (5-21b) into (5-25) yields

$$A_{CL} = \frac{V_{RF} + V_i}{V_i} \tag{5-26}$$

Substitution of Eqs. (5-18) and (5-19) into (5-26) gives the voltage-gain expression

$$A_{CL} = \frac{I_F R_F + I_i R_i}{I_i R_i} = \frac{I_i R_F + I_i R_i}{I_i R_i}$$

$$= \frac{R_F + R_i}{R_i} \tag{5-27}$$

Another form of the gain expression for a noninverting amplifier is obtained by a rearrangement of Eq. (5-27).

$$A_{CL} = 1 + \frac{R_F}{R_i} \tag{5-28}$$

Equation (5-28) reveals that the gain of a noninverting amplifier is always greater than 1.

EXAMPLE 5-6

(a) Solve for both A_{CL} and V_o for the circuit shown in Fig. 5-7. (b) Solve for V_o if V_i is changed to -0.5 V.

SOLUTION (a) Using Eq. (5-27) gives us

$$A_{CL} = \frac{R_F + R_i}{R_i} = \frac{100\text{ k}\Omega + 10\text{ k}\Omega}{10\text{ k}\Omega} = 11$$

From Eq. (5-25),

$$V_o = A_{CL} V_i = (11)(1.0V) = 11V$$

(b) The gain remains unchanged at +11. With $V_i = -0.5$ V, the output voltage becomes $V_o = 11(-0.5\text{ V}) = -5.5$ V. Note that there is no phase shift between V_o and V_i for a noninverting amplifier.

5-4.3 Input and Output Resistance

Figure 5-8 illustrates that the input voltage V_i is connected directly to the noninverting input terminal of the op amp. From assumption 3 we know that the (+) terminal draws no current from V_i and therefore must assume that input resistance is extremely high. With negative feedback, we can indicate that the input resistance R_{in} of a noninverting amplifier is infinite.

The noninverting amplifier has *no loading effect* on any voltage source. It is preferred to an inverting amplifier when little or no current is to be drawn from a circuit. Output resistance R_o is essentially zero ohms (identical to the inverting amplifier).

5-4.4 V_o versus V_i Transfer Function

Equation (5-27) predicts the closed-loop gain A_{CL} for Fig. 5.9a to be

$$A_{CL} = \frac{R_F + R_i}{R_i} = \frac{90\text{ k}\Omega + 10\text{ k}\Omega}{10\text{ k}\Omega} = \frac{100\text{ k}\Omega}{10\text{ k}\Omega} = 10$$

Therefore, when an ac signal of $V_i = +1$ V (peak) is applied to the input, $V_o = A_{CL}V_i = 10(+1\text{ V}) = +10$ V and when $V_i = -1$ V (peak), $V_o = -10$ V (peak). Both V_i versus t and V_o versus t are plotted in Fig. 5.9a. Clearly, the input signal V_i is amplified by the circuit and V_o is a faithful reproduction of V_i without any phase shift.

The solid line shown in Fig. 5.9b indicates that V_o varies from +10 V through 0 to −10 V as V_i changes from +1 V through 0 to −1 V. Circuit gain is represented by this solid line and given as

$$A_{CL} = \frac{\text{rise}}{\text{run}} \tag{5-29*}$$

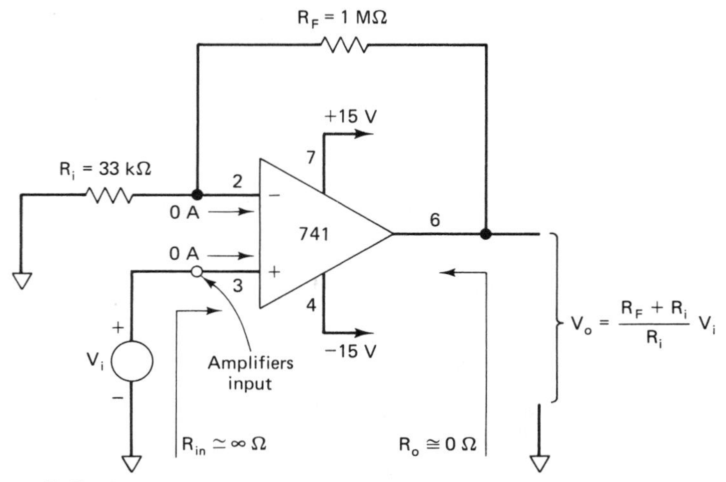

Figure 5-8 Negative feedback increases the input resistance seen by V_i. R_{in} is assumed to be extremely large, or simply infinite ohms.

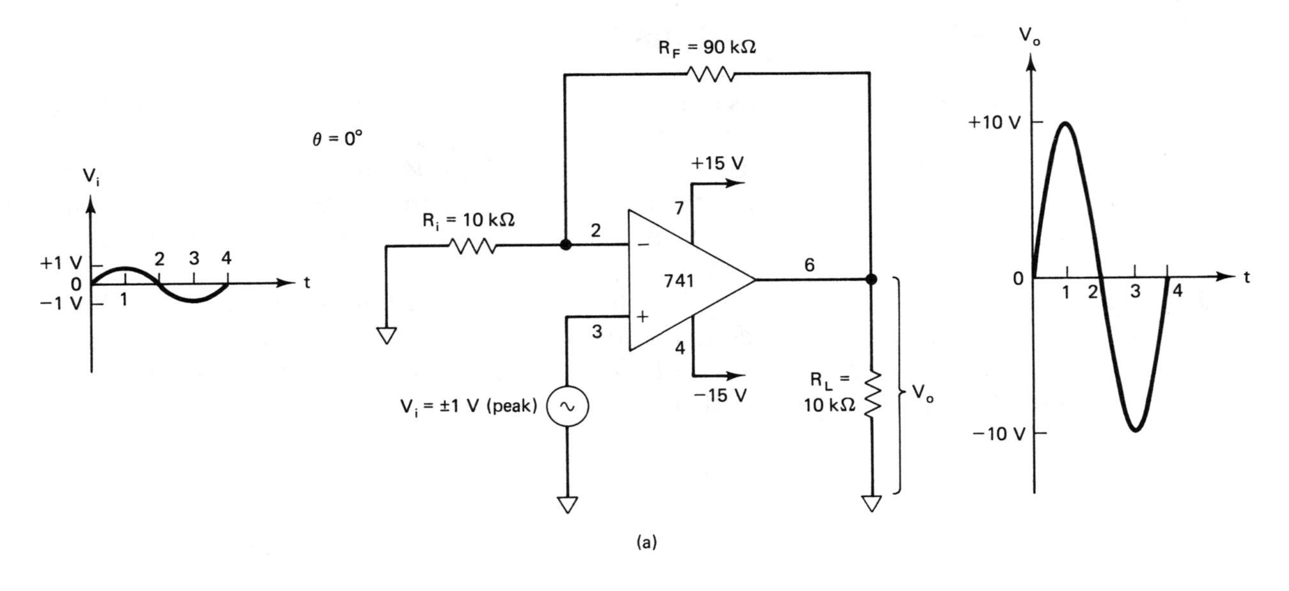

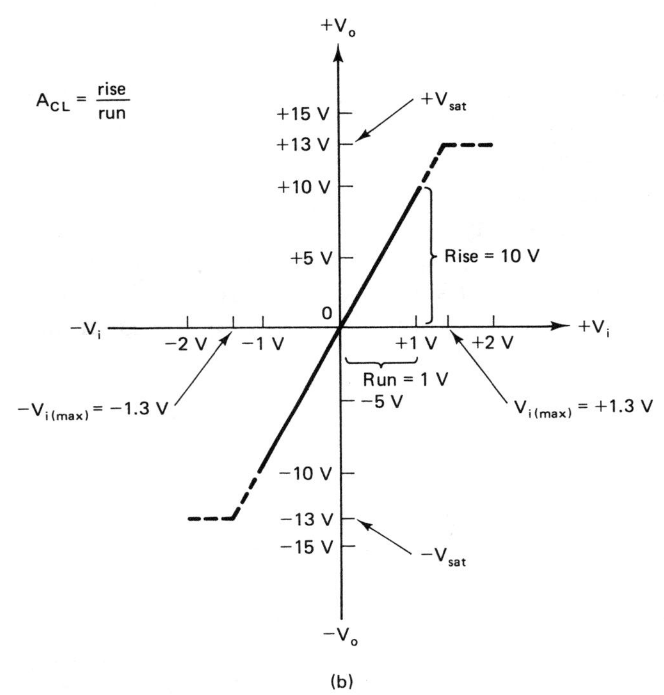

Figure 5-9 Plots of V_i versus t, V_O versus t, and V_O versus V_i for a noninverting amplifier. (a) An ac signal applied to a noninverting amplifier exhibits amplification with no phase shift; (b) the transfer function for a noninverting amplifier has a slope that is up and to the right.

EXAMPLE 5-7

Calculate the gain of Fig. 5-9a from its transfer function shown in Fig. 5-9b.

SOLUTION From Eq. (5-29) the closed-loop gain is found to be

$$A_{CL} = \frac{\text{rise}}{\text{run}} = \frac{+10\text{ V}}{+1\text{ V}} = 10$$

The presence of a plus sign indicates 0° phase shift between V_o and V_i. As long as V_i is below ± 1.3 V, V_o will be less than $\pm V_{sat}$ and exist in the linear region of amplification. If V_i is increased above ± 1.3 V, V_o will be clipped at $\pm V_{sat} = \pm 13$ V.

5-5 PROGRAMMABLE GAIN AMPLIFIER

The usefulness of the noninverting amplifier configuration can be extended with the addition of a few more resistors and a digitally controlled set of analog switches. The result is called PGA or *programmable-gain amplifier* whose gain can be adjusted under software control by a computer. Analog signals inputed to the amplifier can then be scaled to any desired level.

5-5.1 Eight-Channel CMOS Analog Switch

The output port (address bus) of a computer can be connected to the control inputs of an analog switch such as the HI-508A. In this way the computer can control the state of the switches. Manufactured by Harris Semiconductor, the HI-508A eight-channel CMOS analog multiplexer is housed in a 16-pin DIP. (Fig. 5-10). It contains eight separate single-pole, single-throw CMOS switches. While shown symbolically as mechanical devices, these switches are made with CMOS transistors. One end of each

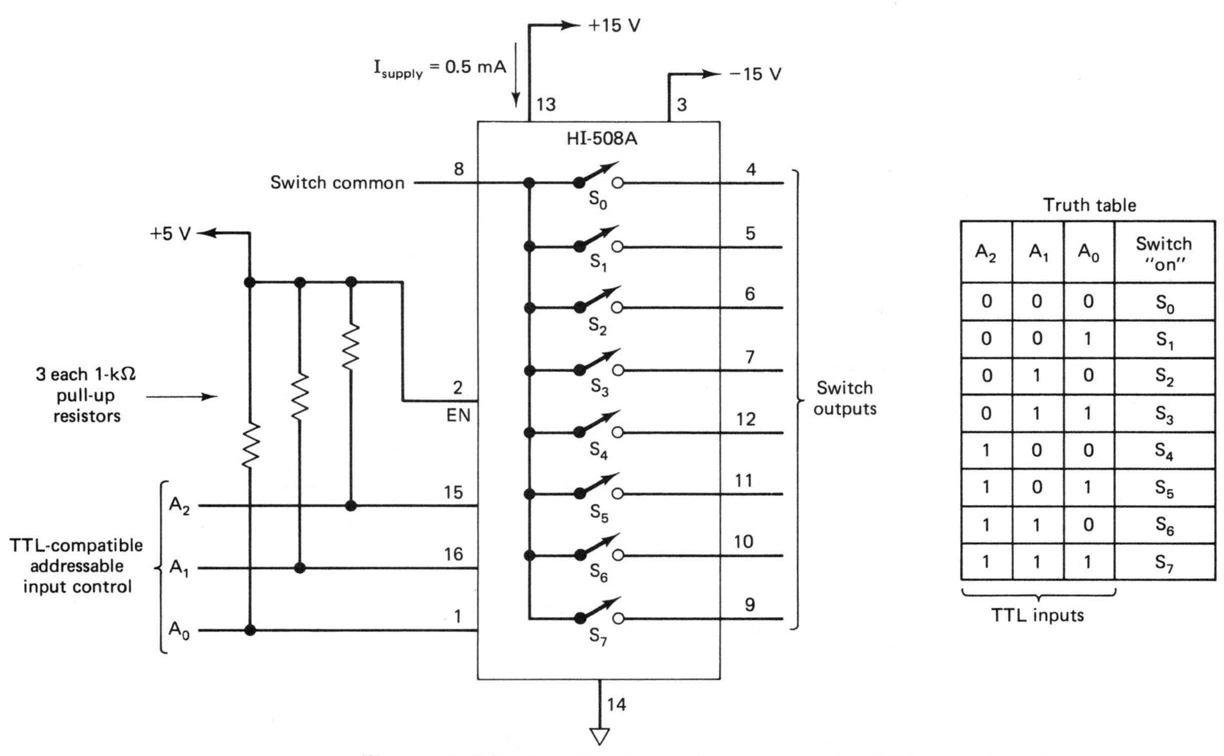

A_2	A_1	A_0	Switch "on"
0	0	0	S_0
0	0	1	S_1
0	1	0	S_2
0	1	1	S_3
1	0	0	S_4
1	0	1	S_5
1	1	0	S_6
1	1	1	S_7

Figure 5-10 The Harris semiconductor HI-508A is a digitally controlled set of CMOS switches that allows direct interface between a computer and the analog world of the op amp.

switch, labeled *switch outputs,* is connected separately to an external output pin. The other end of each switch is connected internally to pin 8 as a *common terminal* or *switch common.* Pin 2 is called *circuit enable.* When grounded (logic 0), all switches are open.

When the EN terminal is enabled by a logic 1, all CMOS switches are placed under the control of the *address inputs* A_0, A_1 and A_2. External 1 kΩ pull-up resistors are recommended as shown in Fig. 5-10.

All possible logic states, at the address lines, are shown in the *truth table* of Fig. 5-10. If a logic 0 appears on all three address lines, only switch S_0 is "on" or "closed." If the computer supplies a logic 1 to A_0 and logic 0 to both A_1 and A_2 (binary equivalent 1), only S_1 is "on" or "closed," and so forth. This truth table indicates that only one switch at a time is activated and all remaining switches are "off" or "open" for the Harris Semiconductor HI508A.

Unlike conventional mechanical switches that are about 0 Ω when "closed," a CMOS switch that is "on" has a resistance R_{on} of typically 1.2 kΩ. When a CMOS switch is "off," the equivalent resistance of the "open" switch is so high that I_{off} is typically 1 nA at room temperature.

5-5.2 Four-Level PGA Amplifier

A four-level PGA is constructed using a portion of the Harris HI-508A analog switch, an op amp, and four resistors, as in Fig. 5-11. The input signal V_i is applied directly to the noninverting input terminal and feedback is provided through whichever analog switch is closed.

The voltage gain setting can be interpreted as follows:

1. R_i' equals the resistance below the switch that is closed.
2. R_F' equals the resistance above the switch that is closed.

Voltage gain A_{CL} becomes

$$A_{CL} = \frac{R_F' + R_i'}{R_i'} \tag{5-30}$$

EXAMPLE 5-8

Determine the gain of the PGA in Fig. 5-11 for all possible binary inputs. $A_2 = 0$ throughout.

SOLUTION When $A_0 = 0$ and $A_1 = 0$, switch S_0 is closed.

$$A_{CL} = \frac{R_F' + R_i'}{R_i'} = \frac{0\ \Omega + 8\ \text{k}\Omega}{8\ \text{k}\Omega} = 1$$

When $A_0 = 1$ and $A_1 = 0$, switch S_1 is closed.

$$A_{CL} = \frac{R_F' + R_i'}{R_i'} = \frac{4\ \text{k}\Omega + 4\ \text{k}\Omega}{4\ \text{K}\Omega} = 2$$

When $A_0 = 0$ and $A_1 = 1$, switch S_2 is closed.

$$A_{CL} = \frac{R_F' + R_i'}{R_i'} = \frac{6\ \text{k}\Omega + 2\ \text{k}\Omega}{2\ \text{k}\Omega} = 4 \quad \text{(shown)}$$

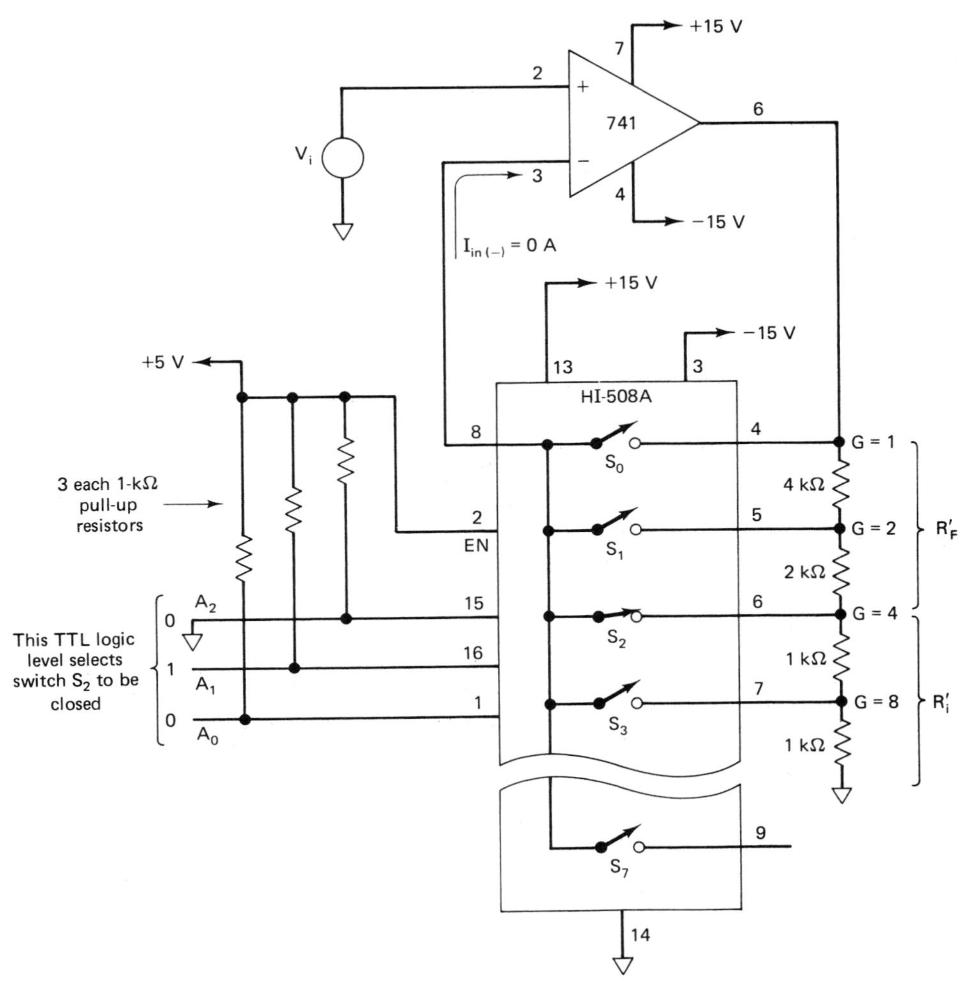

Figure 5-11 Simple noninverting amplifier configured as a programmable gain amplifier. A_{CL} is controlled by a binary code applied from a computer to the addressable inputs of the HI-508A analog switch.

When $A_0 = 1$ and $A_1 = 1$, switch S_3 is closed.

$$A_{CL} = \frac{R'_F + R'_i}{R'_i} = \frac{7\ \text{k}\Omega + 1\ \text{k}\Omega}{1\ \text{k}\Omega} = 8$$

An advantage of this configuration is that the 1.2 kΩ "on" resistance of the analog switch is in series with the inverting input terminal of the op amp. Since $I_{in(-)} = 0$ A, the switch's large on resistance has no effect on the circuit's performance.

5-6 VOLTAGE-FOLLOWER AMPLIFIER

5-6.1 Circuit Analysis

Figure 5-12 shows the basic configuration for a voltage-follower amplifier. This circuit is also referred to as a *buffer, isolation,* or *unity-gain amplifier.* With R_i removed and R_F replaced by a short circuit, V_i is applied directly

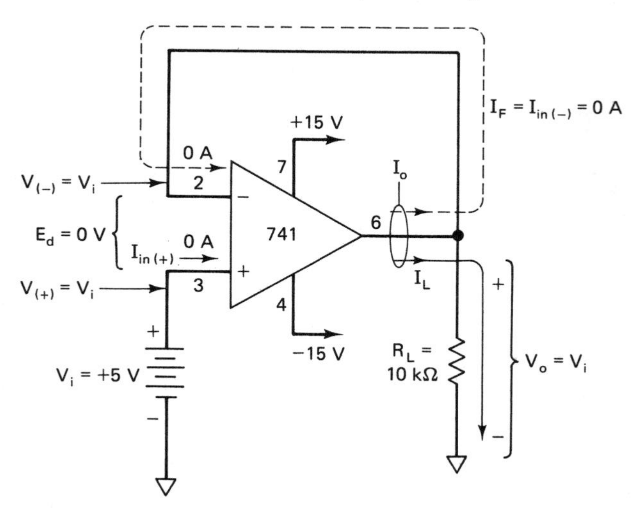

Figure 5-12 A voltage follower amplifier has a closed-loop gain $A_{CL} = 1$ and no phase shift between input and output voltages.

to the noninverting input terminal. One hundred percent negative feedback exists in this circuit. The entire output voltage V_o is fed directly back to the inverting input terminal of the op amp.

A positive voltage V_i is applied directly to the noninverting input terminal of the voltage follower. With negative feedback, $E_d = 0$ V (assumption 1) and $V_{(+)} = V_{(-)}$ (assumption 2). Therefore, the input voltage V_i is also equal to the voltage $V_{(-)}$. Since the (−) input is connected directly to the output terminal, V_i must also equal V_o.

$$V_o = V_i \tag{5-31}$$

Note that the output voltage follows the input voltage in both magnitude and phase. There is no phase shift between input and output for the voltage follower.

$$\theta = 0° \tag{5-32}$$

When V_o is positive, the load current I_L is sourced from the output terminal of the op amp as shown in Fig. 5-12. The value of I_L is given as

$$I_L = \frac{V_o}{R_L} \tag{5-33}$$

In addition to I_L, the output terminal of the op amp must also supply I_F. But unlike the two previous amplifiers discussed in this chapter, I_F for the follower is approximately 0 A; therefore,

$$I_O = I_L \tag{5-34}$$

EXAMPLE 5-9

Solve for the output voltage V_o and current I_L in the circuit of Fig. 5-12.

SOLUTION From Eq. (5-31), $V_o = V_i = +5$ V. The load current $I_L = V_o/R_L = 5\text{ V}/10\text{ k}\Omega = 0.5$ mA and I_O also equals 0.5 mA.

5-6.2 Closed-Loop Voltage Gain

The closed-loop voltage gain for Fig. 5-12 is expressed by

$$A_{CL} = \frac{V_o}{V_i} \tag{5-35}$$

Substitution of Eq. (5-31) into Eq. (5-35) yields the following:

$$A_{CL} = \frac{V_i}{V_i} = 1 \tag{5-36}$$

5-6.3 Input and Output Resistance

As with the amplifiers of Section 5-4, the input voltage V_i is connected directly to the noninverting input terminal (Fig. 5-13). Since pin 3 of the 741 is assumed to draw no current, the input resistance R_{in} is considered to be very large, approaching an infinite value. For completeness, note that R_o for the voltage follower (like R_o for both the inverting and noninverting amplifiers) is considered to be 0 Ω.

5-6.4 V_o versus V_i Transfer Function

An ac voltage of 5 V (peak) is applied to the input terminal of the voltage-follower amplifier shown in Fig. 5-14a. V_o is a faithful reproduction of V_i since $A_{CL} = 1$. Figure 5-14a also shows that there is no phase shift between input and output.

Figure 5-14b depicts the transfer function. The solid line indicates that V_o varies from +5 V through 0 to −5 V as V_i changes from +5 V through 0 to −5 V. Circuit gain can be calculated from the transfer function by determining the slope of the solid line and using equation

$$A_{CL} = \frac{\text{rise}}{\text{run}} \tag{5-37}$$

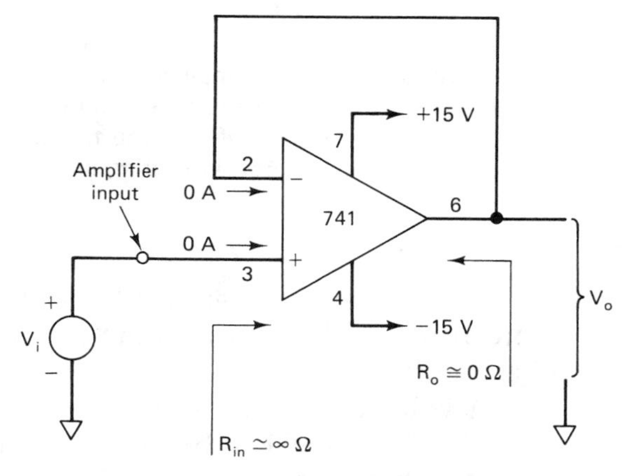

Figure 5-13 The input resistance of a voltage-follower amplifier is assumed to be infinite and its output resistance equal to 0 Ω.

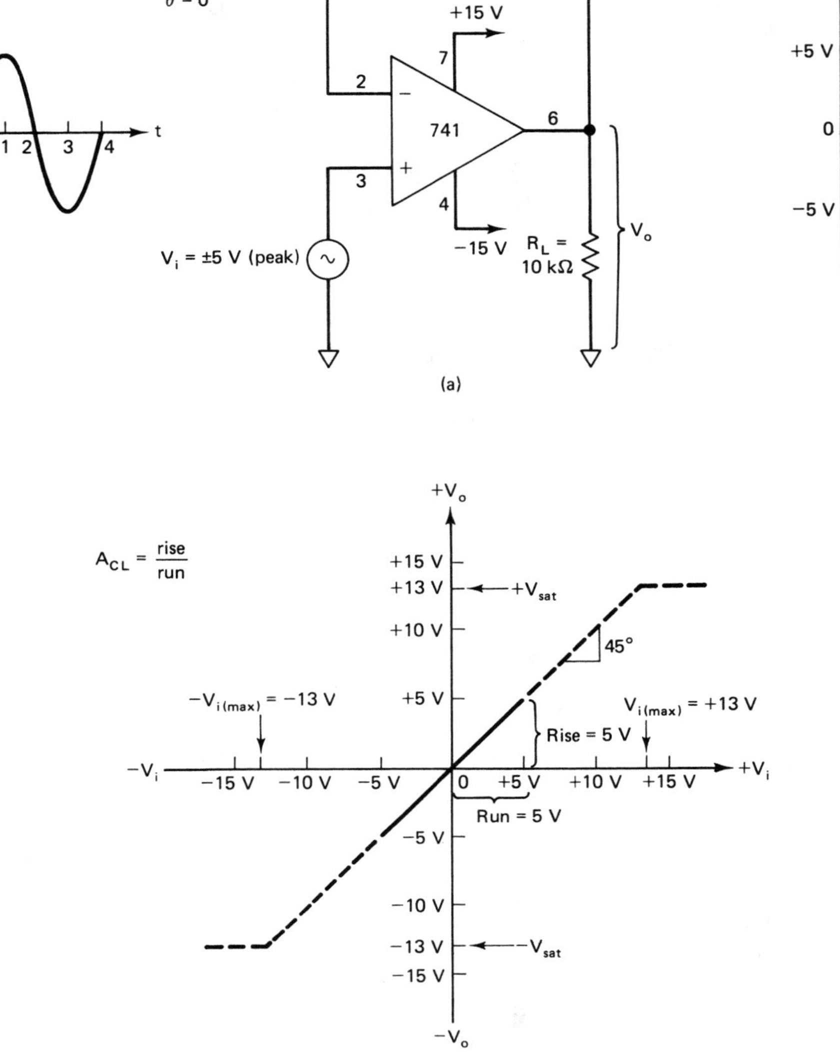

Figure 5-14 The V_i versus t, V_O versus t, and V_O versus V_i plots for a voltage-follower amplifier. (a) An ac signal applied to a voltage-follower amplifier produces an output voltage identical in both amplitude and phase shift; (b) the transfer function for a voltage follower has a slope of 45°, indicating a closed-loop gain of 1.

Since $V_o = V_i$, the gain is 1 and the slope of this line is at a 45° angle. Note also that the transfer function is up and to the right, indicating no phase shift.

A voltage follower is capable of linearly amplifying input signals up to $\pm V_{sat}$. That is, if V_i is below ± 13 V (for a ± 15-V supply), V_o will follow V_i and be below $\pm V_{sat} = \pm 13$ V.

5-7 MAKING A HIGH-INPUT RESISTANCE VOLTMETER

Even though the voltage-follower amplifier of Section 5-6 has no phase shift and a closed-loop voltage gain of only 1, it is a very widely used op-amp configuration. Its usefulness stems from the fact that it has very high input resistance and very low output resistance. These features make it ideal as an interfacing circuit to avoid circuit loading.

Refer to the simple series circuit of Fig. 5-15a. V_2 is predicted from the voltage-divider rule to be

$$V_2 = \frac{V_{CC}\,(R_2)}{R_1 + R_2} = \frac{15\text{ V }(100\text{ k}\Omega)}{400\text{ k}\Omega + 100\text{ k}\Omega} = 3\text{ V}$$

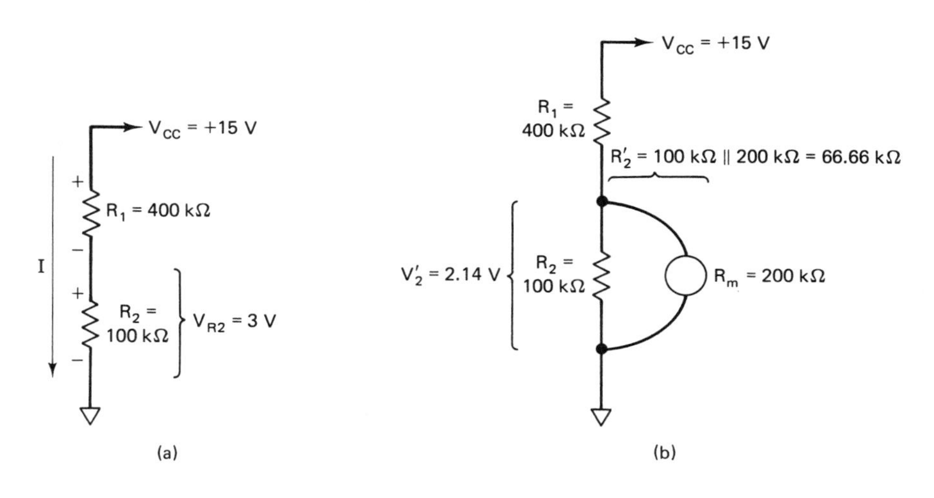

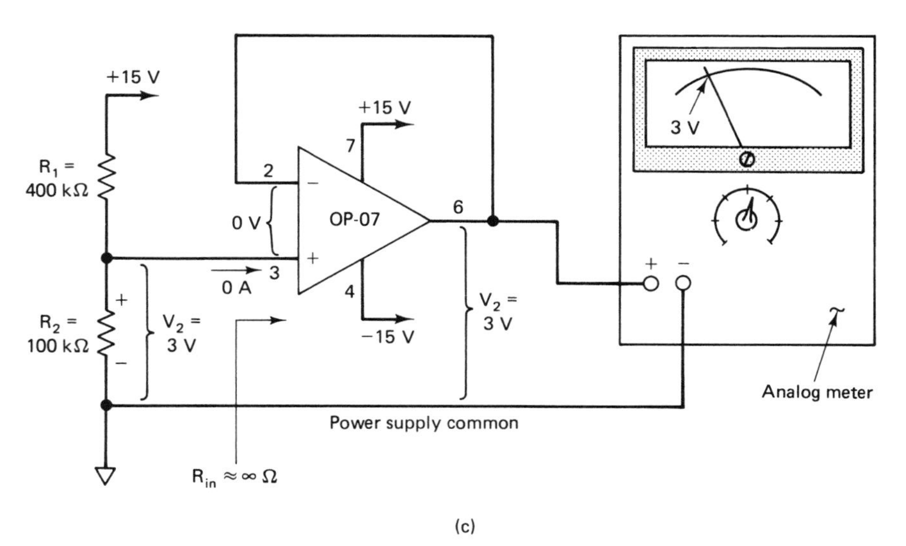

Figure 5-15 Using a high-input resistance voltmeter eliminates meter loading. (a) For this simple series circuit V_{R2} is determined to be 3 V; assume this to be the standard value. (b) V_{R2} equals only 2.14 V when measured with a Simpson 260 multmeter on the 10-V range. (c) Effectively increasing the Simpson 260's meter resistance with the ∞-Ω input resistance of a voltage follower.

Consider this to be the *standard value* for V_2. If we were to measure V_2 with a basic Simpson 260 multimeter on the 10-V range, we would find that the *actual value* of voltage measured was only 2.14 V. This discrepancy between standard and actual values can be explained as follows. The Simpson 260 has a meter resistance of 200,000 Ω on its 10-V range. When this meter is placed in parallel with the 100-kΩ resistance of R_2 to measure V_2, it loads down the circuit. The new resistance across which V_2' is to be measured is $R_2' = 100\ \text{k}\Omega \parallel 200\ \text{k}\Omega = 66.66\ \text{k}\Omega$. The actual voltage, using the voltage-divider rule, is

$$V_2 = \frac{V_{CC}\,(R_2')}{R_1 + R_2'} = \frac{(15\ \text{V})(66.66\ \text{k}\Omega)}{66.66\ \text{k}\Omega + 400\ \text{k}\Omega} = 2.14\ \text{V}$$

To overcome this problem, the 200-kΩ meter resistance of the Simpson 260 is isolated from R_2 by a voltage follower, as shown in Fig. 5-15c. R_2 is in parallel with the infinite input resistance of the op amp. The voltage V_2 measured equals precisely 3.0 V.

End of Chapter Exercises

Name: ____________________

Date: __________ **Grade:** ___

A. FILL-IN THE BLANKS

Fill in the blanks with the best answer.

1. Negative feedback is present when there is a current path from __________ to the __________ input.
2. When an input signal is applied to the (+) input of an op amp with negative feedback, the resulting circuit is called a(n) __________ amplifier.
3. The output current of an op amp with negative feedback is the sum of load current and __________ current.
4. The plot of V_o vs. V_i for an inverting amplifier has a slope that rises up and to the __________.
5. In either a non-inverting or inverting amplifier, voltage at the (+) input equals voltage at the __________ provided the amplifier is not driven into __________.
6. The input resistance of a non-inverting amplifier equals __________.
7. The gain of a non-inverting amplifier is always greater than __________.
8. The gain of a voltage follower always equals __________.
9. A programmable gain amplifier is made with an op amp, resistors and a digitally controlled set of __________ switches.
10. The HI-508A, 8-channel CMOS Analog Multiplexer has __________ digital inputs.

B. TRUE/FALSE

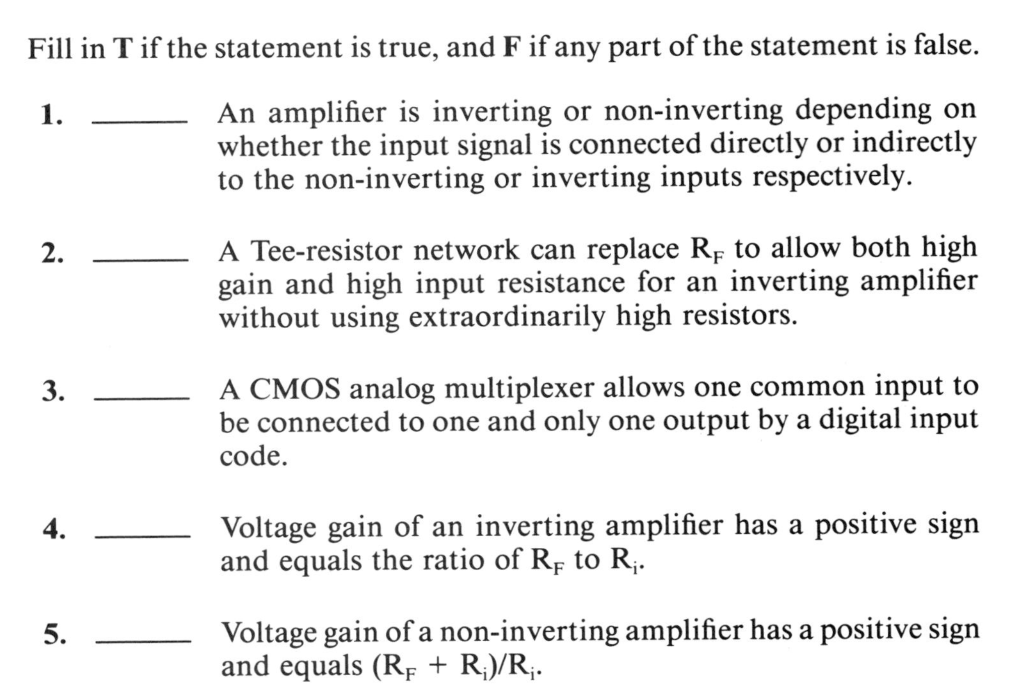

Fill in **T** if the statement is true, and **F** if any part of the statement is false.

1. ______ An amplifier is inverting or non-inverting depending on whether the input signal is connected directly or indirectly to the non-inverting or inverting inputs respectively.

2. ______ A Tee-resistor network can replace R_F to allow both high gain and high input resistance for an inverting amplifier without using extraordinarily high resistors.

3. ______ A CMOS analog multiplexer allows one common input to be connected to one and only one output by a digital input code.

4. ______ Voltage gain of an inverting amplifier has a positive sign and equals the ratio of R_F to R_i.

5. ______ Voltage gain of a non-inverting amplifier has a positive sign and equals $(R_F + R_i)/R_i$.

C. CIRCLE THE CORRECT ANSWER

Circle the correct answer for each statement.

1. The input resistance of an inverting amplifier is equal to (input resistor R_i, infinity ohms).
2. Voltage gain of an amplifier with negative feedback is determined by (the op amp, external resistors) only.
3. The output resistance of an inverting or non-inverting amplifier equals (0Ω, $\infty\Omega$).
4. V_o is (positive, negative) when V_i is positive in an inverting amplifier.
5. The slope of V_o vs V_i for a non-inverting amplifier rises up and to the (left, right).
6. The slope of a V_o vs V_i amplifier characteristic equals its (gain, reciprocal of gain).
7. The mechanical equivalent of an electronic 8-channel analog multiplexer is a(n) (8-pole single-throw switch, single-pole 8-position switch).
8. A transfer function is a plot of (V_i and V_o vs time, V_i vs V_o, V_o vs V_i).
9. Non-inverting amplifier gain is given by ($-R_F/R_i$, R_F/R_i, $1 + \frac{R_F}{R_i}$).

D. MATCHING EXERCISE

Match the name or symbol in column **A** with the statement that matches best in column **B**.

	Column A		Column B
1. ______	1	**a.**	inverting amplifier
2. ______	$-\frac{R_F}{R_i}$	**b.**	$I_L + I_F$
3. ______	$1 + \frac{R_F}{R_i}$	**c.**	voltage follower
4. ______	V_o vs V_i	**d.**	non-inverting amplifier
5. ______	I_o	**e.**	Transfer function

PROBLEMS

5-1. In Fig. 5-2, $R_i = 47\ k\Omega$ and $R_F = 680\ k\Omega$.
- **(a)** Solve for all circuit currents and voltages.
- **(b)** Solve for A_{CL} and θ.
- **(c)** Estimate both R_{in} and R_o.

5-2. **(a)** Draw the transfer function, V_o versus V_i, for the circuit in Problem 5-1 if V_i varies from -1 V through 0 to $+1$ V.
- **(b)** Calculate A_{CL} from this transfer function.

5-3. Repeat Example 5-4 with R_{in} equal to 2.2 $M\Omega$ and $A_{CL} = -200$.

5-4. In Fig. 5-7, $R_{in} = 22\ k\Omega$ and $R_F = 180\ k\Omega$.
- **(a)** Solve for I_i, I_F, I_o, V_{Ri}, V_{RF}, and V_o.
- **(b)** Calculate A_{CL} and θ.
- **(c)** Estimate both R_{in} and R_o.

5-5. **(a)** Draw the transfer function, V_o versus V_i, for the circuit in Problem 5-4, if V_i varies from -1 V through 0 to $+1$ V.
- **(b)** Calculate A_{CL} from this transfer function.

5-6. Repeat Example 5-8 if all resistors are doubled in Fig. 5-11.

5-7. **(a)** Repeat Example 5-9 for an $R_L = 100\ k\Omega$ and $V_i = 10$ V.
- **(b)** Solve for A_{CL} and θ.
- **(c)** Estimate both R_{in} and R_o.

5-8. **(a)** Draw the transfer function, V_o versus V_i, for the circuit in Problem 5-7 if V_i varies from -15 V through 0 to $+10$ V.
- **(b)** Estimate A_{CL} from the slope of this curve.

5-9. Show the complete schematic (two op amps) of an inverting amplifier with $A_{CL} = -10$ and input resistance $R_{in} = \infty\ \Omega$.

5-10. Draw the schematic of an inverting amplifier with an input resistance of 10 kΩ and $A_{CL} = -10$.

Laboratory Exercise 5

Name: ____________________

Date: __________ **Grade:** __

AMPLIFIERS

OBJECTIVES: Upon completion of this laboratory exercise on op amps used as linear amplifiers, you will be able to (1) design and analyze inverting amplifiers; (2) measure the closed-loop voltage gain A_{CL}, phase shift θ between V_i and V_o plus input resistance R_{in}; (3) design and analyze a noninverting amplifier, (4) measure A_{CL} and θ; (5) analyze a voltage-follower amplifier; and (6) measure A_{CL} and θ.

REFERENCE: Chapter 5, 741 Data Sheet (Chapter 1)

PARTS LIST

1	741 op amp	1	33-kΩ resistor
1	10-kΩ resistor	1	100-kΩ resistor
1	15-kΩ resistor	1	120-kΩ resistor

Procedure A: Design and Analysis of an Inverting Amplifier

1. Figure L5-1 illustrates a basic inverting amplifier that is to be designed for a closed-loop voltage gain A_{CL} of -10 and an input resistance of R_{in} of 10 kΩ.

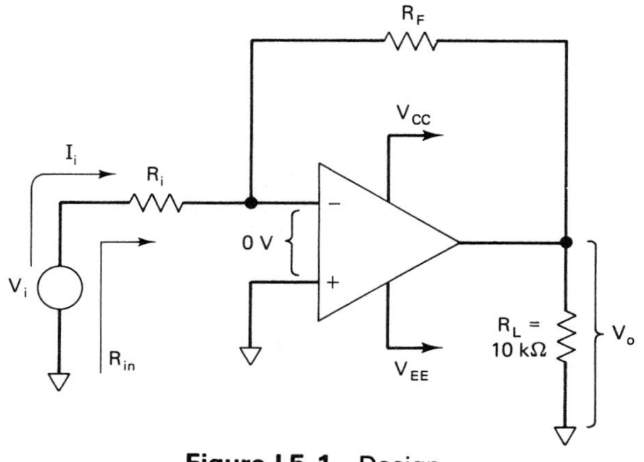

Figure L5-1 Design.

2. Since R_{in} for an inverting amplifier equals R_i, make R_i = __________. Calculate R_F from the gain equation: $A_{\text{CL}} = -\dfrac{R_F}{R_i}$ then $R_F = A_{\text{CL}}\, R_i$. R_F = __________.
3. What is the expected phase shift between V_i and V_o? θ = __________.
4. Figure L5-2 is to be analyzed by first selecting two resistors with color code values of 33 kΩ for R_i and 100 kΩ for R_F. Measure and record these resistor values using a DMM. R_i = __________; R_F = __________. Use these measured values for all calculations involving Fig. L5-2. Include these measured values on Fig. L5-2.
5. Using the measured resistor values found in step 3, solve for the following:

   (a) R_{in} = __________; R_F = __________;

   (b) the closed-loop voltage gain A_{CL} = __________;

   (c) phase shift θ = __________.

   (d) Solve for the expected value of peak output V_o. V_o = __________.

Procedure B: Measuring A_{CL}, θ, and R_{IN}

6. Wire Fig. L5-2. Set V_i equal to a sine wave of ±2 V (peak) at a test frequency of 100 Hz. Using channel 1 of the CRO, obtain the input voltage waveform shown in Fig. L5-3. Measure and sketch V_o versus t on Fig. L5-3.
7. Using Fig. L5-3, measure the phase shift from V_i and V_o. θ = __________.

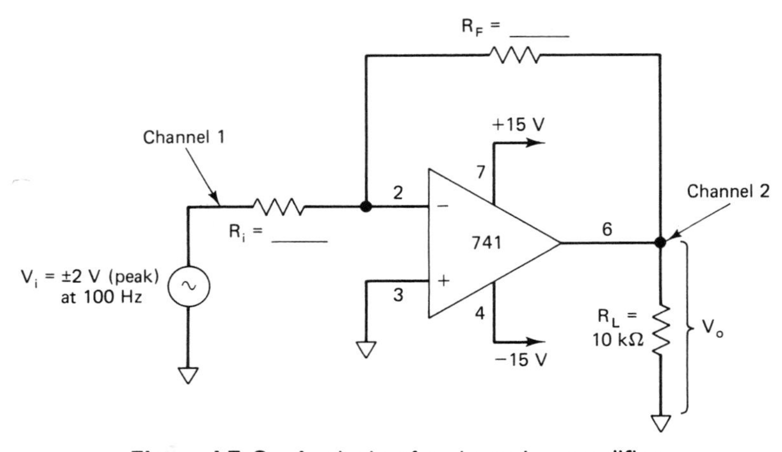

Figure L5-2 Analysis of an inverting amplifier.

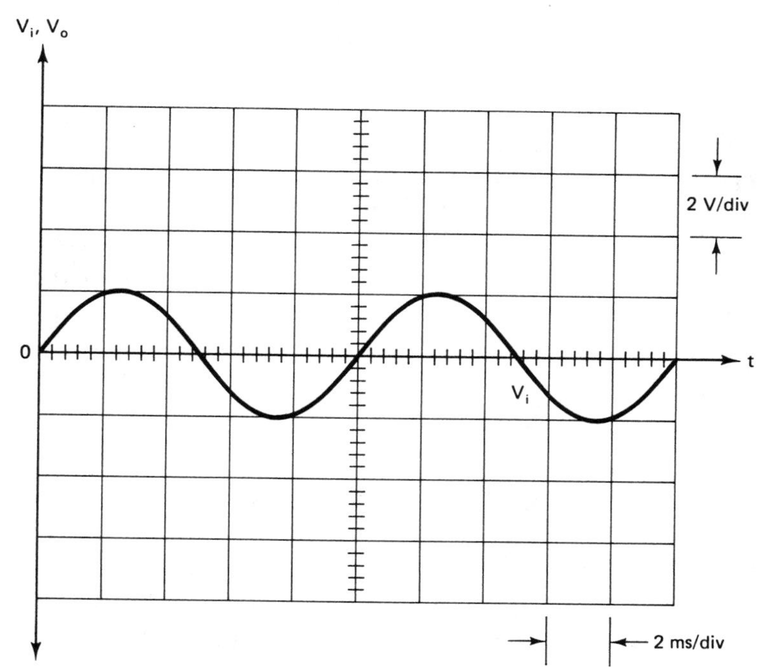

Figure L5-3 V_i versus t and V_o versus t.

8. Using Fig. L5-3, measure the closed-loop voltage gain A_{CL} from the equation

$$A_{CL} = \frac{V_o(\text{peak})}{V_i(\text{peak})} = \frac{V_o(\text{peak})}{2\ V(\text{peak})}$$

A_{CL} = __________.

9. Does the expected value of phase shift in step 5c compare with the measurement of step 7? __________. (b) Does A_{CL} calculated in step 5b compare favorably with the measured value of step 8? __________.

10. Set channel 2 of the CRO to 5 V/div. Increase V_i to +5 V (peak) while monitoring V_o with channel 2. Explain what happened to V_o. ____.

11. Again, refer to Fig. L5-2 and adjust V_i to precisely 1.00 V (rms). Make this measurement with the DMM on ac volts. Next, measure the precise value of rms voltage at the inverting input terminal of the 741 op amp. V_{rms} = __________. (*Note:* $V_{(-)}$ should be approximately equal to 0 V with negative feedback.)

12. The value of R_{in} can be determined from the Ohm's law equation given here and using the measured value of R_i from step 4.

$$R_{in} = \frac{V_i}{I_i} = \frac{V_i}{\dfrac{V_i - V_{(-)}}{R_i}}$$

R_{in} = __________.

13. Is the input resistance just measured approximately equal to the expected value of step 5a? __________.

14. The transfer curve of an inverting amplifier can be displayed on the CRO by first setting $V_i = \pm 2$ V (peak). Use channel 1 at 2 V/div to monitor V_i. Set channel 2 at 5 V/div to monitor V_o. Switch to (XY). Ground both channels and center the spot as shown in Fig. L5-4.

15. Switch both channels to *dc coupling* and sketch the waveform on Fig. L5-2 as a *solid line*. Calculate A_{CL} using the equation $A_{CL} = \frac{\text{rise}}{\text{run}}$.

$A_{CL} =$ __________.

16. Is V_o in phase or 180° out of phase with the input voltage V_i? __________. (*Note:* 0° phase shift, up and to the right; 180° phase shift, down and to the right.)

17. Increase V_i to ± 5 V (peak). Sketch the results on Fig. L5-4 using *dashed lines* to extend the waveshape. Explain why the transfer curve flattens. ____

____.

Procedure C: Design and Analysis of a Noninverting amplifier

18. Figure L5-5 illustrates a basic noninverting amplifier that is to be designed for a closed-loop voltage gain $A_{CL} = 2$.

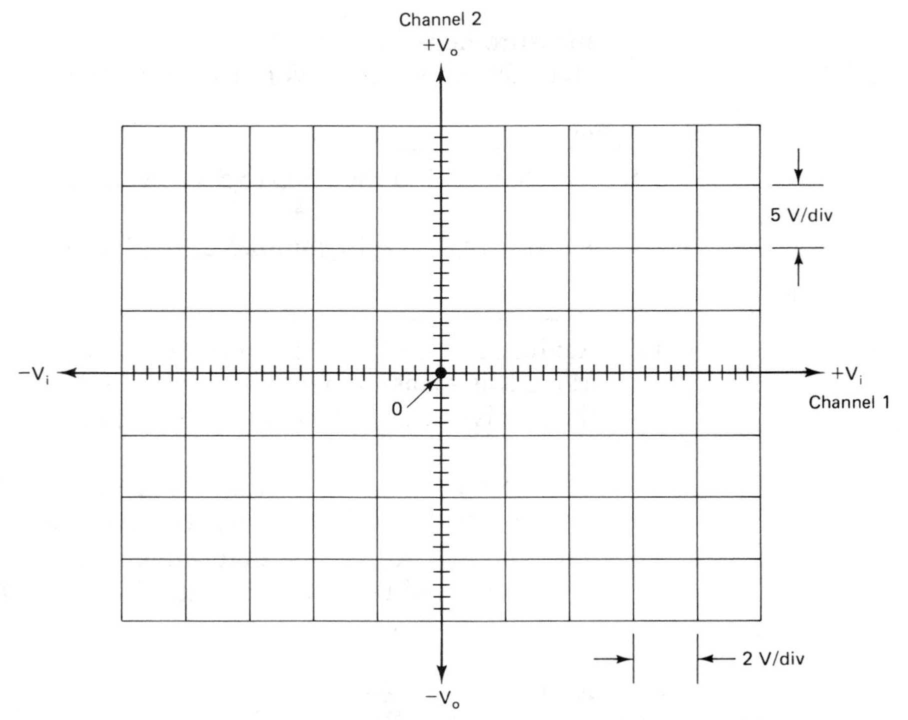

Figure L5-4 Transfer characteristics of an inverting amplifier.

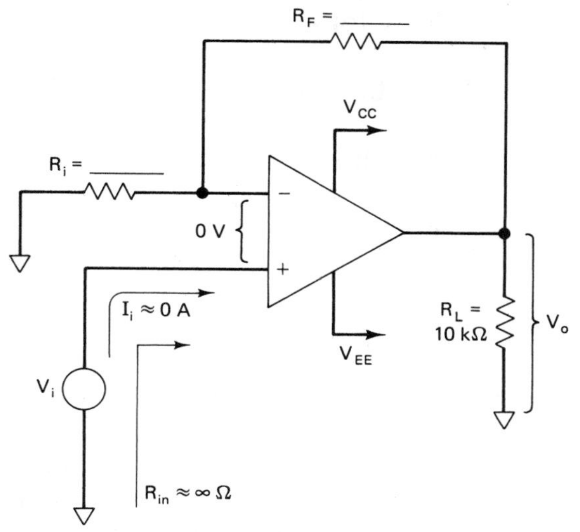

Figure L5-5 Design.

19. Since R_{in} for the noninverting amplifier is very large, we can assume it to be infinite ohms (∞ Ω) for this design.

20. Design Fig. L5-5 for a closed-loop gain A_{CL} of 2

(a) Select a value of R_i between 10 and 100 kΩ.

(b) Solve for R_F using the gain equation: $A_{CL} = \frac{R_F + R_i}{R_i}$; therefore,

$R_F = (A_{CL} - 1)R_i$. R_F = ______.

21. What is the expected phase shift between V_i and V_o? θ = ______.

22. Figure L5-6 is to be analyzed by first selecting two resistors with color-code values of 15 kΩ for R_i and 120 kΩ for R_F. Measure and record these resistor values on Fig. L5-6. Use these measured values for all calculations involving Fig. L5-6.

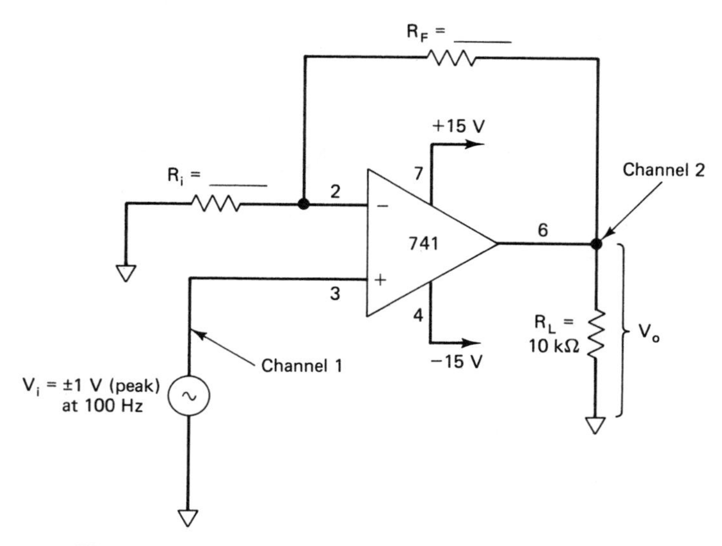

Figure L5-6 Analysis of a noninverting amplifier.

23. Solve for the following:

(a) the estimated value of R_{in} = ________;

(b) closed-loop gain A_{CL} = ________;

(c) phase shift θ = ________.

(d) Solve for the expected value of peak output voltage V_o. V_o = ________.

Procedure D: Measuring A_{CL} and θ

24. Wire Fig. L5-6. Set V_i equal to a sinewave of ± 1 V (peak) at a test frequency of 100 Hz. Use channel 1 of the CRO to obtain the input waveform shown in Fig. L5-7. Measure and sketch V_o versus t on Fig. L5-7.

25. Using Fig. L5-7, measure the phase shift from V_i and V_o. θ = ________.

26. Using Fig. L5-7, measure the closed-loop voltage gain A_{CL}. $A_{CL} = \frac{V_{o(peak)}}{V_{i(peak)}}$ = ________.

27. **(a)** Does the expected value of phase shift in step 23c compare with the measured value of step 25? ________.

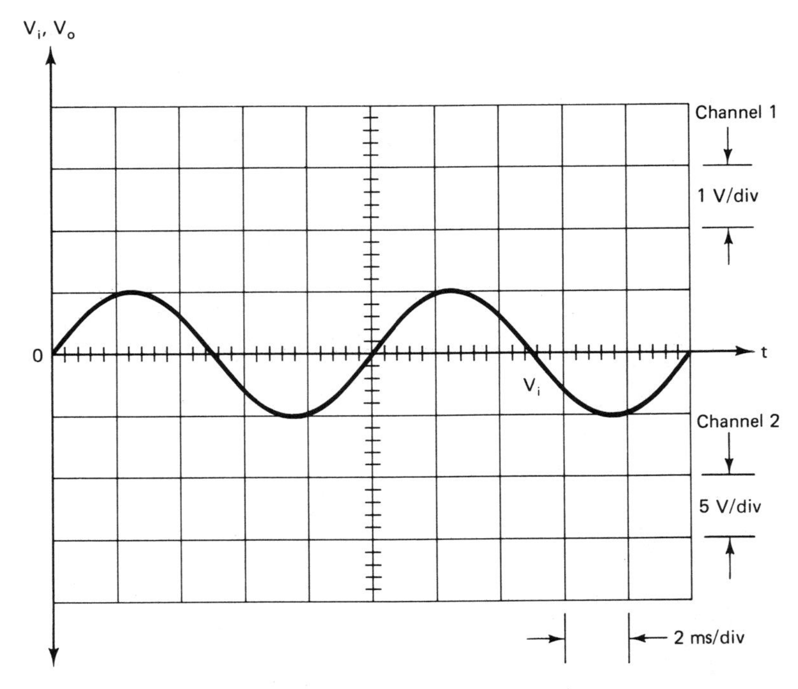

Figure L5-7 V_i versus t and V_o versus t.

(b) Does A_{CL} calculated in step 23b compare favorably with the measured value of step 26? __________.

28. Increase V_i slowly until V_o just begins to clip. Record the maximum peak value of V_i just before distortion begins. $V_{i(max)}$ = __________.

29. The transfer curve of a noninverting amplifier can be displayed on the CRO by first setting V_i to ± 1 V (peak). Use channel 1 set at 0.5 V/div to monitor V_i and channel 2 set at 5 V/div to monitor V_o. Switch the CRO to (XY). Ground both channels and center the spot as shown in Fig. L5-8.

30. Switch both channels to dc coupling and sketch the waveform on Fig. L5-8 as a *solid line.* Calculate the closed-loop gain A_{CL}, using the equation $A_{CL} = \frac{\text{rise}}{\text{run}}$. A_{CL} = __________.

31. Is V_o in phase or 180° out of phase with input voltage? __________.

32. Increase V_o to ± 2 V (peak). Sketch the results on Fig. L5-8, using dashed lines to extend the waveshape. Explain why the transfer curve flattens. ______________________________

______________________________.

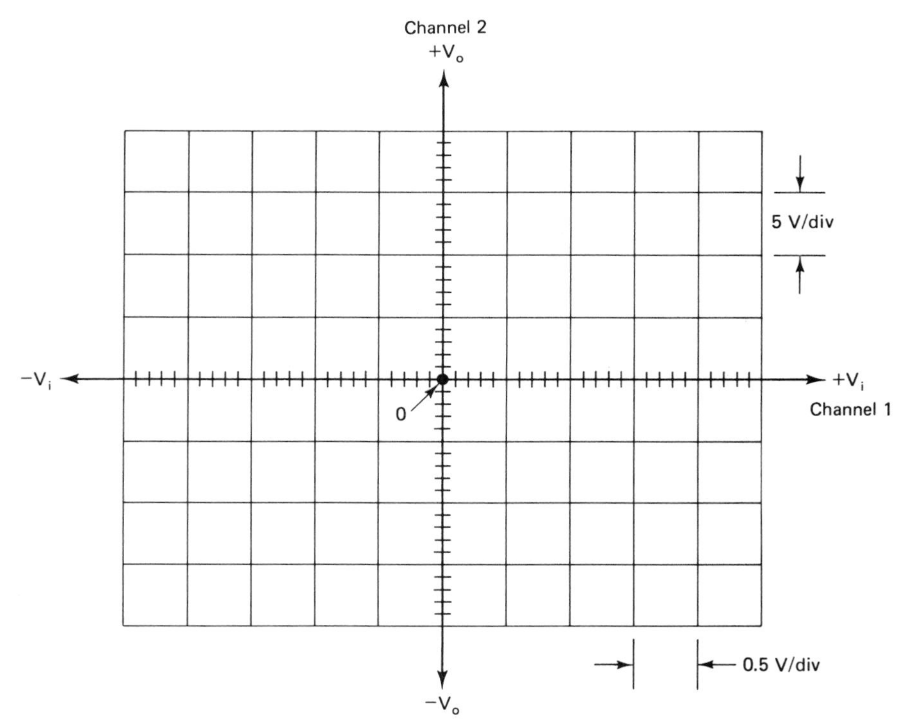

Figure L5-8 Transfer characteristics of a noninverting amplifier.

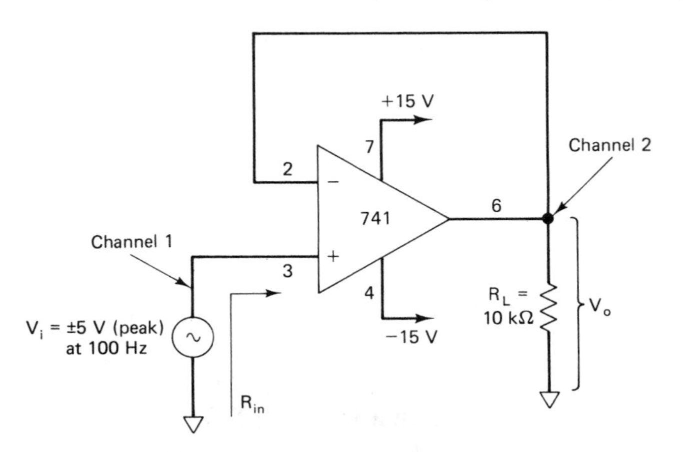

Figure L5-9 Voltage follower.

What is the $\pm V_{sat}$ as measured on Fig. L5-8? $+V_{sat}$ = ________;

$-V_{sat}$ = ________.

Procedure E: Voltage-Follower Amplifier

33. Use the voltage-follower amplifier shown in Fig. L5-9 to predict the following information.

(a) Estimate R_{in}. R_{in} = ________.

(b) Estimate A_{CL}. A_{CL} = ________.

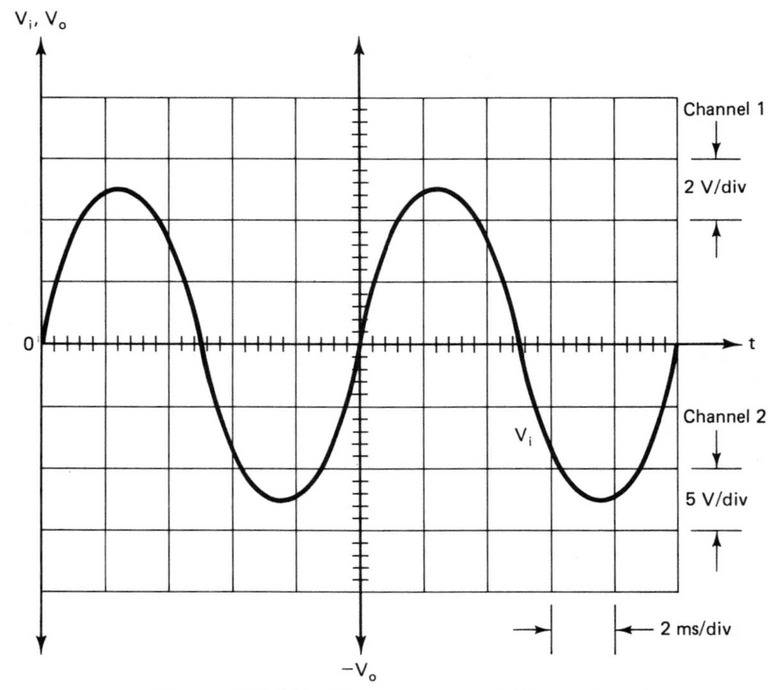

Figure L5-10 V_i versus t, and V_o versus t.

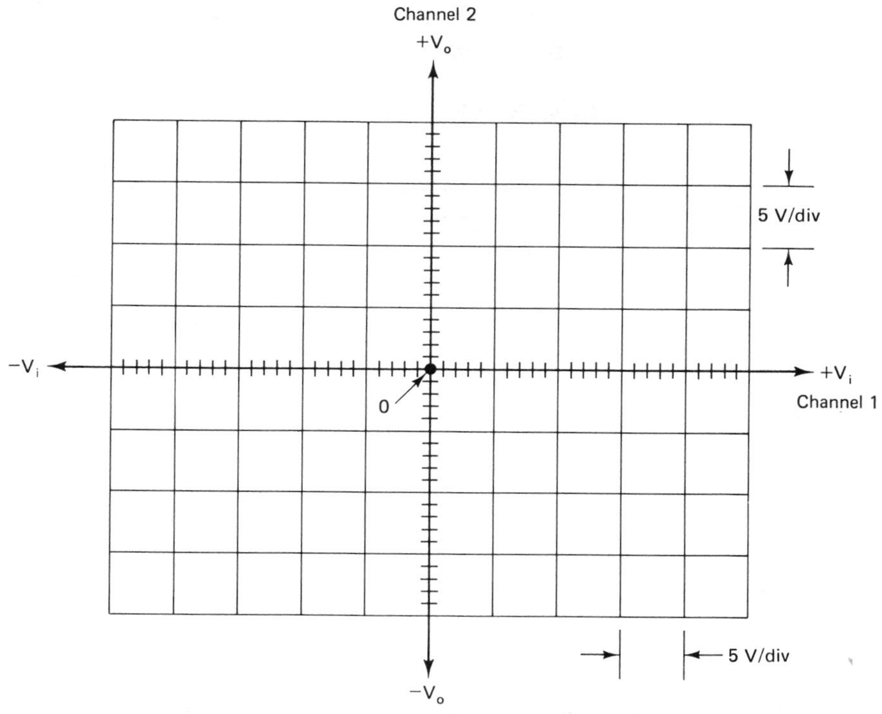

Figure L5-11 Transfer characteristics of a voltage follower.

(c) What is the phase shift between V_i and V_o? $\theta =$ __________.

(d) Determine the expected value of V_o. $V_o =$ __________.

34. Wire the circuit shown in Fig. L5-9. Set V_i equal to a sine wave of ± 5 V (peak) of a test frequency of 100 Hz. Use channel 1 set at 2 v/div to monitor V_i and obtain the input waveform shown in Fig. L5-10. Measure and sketch V_o versus t on Fig. L5-10. Channel 2 should be set to 5 V/div.

35. Using Fig. L5-10, measure the closed-loop gain A_{CL}: $A_{CL} = \frac{V_{o(peak)}}{V_{i(peak)}}$. A_{CL} = __________.

36. Do the expected values of A_{CL} and θ compare favorably with the measured values? __________.

37. Display the transfer curve of a voltage follower by first setting V_i to ± 5 V (peak). With both channels 1 and 2 set to 5 V/div, switch the CRO to (XY). Ground both channels and center the spot as shown in Fig. L5-11.

38. Switch both CRO channels to dc coupling and sketch the waveform on Fig. L5-11 with a solid line. Calculate the closed-loop voltage gain A_{CL} using the equation $A_{CL} = \frac{\text{rise}}{\text{run}}$. $A_{CL} =$ __________.

39. Is V_o in phase or 180° out of phase with the input voltage? __________.

40. Increase V_i to its maximum output. Sketch the results in Fig. L5-11 using dashed lines to extend the waveform. Can you measure $+V_{sat}$?

Explain why. ____

____.

CONCLUSION:

Mathematical Operations with Op Amps and Linear Integrated Circuits

6

LEARNING OBJECTIVES

Upon completion of this chapter on mathematical operations, you will be able to:

- Change the polarity of an analog voltage (signed numbers) without changing its magnitude.
- Sum or add signed numbers with a simple inverting adder.
- Convert a simple inverting adder into a noninverting adder using a sign-changer op amp.
- Take the average of a group of signed numbers and display the results in both inverted or noninverted form.
- Combine an inverting adder and a sign changer op amp into a subtractor.
- Explain the features of both two-quadrant and four-quadrant multipliers.
- Discuss the important specifications of the AD534 analog multiplier IC.
- Use the AD534 analog multiplier IC to perform such functions as multiplication, division, square, and square root.
- Scale the output level of the AD534 IC to any desired valued.
- Provide an offset to the output of the AD534.

6-0 INTRODUCTION

As stated in Chapter 1, the name "operational amplifier" was adopted because the first devices were used to perform mathematical operations. In this chapter we show how op amps and other linear ICs can solve mathematical problems. We begin our discussion with a review of signed numbers and their equivalent, electronic voltage. Then we show how the op amp can accomplish sign changing, addition, averaging, and subtraction.

To perform the more complex operation of multiplication, the

AD534 linear IC multiplier chip (see data sheet pp. 211-16) from Analog Devices Incorporated has been selected. This versatile device requires no external trimming and performs multiplication of signed numbers with a high degree of accuracy. In addition to multiplication, the AD534 can also be wired to perform division, square a number, or take the square root of a number.

6-1 SIGNED NUMBERS

One descriptive method to represent signed numbers is shown in Fig. 6-1a. A graduated vertical line shows positive numbers above the zero point and negative numbers below. As you move farther away from zero, the numbers get progressively larger in either the positive (up) or negative (down) direction.

Figure 6-1b illustrates that signed numbers can be expressed electronically simply as a voltage. The zero point of the voltage scale is equivalent to 0 V or ground. The magnitude of voltage represents the value of the signed number and can be either equal to the number itself (Fig. 6-1b) or a scaled equivalent (Fig. 6-1c). Signed numbers are either positive or negative values. The electronic analog of the sign is voltage polarity and is either a + or − value *measured with respect to ground.*

6-2 SIGN CHANGER

It is sometimes necessary to change the sign of a number from positive to negative or from negative to positive. This is accomplished, in an electrical sense, by changing the polarity of the voltage used to represent the number without altering its magnitude. This can be done by an inverting amplifier with a gain of −1.

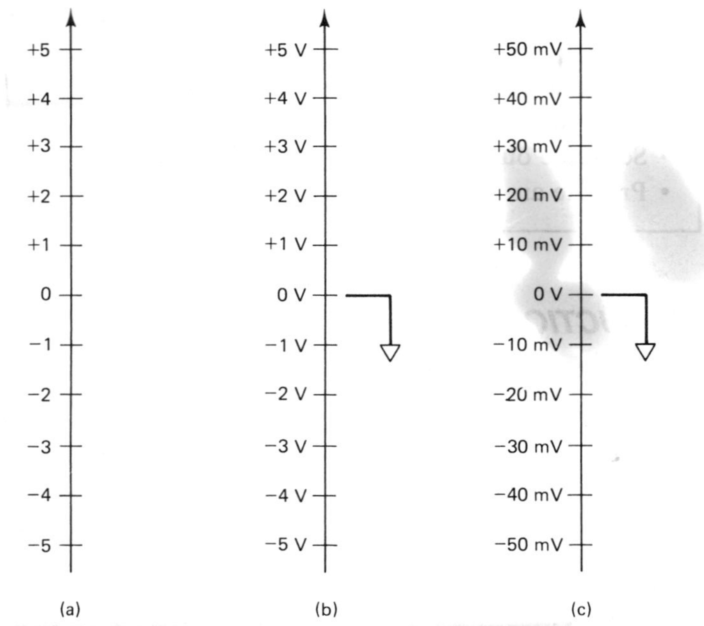

Figure 6-1 Signed numbers can be represented (a) by a vertical scale (b) a direct voltage equivalent, or (c) a scaled equivalent.

Figure 6-2a illustrates a basic sign-changer circuit. Both R_F and R_i are equal to 10-kΩ 1% resistors. The gain is equal to -1 and V_o is

$$V_o = -(V_i) \tag{6-1}$$

Observe that the minus sign outside the parentheses results from the amplifier's 180° phase shift. The value of V_i inside the parentheses can be of either polarity.

EXAMPLE 6-1

Solve for output V_o of Fig. 6-2a if V_i is (a) +3 V; (b) −25 mV.

SOLUTION Using Eq. (6-1), we obtain:

(a) $V_o = -(V_i) = -(3\text{ V}) = -3\text{ V}$

(b) $V_o = -(-25\text{ mV}) = +25\text{ mV}$.

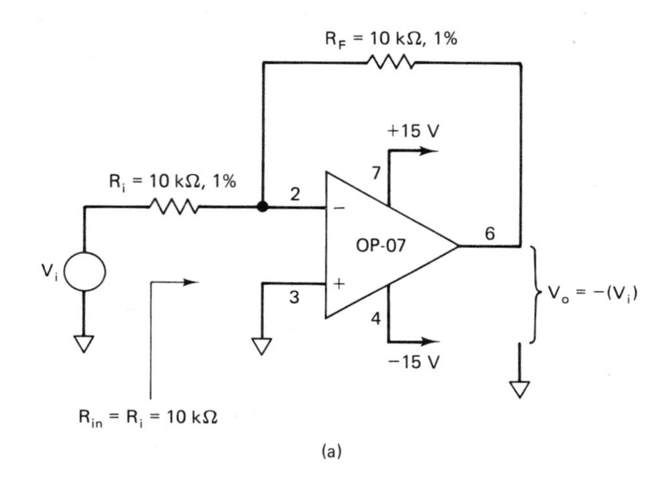

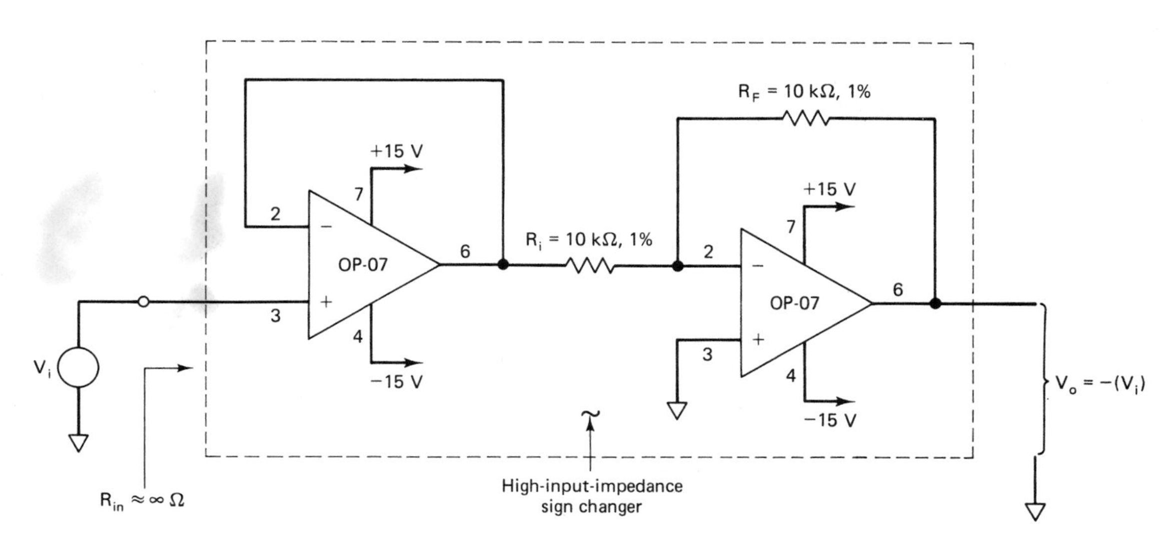

Figure 6-2 Sign-changer circuits. (a) A basic sign-changer circuit, output voltage V_O equals the magnitude of input E_i with a polarity reversal. (b) The kilohm input resistance of the basic sign-changer circuit can be increased to approximately infinite ohms with the addition of a voltage-follower circuit.

The sign changer of Fig. 6-2a has a low input resistance of 10 kΩ. This low value can be improved substantially by adding a buffer amplifier as in Fig. 6-2b. Since the input resistance of the buffer is very high (ideally, ∞ Ω), it can be connected to any voltage source without loading.

6-3 ADDITION OF ANALOG NUMBERS

6-3.1 Basic Inverting Adder

Figures 6-3 to 6-5 show how to add signed numbers. The basic inverting adder in Fig. 6-3 has three input voltages. However, there is no restriction on the number of inputs.

The inverting (−) input terminal is called the *summing node.* It is at virtual ground. Each input current is then set by a separate input voltage and a separate series input resistor. I_1, I_2, and I_3 are determined by Ohm's law as

$$I_1 = \frac{V_1}{R_1} \qquad I_2 = \frac{V_2}{R_2} \qquad I_3 = \frac{V_3}{R_3} \qquad \text{etc.} \tag{6-2}$$

Since the inverting terminal of the op amp draws no current, I_1, I_2, and I_3 flow through the feedback resistor R_F. This summation of currents in R_F yields a voltage across R_F equal to V'_o, or

$$V'_o = -(I_1 + I_2 + I_3)R_F \tag{6-3}$$

The minus sign of Eq. (6-3) is produced by the 180° phase shift of the inverting configuration. Substitution of Eq. (6-2) into (6-3) yields the following *general expression* for an inverting adder:

$$V'_o = -\left(\frac{R_F}{R_1}V_1 + \frac{R_F}{R_2}V_2 + \frac{R_F}{R_3}V_3\right) \tag{6-4}$$

Equation (6-4) shows that each input voltage is amplified by an *independently* controlled *channel gain* that is adjustable simply by changing the value of input resistors.

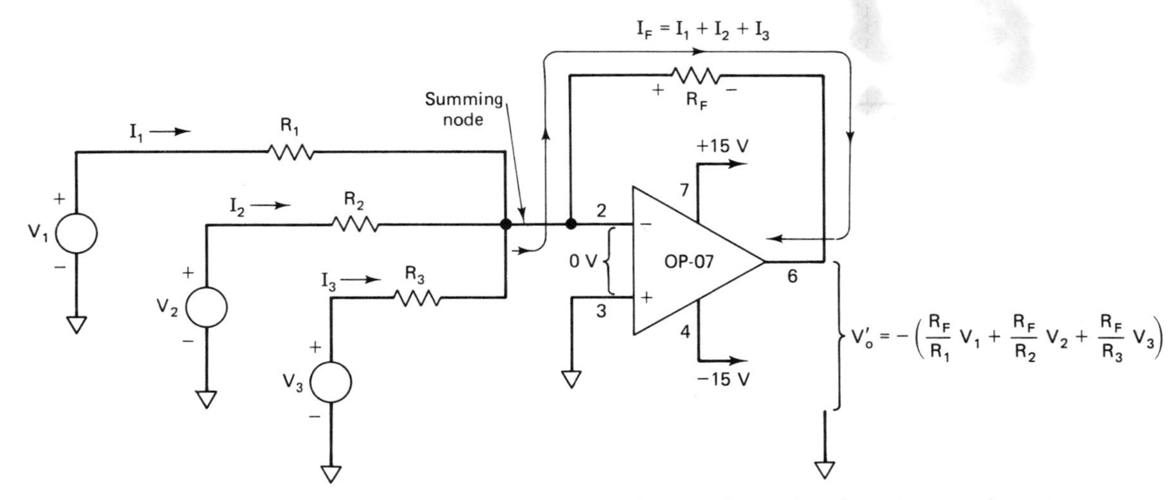

Figure 6-3 basic inverting adder configuration. Each input voltage is amplified by its own channel gain.

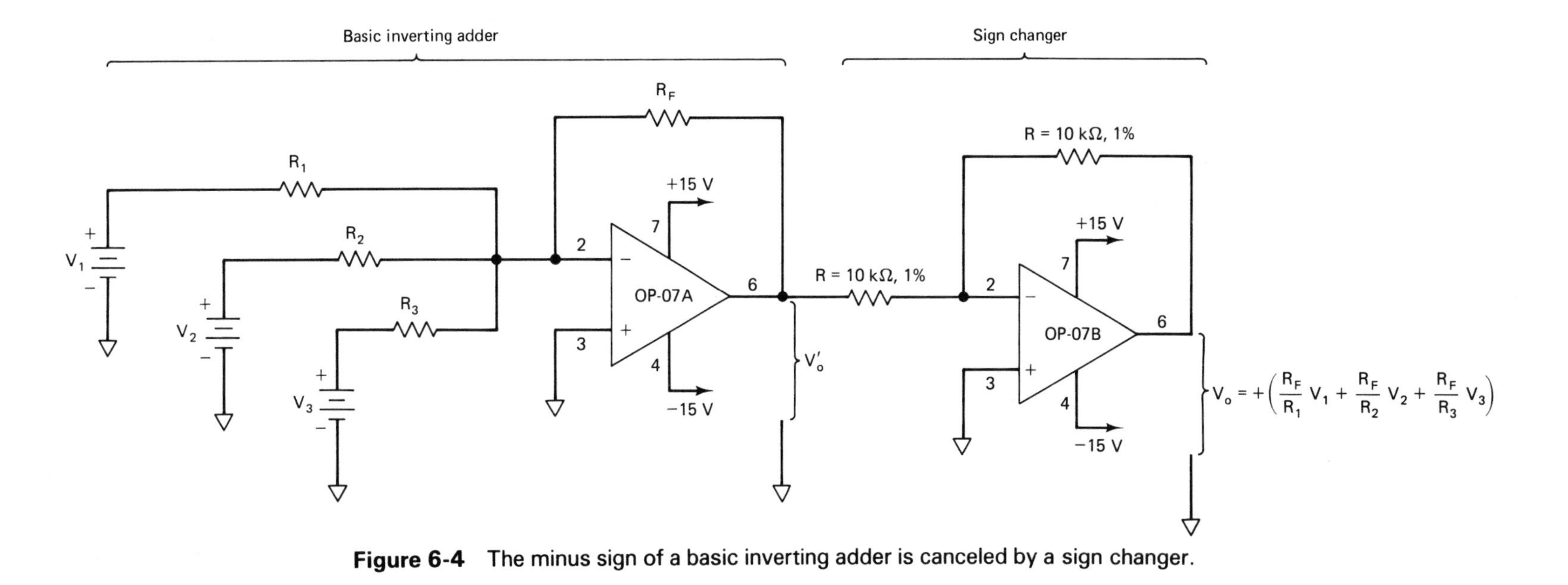

Figure 6-4 The minus sign of a basic inverting adder is canceled by a sign changer.

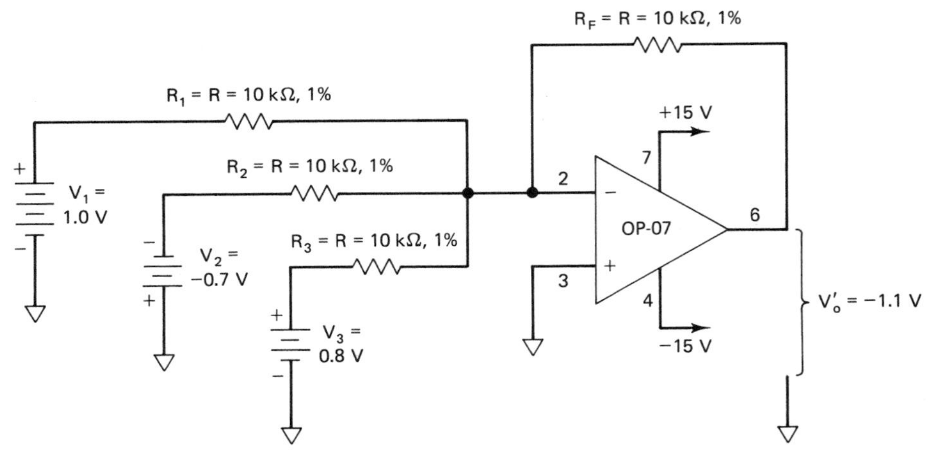

Figure 6-5 This inverting summing amplifier can be converted to a noninverting summing amplifier by adding an output sign changer as in Fig. 6-6.

One obvious difficulty with the basic adder of Fig.6-3 is the minus (−) sign associated with V_o'. The minus sign can be cancelled by adding a sign changer to the output as in Fig. 6-4. Its output V_o can then be expressed as:

$$V_o = +\left(\frac{R_F}{R_1}V_1 + \frac{R_F}{R_2}V_2 + \frac{R_F}{R_3}V_3\right) \tag{6-5}$$

Three adding-type circuits are possible using the basic adder designs of Fig. 6-3 and 6-4. They are the *summing, averaging,* and *scaled* adder circuits that we discuss in the following three sections.

6-3.2 Summing Amplifiers

Our first problem is to construct an inverting summing amplifier that will instantaneously add input voltages directly. We must convert the general expression of Eq. (6-4) to the equation

$$V_o' = -(V_1 + V_2 + V_3) \tag{6-6}$$

To accomplish this, make all resistors equal. If $R_F = R_1 = R_2 = R_3 = R$, the gain for each channel is −1 and V_o' will equal the (negative) sum of all input voltages. See Fig. 6-5.

EXAMPLE 6-2

Design an inverted summing amplifier to add $V_1 = 1.0$ V, $V_2 = -0.7$ V, and $V_3 = 0.8$ V. The input resistance of each channel is specified to be 10 kΩ.

SOLUTION Set all input resistors to 10 kΩ, 1% to satisfy the input resistance requirement. Making R_F also equal to a 10-kΩ 1% resistor will set the gain of each channel to −1 as in Fig. 6-5. From Eq. (6-6),

$$V_o' = -(V_1 + V_2 + V_3) = -(1.0\text{ V} - 0.7\text{ V} + 0.8\text{ V}) = -(+1.1\text{ V}) = -1.1\text{ V}$$

6-3.3 Averaging Circuit

An averaging circuit sums all inputs and then divides that sum by the number of inputs. To produce an output voltage that is proportional to the average of the input voltages, we must first convert Eq. (6-4) to the following expression:

$$V_o' = -\left(\frac{V_1 + V_2 + V_3}{n}\right) \tag{6-7}$$

where n is the number of inputs. This conversion is accomplished by first making the input resistors of each channel equal in value.

$$R_1 = R_2 = R_3 = R \text{ (etc.)} \tag{6-8}$$

To average, make the feedback resistor R_F equal to the input resistor R divided by the number of inputs n, or

$$R_F = \frac{R}{n} \tag{6-9}$$

The minus sign of Eq. (6-7) can be removed by adding a sign changer as in Fig. 6-6.

EXAMPLE 6-3

Design an averaging amplifier to average inputs V_1 and V_2. Assume, for example, that $V_1 = +5$ V and $V_2 = +3$ V. The input resistance of each channel as seen by V_1 and V_2 is to be 20 kΩ.

SOLUTION Refer to the two-input average of Fig. 6-6. Use Eq. (6-8) to set all input resistors to equal 20 kΩ, 1%. That is, $R_1 = R_2 = R = 20$ kΩ. From Eq. (6-9), R_F becomes $R_F = R/n = 20$ kΩ/2 = 10 kΩ. From Eq. (6-7),

$$V_o' - \left(\frac{V_1 + V_2}{n}\right) = -\left(\frac{5\text{ V} + 3\text{ V}}{2}\right) = -\left(\frac{8\text{ V}}{2}\right) = -(4\text{ V})$$

In Fig. 6-6 the sign changer cancels the minus sign and the output becomes $V_o = +4$ V.

6-3.4 Scaled Addition

The inverting adder of Fig. 6-3 can also be used to add input voltages that are also amplified (scaled) during the summing process. The design of a scaled adder begins by (1) selecting a 10-kΩ input resistor for that channel with the highest gain; (2) R_F will then equal 10 kΩ times the highest gain; and (3) the other channel input resistors are equal to R_F divided by the channel's gain.

Thus the input resistance for each channel is calculated from Eq. (6-10) once the channel gain is determined.

$$R_1 = \frac{R_F}{A_1} \qquad R_2 = \frac{R_F}{A_2} \qquad R_3 = \frac{R_F}{A_3} \qquad \text{etc.} \tag{6-10}$$

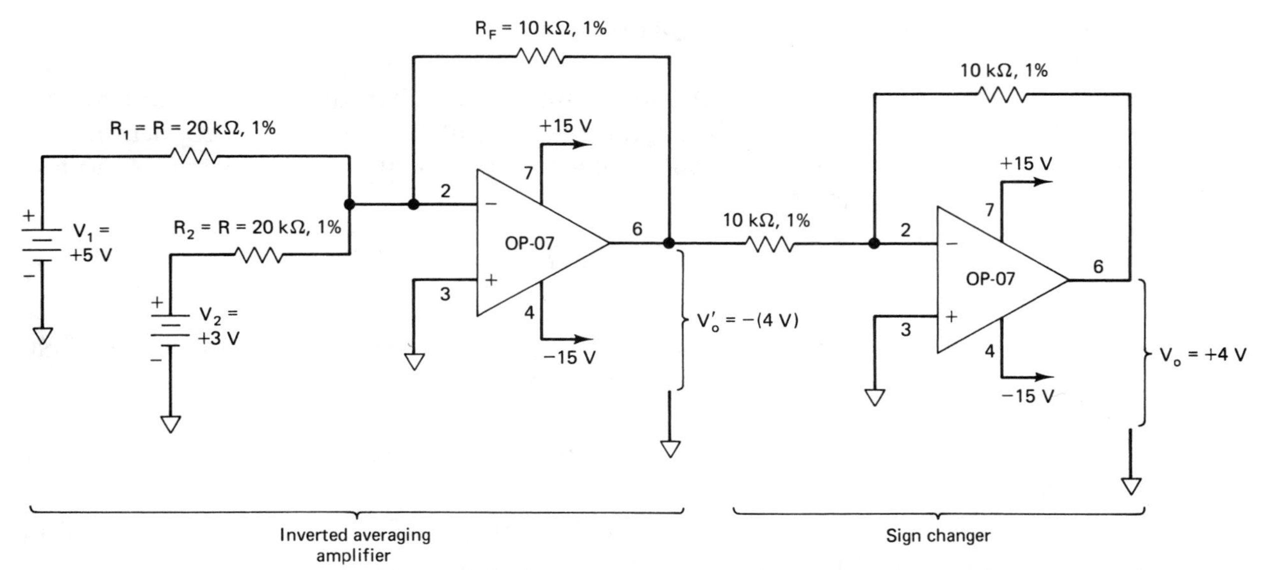

Figure 6-6 Instantaneous averaging with an inverted averager and sign changer.

EXAMPLE 6-4

Design a scaled adder circuit that will amplify $V_1 = +300$ mV by a channel 1 gain of 10 and $V_2 = 1.00$ V by a channel 2 gain of 2 and sum the results.

SOLUTION Pick $R_1 = 10\ \text{k}\Omega$ for the high-gain channel 1. From Eq. (6-10),

$$R_F = 10\ R_1 = 10\ (10\ \text{k}\Omega) = 100\ \text{k}\Omega$$

From Eq. (6-10),

$$R_2 = \frac{100\ \text{k}\Omega}{2} = 50\ \text{k}\Omega$$

From the general equation (6-4),

$$V'_o = -\left(\frac{R_F}{R_1}V_1 + \frac{R_F}{R_2}V_2\right) = -\left(\frac{100\ \text{k}\Omega}{10\ \text{k}\Omega}0.3\ \text{V} + \frac{100\ \text{k}\Omega}{50\ \text{k}\Omega}1\ \text{V}\right)$$

$$V_o = -(3.0\ \text{V} + 2\ \text{V}) = -(5\ \text{V})$$

Figure 6-7 shows the solution.

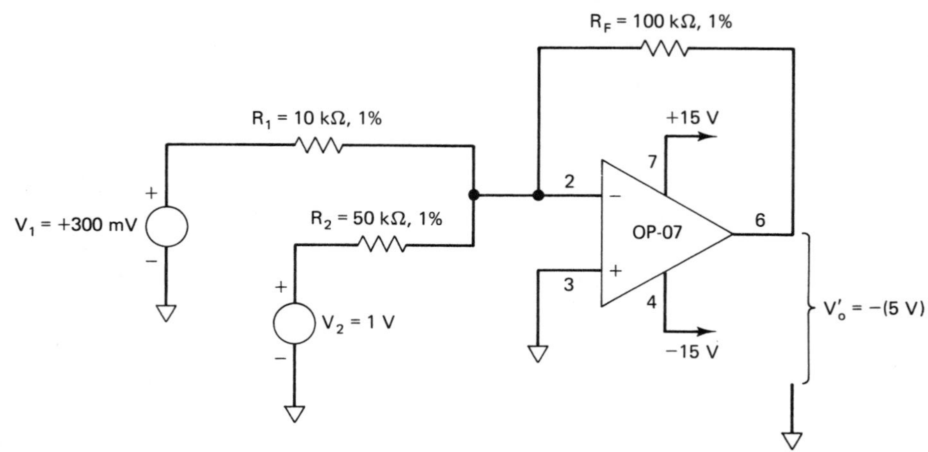

Figure 6-7 Scaled addition. The gain of channel 1 is set to $A_1 = 10$ and channel 2 is set to $A_2 = 2$.

6-4 SUBTRACTION OF ANALOG NUMBERS

6-4.1 Subtraction Operation

In subtraction, the first number is called the *minuend* and the second number the *subtrahend.* The answer is the *difference* and is found by first changing the sign of the subtrahend and then adding.

EXAMPLE 6-5
Subtract from +5 the value +2.

SOLUTION $5 - (+2) = 5 + (-2) = 3.$

Note that subtraction is *addition with* a *sign change on the subtrahend.*

6-4.2 Two-Op-Amp Subtractor

Analog subtraction is shown in Fig. 6-8, where V_2 is to be subtracted from V_1. One op amp performs addition and another produces a sign change for the subtrahend.

V_1 is applied to the sign changer and inverted at output $V_{oA} = -V_1$. V_{oA} is re-inverted by channel 1 of the adder. Thus V_1 appears at output V_o as V_1. V_2 is inverted and added by channel 2 and appears at output V_o as $-V_2$ is given by

$$V_o = V_1 - V_2 \tag{6-11}$$

EXAMPLE 6-6
Design a subtractor circuit that will subtract 3 V from a constant 8 V.

SOLUTION Use Fig. 6-8. Apply the constant 8 V to the sign-changer

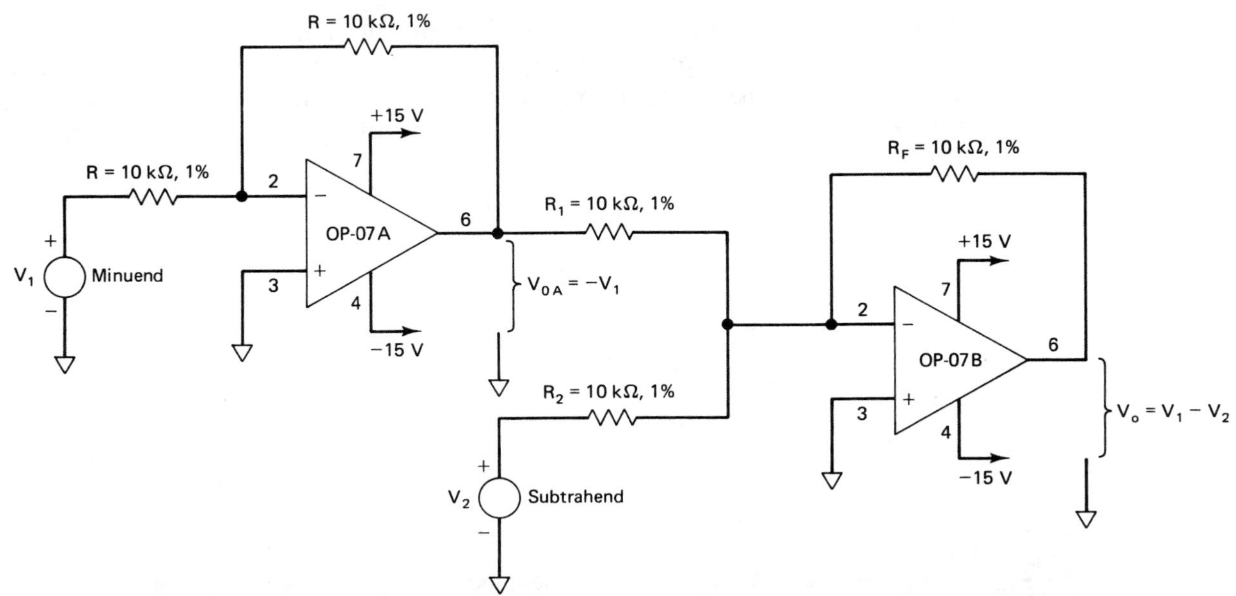

Figure 6-8 Subtraction using a sign changer and an inverting summing amplifier.

circuitry (i.e., $V_1 = 8$ V). Apply $V_2 = 3$ V directly to channel 2 of the inverting adder.

$$V_o = V_1 - V_2 = 8\text{ V} - 3\text{ V} = 5\text{ V}$$

6-5 MULTIPLICATION OF ANALOG NUMBERS

6-5.1 Analog Multiplier ICs

The multiplication of two input signals will be accomplished by a single analog multiplier IC. These precision ICs are packaged in conventional cases and are available from manufacturers such as Analog Devices, Burr-Brown, and Motorola. Figure 6-9 symbolically represents a basic analog multiplier. Two terminals, labeled *X* and *Y*, accept input voltages V_x and V_y. Output voltage V_o is the product of V_x and V_y reduced by a scale factor SF. Expressed mathematically, we have

$$V_o = \frac{V_x \cdot V_y}{\text{SF}} \tag{6-12}$$

The scale factor is normally 10 V and is discussed in Section 6-5.3.

Analog multiplier ICs are available as either two-quadrant or four-quadrant devices. They are distinguished by polarity limitations on the input terminals and the corresponding output voltage response.

6-5.2 Two- and Four-Quadrant Analog Multipliers

Assume that an analog multiplier operates with a scale factor SF equal to 1 V. Equation (6-12) reduces to

$$V_o = \frac{V_x \cdot V_y}{1\text{ V}} \tag{6-13}$$

Also assume that there is no limitation on the magnitude of V_o.

Figure 6-10 is a plot of multiplier output voltage V_o along the vertical axis. Input voltage V_x is plotted along the horizontal axis. Note that the plot is given for a variety of second input voltages V_y, shown as *lines of constant* V_y.

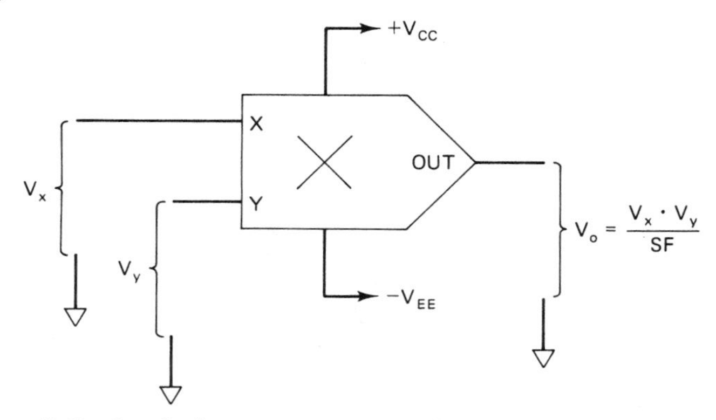

Figure 6-9 Symbolic representation of a generic analog multiplier integrated circuit.

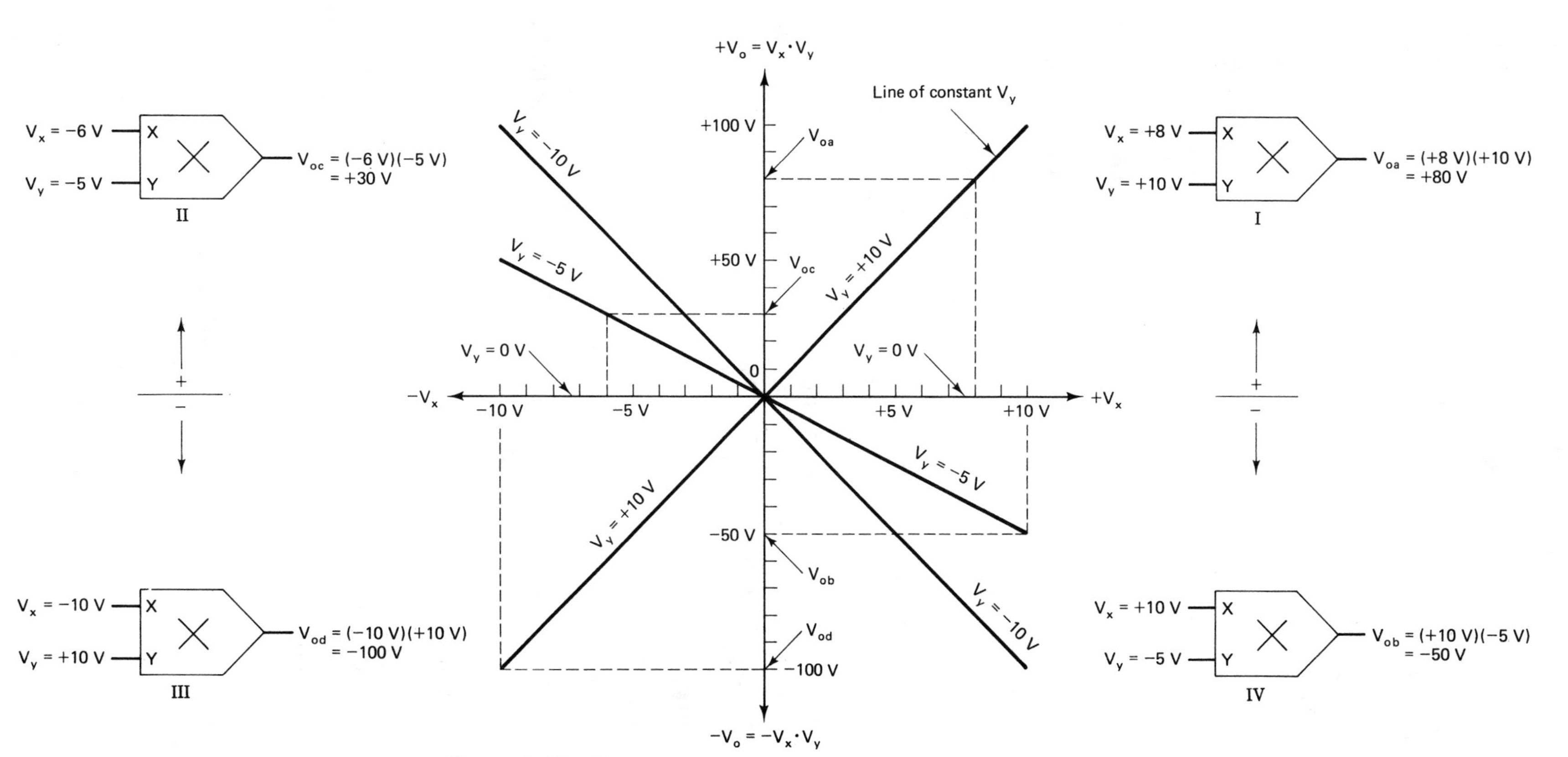

Figure 6-10 Characteristics of both two- and four-quadrant multipliers.

As the name implies, a two-quadrant multiplier operates properly only in quadrants I and IV. V_y can be either an ac signal or a positive or negative dc signal. But V_x must be restricted to positive values in the range 0 to 10 V. Output V_o of a two-quadrant multiplier will have the same sign as V_y.

EXAMPLE 6-7

In the multiplier circuits graphed in Fig. 6-10, find V_o if (a) $V_x = +8$ V and $V_y = +10$ V; (b) $V_x = +10$ V and $V_y = -5$ V.

SOLUTION (a) Using Fig. 6-10, first find $V_x = +8$ V and project up vertically until line $V_y = +10$ V is intersected in quadrant I. Next, project across horizontally to $V_o = +80$ V. From Eq. 6-13,

$$V_o = \frac{8\text{ V} \cdot 10\text{ V}}{1\text{ V}} = +80\text{ V}$$

(b) Find $V_x = +10$ V and project down to the line $V_y = -5$ V in quadrant IV. V_o is found horizontally to be -50 V, or

$$V_o \; \frac{(10\text{ V})(-5\text{ V})}{1\text{ V}} = -50\text{ V}$$

Note that the polarity of V_o follows the polarity of V_y. Also it should be noted that $V_o = 0$ V if either V_x or V_y equals 0 V.

A four-quadrant multiplier eliminates the restriction on the sign of V_x. Both V_x and V_y can be any + or − voltage from 0 to ± 10 V. Operation of a four-quadrant multiplier is identical to that of a two-quadrant device in quadrants I and IV. For an explanation of operation in quadrants II and III, refer to Example 6-8.

EXAMPLE 6-8

(a) Graphically, solve for V_o if $V_x = -6$ V and $V_y = -5$ V. (b) Solve for V_o if $V_x = -10$ V and $V_y = +10$ V.

SOLUTION (a) In Fig. 6-10, the intersection of $V_x = -6$ V and $V_y = -5$ V is in the second quadrant with $V_o = +30$ V. Note that the product of two negative numbers is a positive value.

(b) The product of $V_x = -10$ V and $V_y = +10$ V is found in the third quadrant as $V_o = -100$ V.

Four-quadrant operation will be the product of the two numbers, always with the correct sign. The output will be positive in quadrants I and II (above horizontal axis) and negative in quadrants III and IV (below horizontal axis).

6-5.3 CHARACTERISTICS OF THE AD534 ANALOG MULTIPLIER IC

The AD534 (See data sheets 211-16) is a precision four-quadrant multiplier IC. It is chosen to illustrate the ease with which multiplier chips can be used. Since the AD534 is internally laser trimmed by the manufacturer, no external trim potentiometers are required to obtain the maximum multiplication error of only +0.15% in all four quadrants.

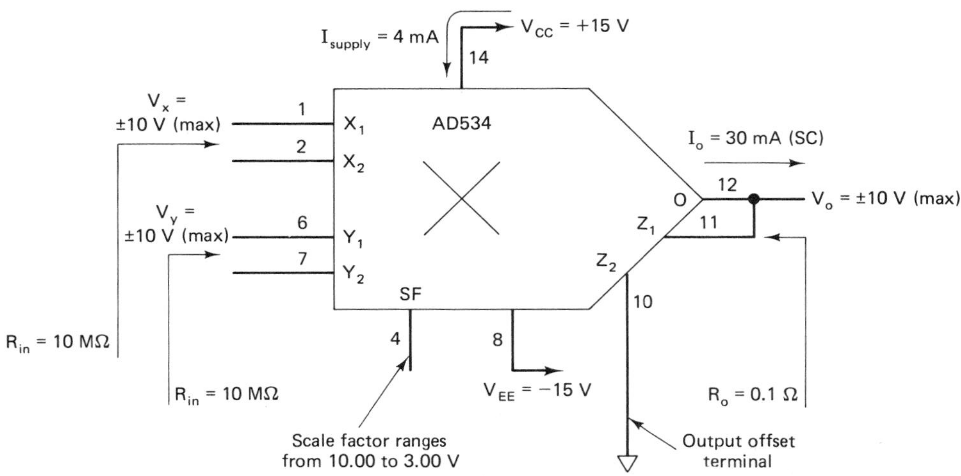

Figure 6-11 Basic characteristics of the AD534 analog multiplier IC.

Figure 6-11 illustrates symbolically some of the more important data sheet information for the AD534. It requires a standard ±15-V bipolar dc power supply (pins 14 and 8) typical of most op amps and only 4 mA of supply current at idle. A single output terminal (pin 12) is internally protected against short circuits to a maximum current of I_o = 30 mA(SC). In normal operation input terminal Z_1 (pin 11) should be connected directly to the output and input terminal Z_2 (pin 10) referenced to ground as shown in Fig. 6-11. For best results, the output voltage V_o should be limited to ±10 V maximum when a ±15-V dc power supply is used.

A pair of X input terminals (pins 1 and 2) and another pair of Y input terminals (pins 6 and 7) are made available to multiply true differential input signals. (See chapter 8 for differential measurements.) For standard single-ended input operation terminals, X_2 and Y_2 are grounded and terminals X_1 and Y_1 become the X and Y input terminals. The input terminal signal swings should also be limited to a maximum of ±10 V.

A scale factor SF of precisely 10.000 V is provided if pin 4 is left open. The scale factor can also be varied from about 3 V to 10.000 V, as will be illustrated later in the squaring circuit of Fig. 6-15.

6-6 APPLICATIONS OF THE AD534

Analog multipliers solve problems such as signal multiplication, squaring signals, volume control, phase-angle detection, and power measurements. We begin with an example of multiplication.

6-6.1 Multiplication of Single-Ended Signals

Figure 6-12 shows the connections required for the multiplication of two single-ended input signals. V_y is a sine wave with peak amplitudes of ±10 V at a frequency of 1000 Hz. It is expressed mathematically as 10 V sin 6280 t. V_x is an adjustable dc voltage that can vary from +10 V through 0 V to −10 V. Since pin 4 is left open, the scale factor is precisely 10 V and the output of Fig. 6-12 responds to Eq. (6-12).

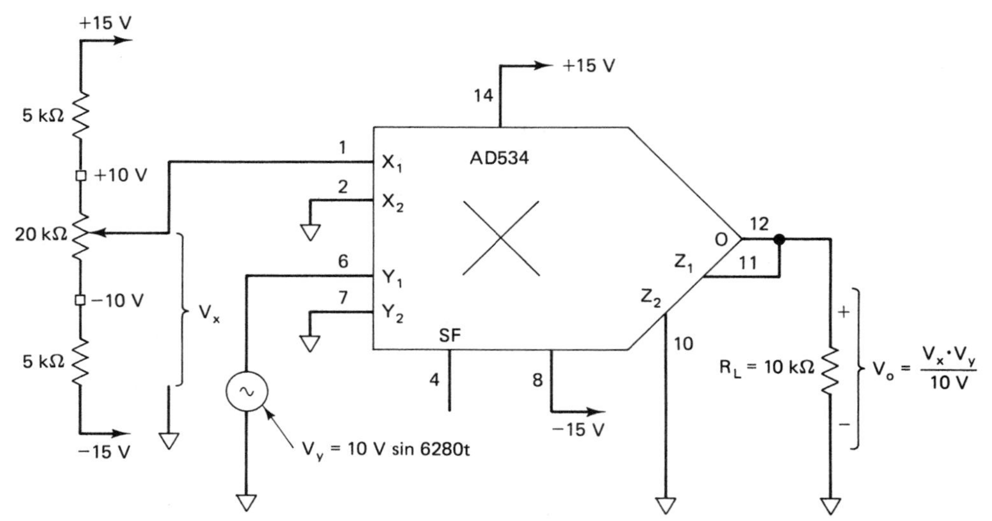

Figure 6-12 Multiplication of two single-ended voltages using the AD534 multiplier chip. V_x can be used as a volume control for V_y or volume control plus sign change for V_y.

EXAMPLE 6-9

Solve for the output voltage of Fig. 6-12, if V_x is (a) +10 V; (b) 0 V; (c) −5 V.

SOLUTION From Eq. (6-12),

$$\text{(a) } V_o = \frac{V_x \cdot V_y}{10 \text{ V}} = \frac{(+10 \text{ V})(10 \text{ V} \sin 6280t)}{10 \text{ V}} = 10 \text{ V} \sin 6280\, t$$

$$\text{(b) } V_o = \frac{(0 \text{ V})(10 \text{ V} \sin 6280t)}{10 \text{ V}} = 0 \text{ V}$$

$$\text{(c) } V_o = \frac{(-5 \text{ V})(10 \text{ V} \sin 6280t)}{10 \text{ V}} = -5 \text{ V} \sin 6280t$$

Note that when V_x is between 0 and +10 V, the 20-kΩ potentiometer acts as a volume control. V_o is a function of the setting of the 20-kΩ pot. When V_x becomes a negative value, it causes V_o to be 180° out of phase with V_y. In the solution of Example 6-9c, $V_o = -5 \text{ V} \sin 6280t$ can also be written as $V_o = 5 \text{ V} \sin (6280t + \pi)$ where π radians = 180° of phase shift.

6-6.2 Multiplication of Two Differential Signals

Figure 6-13 shows the AD534 multiplier wired to multiply differential inputs. Equation (6-12) must be modified to include differential inputs:

$$V_o = \frac{(V_{x1} - V_{x2}) \cdot (V_{y1} - V_{y2})}{\text{SF}} \tag{6-14}$$

EXAMPLE 6-10

Solve for the output voltage V_o of Fig. 6-13 if both bridges are unbalanced with the values shown.

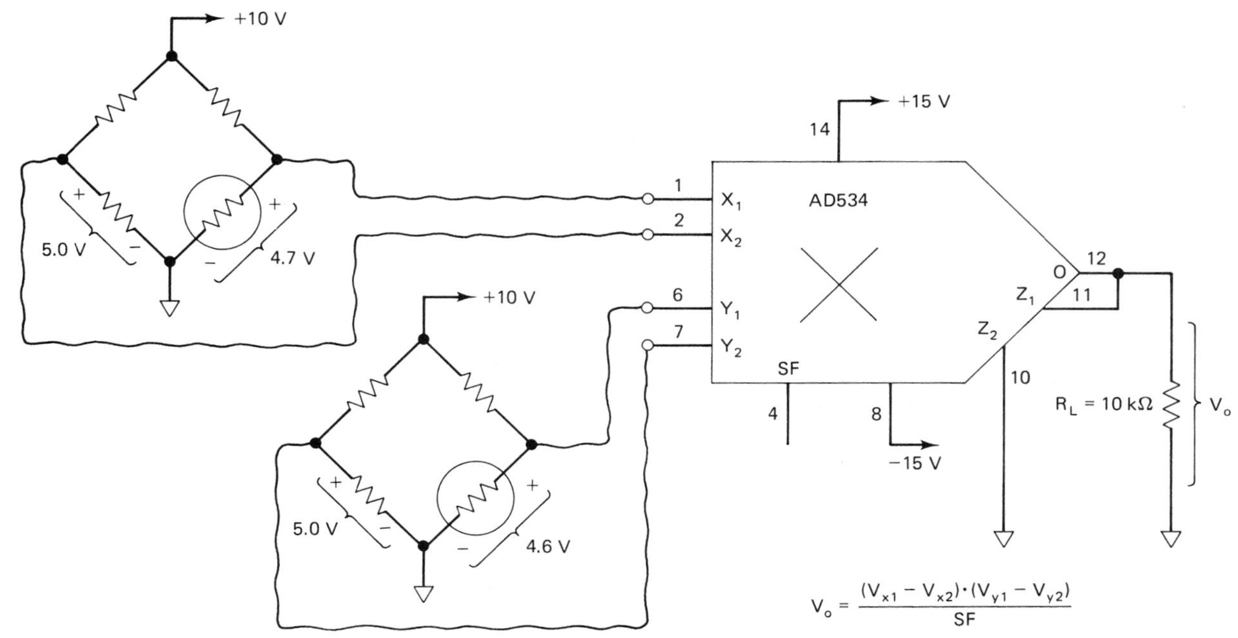

Figure 6-13 The multiplication of two differential input signals is accomplished by using four input terminals.

SOLUTION Since pin 4 is left open, SF = 10 V. Using Eq. (6-14) yields

$$V_o = \frac{(5.00\text{ V} - 4.70\text{ V})(5.00\text{ V} - 4.60\text{ V})}{10\text{ V}}$$

$$= \frac{(0.3\text{ V})(0.4\text{ V})}{10\text{ V}} = \frac{0.12\text{ V}}{10\text{ V}} = 0.012\text{ V}$$

6-6.3 Adding Output Offset

In order to shift, or offset, the output voltage, V_o of an AD534, a dc voltage may be applied to the Z_2 input terminal of pin 10 (Fig. 6-14). Equation (6-12) is modified to include the offset voltage V_{z2} as shown in Eq. (6-15).

$$V_o = \frac{V_x \cdot V_y}{\text{SF}} + V_{z2} \tag{6-15}$$

This terminal arrangement is useful when V_o of the multiplier drives the pen of a chart recorder. V_{z2} can then be used to initially position the pen of a chart recorder.

EXAMPLE 6-11

Solve for V_o in Fig. 6-14 using the information given on the schematic:

SOLUTION Assume that $V_{z2} = 0$ V (pin 10 at ground). Then, using Eq. (6-12), V_o becomes

$$V_o = \frac{(2\text{ V})(5\text{ V-sin }\omega t)}{10\text{ V}} = 1\text{ V sin }\omega t$$

V_o would be a ±1-V (peak) sine wave. To offset this output so that the

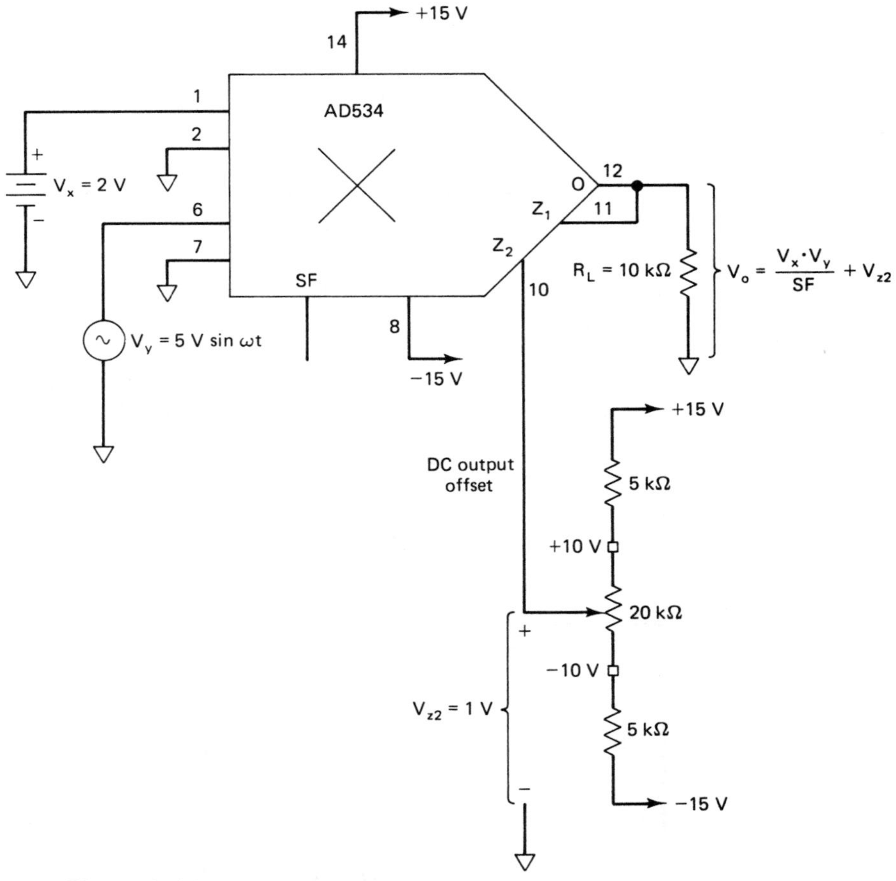

Figure 6-14 Output offset applied to pin 10 of the AD534 multiplier.

sine wave is always positive, the 20-kΩ pot is set so that $V_{z2} = +1$ V. Now using Eq. (6-15), we obtain

$$V_o = \frac{V_x \cdot V_y}{\text{SF}} + V_{z2} = \frac{(2\text{ V})(5\text{ V}\sin \omega t)}{10\text{ V}} + 1\text{ V} = 1\text{ V}\sin \omega t + 1\text{ V}$$

V_o will rise to +2 V at its peak positive point down to 0 V at its peak negative point. With the 20-kΩ pot at pin 10, the output could be offset between a maximum of +10 V above ground and −10 V below ground.

6-6.4 Using the AD534 as a Squarer

The square of any number is calculated if you multiply the number by itself. Figure 6-15 is wired to square a single-ended voltage. Since $V_x = V_y = V$, Eq. (6-12) can be rewritten as

$$V_o = \frac{V^2}{\text{SF}} \tag{6-16}$$

Also note that a resistor R_{SF} has been placed in series with pin 4 and the negative supply. The R_{SF} can be adjusted to alter the scale factor SF according to

$$R_{SF} = 5.4\text{ k}\Omega\left(\frac{\text{SF}}{10\text{ V} - \text{SF}}\right) \tag{6-17}$$

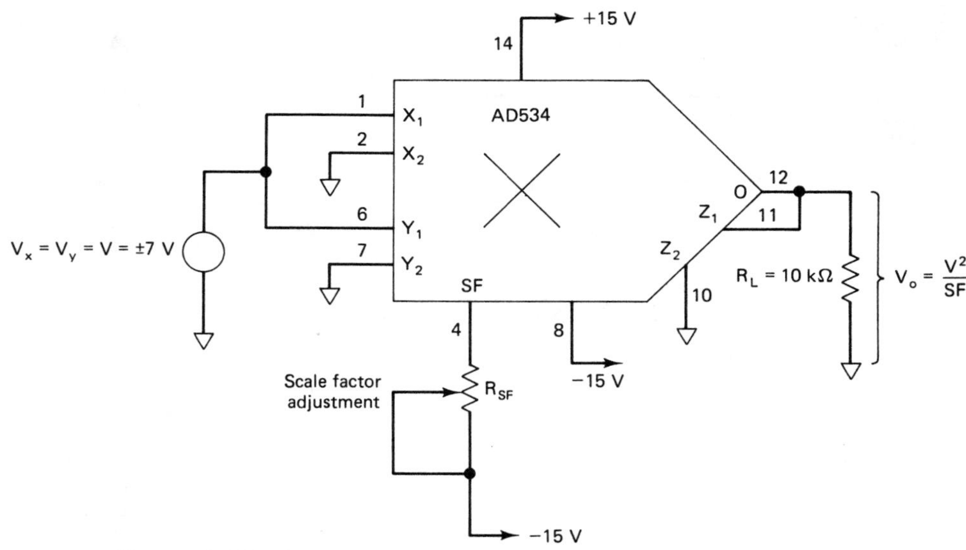

Figure 6-15 Basic circuit to square single-ended voltages. R_{SF} is used to adjust the scale factor from about 3 V to 10 V.

EXAMPLE 6-12

(a) Solve for the output voltage V_o of Fig. 6-15 if $V_x = V_y = +7$ V and SF = 5 V. (b) Solve for V_o if $V_x = V_y = -7$ V and SF = 10 V.

SOLUTION (a) Use Eq. (6-17) to set R_{SF}.

$$R_{SF} = 5.4\text{ k}\Omega\left(\frac{\text{SF}}{10\text{ V} - \text{SF}}\right) = 5.4\text{ k}\Omega\left(\frac{5\text{ V}}{10\text{ V} - 5\text{V}}\right) = 5.4\text{ k}\Omega$$

Using Eq. (6-16) yields

$$V_o = \frac{V^2}{\text{SF}} = \frac{(+7\text{ V})^2}{5\text{ V}} = 9.8\text{ V}$$

(b) For a scale factor of 10 V, Eq. (6-17) becomes

$$R_{SF} = 5.4\text{ k}\Omega\left(\frac{10\text{ V}}{10\text{ V} - 10\text{ V}}\right) = 5.4\text{ k}\Omega\left(\frac{10\text{ V}}{0\text{ V}}\right)$$

$$= \infty\Omega \qquad \text{(open circuit at pin 4)}$$

From Eq. (6-16),

$$V_o = \frac{V^2}{\text{SF}} = \frac{(-7\text{ V})^2}{10\text{ V}} = 4.9\text{ V}$$

Observe that the square of negative number is correctly positive.

6-6.5 Performing Division

In division, the ratio of a *numerator* to its *denominator* results in an answer called the *quotient.* An analog divider can be realized by placing a multiplier, such as the AD534, in the feedback loop of an inverting amplifier with a gain of −1 (sign changer) as in Fig. 6-16.

To analyze this circuit, note that negative feedback forces the same current I through both 10-kΩ resistors. Therefore, V equals V_z but has the opposite polarity:

$$V_z = -V \tag{6-18}$$

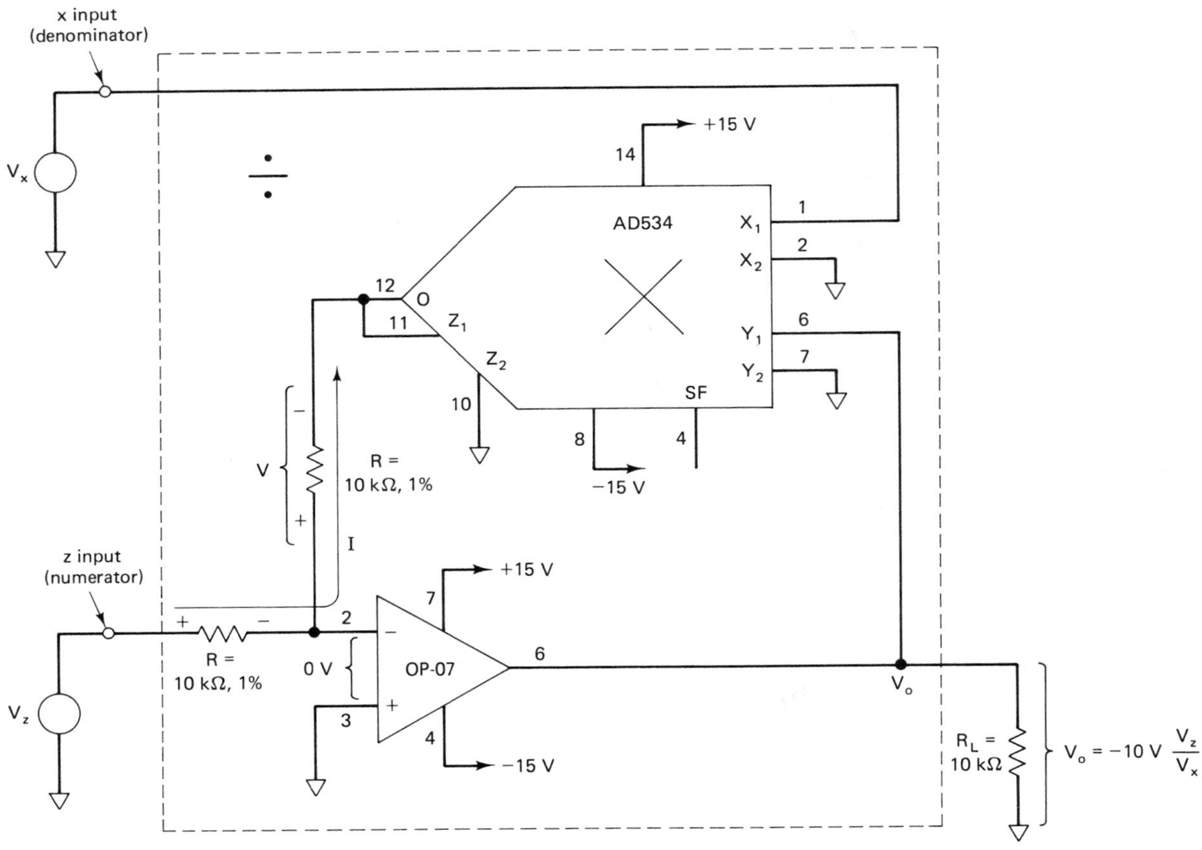

Figure 6-16 The AD534 multiplier used to perform analog division requires only the addition of one op amp and two precision 10 kΩ 1% resistors. V_x must be positive and not equal to zero.

But from Eq. (6-12),

$$V = \frac{V_x \cdot V_o}{\text{SF}} \tag{6-19}$$

Substituting Eq. (6-19) into Eq. (6-18) yields

$$V_z = \frac{-V_x \cdot V_o}{10\text{ V}} \tag{6-20}$$

Where SF = 10 V. Solving for V_o gives us

$$V_o = -10\text{ V}\left(\frac{V_z}{V_x}\right) \tag{6-21}$$

Equation (6-21) shows that output voltage V_o equals a scale factor of −10 V times the ratio, V_z/V_x. V_z can be either a positive or a negative value, but V_x must be restricted to positive values only.

EXAMPLE 6-13

Solve for V_o in Fig. 6-16 if (a) $V_x = +3$ V, $V_z = +1$ V; (b) $V_x = +5$ V, $V_z = -2$ V.

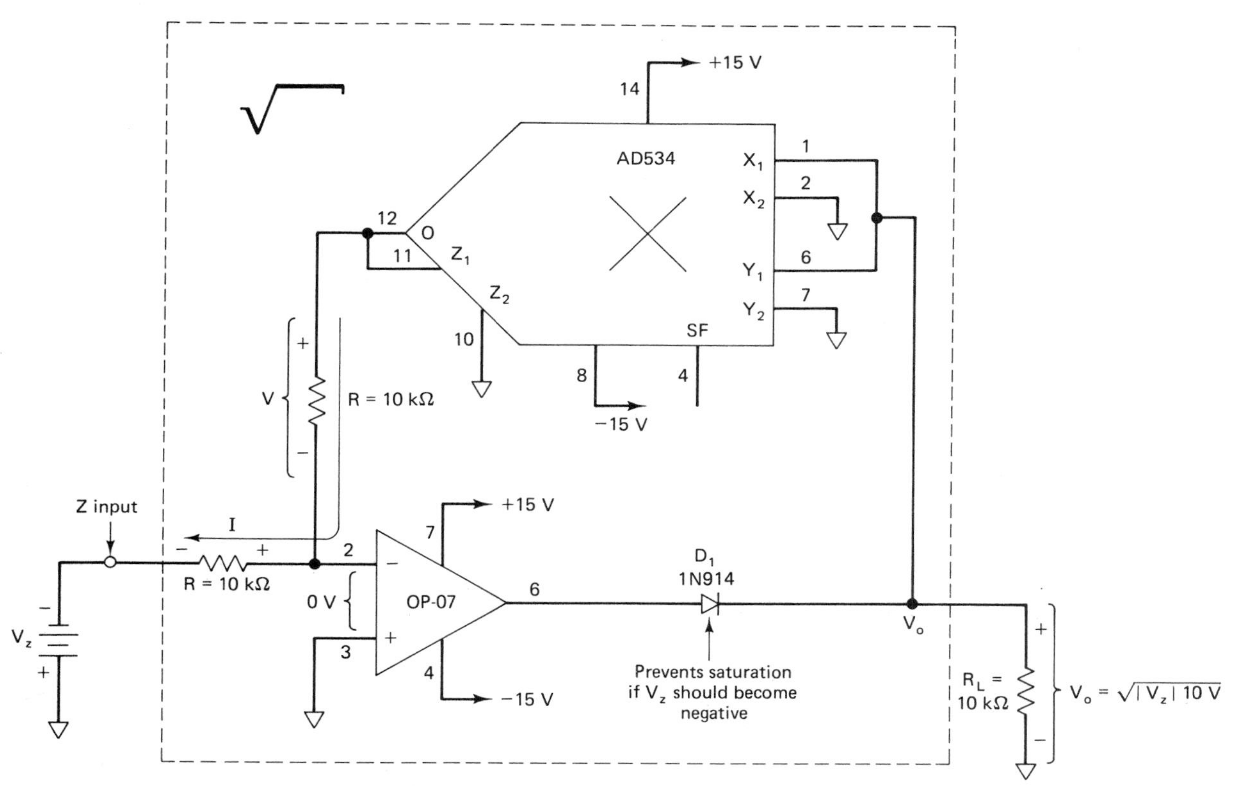

Figure 6-17 The AD534 multiplier is wired as a squaring circuit and placed in the feedback loop of an op amp to build a square rooter.

SOLUTION Use Eq. (6-21), $V_o = -10\text{ V}\left(\frac{V_z}{V_x}\right)$.

(a) $V_o = -10\text{ V}\left(\frac{1\text{ V}}{3\text{V}}\right) = -3.33\text{ V}$

(b) $V_o = -10\text{ V}\left(\frac{-2\text{ V}}{5\text{ V}}\right) = +4.00\text{ V}$

6-6.6 Square Root

An analog square root circuit is shown in Fig. 6-17. The AD534 is configured as a squarer circuit and placed in the feedback loop of an op amp. Equation (6-18) is still valid if $V_z = -V$, is given by

$$V = \frac{V_o^2}{10\text{ V}} \tag{6-22}$$

Substituting Eq. (6-22) into Eq. (6-18) yields

$$V_z = -\frac{V_o^2}{10\text{ V}} \tag{6-23}$$

Solving for V_o, we obtain

$$V_o = \sqrt{-10\text{ V}(V_z)} \tag{6-24}$$

Equation (6-24) shows that V_z must be negative. Diode D_1 has been added to prevent saturation if V_o becomes negative.

End of Chapter Exercises

Name: ____________________

Date: __________ **Grade:** __

A. FILL-IN THE BLANKS

Fill in the blanks with the best answer.

1. A circuit that adds three numbers and inverts their sum is called a three channel __________ __________.
2. Input currents of an inverting adder are determined only by __________ and __________.
3. Signal input currents at an inverting adder all flow together through the __________.
4. An averaging circuits adds the voltages at each input and __________ their sum by __________.
5. A two channel non-inverting adder requires __________ resistors, __________ resistors for the inverting adder and __________ resistors for the sign changer.
6. An analog __________ has an output equal to the difference between two inputs.
7. Typical power supply voltages for an analog multiplier are __________, with output current capability similar to that of an op amp.
8. An analog divider is made by connecting a multiplier within the __________ of an op amp.
9. If no connection is made to pin 4 of an AD584 analog multiplier, its scale factor equals __________.

B. TRUE/FALSE

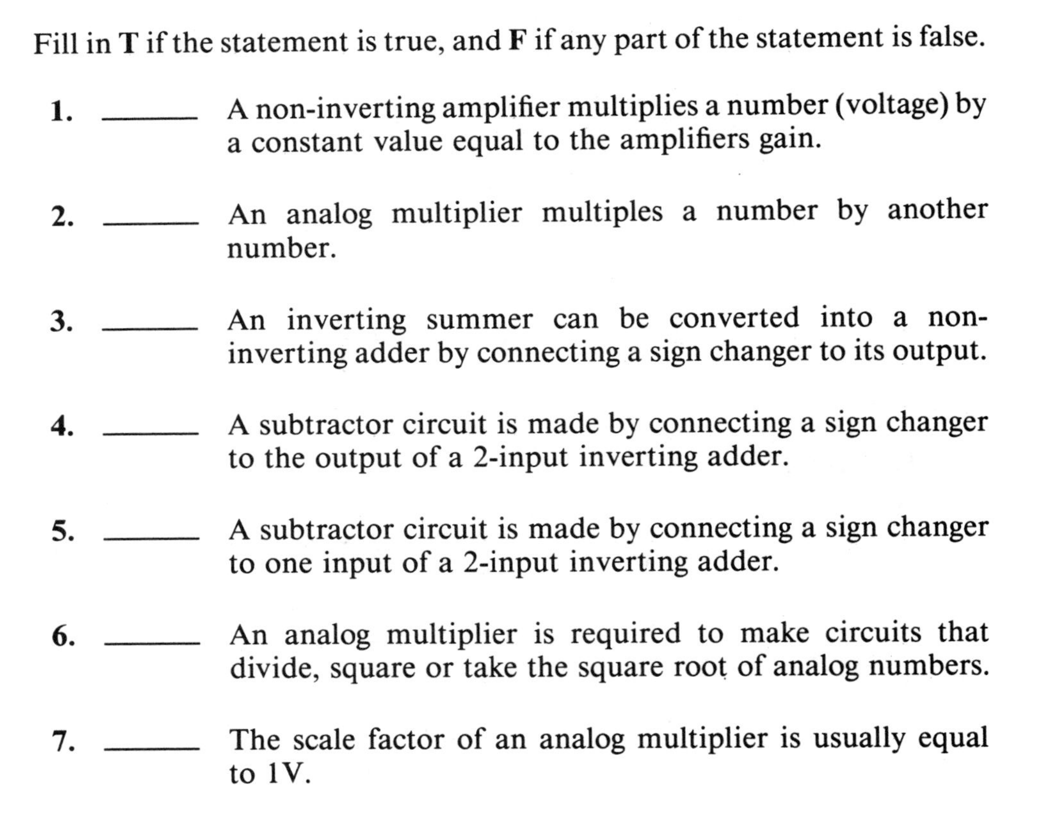

Fill in **T** if the statement is true, and **F** if any part of the statement is false.

1. ______ A non-inverting amplifier multiplies a number (voltage) by a constant value equal to the amplifiers gain.

2. ______ An analog multiplier multiples a number by another number.

3. ______ An inverting summer can be converted into a non-inverting adder by connecting a sign changer to its output.

4. ______ A subtractor circuit is made by connecting a sign changer to the output of a 2-input inverting adder.

5. ______ A subtractor circuit is made by connecting a sign changer to one input of a 2-input inverting adder.

6. ______ An analog multiplier is required to make circuits that divide, square or take the square root of analog numbers.

7. ______ The scale factor of an analog multiplier is usually equal to 1V.

C. CIRCLE THE CORRECT ANSWER

Circle the correct answer for each statement.

1. A scaled addition amplifier has channels whose gains are set by the ratio of R_F to the channels input resistor. To make one begin by choosing 10 kΩ for (R_F, R_i of the lowest gain channel, R_i of the highest gain channel).
2. A sign changer circuit is identical to a (voltage follower, inverting amplifier with $R_F = R_i$).
3. Input currents of an inverting adder (do, do not) interact with one another.
4. The input voltage to an analog square rooter circuit must always be (negative, positive).
5. Assume audio signals are applied to the V_x input of an analog multiplier with SF=10 V. If V_y was varied from 0 to 10 V the audio volume of V_o would (increase, decrease) linearly.
6. The input voltage limits for analog multipliers are typically (±1 V, ±10 V), if SF=10 V.

D. MATCHING

Match the name or symbol in column **A** with the statement that matches best in column **B**.

	COLUMN A		COLUMN B
1. ______	$\frac{V_xV_y}{10}$	**a.**	offset voltage
2. ______	$-10V\frac{Y_z}{V_x}$	**b.**	summer
3. ______	Y_{z2}	**c.**	multiplier
4. ______	V_1-V_2	**d.**	divider
5. ______	V_1+V_2	**e.**	subtractor

PROBLEMS

6-1. (a) Draw the complete circuit schematic for a sign changer with an input resistance of 100 kΩ.
(b) Improve the design in part (a) so that R_{in} approaches ∞ Ω.
(c) What is V_o if $V_i = -4.5$ V is applied to either circuit (a) or circuit (b) of Fig. 6-2?

6-2. $V_i = 3$ V sin $6280t$ in Fig. 6-2a. Solve for V_o.

6-3. Write the general expression for V_o in Fig. 6-3 if only two inputs are present.

6-4. Write the general expression for
(a) V_o' and
(b) V_o of Fig. 6-4 if four inputs are present.

6-5. Design an inverting summing amplifier like Fig. 6-5 to add $V_1 = -0.6$ V and $V_2 = -0.9$ V. Input resistance is to be 25 kΩ.

6-6. Design a noninverting summing amplifier to add $V_1 = +3$ V, $V_2 = 0.5$ V and $V_3 = -1.6$ V. Input resistance is to be 50 kΩ.

6-7. Draw the complete circuit to average $V_1 = 4$ V, $V_2 = +1.5$ V, and $V_3 = +3.5$ V. R_{in} must equal 100 kΩ. V_o should have no inversion.

6-8. Repeat Example 6-4 but set A_1 to 0.5 and A_2 to 3 (see Fig. 6-7).

6-9. Design an analog subtractor like Fig. 6-8 to subtract $V_2 = -2$ V from $V_1 = +3$ V. Show the complete circuit schematic with all voltage levels.

6-10. The following questions relate to the AD534 four-quadrant multiplier chip.
(a) How much supply current is required by pin 14 at idle?
(b) At what current level is the output terminal (pin 12) protected against short circuit?

(c) What is the scale factor SF if pin 4 is left open?
(d) Which pin can be used to provide an output offset?

6-11. Write the complete expression for V_o to multiply two differential input signals and provide an output offset with the AD534.

6-12. What value of R_{SF} must be added to pin 4 to change the scale factor to

(a) 4 V;
(b) 6 V;
(c) 8 V.

6-13. Repeat Example 6-9 for V_y = 5 V sin 377*t* and SF = 5V.

6-14. Repeat Example 6-11 if V_y is changed to 10 V sin 377*t* and V_{z2} = −3 V.

6-15. Draw the complete AD534 circuit to square two numbers. The scale factor SF must equal 6 V and an output offset of −2 V must be provided.

6-16. Solve for V_o in Fig. 6-16 if V_z = 3.5 V sin 1000*t* and V_x = 10 V.

Laboratory Exercise 6

Name: ____________________

Date: __________ **Grade:** ___

ADDITION, SUBTRACTION, MULTIPLICATION, AND DIVISION

OBJECTIVES: Upon completion of this laboratory exercise on mathematical operations, you will be able to (1) design, build, and test several adder/averaging circuits; (2) design, build, and test a two-op-amp subtractor; (3) use the AD534 as a multiplier; and (4) use the AD534 as a squaring circuit.

REFERENCE: Chapter 6

PARTS LIST

- 1 LM324 quad op amp
- 2 OP-07 op amps (see data sheets 207-210)
- 1 AD534 IC Multiplier (see data sheets 211-16)
- 2 10-kΩ potentiometers
- 2 3.9-kΩ resistor
- 1 10-kΩ resistor
- 1 12-kΩ resistor
- 1 4.99-kΩ 1% resistor
- 2 10-kΩ 1% resistors
- 4 15-kΩ 1% resistors
- 2 100-kΩ 1% resistors

Procedure A: Addition, Averaging Circuits

1. The buffered voltage-divider circuit shown in Fig. L6-1 will be used to test several of the circuits in this lab. Wire Fig. L6-1 using three of the four op amps (in an LM324 package) as voltage followers to "buffer" the divider network. As can be seen in Fig. L6-2, pin 4 should be connected to $V_{CC} = +15$ V and pin 11 is connected to $V_{EE} = -15$ V.
2. With power applied, accurately measure (with respect to ground) and record V_1, V_2, and V_3 on Fig. L6-1. Use these values in all calculations to follow.
3. Build the basic three-input inverting and noninverting adder circuit shown in Fig. L6-3. Inputs V_1, V_2, and V_3 to this circuit are developed by the circuit shown in Fig. L6-1. With all resistors of the inverting

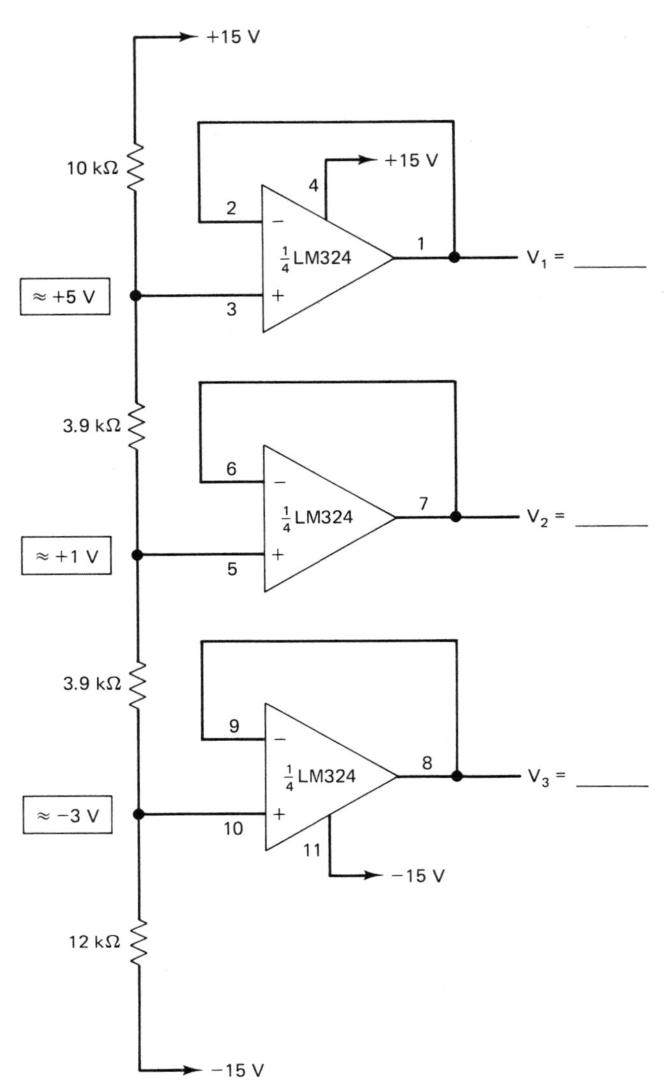

Figure L6-1 Buffered Voltage Divider

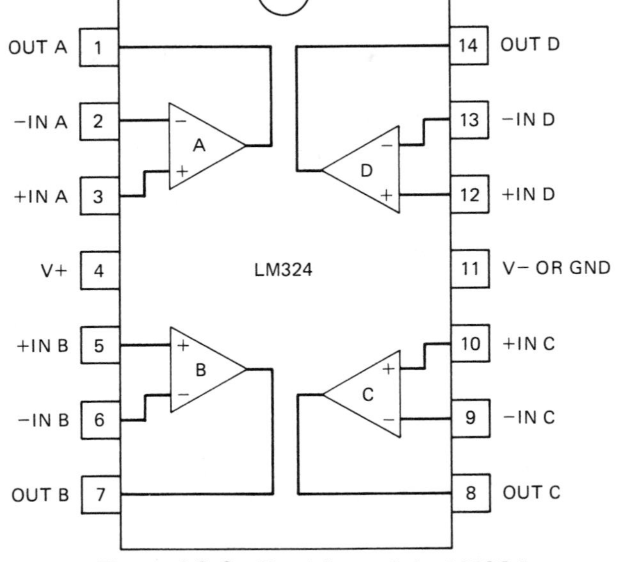

Figure L6-2 Top View of the LM324

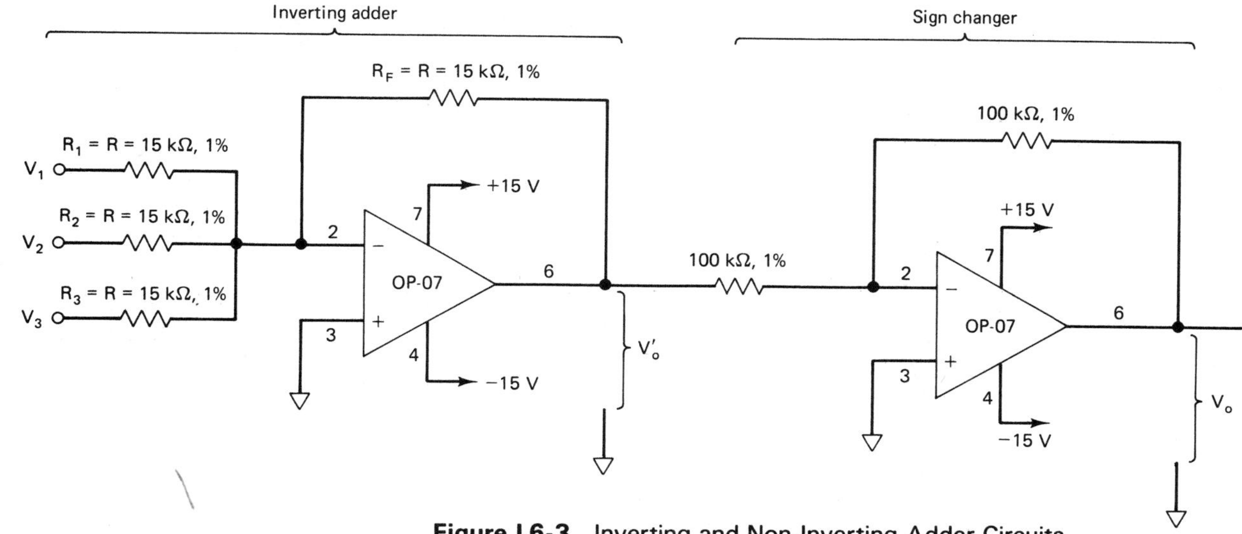

Figure L6-3 Inverting and Non-Inverting Adder Circuits

adder equal to 15 kΩ, the output voltage V_o' can be predicted by the equation $V_o' = -(V_1 + V_2 + V_3)$. $V_o' =$ __________.

4. Measure V_o'. $V_o' =$ __________. Do the calculated and measured values compare favorably? __________.
5. When the sign changer is added as shown in Fig. L6-3, the output voltage V_o is predicted by the equation $V_o = (V_1 + V_2 + V_3)$. $V_o =$ __________.
6. Measure V_o. $V_o =$ __________. Do both calculated and measured values of V_o compare favorably? __________.
7. Figure L6-3 can be redesigned into a noninverting *averager* by changing the 15-kΩ feedback resistor to R_F to $R_F = R/_n = 15\ k\Omega/3 = 5\ k\Omega$. Replace R_F with a 4.99-kΩ 1% resistor.
8. The output voltage V_o of a noninverting averager is calculated using the equation $V_o = \dfrac{(V_1 + V_2 + V_3)}{3}$. $V_o =$ __________.
9. Measure V_o. $V_o =$ __________. Do both calculated and measured values of V_o compare favorably? __________.

Procedure B: Two-Op-Amp Subtractor

10. Build the two-op-amp subtractor shown in Fig. L6-4. The inputs V_1 and V_2 to this circuit are developed by the circuit shown in Fig. L6-1. The output voltage of A_1 should be precisely $-(V_1)$ when the input

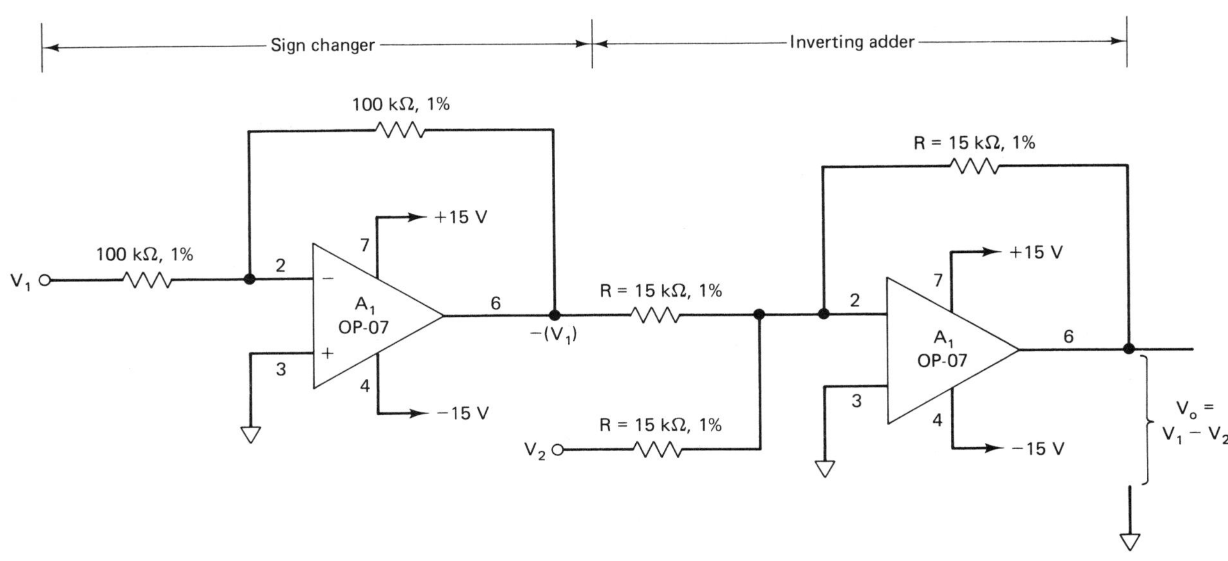

Figure L6-4 Two Op Amp Subtractor

to this sign changer is $+V_1$. Measure both input and output voltages.

$V_1 =$ __________; $-(V_1) =$ __________.

11. With all resistors of the inverting adder section (A_2) equal to 15 kΩ, the output of A_2 at pin 6 can be predicted from the equation

$V_o = V_1 - V_2$. Measure and record V_2. $V_2 =$ __________.

Calculate the value of V_o. $V_o =$ __________.

12. Measure V_o. $V_o =$ __________. Do the calculated and measured

values of V_o compare favorably? __________.

13. Replace V_2 of Fig. L6-4 with V_3 (from Fig. L6-1). Measure and record

this value. $V_3 =$ __________. Predict V_o from the equation $V_o = V_1$

$- V_3$. $V_o =$ __________.

14. Measure V_o. $V_o =$ __________. Do the calculated and measured

values of V_o compare favorably? __________.

Procedure C: Using the AD534 as a Multiplier

15. Build the multiplier circuit shown in Fig. L6-5. Apply power and adjust V_y to 5 V (peak) at 1000 Hz (i.e., $V_y = 5$ V sin $6280t$). Next adjust R_{vol} until $V_x = 0$ V. Then connect point A to ground (pin 10 grounded), and finally connect both CRO, channels as shown.

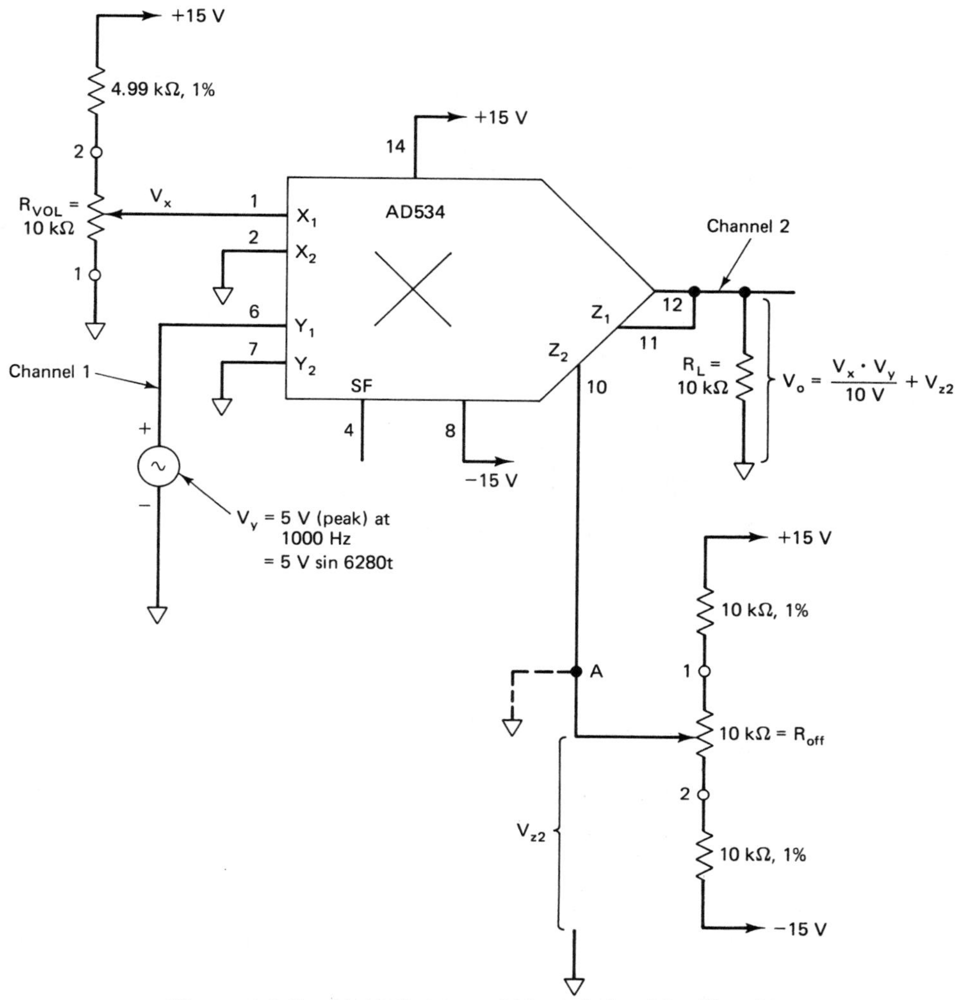

Figure L6-5 Multiplication of V_x and V_y with offset V_z.

16. With pin 10 grounded (V_{z2} = 0 V), the output voltage of the multiplier is predicted by the equation

$$V_o = \frac{V_x \cdot V_y}{10\text{ V}}$$

Adjust R_{vol} until $V_x = 1.0$ V. Use the equation above to predict V_o. V_o = ________. Measure the peak output voltage at pin 12 with channel 2 of the CRO. Is V_o measured the same as V_o calculated? ________

17. Increase R_{vol} until $V_x = 10$ V (or as close as possible). Again predict V_o. V_o = ________. Measure V_o and compare results. V_o = ________. Explain the function of R_{vol}. ________

____

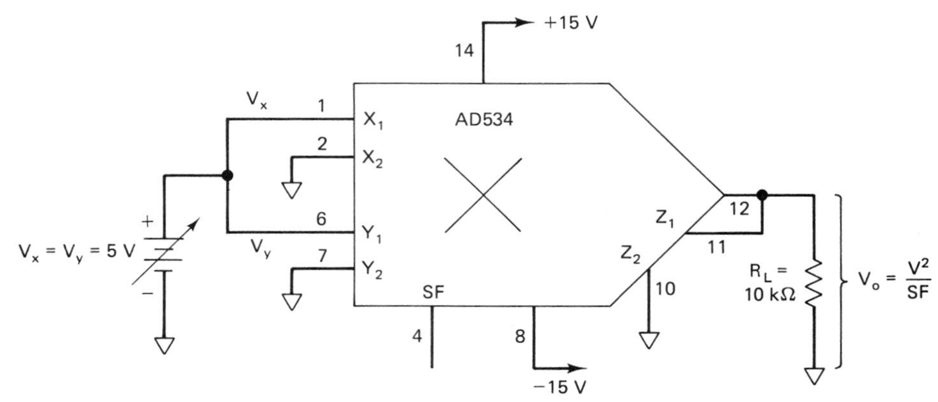

Figure L6-6 Squaring Circuit

18. Remove Z_2 (pin 10) from ground. While monitoring V_o, slowly adjust R_{off} until $V_{z2} = +3$ V. Readjust R_{off} until $V_{z2} = -3$ V.

19. Explain the function of R_{off}. ______________________________

____

Procedure D: Using the AD534 as a Squaring Circuit

20. Build Fig. L6-6. Assume that $V_x = V_y = 5$ V and predict V_o using the equation $V_o = \frac{V^2}{SF}$. $V_o =$ __________.

21. Apply power to Fig. L6-6. Adjust $V_x = V_y$ to 5.0 V. Measure and record V_o. $V_o =$ __________.

22. Reverse the polarity of $V_x = V_y$. Predict V_o. $V_o =$ __________. Measure and record V_o. $V_o =$ __________. Is Fig. L6-6 a true squaring circuit? __________.

CONCLUSION:

Ultra-Low Offset Voltage Op Amp

AD OP-07

FEATURES
Ten Times More Gain Than Other OP-07 Devices (3.0M min)
Ultra-Low Offset Voltage: 10μV
Ultra-Low Offset Voltage Drift: 0.2μV/°C
Ultra-Stable vs. Time: 0.2μV/month
Ultra-Low Noise: 0.35μV p-p
No External Components Required
Monolithic Construction
High Common Mode Input Range: ±14.0V
Wide Power Supply Voltage Range: ±3V to ±18V
Fits 725, 108A/308A, 741 Sockets

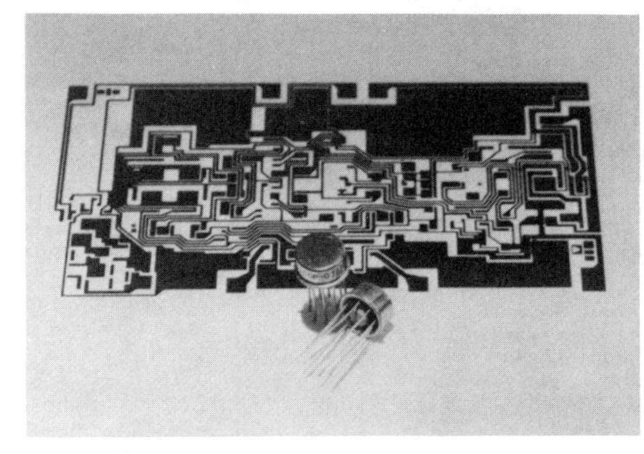

PRODUCT DESCRIPTION

The AD OP-07 is an improved version of the industry-standard OP-07 precision operational amplifier. A guaranteed minimum open-loop voltage gain of 3,000,000 (AD OP-07A) represents an order of magnitude improvement over older designs; this affords increased accuracy in high closed loop gain applications. Input offset voltages as low as 10μV, bias currents of 0.7nA, internal compensation and device protection eliminate the need for external components and adjustments. An input offset voltage temperature coefficient of 0.2μV/°C and long-term stability of 0.2μV/month eliminate recalibration or loss of initial accuracy.

A true differential operational amplifier, the AD OP-07 has a high common mode input voltage range (±14V) high common mode rejection ratio (up to 126dB) and high differential input impedance (50MΩ); these features combine to assure high accuracy in noninverting configurations. Such applications include instrumentation amplifiers, where the increased open-loop gain maintains high linearity at high closed-loop gains.

The AD OP-07 is available in five performance grades. The AD OP-07E, AD OP-07C and AD OP-07D are specified for operation over the 0 to +70°C temperature range, while the AD OP-07A and AD OP-07 are specified for −55°C to +125°C operation. Full processing to the requirements of MIL-STD-883, Class B, is available on the AD OP-07 and AD OP-07A. All devices are packaged in TO-99 hermetically-sealed metal cans.

PRODUCT HIGHLIGHTS

1. Increased open-loop voltage gain (3.0 million, min) results in better accuracy and linearity in high closed-loop gain applications.
2. Ultra-low offset voltage and offset voltage drift, combined with low input bias currents, allow the AD OP-07 to maintain high accuracy over the entire operating temperature range.
3. Internal frequency compensation, ultra-low input offset voltage and full device protection eliminate the need for additional components. This reduces circuit size and complexity and increases reliability.
4. High input impedances, large common mode input voltage range and high common mode rejection ratio make the AD OP-07 ideal for noninverting and differential instrumentation applications.
5. Monolithic construction along with advanced circuit design and processing techniques result in low cost.
6. The input offset voltage is trimmed at the wafer stage. Unmounted chips are available for hybrid circuit applications.

Route 1 Industrial Park; P.O. Box 280; Norwood, Mass. 02062
Tel: 617/329-4700 **TWX: 710/394-6577**
West Coast 714/842-1717 **Mid-West 312/894-3300** **Texas 214/231-5094**

*Courtesy of **Analog Devices Inc.**

SPECIFICATIONS (T_A = +25°C, V_S = ±15V, unless otherwise specified)

MODEL		AD OP-07EH			AD OP-07CH			AD OP-07DH		
PARAMETER	SYMBOL	MIN	TYP	MAX	MIN	TYP	MAX	MIN	TYP	MAX
OPEN LOOP GAIN	A_{VO}	2,000	5,000		1,200	4,000		1,200	4,000	
		1,800	4,500		1,000	4,000		1,000	4,000	
		300	1,000		300	1,000		300	1,000	
OUTPUT CHARACTERISTICS										
Maximum Output Swing	V_{OM}	±12.5	±13.0		±12.0	±13.0		±12.0	±13.0	
		±12.0	±12.8		±11.5	±12.8		±11.5	±12.8	
		±10.5	±12.0			±12.0				
		±12.0	±12.6		±11.0	±12.6		±11.0	±12.6	
Open-Loop Output Resistance	R_O		60			60			60	
FREQUENCY RESPONSE										
Closed Loop Bandwidth	BW		0.6			0.6			0.6	
Slew Rate	SR		0.17			0.17			0.17	
INPUT OFFSET VOLTAGE										
Initial	V_{OS}		30	75		60	150		60	150
			45	130		85	250		85	250
Adjustment Range			±4			±4			±4	
Average Drift										
No External Trim	TCV_{OS}		0.3	1.3		0.5	(Note 3) 1.8		0.7	(Note 3) 2.5
With External Trim	TCV_{OSN}		0.3	1.3		0.4	1.6 (Note 3)		0.7	2.5 (Note 3)
Long Term Stability	V_{OS}/Time		0.3	1.5		0.4	2.0		0.5	3.0
INPUT OFFSET CURRENT										
Initial	I_{OS}		0.5	3.8		0.8	6.0		0.8	6.0
			0.9	5.3		1.6	8.0		1.6	8.0
Average Drift	TCI_{OS}		8	(Note 3) 35		12	(Note 3) 50		12	(Note 3) 50
INPUT BIAS CURRENT										
Initial	I_B		±1.2	±4.0		±1.8	±7.0		±2.0	±12
			±1.5	±5.5		±2.2	±9.0		±3.0	±14
Average Drift	TCI_B		13	(Note 3) 35		18	(Note 3) 50		18	(Note 3) 50
INPUT RESISTANCE										
Differential	R_{IN}	15	50		8	33		7	31	
Common Mode	R_{IN} CM		160			120			120	
INPUT NOISE										
Voltage	e_n p-p		0.35	0.6		0.38	0.65		0.38	0.65
Voltage Density	e_n		10.3	18.0		10.5	20.0		10.5	20.0
			10.0	13.0		10.2	13.5		10.2	13.5
			9.6	11.0		9.8	11.5		9.8	11.5
Current	i_n p-p		14	30		15	35		15	35
Current Density	i_n		0.32	0.80		0.35	0.90		0.35	0.90
			0.14	0.23		0.15	0.27		0.15	0.27
			0.12	0.17		0.13	0.18		0.13	0.18
INPUT VOLTAGE RANGE										
Common Mode	CMVR	±13.0	±14.0		±13.0	±14.0		±13.0	±14.0	
		±13.0	±13.5		±13.0	±13.5		±13.0	±13.5	
Common Mode Rejection Ratio	CMRR	106	123		100	120		94	110	
		103	123		97	120		94	106	
POWER SUPPLY										
Current, Quiescent	I_Q		3.0	4.0		3.5	5.0		3.5	5.0
Power Consumption	P_D		90	120		105	150		105	150
			6.0	8.4		6.0	8.4		6.0	8.4
Rejection Ratio	PSRR	94	107		90	104		90	104	
		90	104		86	100		86	100	
OPERATING TEMPERATURE RANGE	T_{min}, T_{max}	0		+70	0		+70	0		+70
PRICES										
(1–24)			$14.65			$10.15			$ 7.35	
(25–99)			$11.70			$ 8.10			$ 5.85	
(100+)			$ 9.75			$ 6.50			$ 4.55	

NOTES:

1) Input offset voltage measurements are performed by automated test equipment approximately 0.5 seconds after application of power. Additionally, AD OP-07A offset voltage is measured five minutes after power supply application at 25°C, –55°C and +125°C.
2) Long Term Input Offset Voltage Stability refers to the averaged trend line of V_{OS} vs Time over extended periods of time and is extrapolated from high temperature test data. Excluding the initial hour of operation, changes in V_{OS} during the first 30 operating days are typically 2.5μV – Parameter is not 100% tested: 90% of units meet this specification.
3) Parameter is not 100% tested; 90% of units meet this specification.
4) The AD OP-07A and AD OP-07 are available fully processed to MIL-STD-883, Class B. Order AD OP-07-AH-883B or AD OP-07-H-883B.

Specifications subject to change without notice.

AD OP-07AH (AD OP-07-AH-883B)[4]			AD OP-07H (AD OP-07-H-883B)[4]				
MIN	TYP	MAX	MIN	TYP	MAX	TEST CONDITIONS	UNITS
3,000	5,000		2,000	5,000		$R_L \geqslant 2k\Omega$, $V_O = \pm 10V$	V/mV
2,000	4,000		1,500	4,000		$R_L \geqslant 2k\Omega$, $V_O = \pm 10V$, T_{min} to T_{max}	V/mV
300	1,000		300	1,000		$R_L \geqslant 500\Omega$, $V_O = \pm 0.5V$, $V_S = \pm 3V$	V/mV
±12.5	±13.0		±12.5	±13.0		$R_L \geqslant 10k\Omega$	V
±12.0	±12.8		±12.0	±12.8		$R_L \geqslant 2k\Omega$	V
±10.5	±12.0		±10.5	±12.0		$R_L \geqslant 1k\Omega$	V
±12.0	±12.6		±12.0	±12.6		$R_L \geqslant 2k\Omega$, T_{min} to T_{max}	V
	60			60		$V_O = 0$, $I_O = 0$	Ω
	0.6			0.6		$A_{VCL} = +1.0$	MHz
	0.17			0.17		$R_L \geqslant 2k$	V/μs
	10	25		30	75	Note 1	μV
	25	60		60	200	Note 1, T_{min} to T_{max}	μV
	±4			±4		$R_P = 20k\Omega$	mV
	0.2	0.6		0.3	1.3	T_{min} to T_{max}	μV/°C
	0.2	0.6		0.3	1.3	$R_P = 20k\Omega$, T_{min} to T_{max}	μV/°C
	0.2	1.0		0.2	1.0	Note 2	μV/Month
	0.3	2.0		0.4	2.8		nA
	0.8	4.0		1.2	5.6	T_{min} to T_{max}	nA
	5	25		8	50	T_{min} to T_{max}	pA/°C
	±0.7	±2.0		±1.0	±3.0		nA
	±1.0	±4.0		±2.0	±6.0	T_{min} to T_{max}	nA
	8	25		13	50	T_{min} to T_{max}	pA/°C
30	80		20	60			MΩ
	200			200			GΩ
	0.35	0.6		0.35	0.6	0.1Hz to 10Hz, Note 3	μV p-p
	10.3	18.0		10.3	18.0	$f_O = 10Hz$, Note 3	$nV/\sqrt{Hz}$
	10.0	13.0		10.0	13.0	$f_O = 100Hz$, Note 3	$nV/\sqrt{Hz}$
	9.6	11.0		9.6	11.0	$f_O = 1kHz$, Note 3	$nV/\sqrt{Hz}$
	14	30		14	30	0.1Hz to 10Hz, Note 3	pA p-p
	0.32	0.80		0.32	0.80	$f_O = 10Hz$, Note 3	$pA/\sqrt{Hz}$
	0.14	0.23		0.14	0.23	$f_O = 100Hz$, Note 3	$pA/\sqrt{Hz}$
	0.12	0.17		0.12	0.17	$f_O = 1kHz$, Note 3	$pA/\sqrt{Hz}$
±13.0	±14.0		±13.0	±14.0			V
±13.0	±13.5		±13.0	±13.5		T_{min} to T_{max}	V
110	126		110	126		$V_{CM} = \pm CMVR$	dB
106	123		106	123		$V_{CM} = \pm CMVR$, T_{min} to T_{max}	dB
	3.0	4.0		3.0	4.0	$V_S = \pm 15V$	mA
	90	120		90	120	$V_S = \pm 15V$	mW
	6.0	8.4		6.0	8.4	$V_S = \pm 3V$	mW
100	110		100	110		$V_S = \pm 3V$ to $\pm 18V$	dB
94	106		94	106		$V_S = \pm 3V$ to $\pm 18V$, T_{min} to T_{max}	dB
-55		+125	-55		+125		°C
	$68.00 ($73.00)			$29.75 ($34.75)			
	$55.00 ($59.50)			$24.00 ($28.50)			
	$45.00 ($49.00)			$19.50 ($23.50)			

ABSOLUTE MAXIMUM RATINGS

Supply Voltage . . . . ±22V
Internal Power Dissipation (Note 1) 500mW
Differential Input Voltage . . . . ±30V
Input Voltage (Note 2) . . . . ±22V
Output Short Circuit Duration . . . . Indefinite
Storage Temperature Range . . . . −65°C to +150°C
Operating Temperature Range
OP-07A, OP-07 . . . . −55°C to +125°C
OP-07E, OP-07C, OP-07D . . . . 0 to +70°C
Lead Temperature Range (Soldering, 60sec) . . . . 300°C

NOTES:
Note 1: Maximum package power dissipation vs. ambient temperature.

Package Type	Maximum Ambient Temperature for Rating	Derate Above Maximum Ambient Temperature
TO-99 (H)	80°C	7.1mW/°C

Note 2: For supply voltages less than ±22V, the absolute maximum input voltage is equal to the supply voltage.

OUTLINE DIMENSIONS

Dimensions shown in inches and (mm).

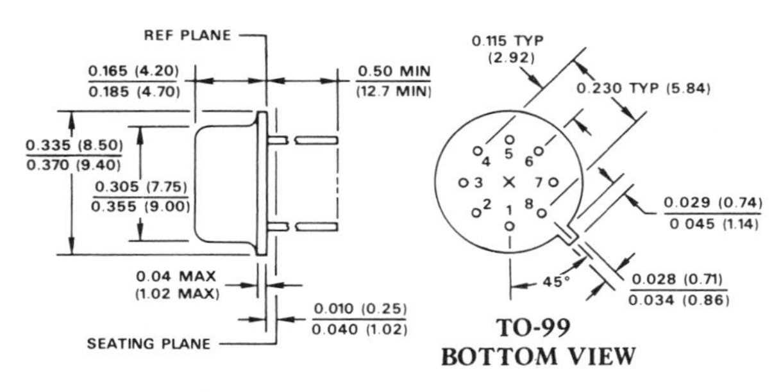

H-PACKAGE

CHIP DIMENSIONS AND BONDING DIAGRAM

Dimensions shown in inches and (mm).

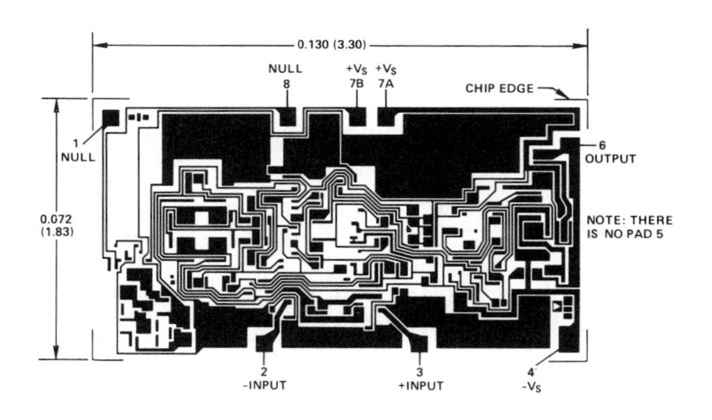

PIN CONFIGURATION TOP VIEW

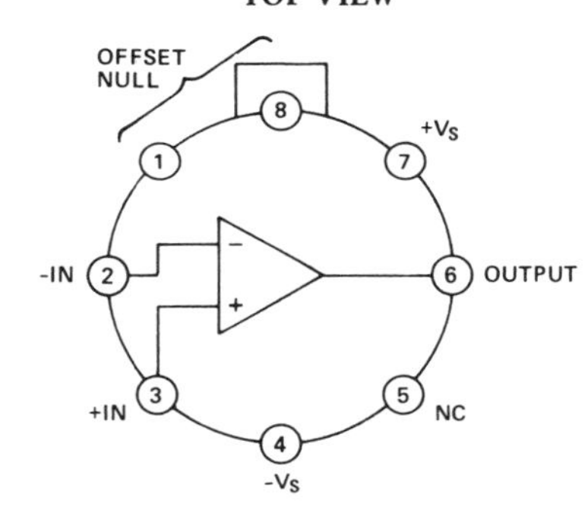

The AD OP-07 is available in wafer-trimmed chip form for precision hybrids. Consult the factory directly for details.

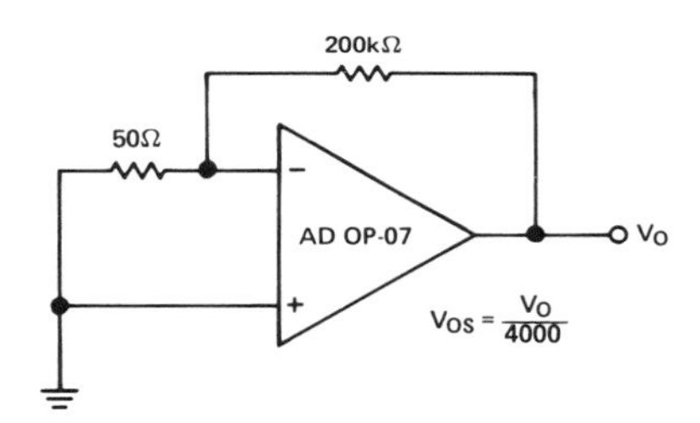

Offset Voltage Test Circuit

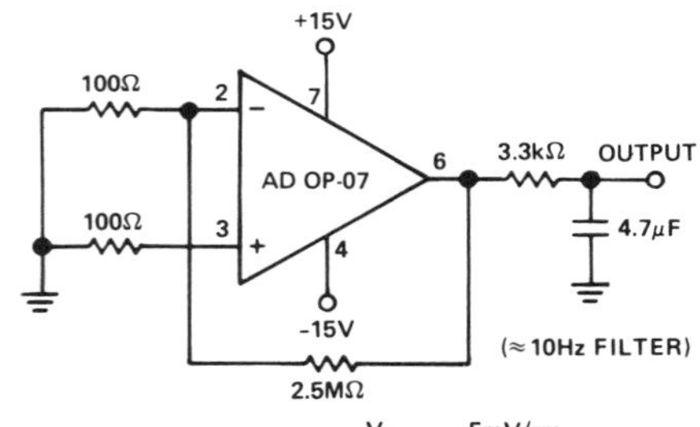

Low Frequency Noise Test Circuit

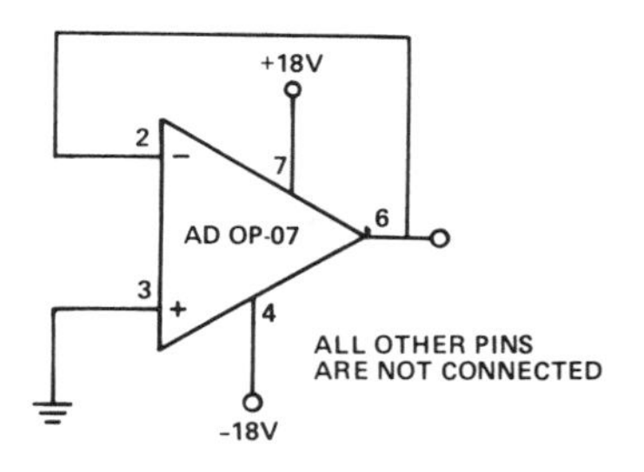

Burn-In Circuit

Internally Trimmed Precision IC Multiplier

AD534

FEATURES

- Pretrimmed to ±0.25% max 4-Quadrant Error (AD534L)
- All Inputs (X, Y and Z) Differential, High Impedance for $[(X_1 - X_2)(Y_1 - Y_2)/10V] + Z_2$ Transfer Function
- Scale-Factor Adjustable to Provide up to X100 Gain
- Low Noise Design: 90μV rms, 10Hz–10kHz
- Low Cost, Monolithic Construction
- Excellent Long Term Stability

APPLICATIONS

- High Quality Analog Signal Processing
- Differential Ratio and Percentage Computations
- Algebraic and Trigonometric Function Synthesis
- Wideband, High-Crest rms-to-dc Conversion
- Accurate Voltage Controlled Oscillators and Filters

AD534 PIN CONFIGURATIONS

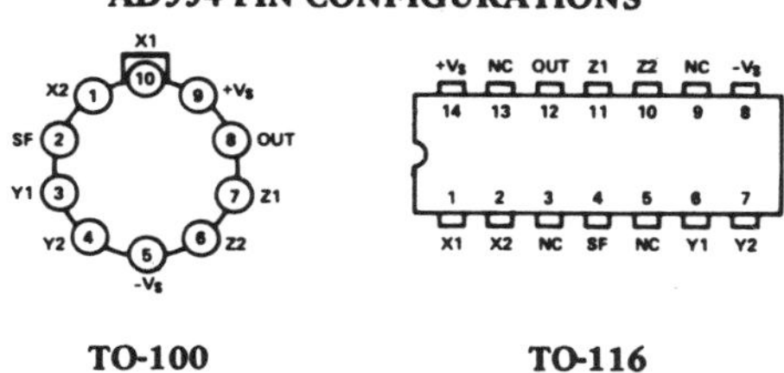

PRODUCT DESCRIPTION

The AD534 is a monolithic laser trimmed four-quadrant multiplier divider having accuracy specifications previously found only in expensive hybrid or modular products. A maximum multiplication error of ±0.25% is guaranteed for the AD534L without any external trimming. Excellent supply rejection, low temperature coefficients and long term stability of the on-chip thin film resistors and buried zener reference preserve accuracy even under adverse conditions of use. It is the first multiplier to offer fully differential, high impedance operation on all inputs, including the Z-input, a feature which greatly increases its flexibility and ease of use. The scale factor is pretrimmed to the standard value of 10.00V; by means of an external resistor, this can be reduced to values as low as 3V.

The wide spectrum of applications and the availability of several grades commend this multiplier as the first choice for all new designs. The AD534J (±1% max error), AD534K (±0.5% max) and AD534L (±0.25% max) are specified for operation over the 0 to +70°C temperature range. The AD534S (±1% max) and AD534T (±0.5% max) are specified over the extended temperature range, -55°C to +125°C. All grades are available in hermetically sealed TO-100 metal cans and TO-116 ceramic DIP packages.

PROVIDES GAIN WITH LOW NOISE

The AD534 is the first general purpose multiplier capable of providing gains up to X100, frequently eliminating the need for separate instrumentation amplifiers to precondition the inputs. The AD534 can be very effectively employed as a variable gain differential input amplifier with high common mode rejection. The gain option is available in all modes, and will be found to simplify the implementation of many function-fitting algorithms such as those used to generate sine and tangent. The utility of this feature is enhanced by the inherent low noise of the AD534: 90μV, rms (depending on the gain), a factor of 10 lower than previous monolithic multipliers. Drift and feedthrough are also substantially reduced over earlier designs.

UNPRECEDENTED FLEXIBILITY

The precise calibration and differential Z-input provide a degree of flexibility found in no other currently available multiplier. Standard MDSSR functions (multiplication, division, squaring, square-rooting) are easily implemented while the restriction to particular input/output polarities imposed by earlier designs has been eliminated. Signals may be summed into the output, with or without gain and with either a positive or negative sense. Many new modes based on implicit-function synthesis have been made possible, usually requiring only external passive components. The output can be in the form of a current, if desired, facilitating such operations as integration.

*Courtesy of **Analog Devices Inc.**

SPECIFICATIONS ($T_A = +25°C$, $\pm V_S = 15V$, $R \geq 2k\Omega$)

Model	AD534J Min	AD534J Typ	AD534J Max	AD534K Min	AD534K Typ	AD534K Max	AD534L Min	AD534L Typ	AD534L Max	Units
MULTIPLIER PERFORMANCE										
Transfer Function		$\frac{(X_1-X_2)(Y_1-Y_2)}{10V}+Z_2$			$\frac{(X_1-X_2)(Y_1-Y_2)}{10V}+Z_2$			$\frac{(X_1-X_2)(Y_1-Y_2)}{10V}+Z_2$		
Total Error[1] ($-10V \leq X, Y \leq +10V$)			**±1.0**			**±0.5**			**±0.25**	%
T_A = min to max		±1.5			±1.0			±0.5		%
Total Error vs Temperature		±0.022			±0.015			±0.008		%/°C
Scale Factor Error (SF = 10.000V Nominal)[2]		±0.25			±0.1			±0.1		%
Temperature-Coefficient of Scaling-Voltage		±0.02			±0.01			±0.005		%/°C
Supply Rejection (±15V ±1V)		±0.01			±0.01			±0.01		%
Nonlinearity, X (X = 20V pk-pk, Y = 10V)		±0.4			±0.2	**±0.3**		±0.10	**±0.12**	%
Nonlinearity, Y (Y = 20V pk-pk, X = 10V)		±0.01			±0.1	**±0.1**		±0.005	**±0.1**	%
Feedthrough[3], X (Y Nulled, X = 20V pk-pk 50Hz)		±0.3			±0.15	**±0.3**		±0.05	**±0.12**	%
Feedthrough[3], Y (X Nulled, Y = 20V pk-pk 50Hz)		±0.01			±0.01	**±0.1**		±0.003	**±0.1**	%
Output Offset Voltage		±5	**±30**		±2	**±15**		±2	**±10**	mV
Output Offset Voltage Drift		200			100			100		µV/°C
DYNAMICS										
Small Signal BW, (V_{OUT} = 0.1 rms)		1			1			1		MHz
1% Amplitude Error (C_{LOAD} = 1000pF)		50			50			50		kHz
Slew Rate (V_{OUT} 20 pk-pk)		20			20			20		V/µs
Settling Time (to 1%, ΔV_{OUT} = 20V)		2			2			2		µs
NOISE										
Noise Spectral-Density SF = 10V		0.8			0.8			0.8		$\mu V/\sqrt{Hz}$
SF = 3V[4]		0.4			0.4			0.4		$\mu V/\sqrt{Hz}$
Wideband Noise f = 10Hz to 5MHz		1			1			1		mV/rms
f = 10Hz to 10kHz		90			90			90		µV/rms
OUTPUT										
Output Voltage Swing	**±11**			**±11**			**±11**			V
Output Impedance (f ≤ 1kHz)		0.1			0.1			0.1		Ω
Output Short Circuit Current (R_L = 0, T_A = min to max)		30			30			30		mA
Amplifier Open Loop Gain (f = 50Hz)		70			70			70		dB
INPUT AMPLIFIERS (X, Y and Z)[5]										
Signal Voltage Range (Diff. or CM		±10			±10			±10		V
Operating Diff.)		±12			±12			±12		V
Offset Voltage X, Y		±5	**±20**		±2	**±10**		±2	**±10**	mV
Offset Voltage Drift X, Y		100			50			50		µV/°C
Offset Voltage Z		±5	**±30**		±2	**±15**		±2	**±10**	mV
Offset Voltage Drift Z		200			100			100		µV/°C
CMRR	**60**	80		**70**	90		**70**	90		dB
Bias Current		0.8	**2.0**		0.8	**2.0**		0.8	**2.0**	µA
Offset Current		0.1			0.1			0.05	**0.2**	µA
Differential Resistance		10			10			10		MΩ
DIVIDER PERFORMANCE										
Transfer Function ($X_1 > X_2$)		$10V\frac{(Z_2-Z_1)}{(X_1-X_2)}+Y_1$			$10V\frac{(Z_2-Z_1)}{(X_1-X_2)}+Y_1$			$10V\frac{(Z_2-Z_1)}{(X_1-X_2)}+Y_1$		
Total Error[1] (X = 10V, $-10V \leq Z \leq +10V$)		±0.75			±0.35			±0.2		%
(X = 1V, $-1V \leq Z \leq +1V$)		±2.0			±1.0			±0.8		%
($0.1V \leq X \leq 10V$, $-10V \leq Z \leq 10V$)		±2.5			±1.0			±0.8		%
SQUARE PERFORMANCE										
Transfer Function		$\frac{(X_1-X_2)^2}{10V}+Z_2$			$\frac{(X_1-X_2)^2}{10V}+Z_2$			$\frac{(X_1-X_2)^2}{10V}+Z_2$		
Total Error ($-10V \leq X \leq 10V$)		±0.6			±0.3			±0.2		%
SQUARE-ROOTER PERFORMANCE										
Transfer Function ($Z_1 \leq Z_2$)		$\sqrt{10V(Z_2-Z_1)}+Y_2$			$\sqrt{10V(Z_2-Z_1)}+X_2$			$\sqrt{10V(Z_2-Z_1)}+X_2$		
Total Error[1] ($1V \leq Z \leq 10V$)		±1.0			±0.5			±0.25		%
POWER SUPPLY SPECIFICATIONS										
Supply Voltage										
Rated Performance		±15			±15			±15		V
Operating	**±8**		**±18**	**±8**		**±18**	**±8**		**±18**	V
Supply Current										
Quiescent		**4**	**6**		**4**	6		**4**	**6**	mA
PACKAGE OPTIONS[6]										
H: TO-100 Package		AD534JH			AD534KH			AD534LH		
D: TO-116 Package (D14A)		AD534JD			AD534KD			AD534LD		

NOTES

[1] Figures given are percent of full scale, ±10V (i.e., 0.01% = 1mV).

[2] May be reduced down to 3V using external resistor between $-V_S$ and SF.

[3] Irreducible component due to nonlinearity: excludes effect of offsets.

[4] Using external resistor adjusted to give SF = 3V.

[5] See functional block diagram for definition of sections.

[6] See Section 19 for package outline information.

Specifications subject to change without notice.

Specifications shown in boldface are tested on all production units at final electrical test. Results from those tests are used to calculate outgoing quality levels. All min and max specifications are guaranteed, although only those shown in boldface are tested on all production units.

Model	AD534S			AD534T			Units
	Min	Typ	Max	Min	Typ	Max	
MULTIPLIER PERFORMANCE							
Transfer Function		$\frac{(X_1 - X_2)(Y_1 - Y_2)}{10V} + Z_2$			$\frac{(X_1 - X_2)(Y_1 - Y_2)}{10V} + Z_2$		
Total Error[1] ($-10V \le X, Y \le +10V$)			**±1.0**			**±0.5**	%
T_A = min to max			**±2.0**			**±1.0**	%
Total Error vs Temperature			±0.02			±0.01	%/°C
Scale Factor Error							
(SF = 10.000V Nominal)[2]		±0.25			±0.1		%
Temperature-Coefficient of							
Scaling-Voltage		±0.02				±0.005	%/°C
Supply Rejection (±15V ±1V)		±0.01			±0.01		%
Nonlinearity, X (X = 20V pk-pk, Y = 10V)		±0.4			±0.2	**±0.3**	%
Nonlinearity, Y (Y = 20V pk-pk, X = 10V)		±0.01			±0.1	**±0.1**	%
Feedthrough[3], X (Y Nulled,							
X = 20V pk-pk 50Hz)		±0.3			±0.15	**±0.3**	%
Feedthrough[3], Y (X Nulled,							
Y = 20V pk-pk 50Hz)		±0.01			±0.01	**±0.1**	%
Output Offset Voltage		±5	**±30**		±2	**±15**	mV
Output Offset Voltage Drift			**500**			**300**	μV/°C
DYNAMICS							
Small Signal BW, (V_{OUT} = 0.1 rms)		1			1		MHz
1% Amplitude Error (C_{LOAD} = 1000pF)		50			50		kHz
Slew Rate (V_{OUT} 20 pk-pk)		20			20		V/μs
Settling Time (to 1%, ΔV_{OUT} = 20V)		2			2		μs
NOISE							
Noise Spectral-Density SF = 10V		0.8			0.8		μV/√Hz
SF = 3V[4]		0.4			0.4		μV/√Hz
Wideband Noise f = 10Hz to 5MHz		1.0			1.0		mV/rms
f = 10Hz to 10kHz		90			90		μV/rms
OUTPUT							
Output Voltage Swing	**±11**			**±11**			V
Output Impedance (f ≤ 1kHz)		0.1			0.1		Ω
Output Short Circuit Current							
R_L = 0, T_A = min to max)		30			30		mA
Amplifier Open Loop Gain (f = 50Hz)		70			70		dB
INPUT AMPLIFIERS (X, Y and Z)[5]							
Signal Voltage Range (Diff. or CM		±10			±10		V
Operating Diff.)		±12			±12		V
Offset Voltage X, Y		±5	**±20**		±2	**±10**	mV
Offset Voltage Drift X, Y		100			150		μV/°C
Offset Voltage Z		±5	**±30**		±2	**±15**	mV
Offset Voltage Drift Z			**500**			**300**	μV/°C
CMRR	**60**	80		**70**	90		dB
Bias Current		0.8	2.0		0.8	2.0	μA
Offset Current		0.1	**2.0**		0.1	**2.0**	μA
Differential Resistance		10			10		MΩ
DIVIDER PERFORMANCE							
Transfer Function ($X_1 > X_2$)		$10V\frac{(Z_2 - Z_1)}{(X_1 - X_2)} + Y_1$			$10V\frac{(Z_2 - Z_1)}{(X_1 - X_2)} + Y_1$		
Total Error[1] (X = 10V, $-10V \le Z \le +10V$)		±0.75			±0.75		%
(X − 1V, $-1V \le Z \le +1V$)		±2.0			±2.0		%
($0.1V \le X \le 10V$, $-10V \le$							
$Z \le 10V$)		±2.5			±1.0		%
SQUARE PERFORMANCE							
Transfer Function		$\frac{(X_1 - X_2)^2 + Z_2}{10V}$			$\frac{(X_1 - X_2)^2 + Z_2}{10V}$		
Total Error ($-10V \le X \le 10V$)		±0.6			±0.3		%
SQUARE-ROOTER PERFORMANCE							
Transfer Function ($Z_1 \le Z_2$)		$\sqrt{10V(Z_2 - Z_1)} + X_2$			$\sqrt{10V(Z_2 - Z_1)} + X_2$		
Total Error[1] ($1V \le Z \le 10V$)		±1.0			±0.5		%
POWER SUPPLY SPECIFICATIONS							
Supply Voltage							
Rated Performance		±15			±15		V
Operating	**±8**		**±22**	**±8**		**±22**	V
Supply Current							
Quiescent		4	6		4	6	mA
PACKAGE OPTIONS[6]							
H: TO-100 Package		AD534SH			AD534TH		
D: TO-116 Package (D14A)		AD534SD			AD534TD		

NOTES

[1]Figures given are percent of full scale, ± 10V (i.e., 0.01% = 1mV).
[2]May be reduced down to 3V using external resistor between $-V_S$ and SF.
[3]Irreducible component due to nonlinearity: excludes effect of offsets.
[4]Using external resistor adjusted to give SF = 3V.
[5]See functional block diagram for definition of sections.
[6]See Section 19 for package outline information.

Specifications subject to change without notice.

Specifications shown in boldface are tested on all production units at final electrical test. Results from those tests are used to calculate outgoing quality levels. All min and max specifications are guaranteed, although only those shown in boldface are tested on all production units.

Using the AD534

ABSOLUTE MAXIMUM RATINGS

	AD534J, K, L	AD534S, T
Supply Voltage	±18V	±22V
Internal Power Dissipation	500mW	*
Output Short-Circuit to Ground	Indefinite	*
Input Voltages, X_1 X_2 Y_1 Y_2 Z_1 Z_2	$\pm V_S$	*
Rated Operating Temperature Range	0 to +70°C	-55°C to +125°C
Storage Temperature Range	-65°C to +150°C	*
Lead Temperature, 60s soldering	+300°C	*

*Same as AD534J specs.

OPTIONAL TRIMMING CONFIGURATION

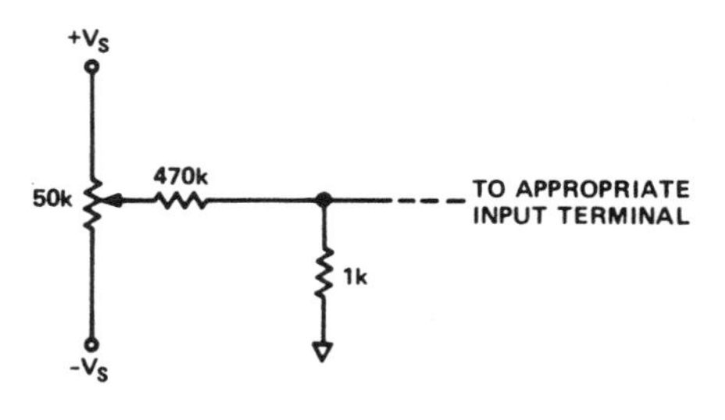

FUNCTIONAL DESCRIPTION

Figure 1 is a functional block diagram of the AD534. Inputs are converted to differential currents by three identical voltage-to-current converters, each trimmed for zero offset. The product of the X and Y currents is generated by a multiplier cell using Gilbert's translinear technique. An on-chip "Buried Zener" provides a highly stable reference, which is laser trimmed to provide an overall scale factor of 10V. The difference between XY/SF and Z is then applied to the high gain output amplifier. This permits various closed loop configurations and dramatically reduces nonlinearities due to the input amplifiers, a dominant source of distortion in earlier designs. The effectiveness of the new scheme can be judged from the fact that under typical conditions as a multiplier the nonlinearity on the Y input, with X at full scale (±10V), is ±0.005% of F.S.; even at its worst point, which occurs when X = ±6.4V, it is typically only ±0.05% of F.S. Nonlinearity for signals applied to the X input, on the other hand, is determined almost entirely by the multiplier element and is parabolic in form. This error is a major factor in determining the overall accuracy of the unit and hence is closely related to the device grade.

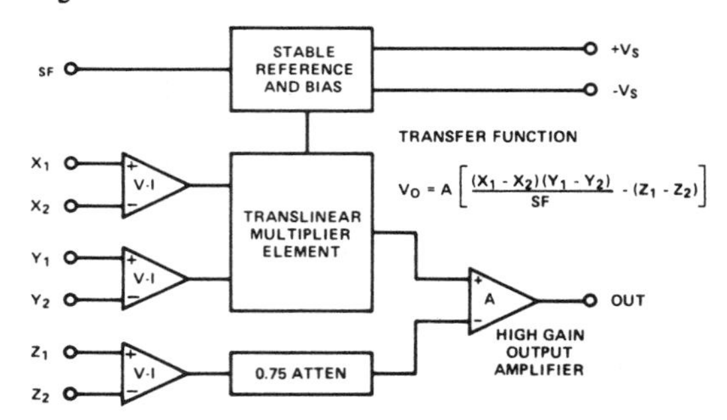

Figure 1. AD534 Functional Block Diagram

The generalized transfer function for the AD534 is given by:

$$V_{OUT} = A\left(\frac{(X_1 - X_2)(Y_1 - Y_2)}{SF} - (Z_1 - Z_2)\right)$$

where A = open loop gain of output amplifier, typically 70dB at dc

X, Y, Z = input voltages (full scale = ±SF, peak= ±1.25SF)

SF = scale factor, pretrimmed to 10.00V but adjustable by the user down to 3V.

In most cases the open loop gain can be regarded as infinite, and SF will be 10V. The operation performed by the AD534, can then be described in terms of equation:

$$(X_1 - X_2)(Y_1 - Y_2) = 10V(Z_1 - Z_2)$$

The user may adjust SF for values between 10.00V and 3V by connecting an external resistor in series with a potentiometer between SF and $-V_S$. The approximate value of the total resistance for a given value of SF is given by the relationship:

$$R_{SF} = 5.4K \frac{SF}{10 - SF}$$

Due to device tolerances, allowance should be made to vary R_{SF} by ±25% using the potentiometer. Considerable reduction in bias currents, noise and drift can be achieved by decreasing SF. This has the overall effect of increasing signal gain without the customary increase in noise. Note that the peak input signal is always limited to 1.25SF (i.e., ±5V for SF = 4V) so the overall transfer function will show a maximum gain of 1.25. The performance with small input signals, however, is improved by using a lower SF since the dynamic range of the inputs is now fully utilized. Bandwidth is unaffected by the use of this option.

Supply voltages of ±15V are generally assumed. However, satisfactory operation is possible down to ±8V (see curve 1). Since all inputs maintain a constant peak input capability of ±1.25SF some feedback attenuation will be necessary to achieve output voltage swings in excess of ±12V when using higher supply voltages.

OPERATION AS A MULTIPLIER

Figure 2 shows the basic connection for multiplication. Note that the circuit will meet all specifications without trimming.

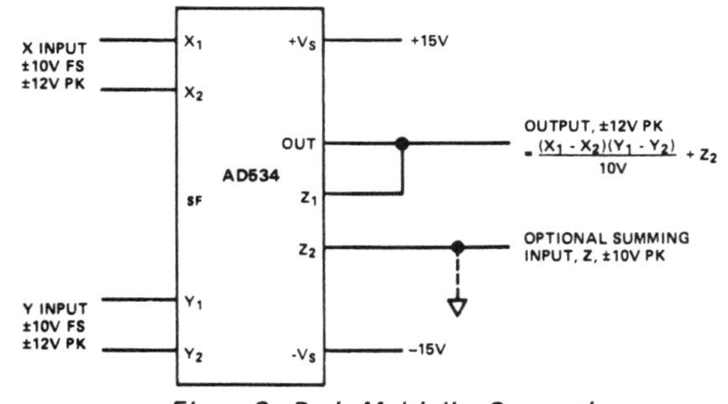

Figure 2. Basic Multiplier Connection

In some cases the user may wish to reduce ac feedthrough to a minimum (as in a suppressed carrier modulator) by applying an external trim voltage (±30mV range required) to the X or Y input (see Optional Trimming Configuration). Curve 4 shows the typical ac feedthrough with this adjustment mode. Note that the Y input is a factor of 10 lower than the X input and should be used in applications where null suppression is critical.

The high impedance Z_2 terminal of the AD534 may be used to sum an additional signal into the output. In this mode the output amplifier behaves as a voltage follower with a 1MHz small signal bandwidth and a 20V/μs slew rate. This terminal should always be referenced to the ground point of the driven system, particularly if this is remote. Likewise the differential inputs should be referenced to their respective ground potentials to realize the full accuracy of the AD534.

A much lower scaling voltage can be achieved without any reduction of input signal range using a feedback attenuator as shown in Figure 3. In this example, the scale is such that $V_{OUT} = XY$, so that the circuit can exhibit a maximum gain of 10. This connection results in a reduction of bandwidth to about 80kHz without the peaking capacitor C_F = 200pF. In addition, the output offset voltage is increased by a factor of 10 making external adjustments necessary in some applications. Adjustment is made by connecting a 4.7MΩ resistor between Z_1 and the slider of a pot connected across the supplies to provide ±300mV of trim range at the output.

Feedback attenuation also retains the capability for adding a signal to the output. Signals may be applied to the high imped-

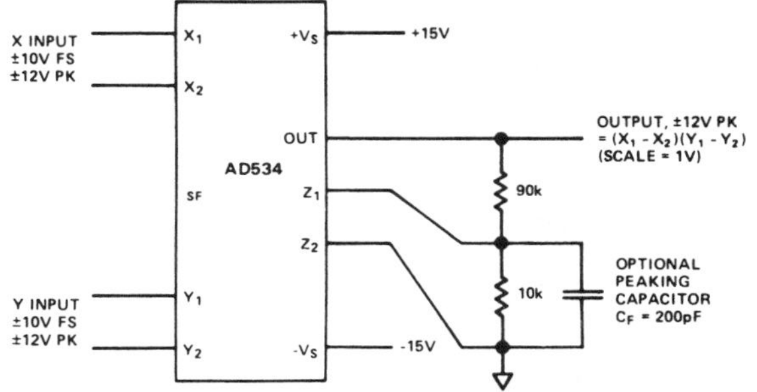

Figure 3. Connections for Scale-Factor of Unity

ance Z_2 terminal where they are amplified by +10 or to the common ground connection where they are amplified by +1. Input signals may also be applied to the lower end of the 10kΩ resistor, giving a gain of -9. Other values of feedback ratio, up to X100, can be used to combine multiplication with gain.

Occasionally it may be desirable to convert the output to a current, into a load of unspecified impedance or dc level. For example, the function of multiplication is sometimes followed by integration; if the output is in the form of a current, a simple capacitor will provide the integration function. Figure 4 shows how this can be achieved. This method can also be applied in squaring, dividing and square rooting modes by appropriate choice of terminals. This technique is used in the voltage-controlled low-pass filter and the differential-input voltage-to-frequency converter shown in the Applications Section.

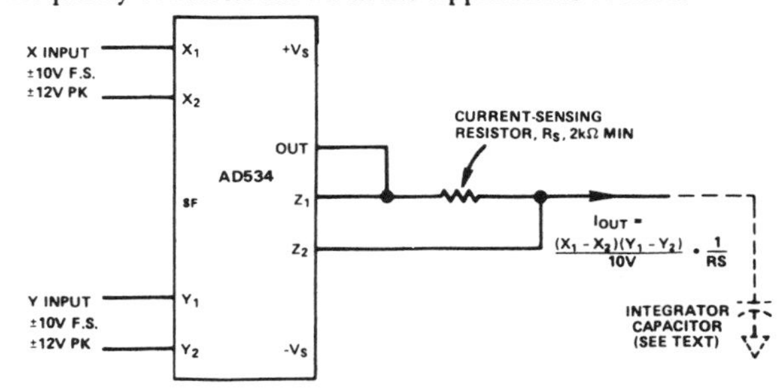

Figure 4. Conversion of Output to Current

OPERATION AS A SQUARER

Operation as a squarer is achieved in the same fashion as the multiplier except that the X and Y inputs are used in parallel. The differential inputs can be used to determine the output polarity (positive for $X_1 = Y_1$ and $X_2 = Y_2$, negative if either one of the inputs is reversed). Accuracy in the squaring mode is typically a factor of 2 better than in the multiplying mode, the largest errors occurring with small values of output for input below 1V.

If the application depends on accurate operation for inputs that are always less than ±3V, the use of a reduced value of SF is recommended as described in the FUNCTIONAL DESCRIPTION section (previous page). Alternatively, a feedback attenuator may be used to raise the output level. This is put to use in the difference-of-squares application to compensate for the factor of 2 loss involved in generating the sum term (see Figure 7).

The difference-of-squares function is also used as the basis for a novel rms-to-dc converter shown in Figure 14. The averaging filter is a true integrator, and the loop seeks to zero its input. For this to occur, $(V_{IN})^2 - (V_{OUT})^2 = 0$ (for signals whose period is well below the averaging time-constant). Hence V_{OUT} is forced to equal the rms value of V_{IN}. The absolute accuracy of this technique is very high; at medium frequencies, and for signals near full scale, it is determined almost entirely by the ratio of the resistors in the inverting amplifier. The multiplier scaling voltage affects only open loop gain. The data shown is typical of performance that can be achieved with an AD534K, but even using an AD534J, this technique can readily provide better than 1% accuracy over a wide frequency range, even for crest-factors in excess of 10.

OPERATION AS A DIVIDER

The AD535, a pin for pin functional equivalent to the AD534, has guaranteed performance in the divider and square-rooter configurations and is recommended for such applications.

Figure 5 shows the connection required for division. Unlike earlier products, the AD534 provides differential operation on both numerator and denominator, allowing the ratio of two floating variables to be generated. Further flexibility results from access to a high impedance summing input to Y_1. As with all dividers based on the use of a multiplier in a feedback loop, the bandwidth is proportional to the denominator magnitude, as shown in curve 8.

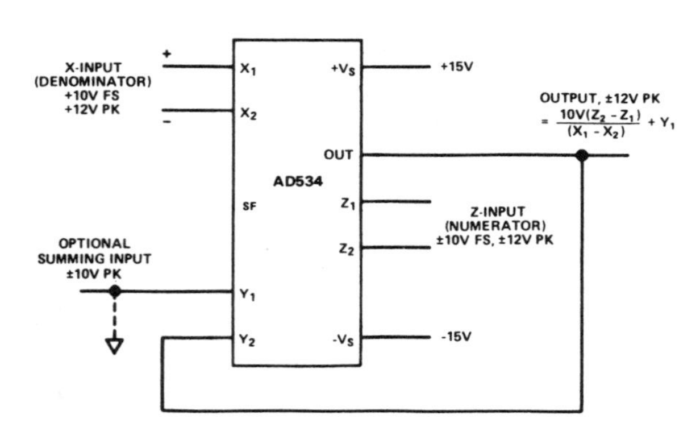

Figure 5. Basic Divider Connection

Without additional trimming, the accuracy of the AD534K and L is sufficient to maintain a 1% error over a 10V to 1V denominator range. This range may be extended to 100:1 by simply reducing the X offset with an externally generated trim voltage (range required is ±3.5mV max) applied to the unused X input (see Optional Trimming Configuration). To trim, apply a ramp of +100mV to +V at 100Hz to both X_1 and Z_1 (if X_2 is used for offset adjustment, otherwise reverse the signal polarity) and adjust the trim voltage to minimize the variation in the output.*

Since the output will be near +10V, it should be ac-coupled for this adjustment. The increase in noise level and reduction in bandwidth preclude operation much beyond a ratio of 100 to 1.

As with the multiplier connection, overall gain can be introduced by inserting a simple attenuator between the output and Y_2 terminal. This option, and the differential-ratio capability of the AD534 are utilized in the percentage-computer application shown in Figure 11. This configuration generates an output proportional to the percentage deviation of one variable (A) with respect to a reference variable (B), with a scale of one volt per percent.

OPERATION AS A SQUARE ROOTER

The operation of the AD534 in the square root mode is shown in Figure 6. The diode prevents a latching condition which could occur if the input momentarily changes polarity. As shown, the output is always positive; it may be changed to a negative output by reversing the diode direction and interchanging the X inputs. Since the signal input is differential, all combinations of input and output polarities can be realized, but operation is restricted to the one quadrant associated with each combination of inputs.

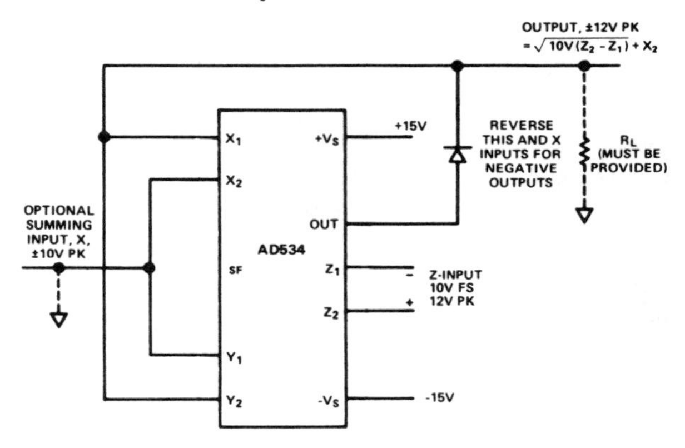

Figure 6. Square-Rooter Connection

In contrast to earlier devices, which were intolerant of capacitive loads in the square root modes, the AD534 is stable with all loads up to at least 1000pF. For critical applications, a small adjustment to the Z input offset (see Optional Trimming Configuration) will improve accuracy for inputs below 1V.

*See the AD535 Data Sheet for more details.

Testing and Measuring with Op Amps

7

LEARNING OBJECTIVES

Upon completion of this chapter on testing and measuring with op amps, you will be able to:

- Explain the operating principles of voltage-to-current converters, current-to-voltage converters, or current-to-current converters.
- Give examples where each converter is the solution to a measuring or testing problem.
- Draw the circuits for transmitters that convert analog or digital signals into a visible (LED) or invisible (IRED) radiant energy output.
- Use a photodiode to convert radiant energy to current and then to voltage.
- Measure temperature with an AD590 temperature-to-current converter.
- Measure solar power density with a solar cell.
- Connect an ordinary dc milliammeter or microammeter into a circuit to measure (1) positive or negative dc voltage, (2) rms ac voltage, or (3) peak ac voltage.
- Show how a function switch allows a voltmeter to measure either dc, rms, or peak ac voltage.
- Show how a range switch is used to allow a voltmeter to measure different ranges of voltages.

7-0 INTRODUCTION

Voltage-to-current (V-to-I) and current-to-voltage (I-to-V) converters can be used to (1) test or select semiconductor devices, (2) make ammeters or voltmeters, and (3) condition semiconductor tranducers to measure physical characteristics of the environment or a process. The applications above are just a small sample of problems that are easily solved by these versatile circuits.

7-1 V-to-I CONVERTERS FOR FLOATING LOADS

7-1.1 Theory of Operation

There are two types of basic *V*-to-*I* converters that have floating loads connected in the negative feedback loop. (A load is "floating" if neither end is grounded.) It is usually intended to hold load current constant in this type of application.

In Fig. 7-1a, input voltage V_i is connected to the op-amp *noninverting* input, so that no current is drawn from V_i. Negative feedback forces the voltage at pin 2 to equal V_i and load current I_L is determined strictly by

$$I_L = \frac{V_i}{R_i} \tag{7-1}$$

Resistor R_i performs the voltage-to-current conversion. In Fig. 7-1b, input voltage V_i also appears across R_i because of the virtual ground at pin 2. Therefore, V_i is converted to load current I_L by the same Eq. (7-1).

Load currents are limited to a maximum of 5 to 10 mA depending on the op amp. Distinctions between the two circuits are presented next.

7-1.2 Advantages and Disadvantages

The *noninverting* *V*-to-*I* converter of Fig. 7-1a draws no current from control voltage V_i. Therefore, V_i can be made from a simple voltage divider and does not have to be buffered with a voltage follower. However, V_o depends on both V_i and load voltage drop $I_L R_L$. Since the limit for V_o is slightly below $\pm V_{sat}$ (± 12 V for prudence with a ± 15-V supply), the magnitude of allowable load voltage drop is limited. For example, if V_i is selected from an available 5-V logic supply, V_{RL} must remain below (12 − 5) V = 7 V.

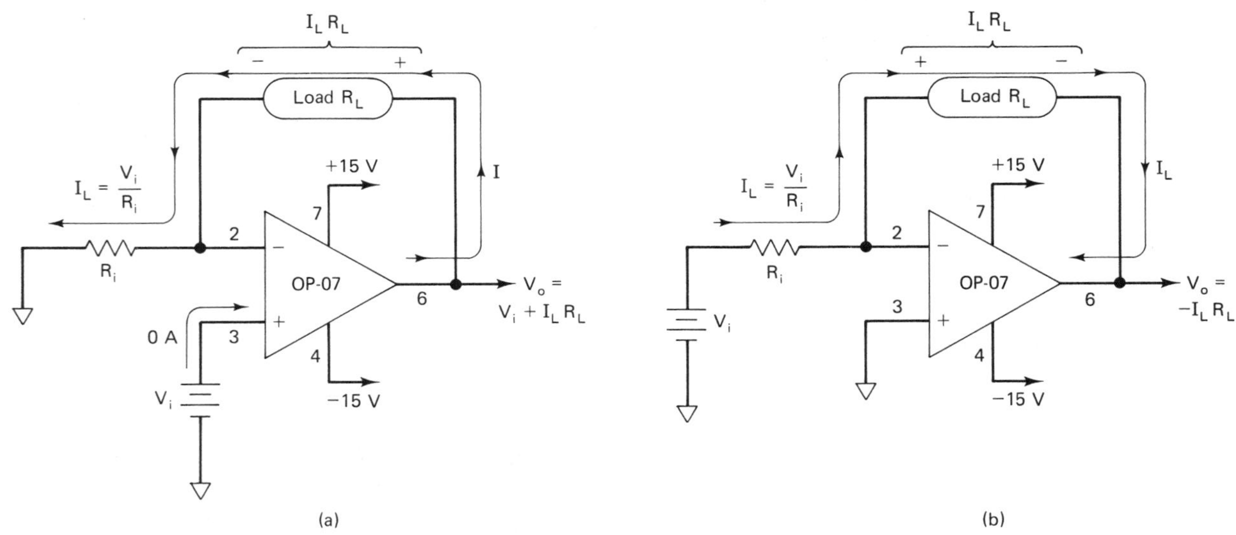

Figure 7-1 Basic voltage-to-current converters for floating loads: (a) no current drawn from voltage source V_i, noninverting; (b) load current drawn from voltage source V_i, inverting.

The *inverting* V-to-I converter in Fig. 7-1b has a virtual ground at one end of the load (pin 2). This is an advantage because the full ± 12 V (prudent saturation voltages) are available for maximum load voltage. Also, output voltage V_o equals the load voltage. This is an advantage where load voltage must be measured at a particular load current.

The only disadvantage for the circuit of Fig. 7-1b is that V_i must furnish all the load current. Note that I_L depends *only* on V_i and R_i for *both* V-to-I converted in Fig. 7-1. I_L does *not* depend on R_L.

7-1.3 Diode and Zener Diode Testers

Silicon diodes make excellent inexpensive temperature sensors. Their voltage *decreases* at a predictable rate of 2.2 mV per degree Celsius as temperature *increases.* This occurs only if the diode's current is held *constant,* usually between 0.1 and 1.0 mA. However, every diode has an unpredictable reference voltage, 580 to 620 mV, at the same reference temperature of 25°C.

To calibrate each diode accurately, you must measure its reference voltage V_F at the same reference temperature, usually 25°C, and at the same constant current. This procedure is shown in Example 7-1.

EXAMPLE 7-1

Calculate a value for R_i in Fig. 7-2a to make $I = 0.1$ mA.

SOLUTION From Eq. (7-1),

$$R_i = \frac{V_i}{I} = \frac{|V_{EE}|}{I} = \frac{|15\text{ V}|}{0.1\text{ mA}} = 150\text{ k}\Omega$$

EXAMPLE 7-2

A supply of 5.1 V $\pm 10\%$ zeners are to be checked by quality control. V_z is to be measured for each zener with a current of exactly 5.0 mA. Find

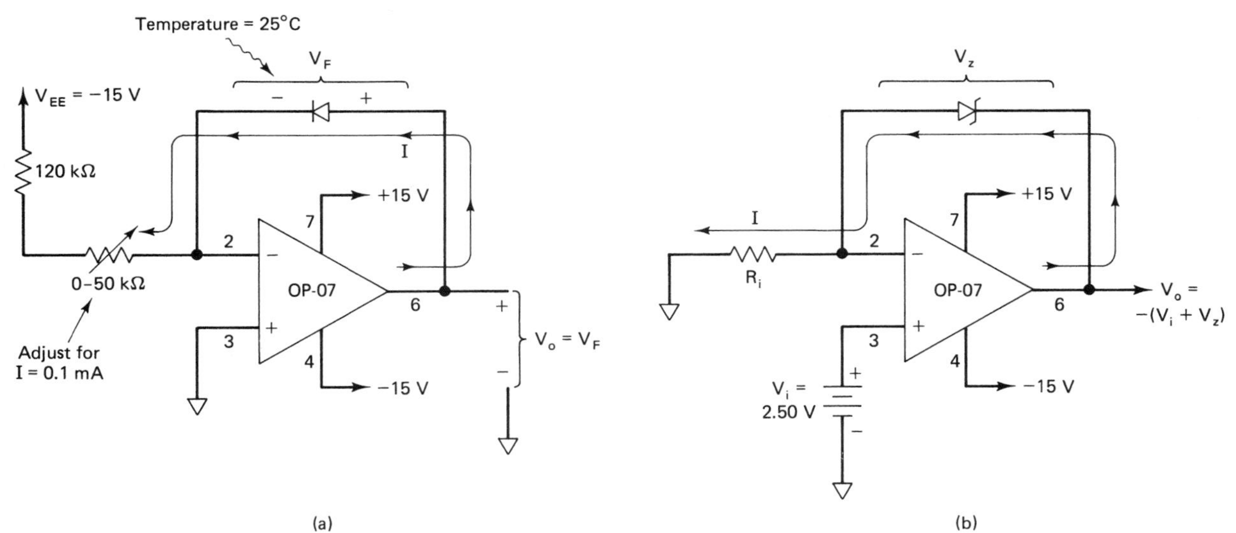

Figure 7-2 *V*-to-*I* converters are used to characterize diodes. (a) temperature-measuring diode calibration circuit; (b) Zener diode voltage measurement.

(a) the required value for R_i in Fig. 7-2b; (b) the value of V_z if V_o = 8.10 V.

SOLUTION (a) From Eq. (7-1),

$$R_i = \frac{2.50\ \text{V}}{5.0\ \text{mA}} = 500\ \Omega$$

(b) From Fig. 7-2b,

$$V_o = (V_i + V_z)$$

Therefore,

$$V_z = (V_o - V_i) = (8.10\ \text{V} - 2.50\ \text{V}) = +5.60\ \text{V}$$

7-2 VOLTAGE-TO-HIGH-CURRENT CONVERTERS

7-2.1 Theory of Operation

The 5 to 10 mA output current limit of an op amp is boosted by the addition of a bipolar junction transistor (BJT), as shown in Fig. 7-3. The BJT's base is connected to the op-amp output and emitter to (−) input. This closes a negative feedback loop by forward-biasing the base-to-emitter junction. Thus voltage between pins 2 and 3 is forced to approximately zero volts, and V_i appears across "current set resistor," R_i.

Emitter current I is still established by Eq. (7-1), where $I = V_i/R_i$. Collector and emitter currents are essentially equal for a BJT in the active state. Therefore, load current through R_L is set by only V_i *and* R_i.

The op amp's output current is determined only by the beta (β) value of the BJT, and is found from

$$I_o = \frac{I}{\beta + 1} \tag{7-2}$$

If BJTs are selected with beta values that exceed 50, load currents are obtainable up to

$$(5\ \text{mA to } 10\ \text{mA}) \times 51 \simeq 250 \text{ to } 500\ \text{mA}$$

7-2.2 Operating Precautions

The BJT in Fig. 7-3 must always have a $V_{CE} > 1$ V to ensure operation in the active region. Therefore, the maximum load resistance is determined by

$$R_{L(\text{max})} = \frac{V_{CC} - 1.0\ \text{V} - V_i}{I} \tag{7-3}$$

If the circuit develops spurious oscillations, add a 100-Ω resistor between pin 6 and base of the transistor or a 50 pF capacitor.

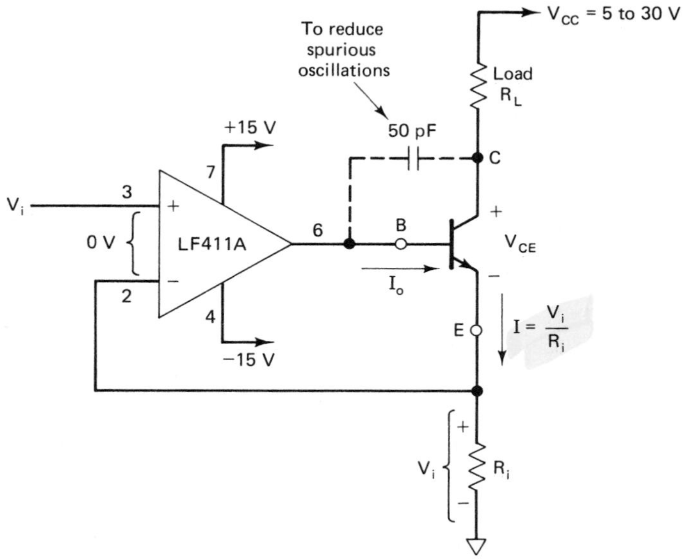

Figure 7-3 Voltage-to-high current converter.

7-2.3 Light-Emitting-Diode Tester

The luminous output of a light-emitting diode (LED) is measured at a constant current, often 20 mA. We connect two LEDs as loads in the circuit of Fig. 7-4a to show that *both* currents are exactly 20 mA and do *not* depend on the different I–V characteristics of each LED. Their light outputs can be compared or measured under the same current test conditions, as shown in the next example.

EXAMPLE 7-3

V_i is a +2.50-V voltage reference in Fig. 7-4a. If LED currents must equal 20.0 mA, find R_i.

SOLUTION From Eq. (7-1),

$$R_i = \frac{V_i}{I} = \frac{2.5\text{ V}}{20\text{ mA}} = 125\ \Omega$$

Note that the collector voltage is 11.8 V and the emitter voltage is 2.5 V. Therefore, $V_{CE} = (11.8 - 2.5)\text{ V} = 9.3\text{ V}$. If $\beta = 49$, op-amp output current would be, from Eq. (7-2),

$$I_o = \frac{I}{\beta + 1} = \frac{20\text{ mA}}{50} = 0.4\text{ mA}$$

7-2.4 Fiber Optic Analog Data Transmitter

Low-frequency analog data can be transmitted over a fiber optic link by the excellent inexpensive circuit of Fig. 7-4b. Op amp B is the V-to-I converter that drives an infrared emitting diode (IRED). Op amp A is an inverting adder (see Section 6-3.2). Its performance is analyzed by an example.

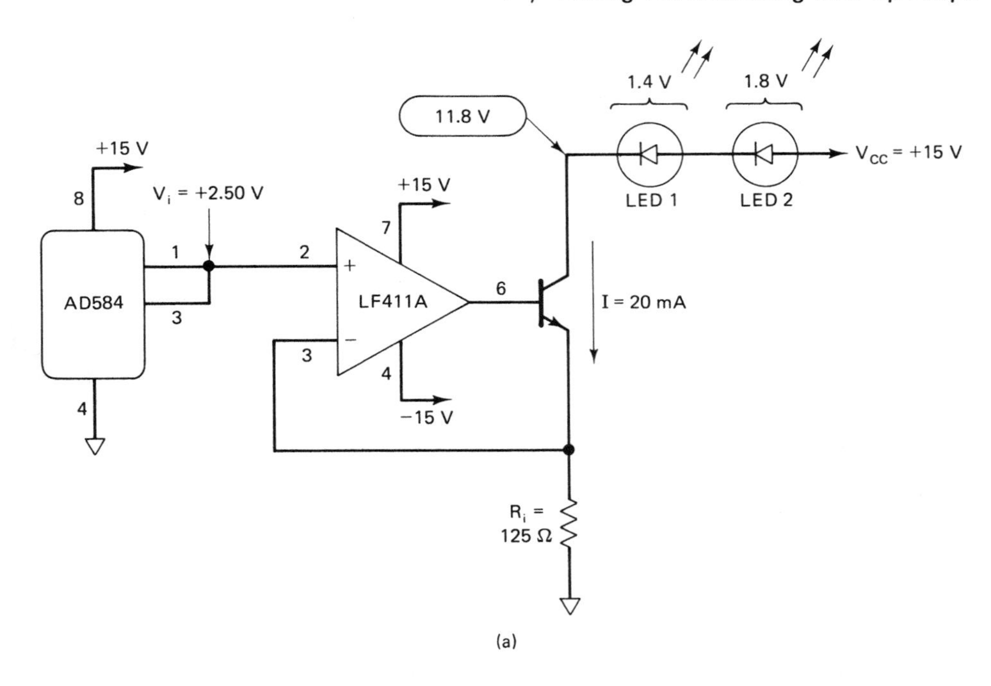

(a)

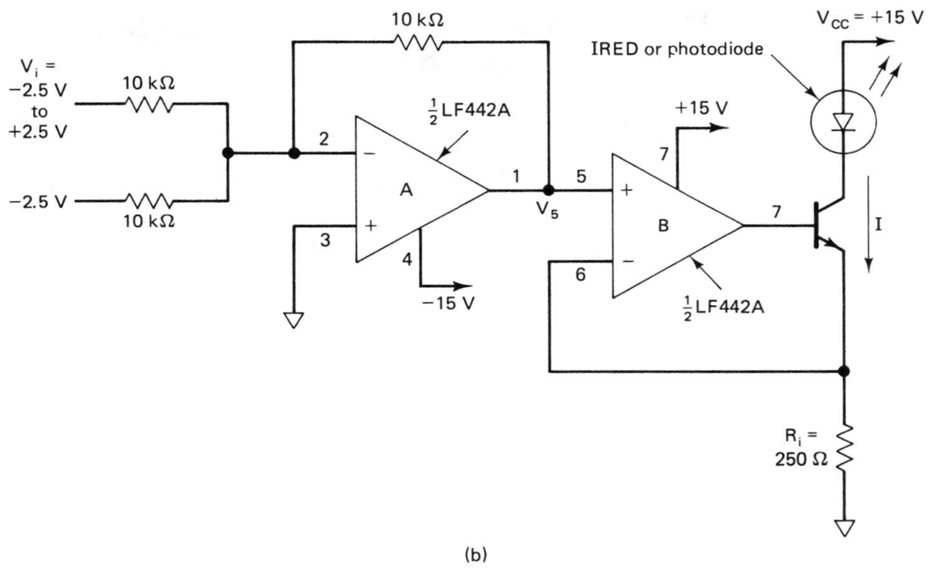

(b)

Figure 7-4 Applications of voltage-to-high current converters. Light outputs of two LEDs are compared or measured in (a) and an analog fiber optic transmitter is shown in (b). (a) Circuit to test output of two LEDs at exactly the same current (BJT is a 2N2222); (b) fiber optic or optoelectronic analog transmitter (dual-FET op amp is an LM 442AN).

EXAMPLE 7-4

Assume that V_i varies sinusoidally from −2.5 V through 0 V to +2.5 V in Fig. 7-4b. Find the corresponding variation in IRED current (and consequently radiant power output).

SOLUTION When $V_i = 0$ V, op amp A's output at V_5 is +2.5 V. Therefore, $I = 2.5$ V/250 Ω = 10 mA. We tabulate the calculations for peak values of V_i as follows:

V_i (V)	V_5 (V)	I (mA)
−2.5	0	20
0	2.5	10
+2.5	5.0	0

Thus as V_i varies sinusoidally, IRED current varies sinusoidally around an average value of 10 mA. The LED radiant power output varies sinusoidally, and these analog data can be detected by a photodiode (see Section 7-6.2 and Fig. 7-11a).

7-2.5 Digital Fiber Optic Transmitter

To make a TTL-compatible fiber optic transmitter, connect the TTL output signal as V_i in Fig. 7-4a and replace the LEDs with an IRED. Add a variable resistor of 0 to 125 Ω in series with R_i. Adjust it for I = 20 mA when the TTL signal is high. The IRED will broadcast full power on a "1" signal and negligible power on a "0" signal.

7-3 VOLTAGE-TO-CURRENT CONVERTER AS A VOLTMETER

7-3.1 Theory of Operation

An excellent dc analog voltmeter can be made with an inexpensive milli- or microammeter, bridge rectifier, op amp, and resistor, as in Fig. 7-5. Assume that measured voltage V_i will range from −1 to +1 V. V_i will see the very high impedance of the op amp's noninverting input.

If V_i is negative as in Fig. 7-5a, diodes D_N steer converted I down through the ammeter to cause an upscale deflection. If V_i is positive as in Fig. 7-5b, diodes D_P steer current in the *same* downward direction through the dc ammeter to cause an upscale deflection.

This dc voltmeter indicates the *magnitude* of input voltage regardless of polarity. Let us call the ammeter's full-scale deflection I_{FS}. In general, we want the meter to indicate full scale when V_i is at the maximum expected value V_{FS}. Since V_i appears across R_i, the full-scale current is set only by R_i.

$$R_i = \frac{V_{FS}}{I_{FS}} \qquad (7\text{-}4)$$

EXAMPLE 7-5

A 0- to 100-μA meter movement is used to construct a dc voltmeter (in the circuit of Fig. 7-5) to measure dc voltages that range from −1.0 to 1.0 V. Find R_i.

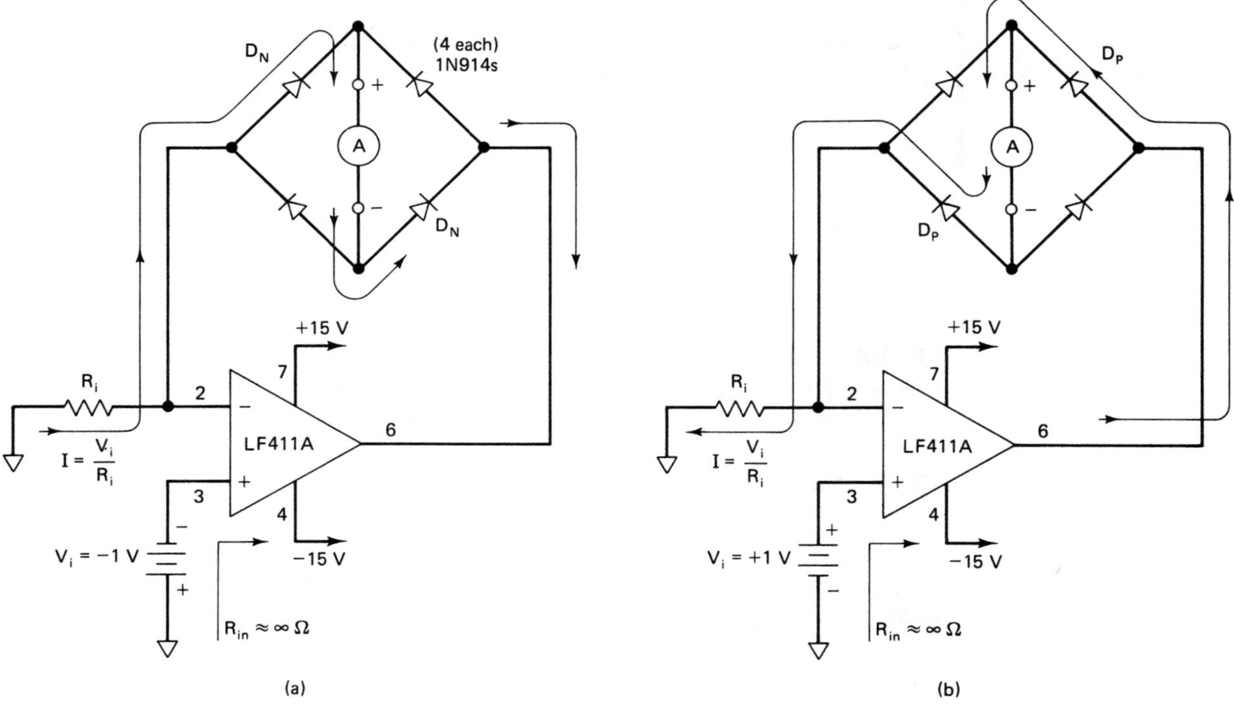

Figure 7-5 Dc ammeter *A* is the output load in the feedback loop. The diodes steer current downward through *A* in the same direction regardless of the polarity of V_i. (a) Negative *V*-to-*I* converter; (b) positive *V*-to-*I* converter.

SOLUTION $V_{FS} = 1$ V magnitude; $I_{FS} = 100\ \mu$A from Eq. (7-4):

$$R_i = \frac{1\text{ V}}{100\ \mu\text{A}} = 10.0\text{ k}\Omega$$

The meter's face would be re-labeled 0 to 1 V to indicate directly the magnitude of V_i.

7-3.2 Function Switch

If V_i were replaced by an ac sine wave in Fig. 7-6, current through R_i would be alternating. However, current through the meter would be rectified ac and the meter would indicate the *average* value of the rectified current.

To make an ac voltmeter, all we have to do is change the value of R_i to

$$R_i = 0.90\ \frac{\text{rms}\, V_{FS}}{\text{dc}\ I_{FS}} \tag{7-5a}$$

To make a peak-indicating ac voltmeter, for sine waves only, R_i is given by

$$R_i = 0.636\ \frac{\text{peak}\ V_{FS}}{\text{dc}\ I_{FS}} \tag{7-5b}$$

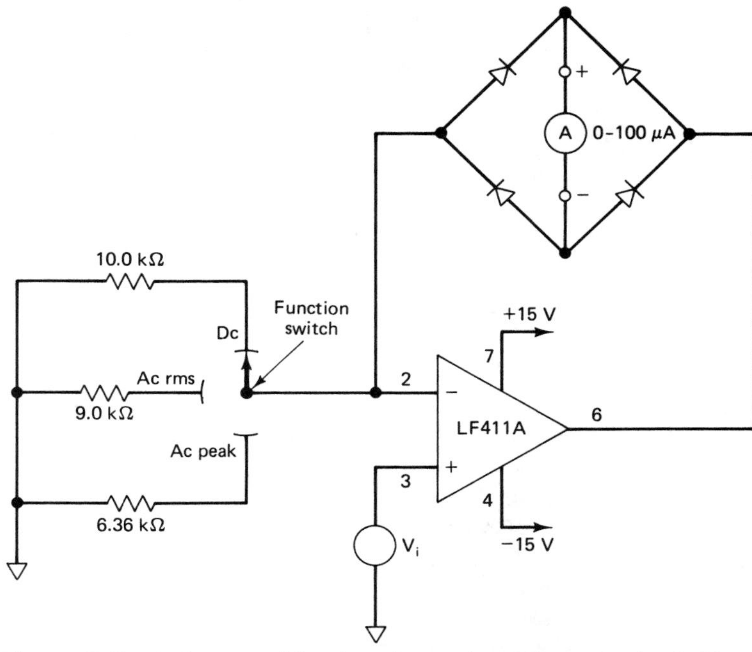

Figure 7-6 A three-position function switch allows the basic *V*-to-*I* converter to measure rms ac, peak ac, or peak dc voltages.

The procedure is illustrated by an example

EXAMPLE 7-6

Given a 100-μA meter movement in Fig. 7-6, find R_i (a) if V_i is a 0- to 1-V (rms) sine wave; (b) if the peak values of V_i are 0 to ±1 V.

SOLUTION (a) From Eq. (7-5a),

$$R_i = 0.90 \frac{1 \text{ V rms}}{100 \ \mu\text{A}} = 9.0 \text{ k}\Omega$$

(b) From Eq. (7-5b),

$$R_i = 0.636 \frac{1 \text{ V peak}}{100 \ \mu\text{A}} = 6.36 \text{ k}\Omega$$

We use the same meter movement with a function switch to measure either dc, ac rms, or peak voltages in Fig. 7-6. The function switch position indicates what type of voltage is being measured.

7-3.3 Range Switch

The basic ac/dc voltmeter of Fig. 7-6 can be made more versatile by adding a range switch. Suppose that you must measure dc voltages between 0 and ±10 V. We can use the same hardware in Fig. 7-6 if we divided the voltage to be measured by 10. Then if $V_i = 9$ V, the ammeter needle would indicate 0.9 mA. The meter's face would then be marked with the symbol 9 under this location of the needle.

This can be accomplished by adding a *range* switch and a voltage

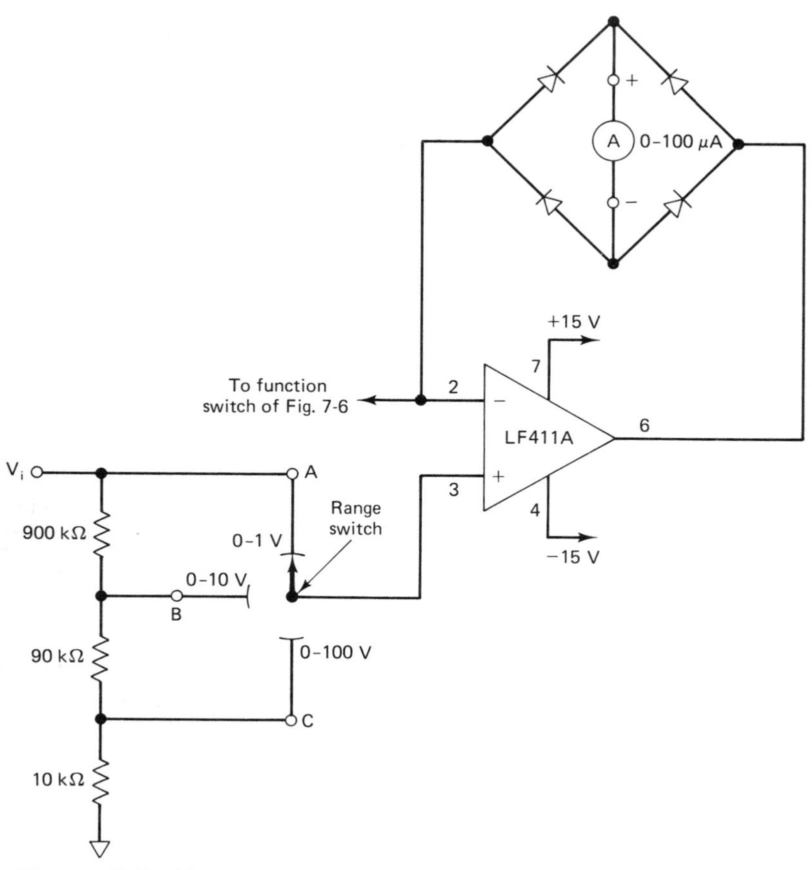

Figure 7-7 Three resistors form a 10-to-1 or 100-to-1 voltage divider for V_i. A range switch allows the 0- to l-V voltmeter of Figure 7-6 to measure $V_i = 0$ to 1 V, $V_i = 0$ to 10 V, or $V_i = 0$ to 100 V.

divider as in Fig. 7-7. V_i can be connected directly to the 0- to 1-V voltmeter. If the range switch is thrown to point B, V_i is divided by 10 and we can now measure voltages in the range 0 to ±10 V. Point *c* selects a 100-to-1 divider to allow the 0- to 1-V voltmeter to measure voltages in the range 1 to ±100 V.

Finally, to measure smaller voltages in the range of 0 to ±100 mV, simply add another position on the range switch plus an amplifier with a voltage gain of 10.

7-4 CURRENT SOURCES AND SINKS

7-4.1 DC Current Source

In Fig. 7-8a, constant current I_L is sourced to load R_L. The R_1 and R_2 voltage divider establishes 2 V across R_1. The op amp forces this constant 2-V drop across R_1 to appear across "load current set resistor" R_S. Therefore, load current is set by

$$I_L = \frac{V_{RS}}{R_S} \tag{7-6}$$

independent of the value of R_L.

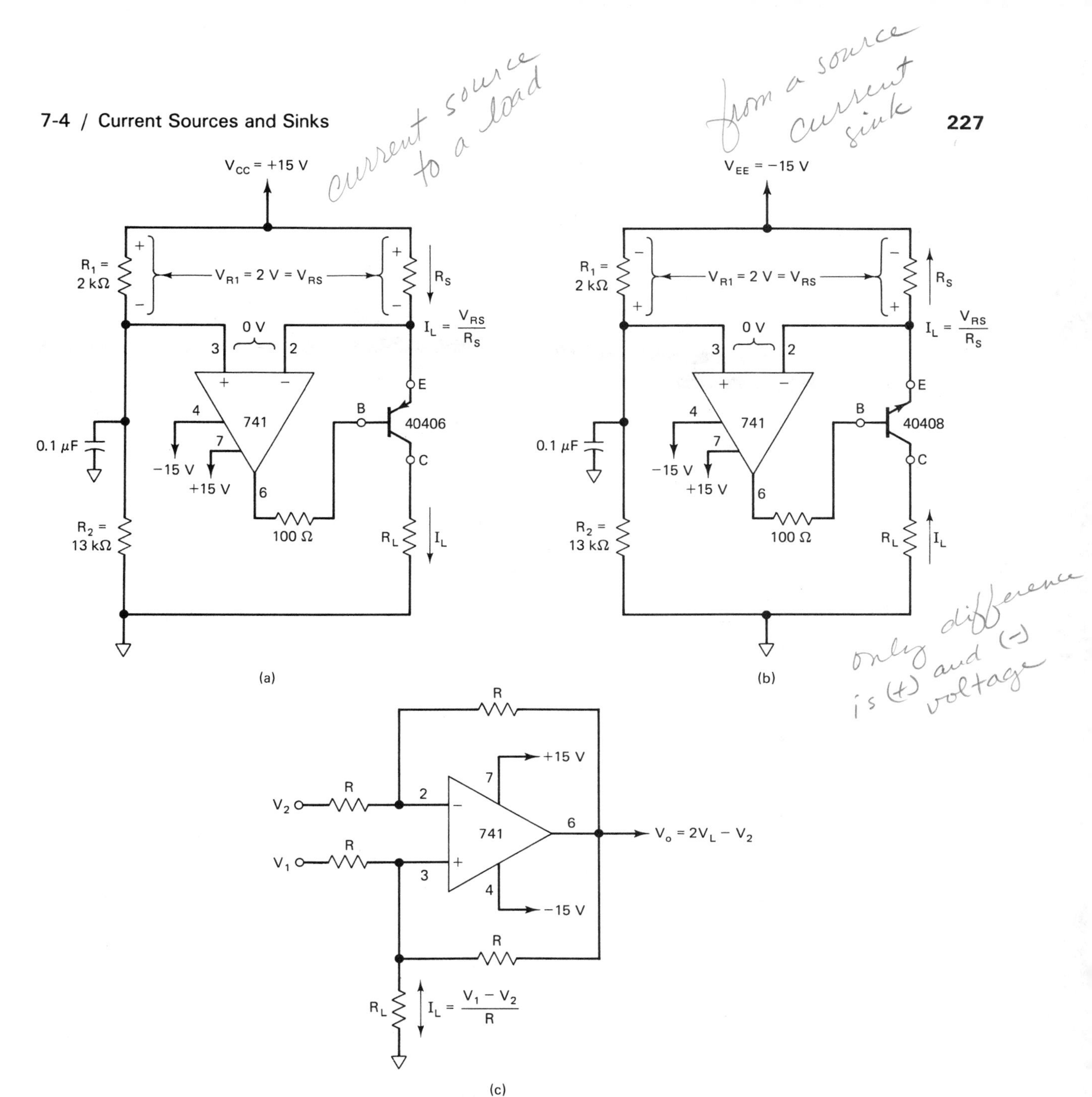

Figure 7-8 Current sources and sinks: (a) dc current source; (b) dc current sink; (c) ac current source.

EXAMPLE 7-7

What value is required for R_S if I_L must equal 20 mA in Fig. 7-8a?

SOLUTION From Eq. (7-6),

$$R_S = \frac{V_{RS}}{I_L} = \frac{2 \text{ v}}{20 \text{ mA}} = 100\ \Omega$$

7-4.2 DC Current Sink

Current flows from load R_L into the collector of the NPN transistor in Fig. 7-8b. Thus the 40408 transistor and op amp form a current sink. I_L is set by the same equation, Eq. (7-6), as in Section 7-4.1. As the value of R_L

increases, the collector voltage will approach the emitter voltage in both Figs. 7-8a and b. You must allow 1 V "headroom" for V_{CE} to keep the transistor out of saturation.

7-4.3 AC Current Source

The circuit of Fig. 7-8c is classified as a differential voltage-to-current converter. All resistors labeled R must be matched to within ±1%. Then current I_L, V_L and V_o are found from

$$I_L = \frac{V_1 - V_2}{R} \tag{7-7a}$$

$$V_L = I_L R_L \tag{7-7b}$$

$$V_o = 2V_L - V_2 \tag{7-7c}$$

EXAMPLE 7-8

In Fig. 7-8, $R = 10\ \text{k}\Omega$ and $R_L = 1\ \text{k}\Omega$. Assume that $V_2 = 0$ V. Find I_L, V_L, and V_o for (a) $V_1 = +1$ V; (b) $V_1 = -1$ V.

SOLUTION From Eq. (7-7),

(a) $I_L = \dfrac{1\ \text{V} - 0}{10\ \text{k}\Omega} = 0.1\ \text{mA}$

$V_L = 0.1\ \text{mA} \times 1\ \text{k}\Omega = 0.1\ \text{V}$

$V_o = 2(0.1\ \text{V}) - 0\ \text{V} = 0.2\ \text{V}$

(b) $I_L = \dfrac{-1\ \text{V} - 0}{10\ \text{k}\Omega} = -\ 0.1\ \text{mA}$

$V_L = -\ 0.1\ \text{mA} \times 1\ \text{k}\Omega = -0.1\ \text{V}$

$V_o = 2(-0.1\ \text{V}) = -0.2\ \text{V}$

Example 7-8 shows that load current through R_L can flow in both directions (source–sink), using Fig. 7-8.

7-5 TRANSDUCERS AS CURRENT SOURCES

Transducers convert energy from one form into another. We select three types of transducers that convert a physical characteristic of our environment into an electric current.

7-5.2 Photodiodes

The first transducer is a photodiode. It converts incoming light or radiant energy into an output light current I_λ. When the photodiode is reverse biased, its output current will increase linearly from essentially zero to typically tens of microamperes as received light or radiation energy, H, increases (Fig. 7-9a).

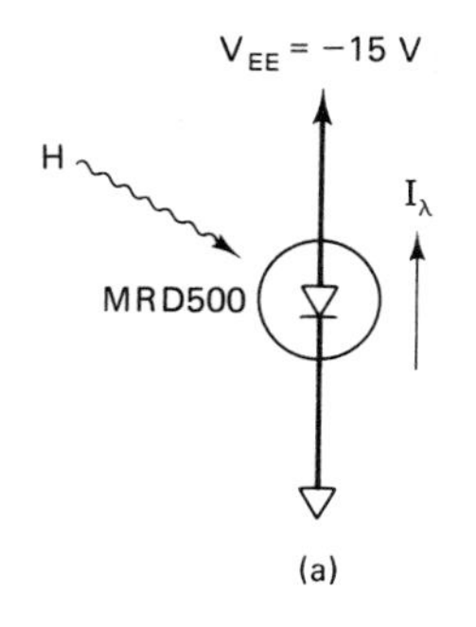

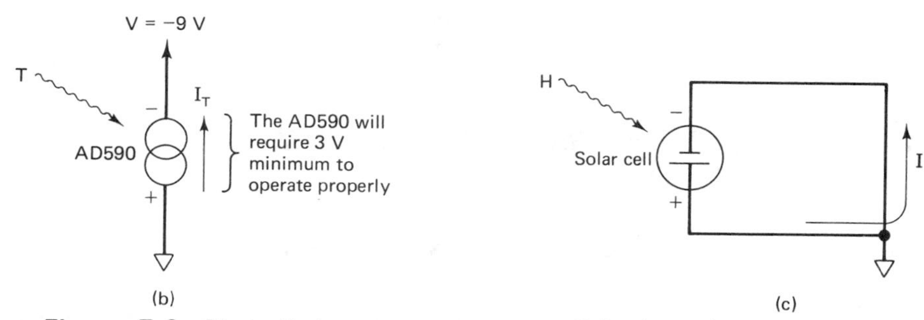

Figure 7-9 Photodiode current measures light intensity in (a). AD590 current measures temperature in (b). A solar cell's short-circuit current measures received solar power density in (c). (a) Photodiode transducer; (b) AD590 temperature transducer; (c) Solar cell transducer.

7-5.2 Temperature Transducer

In Fig. 7-9b, an AD590 transducer converts ambient temperature T to a current I_T. If you measure I_T, it tells the ambient temperature in either Kelvin, Celsius, or Fahrenheit degrees from the appropriate equation:

$$I_T = \left(\frac{1\ \mu\text{A}}{\text{K}}\right)T_K \tag{7-8a}$$

$$I_T = \left(\frac{1\ \mu\text{A}}{°\text{C}}\right)T_C + 273.2\ \mu\text{A} \tag{7-8b}$$

$$I_T = \left(\frac{5\ \mu\text{A}}{9°\text{F}}\right)T_F + 255.4\ \mu\text{A} \tag{7-8c}$$

The AD590 must be biased properly with a supply voltage.

7-5.3 Solar Cell Transducer

Unlike a photodiode, the solar cell needs no biasing. It converts incoming solar power directly into electrical power. All solar cells output an open-circuit voltage of about 0.5 V. This voltage does *not* indicate the amount of incoming solar power.

The short-circuit current of a solar cell is a direct measure of received solar power density. For example, a 3-inch circular cell will output a maximum of typically 1 A, when directly exposed to a noonday sun and receives solar power at the rate of about 100 mW/cm^2. Call this power level, 1 sun.

Suppose that clouds partially obscure the sun so that the cell receives only $\frac{1}{2}$ sun or 50 mW/cm^2. Its short-circuit current output will exactly halve, to 0.5 A.

7-5.4 Measuring with Current Output Transducers

All the transducers in this section share one common characteristic. They act like current sources. Thus we must select a circuit type that will accept *all* of the transducer's current and convert it into a voltage. This means that the transducer's current must flow into or out of a short circuit or virtual ground. We learn how easy it is to do this in the next section.

7-6 PRACTICAL CURRENT-TO-VOLTAGE CONVERTERS

7-6.1 Theory of Operation

The current source in Fig. 7-10a thinks it is working into a short circuit. Negative feedback forces the voltage between (−) and (+) inputs of the op amp to equal zero volts. Thus no current is lost through R_N. *All* of I_{SC} flows through R_F and is converted to a voltage. If the current source's output exceeds about 5 to 10 mA, a current boost transistor is added as in Fig. 7-10b.

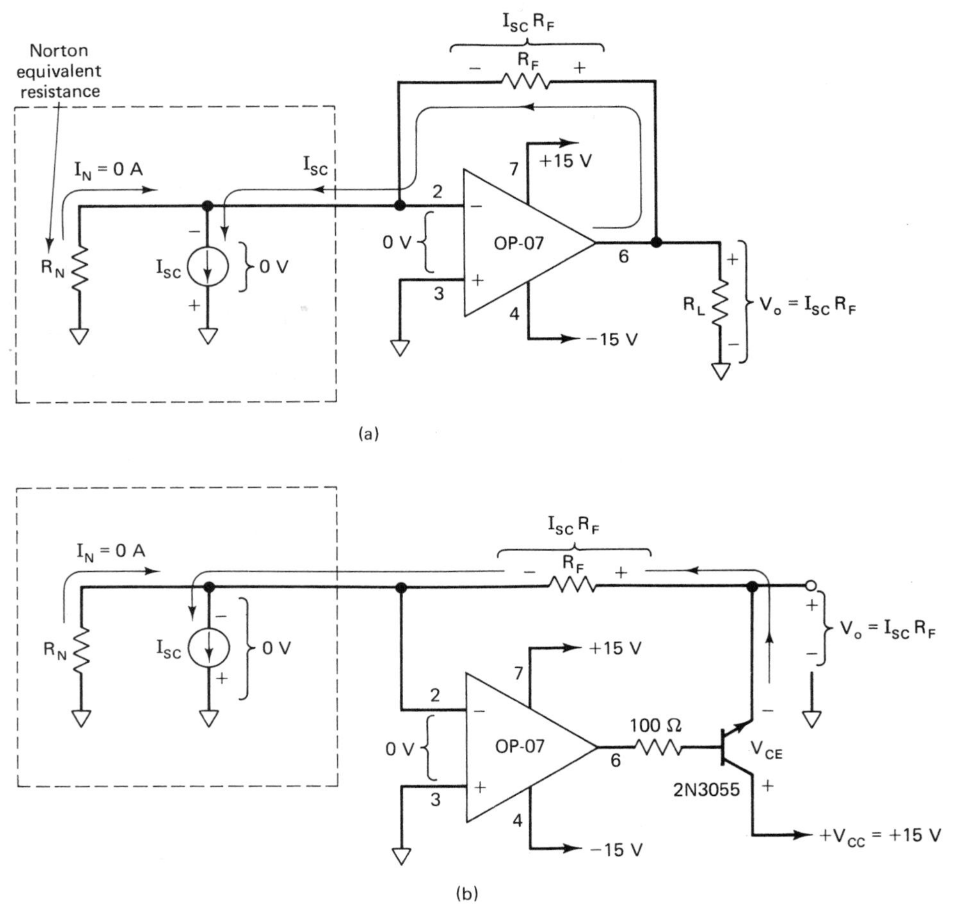

Figure 7-10 Both figures show that the current sources work into an apparent short circuit. All of the short-circuit current flows through R_f to be converted into an output voltage. (a) Current source short-circuit current is converted to an output voltage by R_f; (b) high current-to-voltage converter.

7-6.2 Light Detector

A PIN silicon MRD500 photodiode is reverse biased in Fig. 7-11 to make a light-to-current transducer. Reverse current I_L increases as received light power H increases. Current I_L is converted to a voltage V_o by R_F.

EXAMPLE 7-9

In the photodetector of Fig. 7-11a, $I_L = 1\ \mu$A if (a) radiation energy $H = 2$ mW/cm^2, and (b) 2 μA if $H = 4$ mW/cm^2. When $R_F = 200$ kΩ, calculate V_o.

SOLUTION From Fig. 7-11, $V_o = I_L R_F$.

(a) $V_o = (1\ \mu\text{A})\,(200\ \text{k}\Omega) = 0.2$ V

(b) $V_o = (2\ \mu\text{A})200\ \text{k}\Omega = 0.4$ V.

This photodetector can act as a receiver for the transmitter of Fig. 7-4b. The pair can form an optoelectronic, low-frequency, analog communications channel. The connecting link can be line-of-sight, with send and receive concentrating lenses, or can be via fiber optic cable.

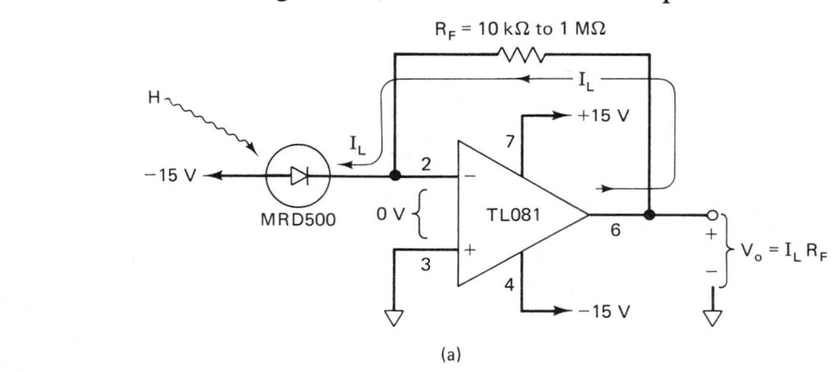

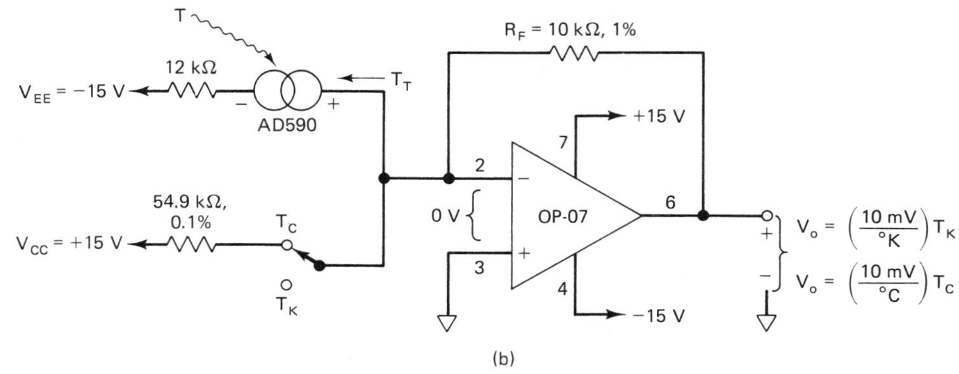

Figure 7-11 The optoelectronic receiver in (a) and temperature detector in (b) are examples of low current-to-voltage converters. (a) Incoming light or infrared or radiant energy is converted into current by a photo or *IR* diode and then into a voltage by R_f; (b) electronic Celsius or Kelvin thermometer. Output V_o equals 10 MV per °K or °C, depending on whether the function switch is positioned to T_K or T_C, respectively.

7-6.3 Electronic Thermometer

The AD590 in Fig. 7-11b is biased on by the -15V supply. If the function switch is thrown to position T_K, each degree Kelvin is converted to a current of 1 μA and then to an output voltage of 1 μA $\times$ 10 kΩ = 10 mV. Thus, from Eq. (7-8a),

$$V_o = \left(\frac{10 \text{ mV}}{°\text{K}}\right) T_K \tag{7-9a}$$

If the function switch is positioned to T_C, a constant current of 273.2 μA is injected into R_F so that current through R_F equals 0 μA when T_K = 273.2°C. Thus V_o will equal zero when ambient temperature equals 0°C or 273.2°K and V_o will be found from

$$V_o = \left(\frac{10 \text{ mV}}{°\text{C}}\right) T_C \tag{7-9b}$$

EXAMPLE 7-10

In Fig. 7-11b, the ambient temperature is 100°C or 373.2°K. Find V_o if the switch is thrown to (a) the °C position; (b) the °K position.

SOLUTION (a) From Eq. (7-9b),

$$V_o = \left(\frac{10 \text{ mV}}{°\text{C}}\right) (100°\text{C}) = 1000 \text{ mV} = 1 \text{ V}$$

(b) From Eq. (7-9a);

$$V_o = \left(\frac{10 \text{ mV}}{°\text{K}}\right) (372.3°\text{K}) = 3.732 \text{ V}$$

7-6.4 Measuring Solar Power or Energy

We select the high-current I-to-V converter of Fig. 7-12 to convert the 0- to 1-A output of a 3-inch circular solar cell, to a 0- to 10-V input required by an anlog-to-digital converter. The virtual ground at the (−) input of the op amp ensures that all the solar cell's short-circuit current flows through R_F.

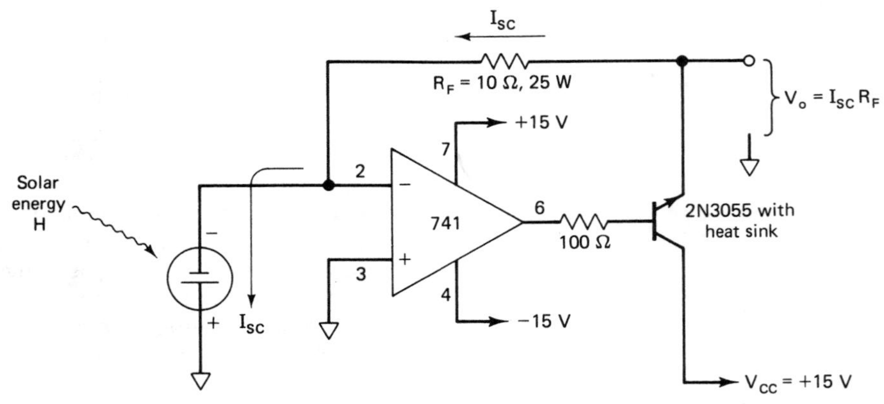

Figure 7-12 Solar power density measuring circuit.

EXAMPLE 7-11

The 3-in. solar cell in the circuit of Fig. 7-12 is exposed to a received radiation or solar power intensity of $\frac{1}{2}$ sun or 50 mW/cm^2. If $I_{SC} = 0.5$ A, find V_o.

SOLUTION Since $V_o = I_{SC}R_F$,

$$V_o = (0.5 \text{ A})(10\ \Omega) = 5.0 \text{ V}$$

Since energy equals power times time, the output of Fig. 7-12 can be connected to a voltage-to-frequency converter. If the output of the *V/F* converter is connected to a digital counter, its total count is a measure of the solar energy received. This count can be translated into the equivalent of free energy dollars by appropriate conversion factors.

7-7 CURRENT-TO-CURRENT CONVERTERS

In Figure 7-13, all input current I flows through AR and causes voltage V_{AR} to equal I(AR). Voltage across load resistor R is given by $V_R = I_L R$ and must equal V_R. If we equate these voltages and solve for I_L;

$$V_R = I_L R = V_{AR} = I \times \text{AR}$$

we get

$$I_L = \text{A} \times \text{I} \tag{7-10}$$

Equation (7-10) shows that the ratio of feedback resistor AR to load resistor R is the current amplification factor A. Thus input current I is amplified by A and delivered to the grounded load resistor R as AI.

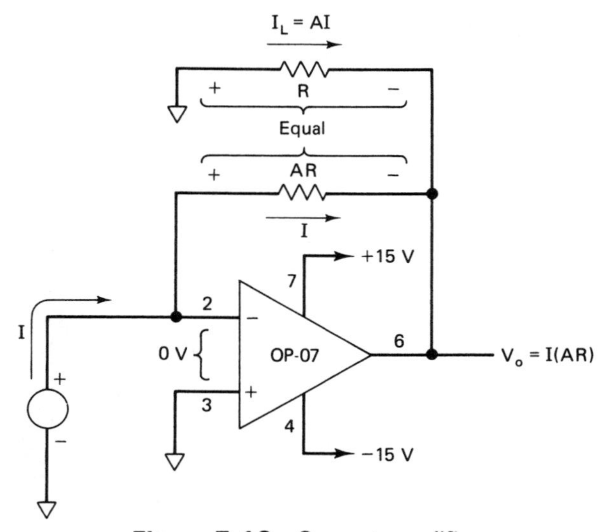

Figure 7-13 Current amplifier.

End of Chapter Exercises

Name: ______________________

Date: __________ **Grade:** ___

A. *FILL-IN THE BLANKS*

Fill in the blanks with the best answer.

1. There are two types of basic, V-to-I converters with floating loads. They are __________ and __________.
2. If load current for a V-to-I converter exceeds the op amp's output current capability, current boast must be added by a __________.
3. Meter current in the analog voltmeter of Fig. 7-3 is determined by __________ and __________.
4. The switch used to change a voltmeter from ac to dc measurements is called a __________ switch.
5. The switch used to change a voltmeter's ability to measure values of 0 to 1 V or 0 to 10 V is called a __________ switch.
6. Transducers convert __________ from one form to another.
7. An AD590 outputs 273.2 μA when its temperature is __________°C.

B. *TRUE/FALSE*

Fill in **T** if the statement is true, and **F** if any part of the statement is false.

1. ______ The switch that changes a voltmeter from ac to dc readings is called a range switch.
2. ______ Load resistance changes will change load current in a V-to-I converter.

3. ______ Light output of an LED is approximately doubled if its current is doubled.

4. ______ Output current of the current source or sink circuit in Fig. 7-4 is a differential voltage-to-current converter.

5. ______ Circuits used with solar cells to measure solar power density are called V-to-I converters.

6. ______ Current flows to the load in a current sink.

7. ______ An AD590 transducer's output current increases as light intensity increases.

C. CIRCLE THE CORRECT ANSWER

Circle the correct answer for each statement.

1. The switch that changes a voltmeter from ac to dc readings is called a (range, function) switch.
2. The op amp's inverting terminal, in an inverting V-to-I detector is at (virtual ground, potential of the input voltage).
3. Current (is, is not) drawn from the source voltage in a non-inverting V-to-I detector.
4. Load current in a V-to-I converter (does, does not) change if load voltage or load resistance changes.
5. In a current-to-voltage converter, output voltage (does, does not) change if input current changes.

D. MATCHING

Match the name or symbol in column **A** with the statement that matches best in column **B**.

	COLUMN A		COLUMN B
1. ______	AC current source	a.	Light detector
2. ______	transducer	b.	Temperature transducer
3. ______	Photo diode	c.	Sun power
4. ______	Solar cell	d.	Energy converter
5. ______	AD 590	e.	Differential V-to-I converter

PROBLEMS

7-1. In Fig. 7-1a and b, $V_i = 5$ V, $R_i = 1$ kΩ, and $R_L = 1$ kΩ. Find load current I_L and output voltage V_o.

7-2. What value is required for R_i in Fig. 7-2a if diode voltage is to be measured at $I = 1$ mA?

7-3. The circuit of Fig. 7-2b is used to measure zener diode voltage at a current of 5 mA. The only available voltage sources is V = −5 V. What changes must be made in the circuit?

7-4. Four LEDS are to be tested at 20 mA in the *V*-to-*I* converter of Fig. 7-4a. The only available voltage for V_i is a 5-V logic supply. Find
(a) the new required value for R_i;
(b) V_{CE} of the transistor if $V_{CC} = 15$ V.

7-5. A 0- to 1-mA dc ammeter is available to make a voltmeter to measure 1.5-V batteries. When the measuring leads are connected to a 1.5-V battery, regardless of polarity, the meter should indicate a full scale. Draw the circuit.

7-6. A 0- to 1-mA meter movement is installed in Fig. 7-6 along with a function switch. Find R_i to measure
(a) ± 1 V dc;
(b) 1 V rms;
(c) ± 1 V peak.

7-7. Draw the schematic for an ac/dc voltmeter that will measure dc rms or peak ac voltages over ranges of 0 to 1 V or 0 to 10 V. (*Hint:* Combine Figs. 7-6 and 7-7.)

7-8. In the current source of Fig. 7-8a, $R_S = 200$ Ω and $R_L = 400$ Ω. Calculate
(a) I_L;
(b) V_L; and
(c) transistor voltage V_{CE}.

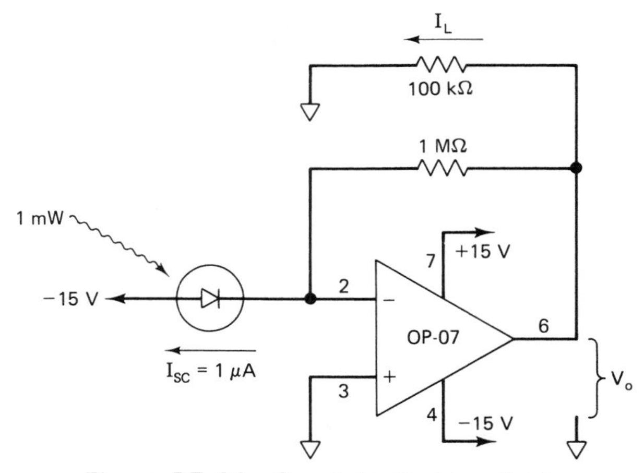

Figure P7-11 Circuit for Problem 7-11.

7-9. All R resistors equal 10 kΩ in the ac current source of Fig. 7-8c. If $V_i = -1$ V, $V_2 = 1$ V and $R_L = 10$ kΩ, find

(a) I_L;

(b) V_L,

(c) Show that $V_o = -5$ V.

7-10. The solar cell of Fig. 7-12 outputs a short-circuit current at 1.0 A when exposed to sunlight with a full sun power density of 100 mW/cm^2. Calculate V_o if the solar cell receives full power.

7-11. The output of a photodiode is converted to a voltage V_o and also a current I_L in Fig. P7-11. Find

(a) V_o;

(b) I_L.

Laboratory Exercise 7.

Name: ____________________

Date: __________ **Grade:** ___

VOLTAGE-TO-CURRENT CONVERTERS

OBJECTIVES: Upon completion of this laboratory exercise on voltage-to-current converters, you will be able to (1) build a zener diode tester; (2) drive two LEDs with equal currents that are independent of the LED voltage drops; (3) make a positive or negative dc voltmeter; (4) make an ac voltmeter; (5) add a function switch to make an ac–dc voltmeter.

REFERENCE: Chapter 7

PARTS LIST

1	OP-07 op amp	1	120-Ω 5%
1	AD584 precision voltage reference (see data sheets 244-48)	1	1 kΩ 5%
		1	3.3 kΩ 5%
1	NPN transistor	1	15 kΩ 5%
4	1N 914 diodes	1	10 kΩ 1%
1	0- to 100-μA meter	1	9.0-kΩ 1% (made from available stock)
2	LEDs		
1	Zener diode, 5V		

Procedure A: Zener Diode Tester

1. Measure resistor R_i that will be used in the circuit of Fig. L7-1. R_i = __________. Build the circuit of Fig. L7-1. Measure the voltage at pin 2 of the op amp. It should be within millivolts of ground potential. Measure V_i. V_i = __________. Predict the value of I_z from $I_z = \frac{V_i}{R_i} \cdot I_z$ = __________.
2. Measure the value of V_o with a DVM to obtain the value of V_z when it conducts a current of I_z. V_o = __________.

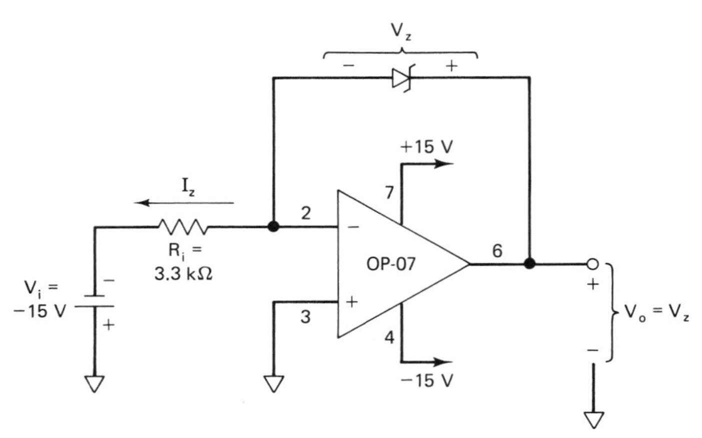

Figure L7-1 Zener diode tester.

3. Reverse the zener diode connections in Fig. L7-1. Measure V_o. V_o = __________. Does V_o equal the typical value of a forward-biased diode? __________. What current value now flows through the zener? __________.

Procedure B: LED Tester

4. Refer to Fig. L7-2. Measure the value of R_i. R_i = __________. Build the circuit and measure V_i with a DMM. Measure the voltage at pin 3 of the OP-07. It should equal V_i within a few millivolts. Predict the value of I from $I = \frac{V_i}{R_i}$. I = __________.
5. Measure the voltage across LED 1, and LED 2 with a DMM. LED

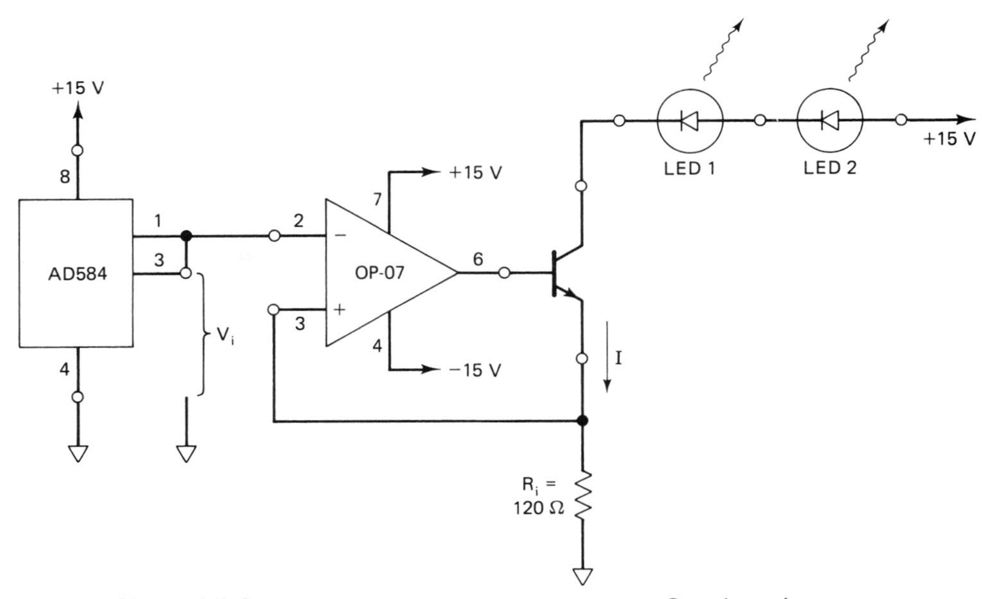

Figure L7-2 Voltage-to-high-current converter. See data sheets pp 244-48.

1, __________; LED 2, __________. (*Note:* The DMM should have floating inputs.) Do the LEDs have equal voltage drops?

__________. Do they conduct equal or unequal currents?

__________. Is the voltage at pin 6 of the op amp higher or lower

than V_i? __________. Measure V_6 to verify your answer. Predict

the voltage drop across the transistor. V_{CE} = __________. Measure V_{CE} to check your answer. If you added another LED in series with

LED 1 and LED 2, would V_{CE} decrease or increase? __________. Suppose that you arbitrarily restrict the minimum V_{CE} to 2.0 V. How many LEDs could you connect in series before V_{CE} dropped

under 2 V? __________. Would every LED conduct the same

current?__________.

Procedure C: Positive or Negative DC Voltmeter

6. Build the circuit of Fig. L7-3. Jumper the 15-kΩ resistor to the

+15V supply. Measure V_i. V_i = __________. Predict meter

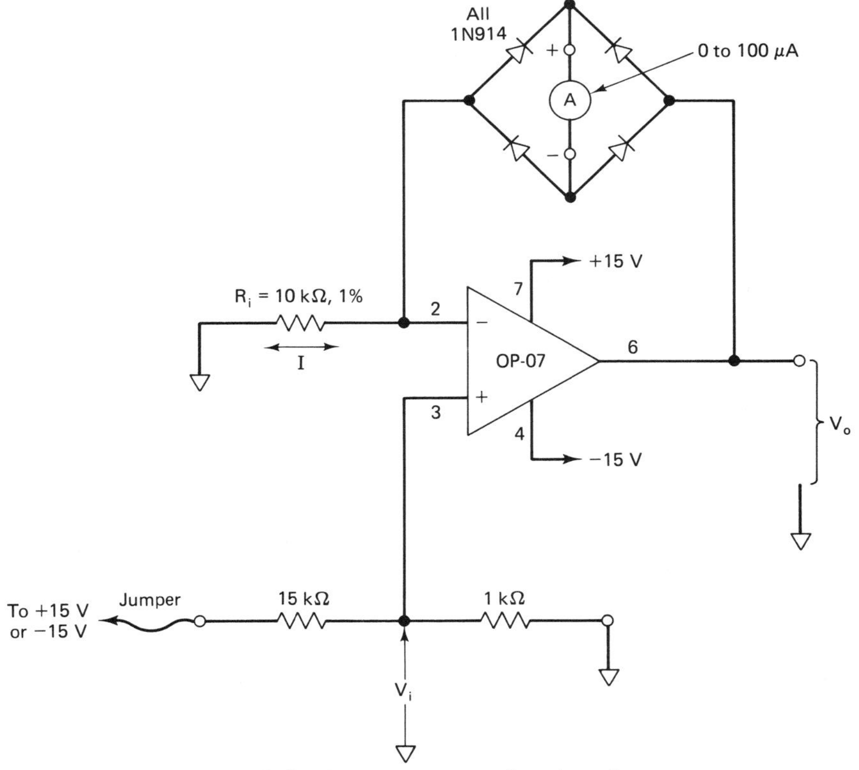

Figure L7-3 Positive or negative dc voltmeter.

current I from $I = \frac{V_i}{R_i}$. Predicted I = __________. What current is indicated by the microammeter? I = __________. (*Note:* The voltages at pins 2 and 3 of the OP-07 should be equal.) Measure the value of V_o. V_o = __________. Explain why V_o is higher than V_i. ____

____

7. Change the jumper wire to power the 15-kΩ resistor from the −15V supply. Measure V_i. V_i = __________. Does the micro ammeter indicate approximately the same value as it did when V_i was positive? __________. Complete the following sentence that refers to Fig. L7-3. When V_i varies from 0 to +1 V or 0 to −1 V, meter current varies from __________.

Procedure D: AC Voltmeter

8. Use any combination of 1% resistors to make up R_i = 9 kΩ for the circuit of Fig. L7-4. Measure R_i. R_i = __________. Build the circuit of Fig. L7-4. Ground pin 3 of the op amp and check that pins 2 and 6 are at ground potential with a dc voltmeter. Remove the ground on pin 3.
9. Connect a sinusoidal voltage from a signal generator to pin 3 as

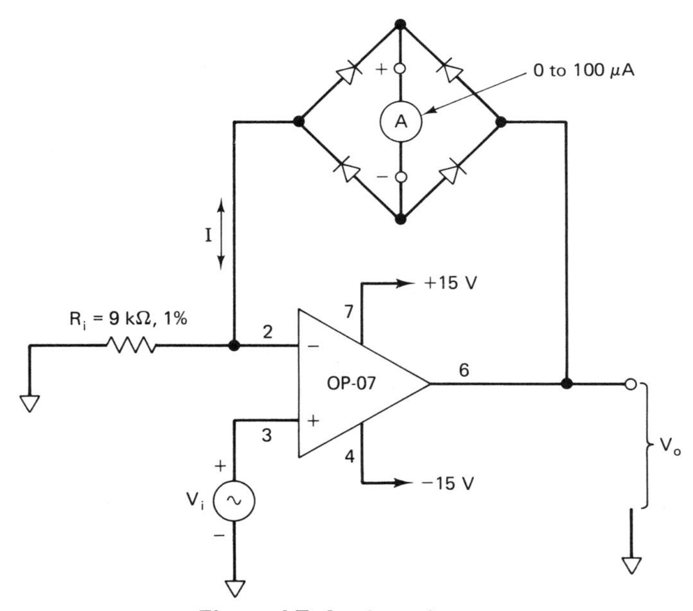

Figure L7-4 Ac voltmeter.

signal V_o. Set the frequency at 100 Hz. Use an ac voltmeter to adjust V_i to 1.0 V rms. What reading is indicated by the microammeter?

$I =$ __________. Reduce V_i by half to $V_i = 0.5$ V rms. What reading is indicated by the microammeter? $I =$ __________. Did the value I halve when V_i was halved? __________.

10. Refer to Fig. 7-6 to solve the following problem. Combine the dc voltmeter circuit of Fig. L7-3 with the ac voltmeter circuit of Fig. L7-4 together with a two-position function switch to make a 0- to 1- V ac or dc voltmeter. Sketch your design in the space below.

CONCLUSION:

Pin Programmable Precision Voltage Reference

AD584*

FEATURES
Four Programmable Output Voltages:
10.000V, 7.500V, 5.000V, 2.500V
Laser-Trimmed to High Accuracies
No External Components Required
Trimmed Temperature Coefficient:
5ppm/°C max, 0 to +70°C (AD584LH)
15ppm/°C max, -55°C to +125°C (AD584TH)
Zero Output Strobe Terminal Provided
Two Terminal Negative Reference
Capability (5V & Above)
Output Sources or Sinks Current
Low Quiescent Current: 1.0mA max
10mA Current Output Capability
Available as a JAN Device

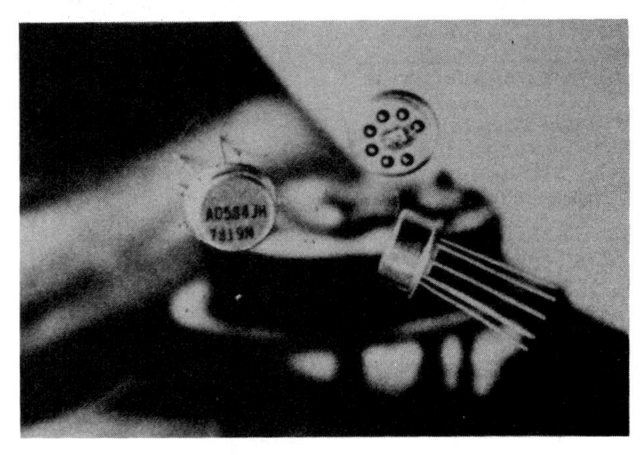

PRODUCT DESCRIPTION

The AD584 is an eight-terminal precision voltage reference offering pin-programmable selection of four popular output voltages: 10.000V, 7.500V, 5.000V and 2.500V. Other output voltages, above, below or between the four standard outputs, are available by the addition of external resistors. Input voltage may vary between 4.5 and 30 volts.

Laser Wafer Trimming (LWT) is used to adjust the pin-programmable output levels and temperature coefficients, resulting in the most flexible high precision voltage reference available in monolithic form.

In addition to the programmable output voltages, the AD584 offers a unique strobe terminal which permits the device to be turned on or off. When the AD584 is used as a power supply reference, the supply can be switched off with a single, low-power signal. In the "off" state the current drain by the AD584 is reduced to about 100μA. In the "on" state the total supply current is typically 750μA including the output buffer amplifier.

The AD584 is recommended for use as a reference for 8-, 10- or 12-bit D/A converters which require an external precision reference. The device is also ideal for all types of A/D converters of up to 14 bit accuracy, either successive approximation or integrating designs, and in general can offer better performance than that provided by standard self-contained references.

The AD584J, K, and L are specified for operation from 0 to +70°C; the AD584S and T are specified for the -55°C to +125°C range. All grades are packaged in a hermetically-sealed eight-terminal TO-99 metal can.

PRODUCT HIGHLIGHTS

1. The flexibility of the AD584 eliminates the need to design-in and inventory several different voltage references. Furthermore one AD584 can serve as several references simultaneously when buffered properly.
2. Laser trimming of both initial accuracy and temperature coefficient results in very low errors over temperature without the use of external components. The AD584LH has a maximum deviation from 10.000 volts of ±7.25mV from 0 to +70°C.
3. The AD584 can be operated in a two-terminal "Zener" mode at 5 volts output and above. By connecting the input and the output, the AD584 can be used in this "Zener" configuration as a negative reference.
4. The output of the AD584 is configured to sink or source currents. This means that small reverse currents can be tolerated in circuits using the AD584 without damage to the reference and without disturbing the output voltage (10V, 7.5V and 5V outputs).
5. The AD584 reference is also available as a JAN device under the part numbers JM38510/12801BGC (similar to the AD584S) and JM38510/12802BGC (similar to the AD584T). Both devices are offered in an eight pin JEDEC TO-99 style header package.

*Covered by U.S. Patent No. 3,887,863; RE 30,586

One Technology Way; Norwood, MA 02062-9106 U.S.A.
Tel: 617/329-4700 **Twx: 710/394-6577**
Telex: 174059 **Cables: ANALOG NORWOODMASS**

*Courtesy of **Analog Devices, Inc.**

SPECIFICATIONS (@ V_{IN} = 15V and 25°C)

Model	AD584J Min	AD584J Typ	AD584J Max	AD584K Min	AD584K Typ	AD584K Max	AD584L Min	AD584L Typ	AD584L Max	Units
OUTPUT VOLTAGE TOLERANCE										
Maximum Error[1] for Nominal										
Outputs of:										
10.000V			**±30**			**±10**			**±5**	mV
7.500V			**±20**			**±8**			**±4**	mV
5.000V			**±15**			**±6**			**±3**	mV
2.500V			**±7.5**			**±3.5**			**±2.5**	mV
OUTPUT VOLTAGE CHANGE										
Maximum Deviation from +25°C										
Value, T_{min} to T_{max}[2]										
10.000, 7.500, 5.000V Outputs			**30**			**15**			**5**	ppm/°C
2.500V Output			**30**			**15**			**10**	ppm/°C
Differential Temperature										
Coefficients Between Outputs		5			3			3		ppm/°C
QUIESCENT CURRENT		0.75	**1.0**		0.75	**1.0**		0.75	**1.0**	mA
Temperature Variation		1.5			1.5			1.5		μA/°C
TURN-ON SETTLING TIME TO 0.1%		200			200			200		μs
NOISE										
(0.1 to 10Hz)		50			50			50		μV p-p
LONG-TERM STABILITY		25			25			25		ppm/1000 Hrs.
SHORT-CIRCUIT CURRENT		30			30			30		mA
LINE REGULATION (No Load)										
$15V \leq V_{IN} \leq 30V$			**0.002**			**0.002**			**0.002**	%/V
$(V_{OUT} + 2.5V) \leq V_{IN} \leq 15V$			**0.005**			**0.005**			**0.005**	%/V
LOAD REGULATION										
$0 \leq I_{OUT} \leq 5mA$, All Outputs		20	**50**		20	**50**		20	**50**	ppm/mA
OUTPUT CURRENT										
$V_{IN} \geq V_{OUT} + 2.5V$										
Source @ +25°C	10			10			10			mA
Source T_{min} to T_{max}	5			5			5			mA
Sink T_{min} to T_{max}	5			5			5			mA
Sink −55°C to +85°C	–			–			–			mA
TEMPERATURE RANGE										
Operating	0		+70	0		+70	0		+70	°C
Storage	−65		+175	−65		+175	−65		+175	°C

NOTES
[1]At Pin 1.
[2]Calculated as average over the operating temperature range.
Specifications subject to change without notice.

Specifications shown in boldface are tested on all production units at final electrical test. Results from those tests are used to calculate outgoing quality levels. All min and max specifications are guaranteed, although only those shown in boldface are tested on all production units.

OUTLINE DIMENSIONS

Dimensions shown in inches and (mm).

TO-99 Package

0.335 (8.50)
0.305 (7.75)
0.185 (4.70)
0.165 (4.19)
0.04 MAX (1.0)
0.5 MIN (12.70)
INSULATION 0.05 MAX (1.27)
0.370 (9.40)
0.335 (8.50)
8 LEADS
0.019 (0.48)
0.016 (0.41) DIA
0.2 TYP (5.1)
45° EQUALLY SPACED
0.045 (1.1)
0.020 (0.51)
0.034 (0.86)
0.027 (0.69)

PIN DESIGNATION

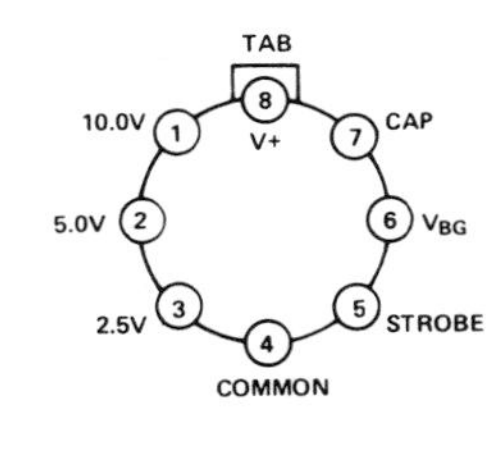

TOP VIEW

SPECIFICATIONS (@ V_{IN} = 15V and 25°C)

Model	AD584S Min	AD584S Typ	AD584S Max	AD584T Min	AD584T Typ	AD584T Max	Units
OUTPUT VOLTAGE TOLERANCE							
Maximum Error[1] for Nominal							
Outputs of:							
10.000V			**±30**			**±10**	mV
7.500V			**±20**			**±8**	mV
5.000V			**±15**			**±6**	mV
2.500V			**±7.5**			**±3.5**	mV
OUTPUT VOLTAGE CHANGE							
Maximum Deviation from +25°C							
Value, T_{min} to T_{max}[2]							
10.000, 7.500, 5.000V Outputs			**30**			**15**	ppm/°C
2.500V Output			**30**			**20**	ppm/°C
Differential Temperature							
Coefficients Between Outputs		5			3		ppm/°C
QUIESCENT CURRENT		0.75	**1.0**		0.75	**1.0**	mA
Temperature Variation		1.5			1.5		μA/°C
TURN-ON SETTLING TIME TO 0.1%		200			200		μs
NOISE							
(0.1 to 10Hz)		50			50		μV p-p
LONG-TERM STABILITY		25			25		ppm/1000 Hrs.
SHORT-CIRCUIT CURRENT		30			30		mA
LINE REGULATION (No Load)							
$15V \leq V_{IN} \leq 30V$			**0.002**			**0.002**	%/V
$(V_{OUT} + 2.5V) \leq V_{IN} \leq 15V$			**0.005**			**0.005**	%/V
LOAD REGULATION							
$0 \leq I_{OUT} \leq 5mA$, All Outputs		20	**50**		20	**50**	ppm/mA
OUTPUT CURRENT							
$V_{IN} \geq V_{OUT} + 2.5V$							
Source @ +25°C	10			10			mA
Source T_{min} to T_{max}	5			5			mA
Sink T_{min} to T_{max}	200			200			μA
Sink −55°C to +85°C	5			5			mA
TEMPERATURE RANGE							
Operating	−55		+125	−55		+125	°C
Storage	−65		+175	−65		+175	°C

NOTES
[1]At Pin 1.
[2]Calculated as average over the operating temperature range.
Specifications subject to change without notice.

Specifications shown in boldface are tested on all production units at final electrical test. Results from those tests are used to calculate outgoing quality levels. All min and max specifications are guaranteed, although only those shown in boldface are tested on all production units.

ABSOLUTE MAX RATINGS

Input Voltage V_{IN} to Ground 40V
Power Dissipation @ +25°C 600mW
Operating Junction Temperature Range . . −55°C to +125°C
Lead Temperature (Soldering 10sec) . . . . +300°C
Thermal Resistance
Junction-to-Ambient . . . . 150°C/W

20-TERMINAL LCC PACKAGE

Dimensions shown in inches and (mm).

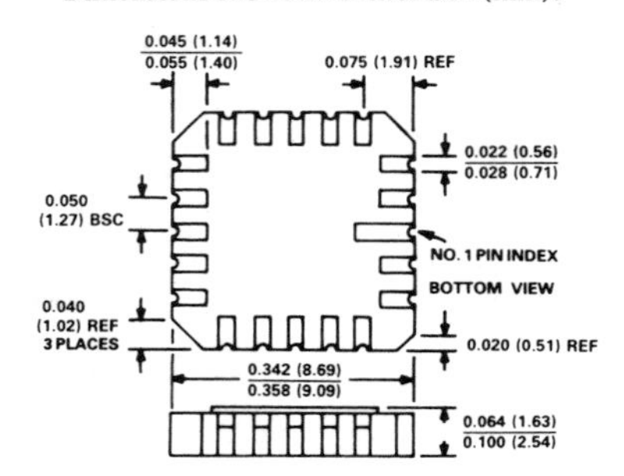

PIN DESIGNATION

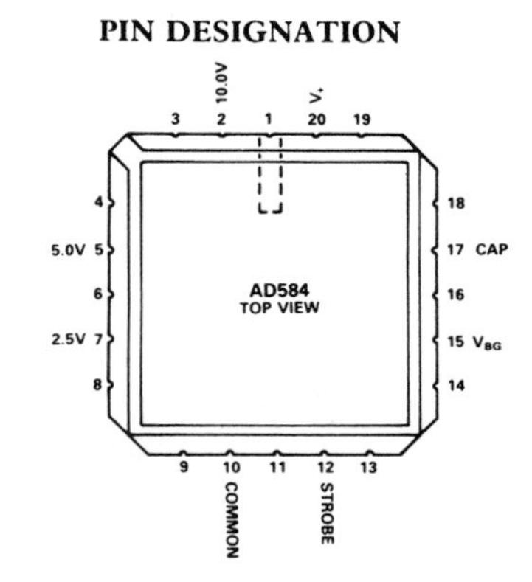

Applying the AD584

APPLYING THE AD584

With power applied to pins 8 and 4 and all other pins open the AD584 will produce a buffered nominal 10.0V output between pins 1 and 4 (see Figure 1). The stabilized output voltage may be reduced to 7.5V, 5.0V or 2.5V by connecting the programming pins as follows:

OUTPUT VOLTAGE	PIN PROGRAMMING
7.5V	Join the 2.5V and 5.0V pins (2) and (3).
5.0V	Connect the 5.0V pin (2) to the output pin (1).
2.5V	Connect the 2.5V pin (3) to the output pin (1).

The options shown above are available without the use of any additional components. Multiple outputs using only one AD584, are also possible by simply buffering each voltage programming pin with a unity-gain noninverting op amp.

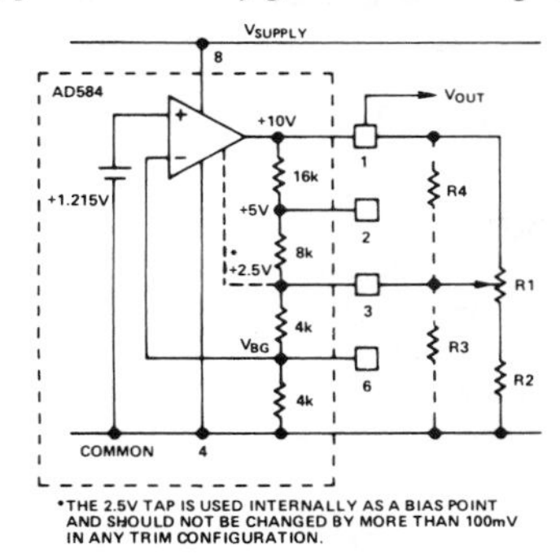

Figure 1. Variable Output Options

The AD584 can also be programmed over a wide range of output voltages, including voltages greater than 10V, by the addition of one or more external resistors. Figure 1 illustrates the general adjustment procedure, with approximate values given for the internal resistors of the AD584. The AD584 may be modeled as an op amp with a noninverting feedback connection, driven by a high stability 1.215 volt bandgap reference (see Figure 3 for schematic).

When the feedback ratio is adjusted with external resistors, the output amplifier can be made to multiply the reference voltage by almost any convenient amount, making popular outputs of 10.24V, 5.12V, 2.56V or 6.3V easy to obtain. The most general adjustment (which gives the greatest range and poorest resolution) uses R1 and R2 alone (see Figure 1). As R1 is adjusted to its upper limit the 2.5V pin 3 will be connected to the output, which will reduce to 2.5V. As R1 is adjusted to its lower limit, the output voltage will rise to a value limited by R2. For example, if R2 is about 6kΩ, the upper limit of the output range will be about 20V even for large values of R1. R2 should not be omitted; its value should be chosen to limit the output to a value which can be tolerated by the load circuits. If R2 is zero, adjusting R1 to its lower limit will result in a loss of control over the output voltage. If precision voltages are required to be set at levels other than the standard outputs, the 20% absolute tolerance in the internal resistor ladder must be accounted for.

Alternatively, the output voltage can be raised by loading the 2.5V tap with R3 alone. The output voltage can be lowered by connecting R4 alone. Either of these resistors can be a fixed resistor selected by test or an adjustable resistor. In all cases the resistors should have a low temperature coefficient to match the AD584 internal resistors, which have a negative T.C. less than 60ppm/°C. If both R3 and R4 are used, these resistors should have matched temperature coefficients.

When only small adjustments or trims are required, the circuit of Figure 2 offers better resolution over a limited trim range. The circuit can be programmed to 5.0V, 7.5V or 10V and adjusted by means of R1 over a range of about ±200mV. To trim the 2.5V output option, R2 (Figure 2) can be reconnected to the bandgap reference (pin 6). In this configuration, the adjustment should be limited to ±100mV in order to avoid affecting the performance of the AD584.

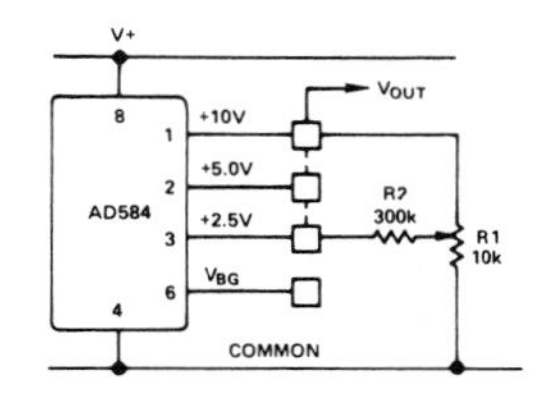

Figure 2. Output Trimming

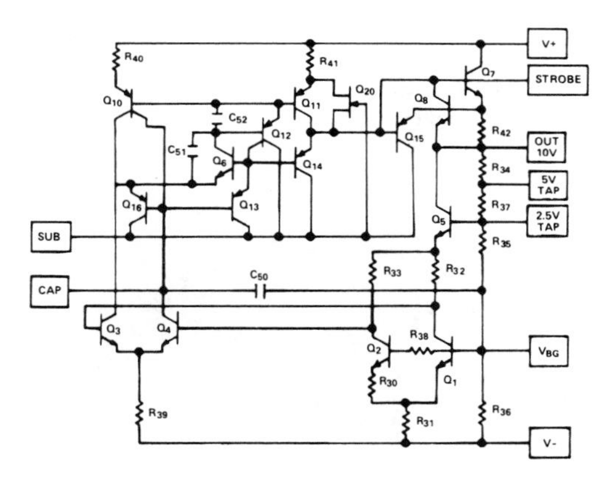

Figure 3. Schematic Diagram

PERFORMANCE OVER TEMPERATURE

Each AD584 is tested at five temperatures over the -55°C to +125°C range to ensure that each device falls within the Maximum Error Band (see Figure 4) specified for a particular grade (i.e., S and T grades); three-point measurement guarantees performance within the error band from 0 to +70°C (i.e., J, K, or L grades). The error band guaranteed for the AD584 is the maximum deviation from the initial value at +25°C. Thus, given the grade of the AD584, the designer can easily determine the maximum total error from initial tolerance plus temperature variation. For example, for the AD584T, the initial tolerance is ±10mV and the error band is ±15mV. Hence, the unit is guaranteed to be 10.000 volts ±25mV from -55°C to +125°C.

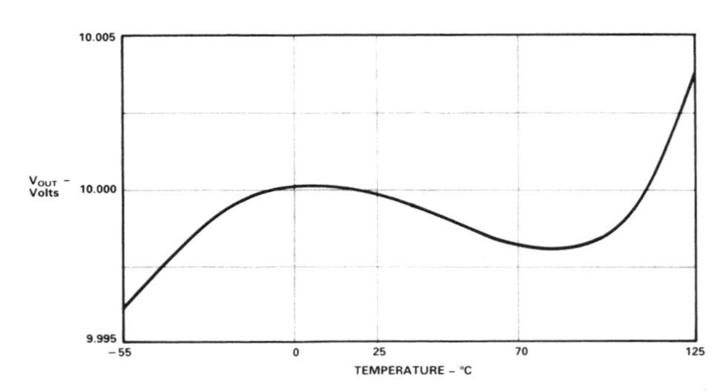

Figure 4. Typical Temperature Characteristic

OUTPUT CURRENT CHARACTERISTICS

The AD584 has the capability to either source or sink current and provide good load regulation in either direction, although it has better characteristics in the source mode (positive current into the load). The circuit is protected for shorts to either positive supply or ground. The output voltage vs. output current characteristics of the device is shown in Figure 5. Source current is displayed as negative current in the figure; sink current is positive. Note that the short circuit current (i.e., zero volts output) is about 28mA; when shorted to +15 volts, the sink current goes to about 20mA.

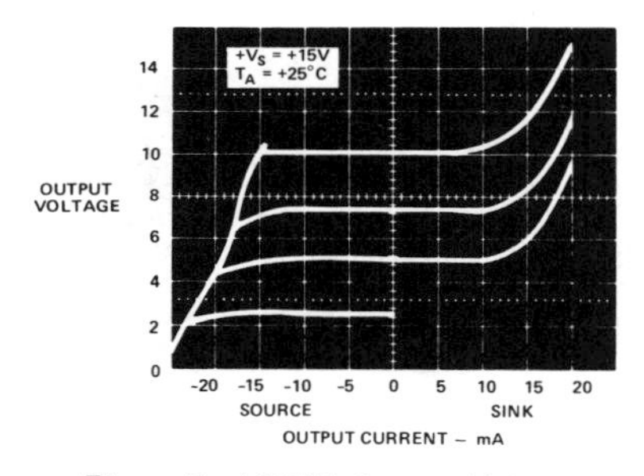

Figure 5. AD584 Output Voltage vs. Sink and Source Current

DYNAMIC PERFORMANCE

Many low power instrument manufacturers are becoming increasingly concerned with the turn-on characteristics of the components being used in their systems. Fast turn-on components often enable the end user to keep power off when not needed, and yet respond quickly when the power is turned on for operation. Figure 6 displays the turn-on characteristic of the AD584. Figure 6a is generated from cold-start operation and represents the true turn-on waveform after an extended period with the supplies off. The figure shows both the coarse and fine transient characteristics of the device; the total settling time to within ±1 millivolt is about 180μs, and there is no long thermal tail appearing after the point.

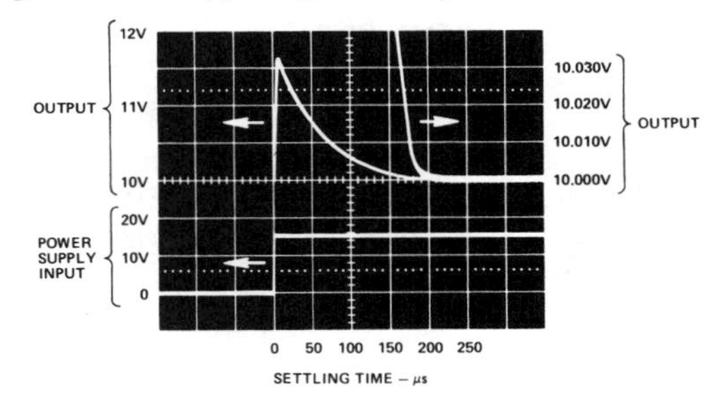

Figure 6. Output Settling Characteristic

NOISE FILTERING

The bandwidth of the output amplifier in the AD584 can be reduced to filter the output noise. A capacitor ranging between 0.01μF and 0.1μF connected between the Cap and V_{BG} terminals will further reduce the wideband and feedthrough noise in the output of the AD584, as shown in Figure 8.

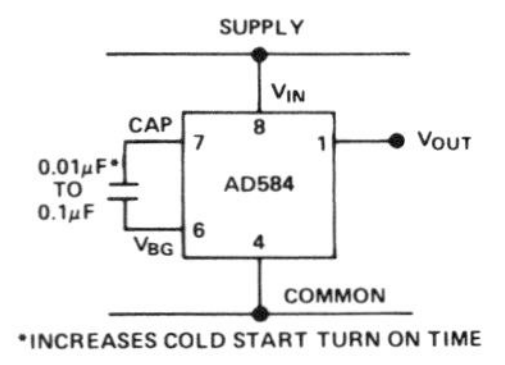

Figure 7. Additional Noise Filtering with an External Capacitor

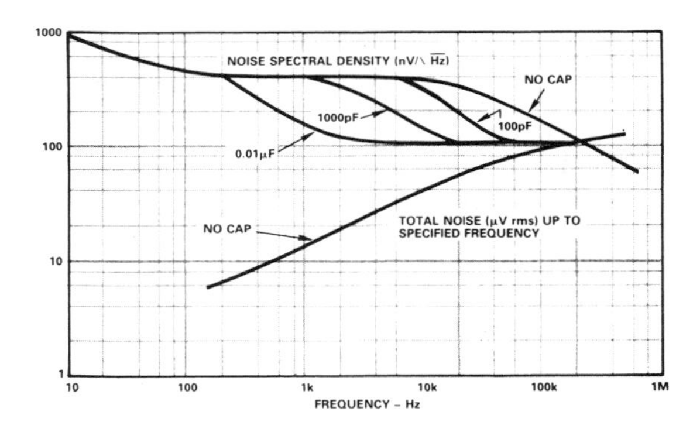

Figure 8. Spectral Noise Density and Total rms Noise vs. Frequency

Instrumentation Amplifiers

8

LEARNING OBJECTIVES

Upon completion of this chapter on the instrumentation amplifier, you will be able to:

- Determine when to use an instrumentation amplifier rather than an inverting or noninverting amplifier.
- Set the differential voltage gain of an instrumentation amplifier.
- Describe what conditions cause the common-mode voltage gain to be greater than zero.
- Use the manufacturer's data sheet value of CMRR to predict circuit performance.
- Wire an IA to amplify differential voltages while simultaneously rejecting common-mode voltages
- Discuss what major improvements over the basic differential amplifier are achieved with the three op-amp instrumentation amplifier.
- Use the reference terminal of an instrumentation amplifier to offset output voltage by any positive or negative voltage level.
- Wire an IA for remote voltage sensing.
- Program the gain of a commercially available INA110 instrumentation amplifier for differential voltage measurements.
- Use a strain gauge and an instrumentation amplifier to measure either tension or compression on a structure.

8-0 INTRODUCTION

This chapter centers on the ability of both differential and instrumentation amplifiers to reject any voltage that is common to both their inputs while simultaneously amplifying the voltage difference between inputs. These characteristics make this amplifier type the correct choice for a large number of industrial problems where a very small signal (differential input) must be amplified in the presence of large amounts of noise or background (common-mode input) voltage.

The difference amplifier is presented first with a discussion of both differential voltage gain, A_{DIFF}, and common-mode voltage gain, A_{CM}. Also included here is the "figure of merit," or CMRR, which describes the ability of these circuits to amplify differential inputs while rejecting common-mode inputs.

Differential amplifier limitations such as low, unequal input resistances are overcome with a three-op-amp type of instrumentation amplifier. This circuit configuration will also allow for the adjustment of A_{DIFF} with the change of only one resistor. This method of gain adjustment preserves good CMRR.

A commercially available instrumentation amplifier from Burr-Brown, the INA110, extends the IA's capabilities to include (1) gain selection by pin programming, (2) remote voltage sensing, and (3) output voltage offset. Finally, the INA110 IA is combined with a strain gauge to solve the practical problem of detecting very small changes in length on a mechanical structure due to stress.

8-1 BASIC DIFFERENTIAL AMPLIFIER

8-1.1 Differential Voltage Gain

A basic differential amplifier is shown in Fig. 8-1. It consists of a single op amp plus four precision 0.1% resistors. It is capable of producing an output signal that is equal to the *difference voltage* appearing at its input terminals. Two single-ended input voltages, V_1 and V_2, are applied to the (+input) and (−input) terminals, respectively. These voltages are connected through equal 100-kΩ input resisters R to the corresponding (+) and (−) op-amp inputs. The circuit is completed with a gain setting feedback resistor gR, plus a terminating resistor gR. The R resistors must be equal and the gR resistors must be equal. This is necessary for

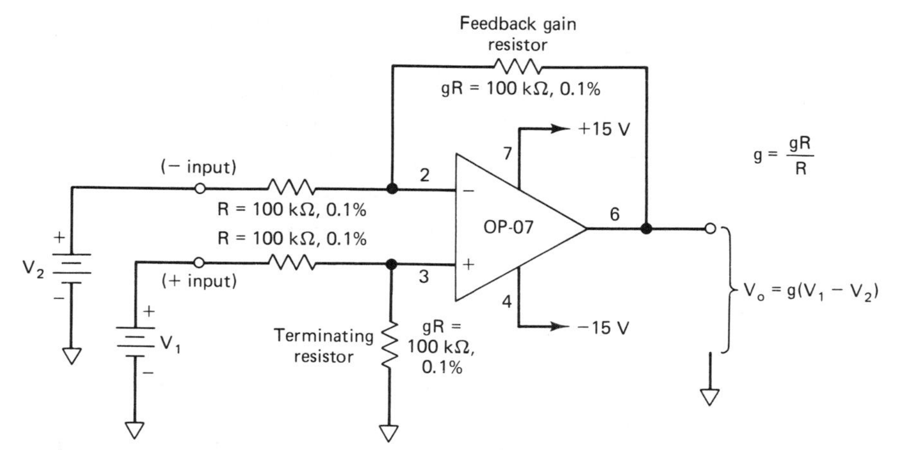

Figure 8-1 A basic differential amplifier circuit has an output voltage V_o that depends on the difference between inputs V_1 and V_2.

maximum CMRR. However, for this circuit, the R and gR resistors are chosen to be equal at 100 kΩ. The circuit *multiplication* or *gain factor* g is then $g = gR/R = 100\ \text{k}\Omega/100\ \text{k}\Omega = 1$.

Let's use the superposition theorem to analyze how the circuit in Fig. 8-1 operates. With V_2 set to 0 V (replaced by a short circuit), the circuit becomes a noninverting amplifier. The output voltage, due only to V_1, is then given as $+gV_1$. With V_1 set to 0 V (replaced by a short circuit) the circuit becomes an inverting amplifier with a closed-loop gain of $A_{CL} = -g$. V_2 is therefore amplified by this gain to produce an output voltage of $-gV_2$.

With *both* V_1 and V_2 applied, output voltage V_o is $gV_1 - gV_2$, or

$$V_o = g(V_1 - V_2) \tag{8-1}$$

Equation (8-1) indicates that the output voltage V_o is equal to the *voltage difference* between V_1 and V_2 multiplied by the gain factor g. The *differential voltage gain* A_{DIFF} can be specified as

$$A_{DIFF} = g = \frac{gR}{R} = \frac{V_o}{V_1 - V_2} \tag{8-2}$$

It should be noted that V_o does not depend on the magnitude of V_1 and V_2, but only on the voltage *difference* between them.

EXAMPLE 8-1

Determine the output voltage V_o of Fig. 8-1 if (a) $V_1 = 835$ mV and $V_2 = 830$ mV; (b) $V_1 = 835$ mV and V_2 is changed to 0 V.

SOLUTION The differential gain $A_{DIFF} = g = gR/R = 100\ \text{k}\Omega/100\ \text{k}\Omega = 1$. From Eq. (8-1) output voltage V_o is

(a) $V_o = g(V_1 - V_2) = 1\ (835\ \text{mV} - 830\ \text{mV}) = 5\ \text{mV}$

(b) $V_o = 1\ (835\ \text{mV} - 0\ \text{V}) = 835\ \text{mV}$

EXAMPLE 8-2

Revise the circuit of Fig. 8-1 for a differential gain of (a) 10; (b) 100.

SOLUTION (a) To increase the differential gain to 10, make the gain factor g equal to 10 [Eq. (8-2)]. If the input resistors R are left at 100 kΩ, then the gR resistors are to be increased to $gR = g(R) = 10(100\ \text{k}\Omega) = 1\ \text{M}\Omega$. Replace both the feedback and terminating resistors with 1 MΩ, 0.1% resistors.

(b) For a gain of 100, first reduce both input resistors to 10 kΩ. Then both gR resistors are equal to $g(R) = 100(10\ \text{k}\Omega) = 1\ \text{M}\Omega$.

8-1.2 Common-Mode Voltage Gain

Equation (8-1) reveals that the output voltage of an ideal differential amplifier is equal to 0 V if the voltage V_1 is made equal to V_2 (i.e., $V_1 = V_2$). $V_o = g(V_1 - V_2) = g(0\ \text{V}) = 0\ \text{V}$. As shown in Fig. 8-2, this is accomplished by wiring both the (+ input) and (− input) terminals to a *common* voltage source. This voltage is called *common-mode* input voltage V_{CM}.

For the ideal amplifier, any common-mode input voltage should be

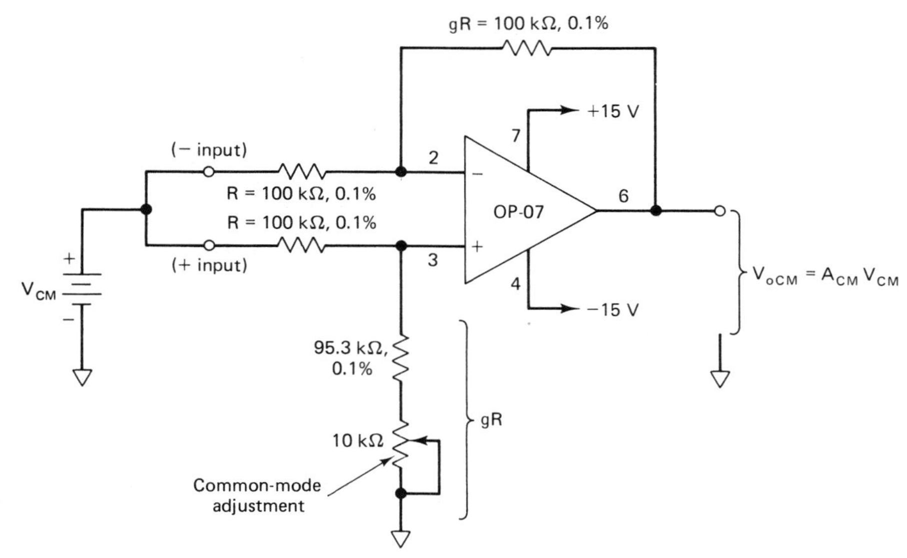

Figure 8-2 Basic differential amplifier circuit in its common-mode configuration. The 25-kΩ potentiometer is adjusted until V_{OCM} is as small as possible.

amplified identically by both the inverting and noninverting sides of the op amp to produce a *common-mode* output voltage V_{OCM} of precisely O V. Common-mode voltage gain A_{CM} is expressed as

$$A_{CM} = \frac{V_{OCM}}{V_{CM}} \tag{8-3}$$

For the ideal differential amplifier A_{CM} is equal to zero.

In a practical differential amplifier, two factors contribute to the presence of a small value of common-mode output voltage. First, it is difficult to manufacture op amps with symmetrical input stages. Second, any slight mismatch between the R/gR ratio in the inverting path and the R/gR ratio in the noninverting path will result in different values of positive and negative gains. Any slight gain mismatch will produce some common-mode output voltage.

A practical compensation technique to minimize resistor mismatch is to build the gR terminating resistor with a resistor–potentiometer combination as shown in Fig. 8-2. Then with a common-mode voltage applied to the input terminals, a trim pot is adjusted until V_{OCM} is reduced to its minimum value. For improved performance the trim pot selected should be a high-quality 10-turn unit.

EXAMPLE 8-3

The common-mode configuration of Fig. 8-2 is tested in the laboratory with an V_{CM} of 1.5 V. It is found that V_{OCM} can be trimmed to a minimum of 2 mV by adjustment of the gR trim pot. Solve for the value of common-mode voltage gain.

SOLUTION From Eq. (8-3),

$$\text{ACM} = \frac{V_{OCM}}{V_{CM}} = \frac{0.002\text{ V}}{1.5\text{ V}} = 0.00133$$

8-1.3 Common-Mode Rejection Ratio

The *common-mode rejection ratio* (CMRR) defines the ability of an op amp to amplify differential input signals while simultaneously rejecting (attenuating) common-mode input signals. CMRR is expressed in equation form as

$$\text{CMRR} = \frac{A_{\text{DIFF}}}{A_{\text{CM}}} \tag{8-4}$$

Typically, differential amplifiers have a large differential gain A_{DIFF} and a small common-mode gain A_{CM}. The value of CMRR is usually an extremely large number. The manufacturers therefore specify on their data sheet, the common-mode rejection ratio in units of decibels. Therefore, CMRR_{dB} is given as

$$\text{CMRR}_{\text{dB}} = 20 \log \text{CMRR} \tag{8-5}$$

It is wise to select a device with a large value of CMRR to ensure that common-mode output voltage will be extremely small.

EXAMPLE 8-4

(a) Solve for the CMRR of a differential amplifier with a $A_{\text{DIFF}} = 10$ and a $A_{\text{CM}} = 0.001333$. (b) Express the results in decibels.

SOLUTION (a) Using Eq. (8-4),

$$\text{CMRR} = \frac{A_{\text{DIFF}}}{A_{\text{CM}}} = \frac{10}{0.001333} = 7500$$

(b) $\text{CMRR}_{\text{dB}} = 20 \log 7500 = 20\,(3.875) = 77.5$ dB.

To predict circuit performance it will be necessary to convert the common-mode rejection ratio given by the manufacturer on their data sheet from decibels to basic units. Solving Eq. (8-5) for CMRR yields

$$\text{CMRR} = 10^{\frac{\text{CMRRdB}}{20}} \tag{8-6}$$

8-1.4 Measurements with a Basic Differential Amplifier

The ability of the differential amplifier to reject a common-mode voltage while simultaneously amplifying difference voltages makes it the correct selection for solving a large variety of industrial problems. To illustrate its usefulness, let's choose (from the biomedical field) the problem of monitoring a patient's cardiac response as a result of heart action. This is quite a difficult task since the typical heart signal is 1 mV or less.

The environment in which the human body exists is exceedingly noisy, with undesired body voltages as high as several hundred millivolts. The solution to this problem is illustrated in Fig. 8-3a. The differential amplifier is attached to the body through two silver-plated electrodes

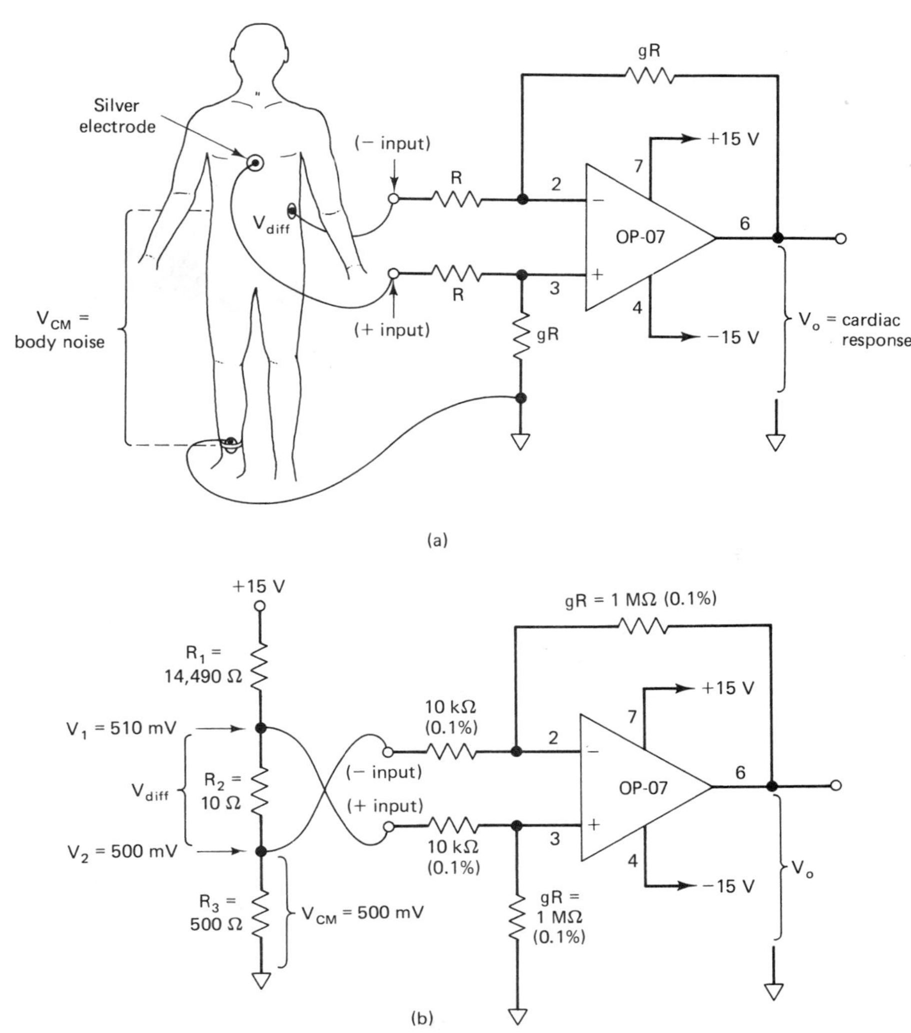

Figure 8-3 The differential amplifier can separate a cardiac signal from body noise signals as in (a). A safe way of duplicating the cardiac measurement is shown in (b). (a) Cardiac signal V_1–V_2 is amplified by the differential amplifier. Common-mode noise voltage is attenuated by CMRR of the amplifier; (b) circuit for Example 8-5. V_2 represents body noise. V_{DIFF} represents the cardiac signal.

(with electrode paste). In this way, the cardiac signal becomes a differential input voltage $V_1 - V_2 = V_{DIFF}$ and will be amplified by A_{DIFF}.

A third electrode, strapped to the right leg, is attached to the ground. This forces body noise to become a common-mode input voltage that will be attenuated by A_{CM} of the differential amplifier. The output voltage V_o will then contain the amplified heart signal and body noise will be all but removed. The electrical principles demonstrated in Fig. 8-3a are investigated by the laboratory test circuit of Fig. 8-3b and the following example.

EXAMPLE 8-5

In this example, the voltage V_{DIFF} across R_2 represents the cardiac signal to be amplified. Voltage V_{CM} across R_3 represents the common-mode voltage to be rejected. Solve for a (a) V_{OCM} if $CMRR_{dB} = 90$ dB; (b) V_{DIFF}; (c) the amplified output voltage due only to the cardiac signal.

SOLUTION (a) From Eq. (8-6), CMRR $= 10^{90/20} = 31{,}623$. The differential gain is

$$A_{DIFF} = g = \frac{gR}{R} = \frac{1\ M\Omega}{10\ k\Omega} = 100$$

From Eq. (8-4),

$$A_{CM} = \frac{A_{DIFF}}{CMRR} = \frac{100}{31{,}623} = 0.00316$$

The output voltage V_{OCM} is then $V_{OCM} = A_{CM} V_{CM}$, where $V_{CM} = V_2$:

$V_{OCM} = 0.00316 \times 500\ mV = 1.58\ mV$

(b) $V_{DIFF} = V_1 = V_2 = 510\ mV - 500\ mV = 10\ mV$
(c) From Eq. (8-1),

$$V_o = A_{DIFF} (V_1 - V_2) = 100(10\ mV) = 1\ V$$

The results of Example 11-5 show that the amplified cardiac signal of 1.0 V exists in the presence of a 1.58 mV noise voltage at the output of the op amp.

8-2 INSTRUMENTATION AMPLIFIER

8-2.1 Disadvantages of the Basic Differential Amplifier

One disadvantage of the basic differential amplifier in Fig. 8-1 is the unequal input resistances presented to V_1 and V_2. This can be overcome by buffering each input of the differential amplifier with a voltage follower as in Fig. 8-4. Input resistances seen by V_1 and V_2 are equal and approach very large values (infinity).

To change the voltage gain of a basic differential amplifier, *both gR* resistors must be adjusted synchronously to maintain a high CMRR. The solution to this second disadvantage is to provide voltage gain adjustment by changing only one resistor, as shown in the three-op-amp instrumentation amplifier of Section 8-2.2.

8-2.2 Three-Op-Amp Instrumentation Amplifier

The three-op-amp instrumentation amplifier in Fig. 8-4 consists of two stages. The buffered input stage uses two OP-07 op amps and three resistors. It provides both high input resistance and gain adjustment, with a single resistance change of mR. A second differential amplifier stage uses Burr-Brown's INA105 unity-gain differential amplifier with four on-chip precision resistors.

We begin the analysis of Fig. 8-4 by first investigating the input stage. A_1 and A_2 are voltage followers to buffer the inputs. This eliminates the problem of low input resistance typical of a basic differential amplifier.

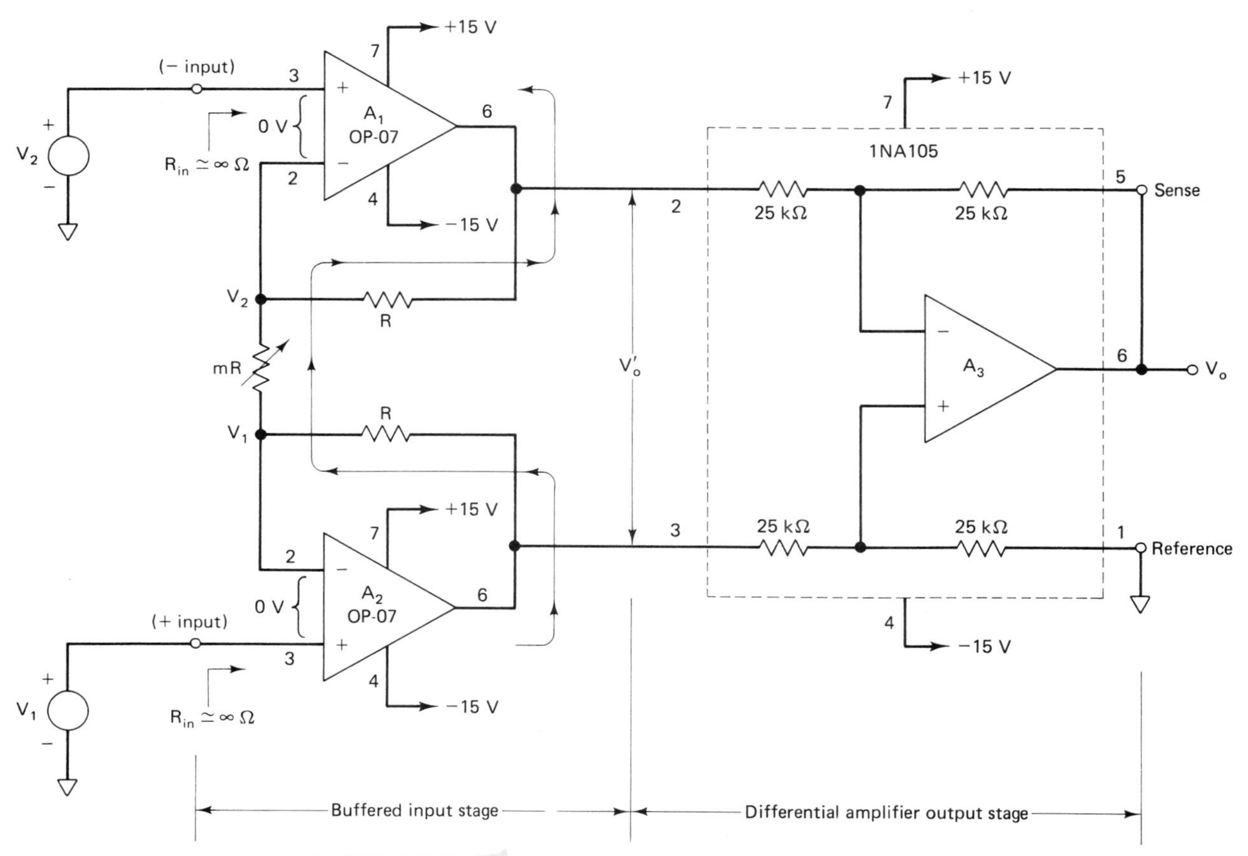

Figure 8-4 The basic instrumentation amplifier has buffered inputs to present high impedance to input V_1 and V_2. Gain is adjusted by a single resistor *mR*. Output V_O is single-ended and responds only to the difference between V_1 and V_2.

Voltage sources V_1 and V_2 "see" essentially infinite input resistance since they are now connected directly to the noninverting input terminals of the OP-07 op amps. The inputs are immune to any small mismatch in source resistance and excellent CMRR can be maintained.

An *mR* potentiometer and two fixed resistors of value *R* provide easy adjustment of voltage gain. The differential input voltage, $V_1 - V_2$, to differential output voltage V_o' (or gain expression) is dependent only on the value of the *mR* resistor, as follows: Assume that $V_1 > V_2$. Negative feedback forces E_d of both A_1 and A_2 to be 0 V. The voltage V_1 appears at the bottom of mR and V_2 appears at the top of *mR*. The voltage developed across the *mR* resistor is then the differential voltage $V_1 - V_2$. Since $V_1 > V_2$, the current *I* must flow in the direction shown and is the same through all three resistors. This current can be expressed as

$$I_{mR} = \frac{V_1 - V_2}{\text{mR}} \tag{8-7}$$

The differential output voltage V_o becomes

$$V_o' = \frac{V_1 - V_2}{mR}(2R + mR) \tag{8-8}$$

The gain expression for the input stage of Fig. 8-4 is

$$A_{\text{DIFF}} = \frac{V_o'}{V_1 - V_2} = 1 + \frac{2}{m} \tag{8-9a}$$

where

$$m = \frac{mR}{R} \tag{8-9b}$$

EXAMPLE 8-6

In Fig. 8-4, $V_1 = 5.025$ V and $V_2 = 5.000$ V. If $R = 20$ kΩ and mR is adjusted to precisely 404 Ω, solve for (a) the differential voltage gain of the input stage; (b) the voltage V_o'.

SOLUTION (a) Using Eq. (8-9b), the resistor ratio $m = mR/R = 404$ Ω/20 kΩ = 0.0202. From Eq. (8-9),

$$A_{\text{DIFF}} = 1 + \frac{2}{m} = 1 + \frac{2}{0.0202} = 100$$

(b) The differential output voltage V_o' is then determined from Eq. (8-9).

$$\begin{aligned} V_o' &= A_{\text{DIFF}}(V_1 - V_2) \\ &= 100(5.025\text{ V} - 5.000\text{ V}) \\ &= 100(0.025\text{ V}) = 2.5\text{ V} \end{aligned}$$

The INA105 unity-gain differential amplifier in Fig. 8-4 converts the differential output V_o' of the input stage to a single-ended output V_o. Housed in a conventional mini-DIP package, the INA105 consists of an op amp A_3 and four resistors, laser trimmed to precisely 25 kΩ. The gain of the second stage is unity. Therefore, the instrumentation amplifier's gain is all achieved in the first stage. The IA's gain from differential input $V_1 - V_2$ to single-ended output V_o is

$$V_o = A_{\text{DIFF}}(V_1 - V_2) \tag{8-10}$$

where A_{DIFF} is found from Eq. (8-9a).

To provide flexibility, the 1NA105's feedback resistor is brought out to pin 5 and labeled the *sense* terminal. Normal differential amplifier operation requires that this terminal be connected to the output at pin 6. Pin 1 is labeled as the *reference* terminal and is usually connected to ground. Both sense and reference terminals are discussed in the next two sections.

8-2.3 Offsetting the Output Voltage

Normally, output voltage of an instrumentation amplifier is the product of the differential input voltage times A_{DIFF} as in Eq. (8-10). It may be desirable to offset V_o by a dc voltage level. For example, when the output of an instrumentation amplifier drives the pen of a chart recorder, it is often convenient to be able to position the pen at a point other than its mechanical zero.

Output offsetting can be accomplished as shown in Fig. 8-5 by applying a dc voltage to reference terminal pin 1. Figure 8-5a illustrates how a +2.5-V offset is established using an AD584 precision reference source. The output voltage of the instrumentation amplifier is now

$$V_o = V_{\text{ref}} + A_{\text{DIFF}}\,(V_1 - V_2) \tag{8-11}$$

When an adjustable bipolar offset is required, V_{ref} is constructed as in Fig. 8-5b. To minimize CMRR error, the effective resistance placed in series with pin 1 must always be a small value. Always connect pin 1 to the output terminal of an op amp.

8-2.4 Remote Voltage Sensing

An instrumentation amplifier is connected to a remote load R_L in Fig. 8-6a. Load current flows through wire resistance R_{wo}. The resulting voltage drop reduces V_o below the value predicted by Eq. (8-10). To eliminate the effect of this voltage drop, the sense and reference terminals are wired to the load (rather than as in Fig. 8-4).

Equal-length wires of identical gauge assure that both 25-kΩ resistors are increased by an equal amount of wire resistance R_w. This maintains good CMRR. The voltage drop across R_{wo} is inside the feedback loop and has no effect on V_o. V_o is again determined from Eq. (8-10). If more load current is required than can be delivered by the instrumentation amplifier, a current boost should be added as in Fig. 8-6b.

8-3 INA110 INSTRUMENTATION AMPLIFIER

8-3.1 Internal Architecture

The internal architecture of Burr-Brown's INA110 instrumentation amplifier (see data sheets 281-85) is shown in Figure 8-7a. Its schematic representation is given in Fig. 8-7b. This design is based on a three-op-amp configuration but with FET buffers added at both inputs to achieve a very high input resistance in the order of teraohms (10^{12} Ω).

Four internal gain-setting resistors are available as part of the input stage (A_1 and A_2). This easily allows the user to change the IA's voltage gain. The output stage (A_3) is a unity-gain differential amplifier with four precision-matched 10-kΩ resistors designed to assure high CMRR.

For the INA110, a minimum CMRR of 90 dB is typical when the instrumentation amplifier is programmed for a gain of 1. The CMRR improves to 104 dB for a gain of 10. For gains about 100, the CMRR equals 110 dB. Versatility is maintained by bringing both *sense* and *reference* terminals out to pins 10 and 6, respectively, to allow remote voltage sensing or output offset.

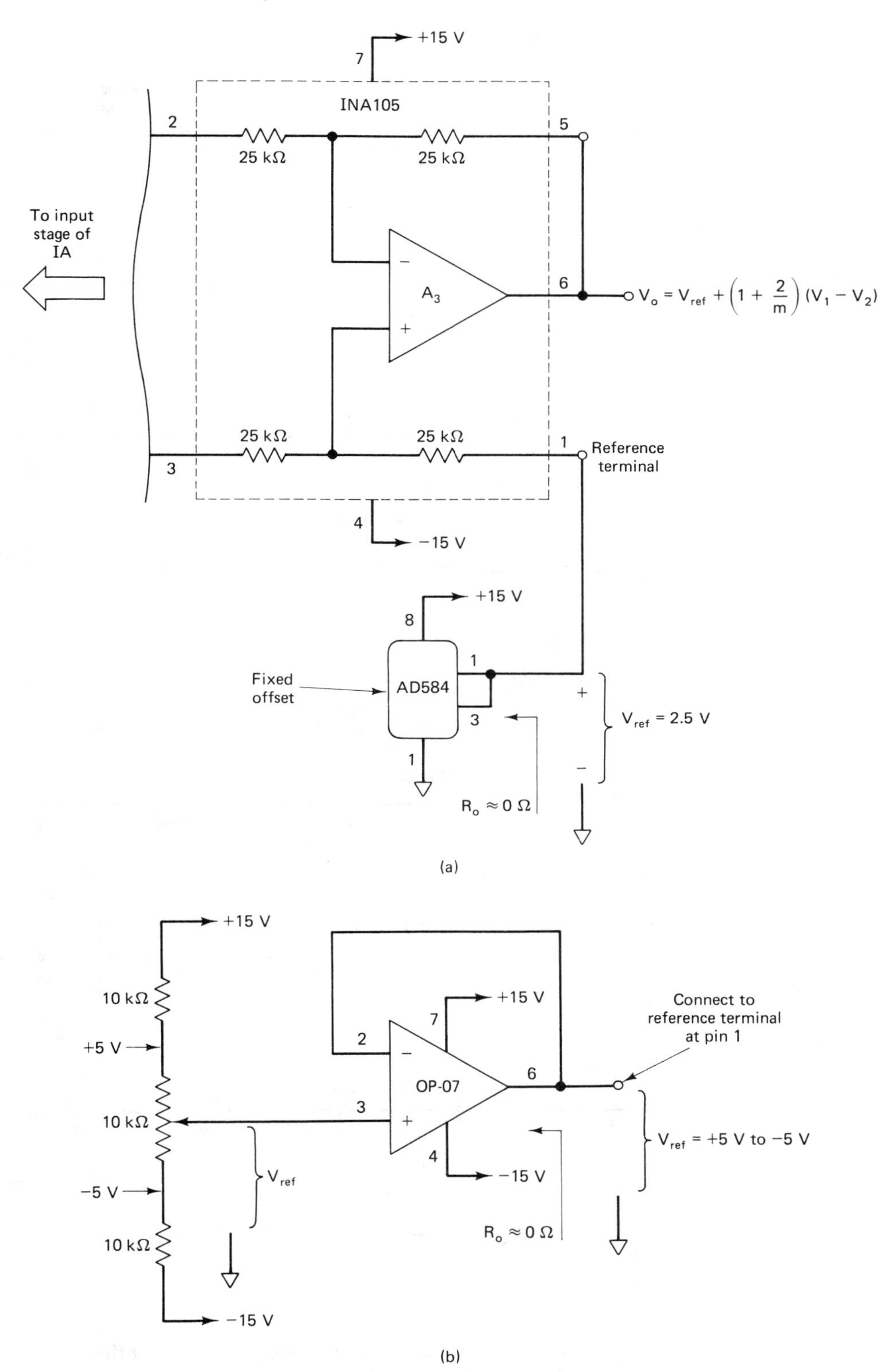

Figure 8-5 The output of an instrumentation amplifier can be offset by connecting a voltage to the reference terminal as in (a). The variable offset voltage in (b) must have zero output impedance. Therefore, the divider is buffered with an op amp. (a) Any voltage applied to the reference terminal offsets output V_O by the same amount; (b) an adjustable offset voltage is made with a voltage divider and a voltage follower.

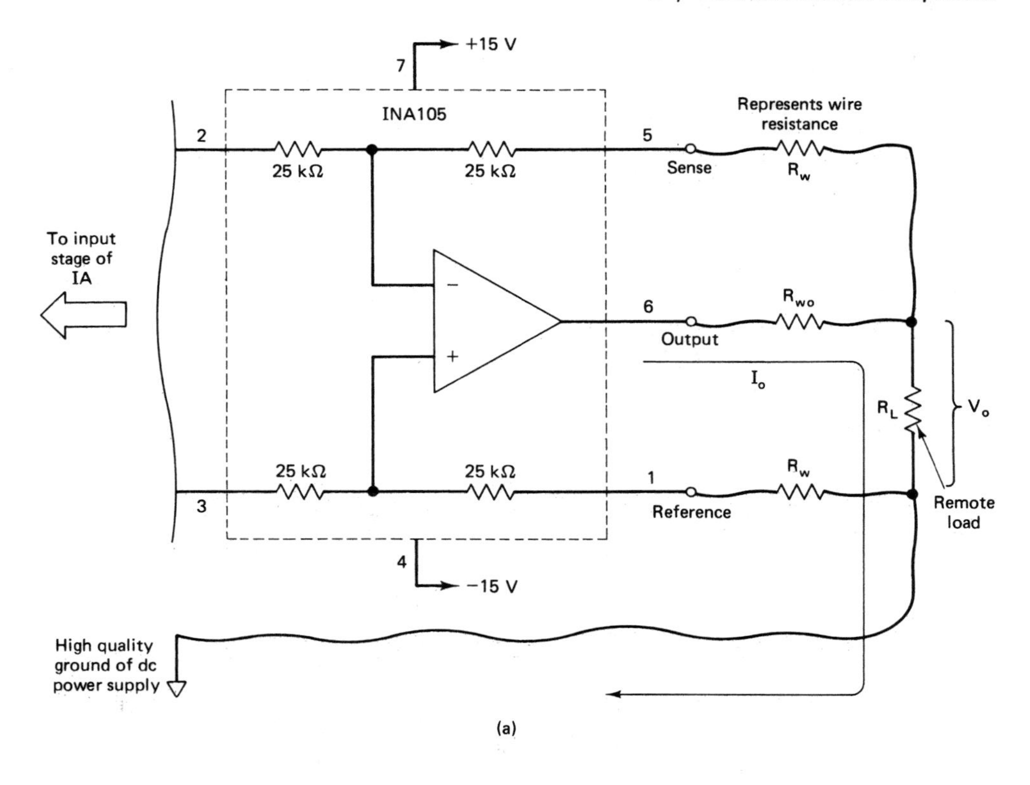

(a)

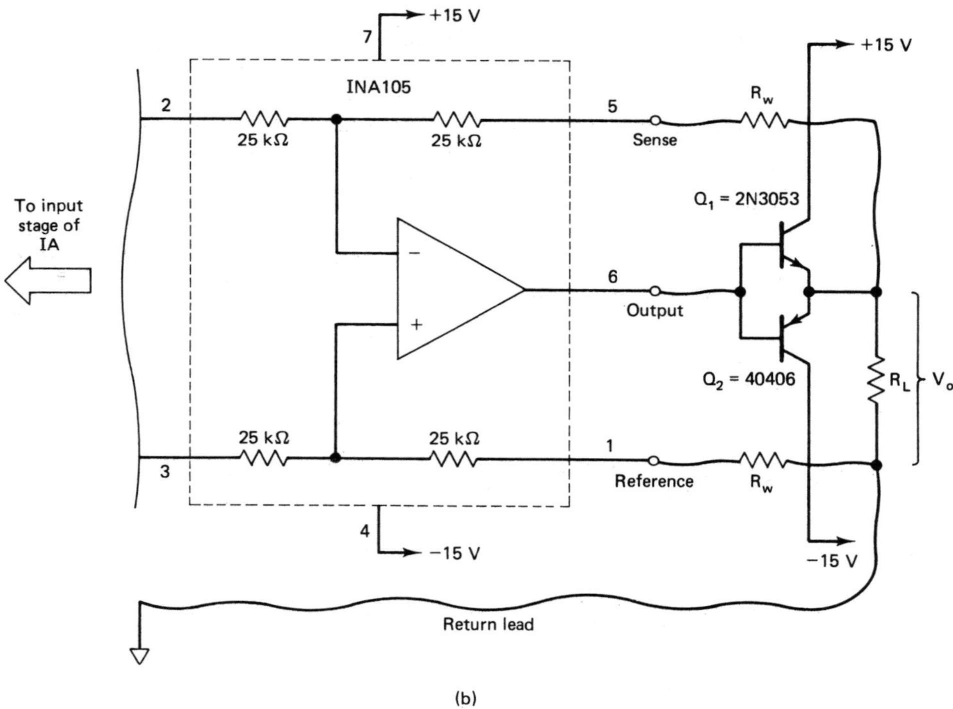

(b)

Figure 8-6 Remote load sensing eliminates the effect of wire resistance in (a) or transfer V_{BE} in (b). (a) The sense and reference terminals are extended to the load and minimizes errors due to wire resistance; (b) load current capacity is increased by boast transisters Q_1 and Q_2. The sense and reference terminals are extended to R_L and eliminate any V_{BE} drop of the transistors.

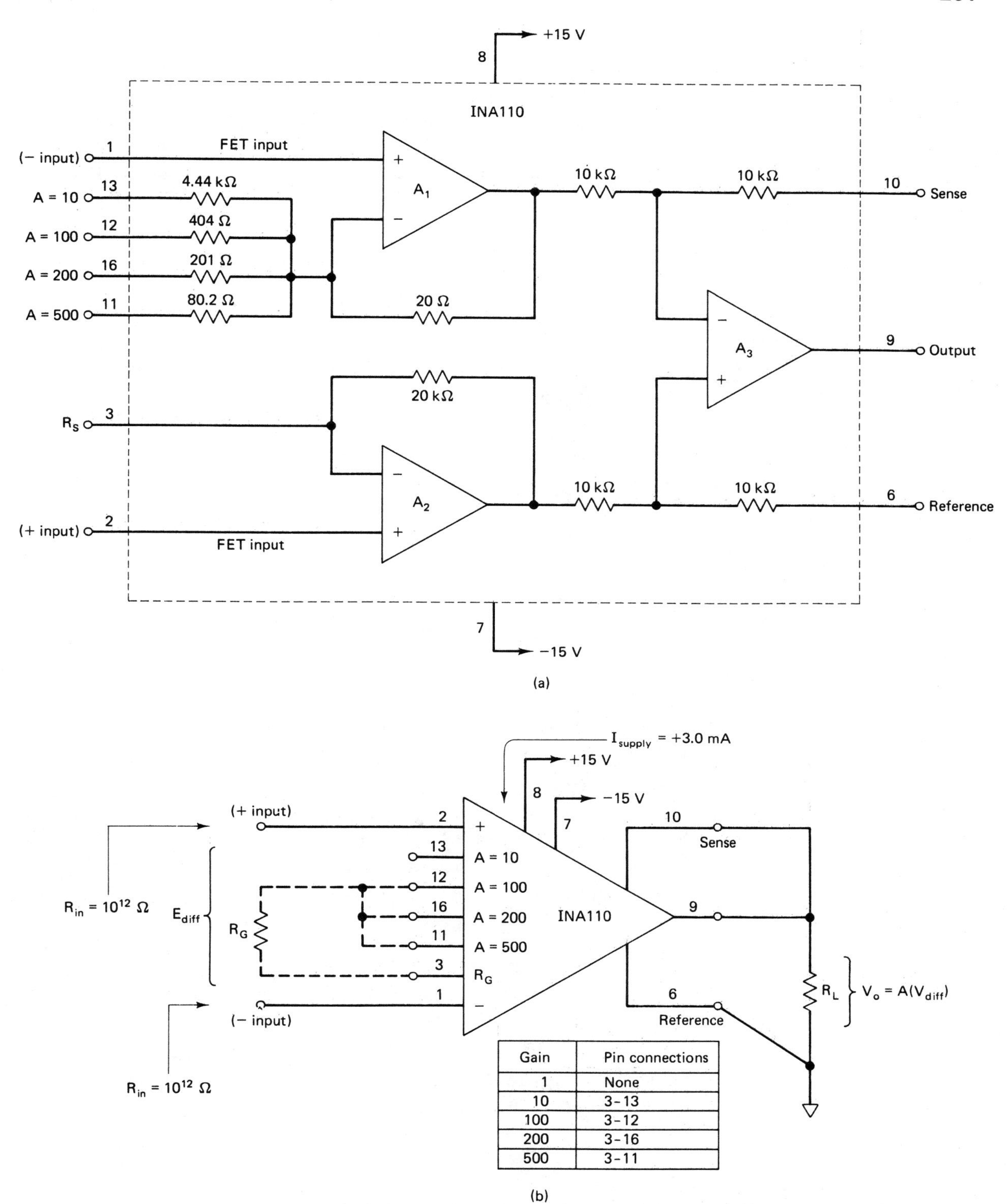

Gain	Pin connections
1	None
10	3-13
100	3-12
200	3-16
500	3-11

Figure 8-7 (a) Schematic diagram of the 1NA110 instrumentation amplifier; (b) circuit symbol and pin connections to program gain.

8-3.2 Pin-Programmable Voltage Gain

Five different pin-selectable gains of 1, 10, 100, 200, and 500 are obtainable. Simply wire 3 of the INA110 to the appropriate internal gain setting resistor as directed by the table in Fig. 8-7b. Gains other than those indicated above are possible by adding an external gain setting resistor R_G between pin 3 and pins 12, 16, and 11 connected together. (Fig. 8-7b). The correct value of R_G is selected with

$$R_G = \frac{40{,}000\ \Omega}{A - 1} - 50\ \Omega \qquad (8\text{-}12)$$

EXAMPLE 8-7

Solve for the value of R_G required to set the following gains, A: (a) 300, (b) 600, and (c) 800.

SOLUTION R_G is calculated from Eq. (8-12) as follows:

$$\text{(a) } R_G = \frac{40{,}000\ \Omega}{300 - 1} - 50\ \Omega = 133.8\ \Omega - 50\ \Omega = 83.8\ \Omega$$

$$\text{(b) } R_G = \frac{40{,}000\ \Omega}{600 - 1} - 50\ \Omega = 66.8\ \Omega - 50\ \Omega = 16.8\ \Omega$$

$$\text{(c) } R_G = \frac{40{,}000\ \Omega}{800 - 1} - 50\ \Omega = 50.0\ \Omega - 50\ \Omega = 0\ \Omega \text{ (short)}$$

Note that the *maximum possible differential* gain is 800 when pin 3 is shorted to pins 12, 16, and 11.

8-3.3 Differential Voltage Measurements with the INA110 Instrumentation Amplifier

Figure 8-8a shows how an INA110 IA is used to measure a voltage drop across the feedback resistor of a noninverting amplifier. The usual way to measure this voltage would be (1) to measure V_1 with respect to ground, (2) to measure V_2 with respect to ground, and (3) to calculate V_{RF} from

$$\begin{aligned} V_{RF} &= V_1 - V_2 \\ &= 195\text{ mV} - 15\text{ mV} = 180\text{ mV rms} \end{aligned} \qquad (8\text{-}13)$$

The instrumentation amplifier can make the differential measurement in one step if its gain A equals 1, V_o will equal V_{RF}. Figure 8-8a also shows how V_{RF} can be amplified by 10 and offset by V_{ref}.

EXAMPLE 8-8

Find V_o in the circuit of Fig. 8-8a. The gain of the instrumentation amplifier is set to 10 and V_{ref} is adjusted to 4 V.

SOLUTION First note that 15 mV (rms) voltage across R_i is a common-mode signal that will be attenuated by the amplifier's CMRR_{dB} of 104 dB. Equation (8-6) is used to solve for the value of CMRR.

$$\text{CMRR} = 10^{\text{CMRR}_{\text{db}}/20} = 10^{104/20} = 10^{5.2} \approx 158{,}500$$

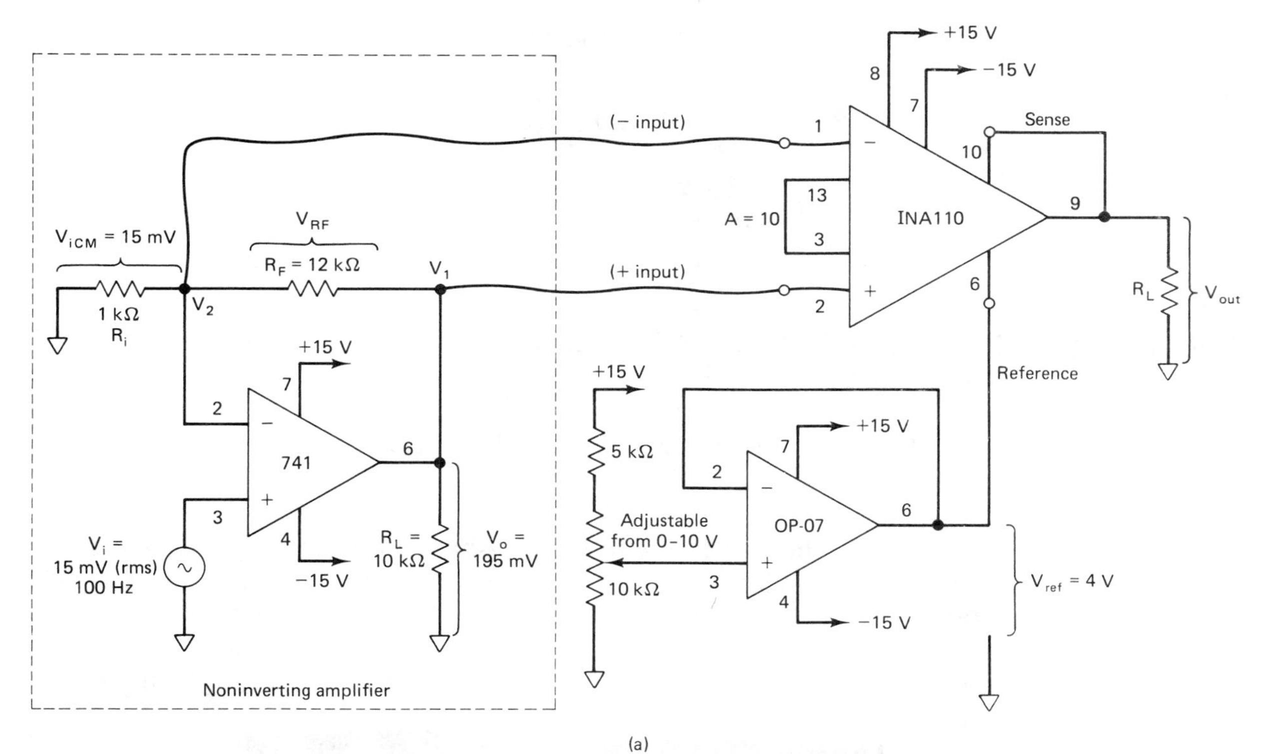

(a)

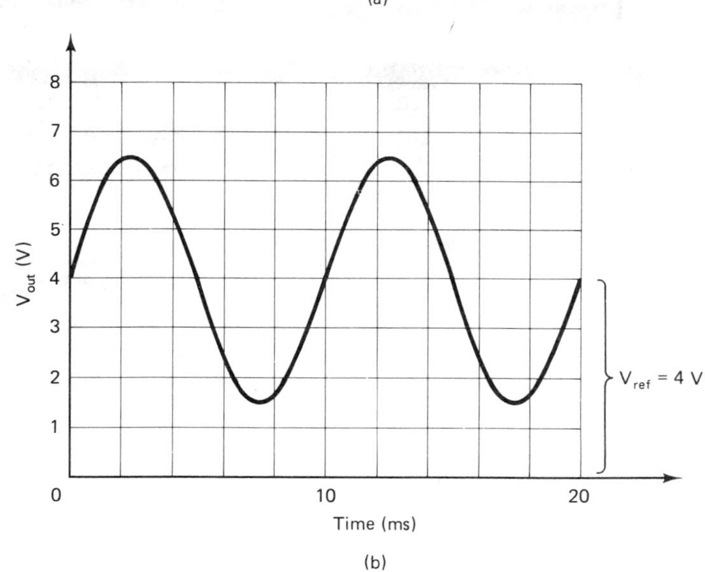

(b)

Figure 8-8 The differential voltage V_{RF} in (a) is amplified by 10 and offset by 4 V. The resulting output is shown in (b). (a) The IA measures the differential voltage across R_F; (b) output voltage V_O is shown for the circuit in (a).

From Eq. (8-4) the common-mode voltage gain becomes

$$A_{CM} = \frac{A_{DIFF}}{CMRR} = \frac{10}{158,500} = 0.000063$$

Therefore, the common-mode voltage appearing at the output is, from Eq. (8-3),

$$V_{OCM} = A_{CM} V_{CM} = 0.000063(15 \times 10^{-3} \text{ V}) \approx 0.95 \ \mu\text{V}$$

This small value can be neglected completely. From Eq. (8-11), find V_{out}:

$$V_{out} = V_{ref} + A_{DIFF}(V_1 - V_2)$$
$$= 4\text{ V} + 10[195 - 15]\text{ mV (ac rms)} = 4\text{ V (dc)} + 1.80\text{ V (ac rms)}$$

See Fig. 8-8b for the shape of V_{out}.

8-4 MEASURING TENSION AND COMPRESSION WITH A STRAIN GAUGE

8-4.1 Problem of Measuring Mechanical Strain

Figure 8-9a shows a *505 round test specimen.* It is used by the mechanical engineer to investigate physical properties of various materials. A typical test would apply a *stress,* in units of pounds per square inch (psi), as an input to the test sample. The output to be measured is the *strain* that results from the stress. Strain causes the original length L of the test specimen to change by an amount ΔL.

If the input stress is a *tension,* the resulting strain is an elongation of the rod and is represented by $+\Delta L$. An input stress that tends to shorted the rod is a *compression* and changes output length by $-\Delta L$. Input stress is plotted along the vertical axis and the output strain along the horizontal axis to give a stress–strain diagram as in Fig. 8-9b. One important feature of this diagram is the slope of the line in its linear region. The mechanical

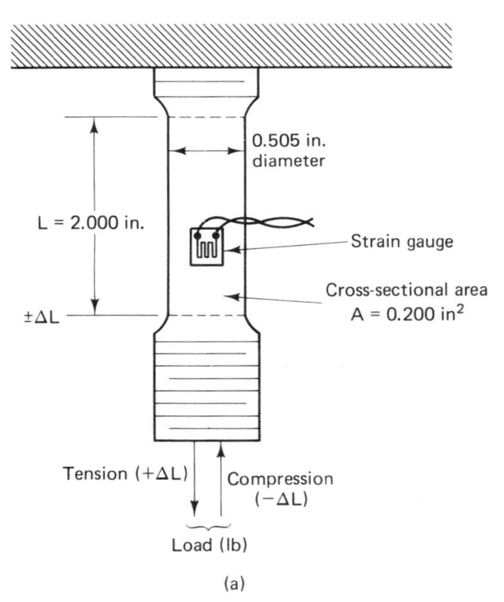

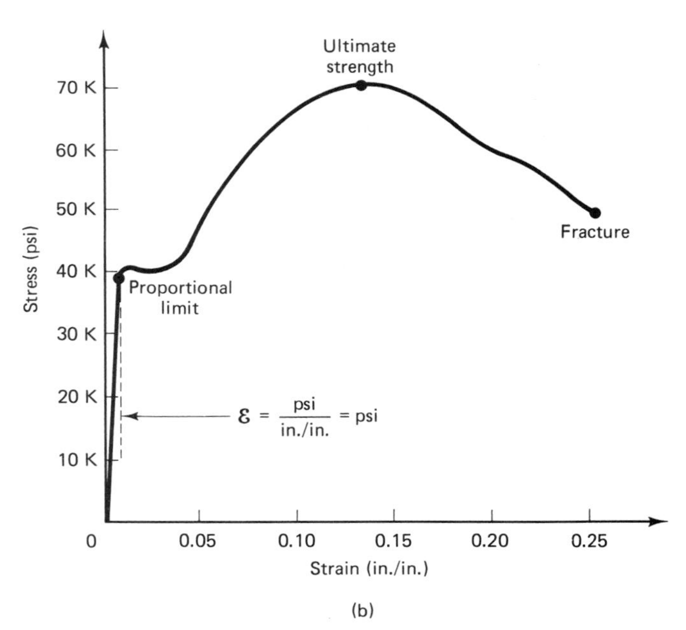

Figure 8-9 Measurement of tension and compression is accomplished with a test specimen and strain gauge. (a) The 505-round test specimen is secured at one end with a known load in pounds connected to the other end; (b) typical stress–strain diagram.

engineer defines this slope as a ratio called the *modulus of elasticity* ε for the material being tested. ε is given by

$$\varepsilon = \frac{\text{stress}}{\text{strain}} = \frac{\text{force/area}}{\pm \Delta L/L} \tag{8-14}$$

The modulus of elasticity, in units of psi, is known for all common materials and is readily available to the mechanical engineer. If we can measure $\Delta L/L$ or strain, we can calculate stress from Eq. (8-14). A strain measurement then allows calculation of weight, force, or pressure.

8-4.2 Basic Strain-Gauge Characteristics

A strain gauge is a transducer that when bonded to a mechanical structure can be used to measure very small changes in its length. A typical foil-type gauge is shown in Fig. 8-10 bonded to a 505 test specimen. The gauge consists of a thin base material on which a resistor foil pattern is deposited.

Under no stress the resistance R of a strain gauge is normally 350 Ω ± 0.1%*. When tension is applied, the length of the test sample is increased by $+\Delta L$, causing gauge resistance to increase by $+\Delta R$.

Compression results in the transducer's resistance decreasing with a change of $-\Delta R$ indicating a length reduction. Length changes in this type of measurement can be as small as a few millionths of an inch and will produce resistance changes in the strain gauge of only a few milliohms. The problem for the electrical engineer is then to be able to measure

*Manufacturer's also make gauges with $R = 120\ \Omega \pm 0.1\%$.

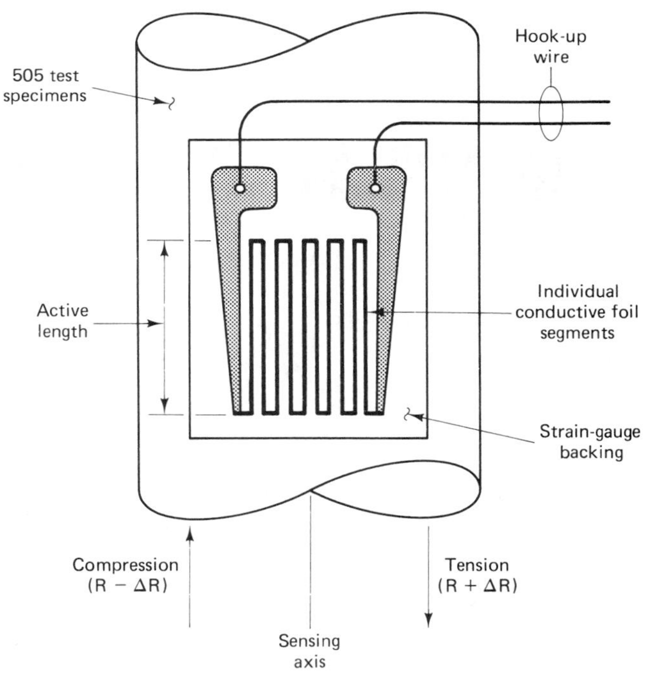

Figure 8-10 The foil-type strain gauge is a conductive foil deposited on a backing. The backing is cemented to provide intimate contact with the test specimine.

precisely a very small resistance $\pm\Delta R$ change in the presence of the large gauge resistance R.

8-4.3 Strain-Gauge Construction

One common form of conductive foil pattern or filament design is shown in Fig. 8-10. This *flat-grid* configuration consists of a "series" of short foil segments oriented along the *sensing axis* of the gauge. The total *active length* is the sum of the individual segments. This type of pattern maximizes the gauge's active length within a minimum area. Resistive foils are made from alloys of nickel such as Constantan, Nichrome V, and Dynalog. For very high temperature applications, strain gauges should be fabricated from platinum wire.

After the active length and foil material is selected, the filament's cross-sectional area is determined by the manufacturer. This sets a standard value for unstrained gauge resistance of typically 120 or 350 Ω. The base material is chosen according to the required temperature range. A paper base is acceptable for temperatures of up to 180°F. Epoxy backing is used for up to 250°F. Use a Bakelite-impregnanted glass fiber for high-temperature applications up to 450°F. Access to the conductive foil pattern of the strain gauge is through insulated instrument strain relief wire via soldering pads.

8-4.4 Mounting Foil-Type Strain Gauges

Specific instructions for installing strain gauges are provided by the manufacturers. In general, the surface should be slightly roughened with either 180-grit sandpaper or a specified chemical precoat liquid. Roughening is followed by an appropriate cleaning solvent.

Cements such as Duco, Eastman 910, and EPY-150 bond the gauge tightly (intimate contact), so that its percentage change in length equals that of the material. A liberal amount of cement is applied to the prepared surface. The gauge is oriented so that its active length aligns with the sensing axis. When clamping is required, a release material such as felt or Teflon is placed over the gauge. Weight is added (as directed by the adhesive manufacturer) for the required curing time.

8-4.5 Measuring Strain-Gauge Resistance

Figure 8-11 shows a working strain gauge mounted to a test specimen. Assume that three ideal 350-Ω resistors are added to form a conventional Wheatstone bridge, and the bridge is driven by an ideal voltage source V_{ref} = 10 V.

If the specimen is under no stress, the gauge resistance is $R \pm \Delta R =$ 350 Ω ± 0 Ω = 350 Ω. With all resistances equal, V_1 and V_2 equal $V_{ref}/2$. The bridge is said to be balanced and the differential voltage $V_1 - V_2 = 0$ V.

If the test specimen is placed under tension, ΔR is positive and V_1

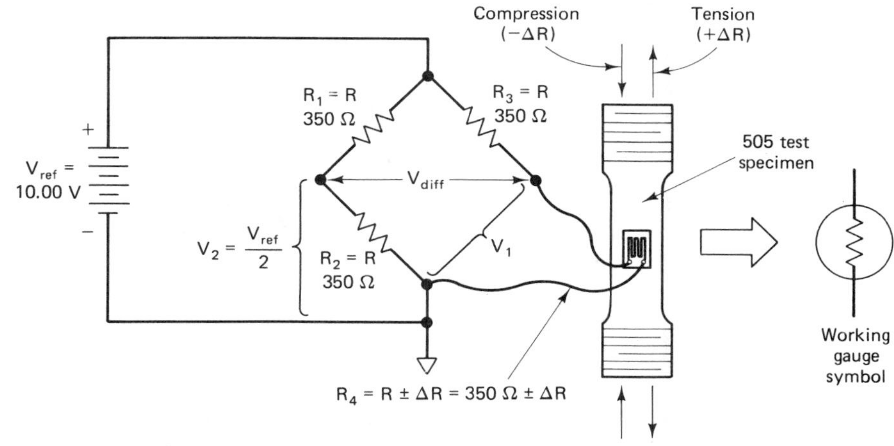

Figure 8-11 A single working gauge is placed in one arm of a wheatstone. Any stress causes a change in R_4 of ΔR and results in a differential output voltage of $V_1 - V_2$.

increases. If compression is applied, ΔR is negative and V_1 decreases. The differential voltage is

$$V_1 - V_2 = V_{\text{DIFF}} \simeq \left(\frac{\Delta R}{4R}\right) V_{\text{ref}} \tag{8-15}$$

EXAMPLE 8-9

(a) Solve for the differential voltage V_{DIFF} of Fig. 8-11 if ΔR changes by $+14$ mΩ. (b) Is the test specimen under tension or compression? (c) Solve for V_2.

SOLUTION (a) From Eq. (8-15),

$$V_1 - V_2 = V_{\text{DIFF}} = \left(\frac{\Delta R}{4R}\right) V_{\text{ref}} = \left(\frac{0.014\ \Omega}{4\ (350\ \Omega)}\right) 10\ \text{V} = 100\ \mu\text{V}$$

(b) A positive ΔR indicates tension.

(c) $V_2 = \text{constant} = V_{\text{ref}}/2 = 10.000\ \text{V}/2 = 5.000\ \text{V}$

We conclude that under no stress the bridge is balanced and $V_{\text{DIFF}} = 0$ V. A strain (either tension or compression) will produce an imbalance in the bridge and a differential output voltage will result.

8-4.6 Measuring $\Delta L/L$ or Strain

Manufacturers of strain gauges specify a *gauge factor* GF for each gauge. The gauge factor is the ratio of a fractional change in gauge resistance $\Delta R/R$ to its fractional change in length $\Delta L/L$. The gauge factor is then

$$\text{GF} = \frac{\Delta R/R}{\Delta L/L} \tag{8-16}$$

The ratio $\Delta L/L$ is called *unit strain.* The strain gauge's $\Delta L/L$ equals the specimens $\Delta L/L$ because they are bonded together.

Gauges made from constantan have a gauge factor of approximately 2. The gauge factor for Nichrome V material is 2.2 and for Dynalog GF = 3.5. We extend Example 8-9 to learn how to measure $\Delta L/L$ or strain.

EXAMPLE 8-10

Assume that the strain gauge in Example 8-9 and Fig. 8-11 has a GF = 2. Tension causes a differential output voltage of 100 μV. Find (a) $\Delta R/R$; (b) the unit strain $\Delta L/L$.

SOLUTION Solve Eq. (8-15) for ΔR:

$$\Delta R = \left(\frac{V_{\text{DIFF}}}{V_{\text{ref}}}\right) 4R = \left(\frac{100\ \mu\text{V}}{10\ \text{V}}\right) 1400\ \Omega = 0.014\ \Omega$$

(b) From Eq. (8-16), the unit strain is

$$\frac{\Delta L}{L} = \frac{\Delta R/R}{\text{GF}} = \frac{0.014\Omega/350\ \Omega}{2.0} = 20\ \mu\text{in./in.}$$

The mechanical engineer can now use this information on strain to determine tensile strength and loading.

8-5 IMPROVING THE STRAIN-GAUGE BRIDGE

8-5.1 Adding an Instrumentation Amplifier

An instrumentation amplifier such as the INA110 from Burr-Brown is added to the bridge circuit of Fig. 8-11 as in Fig. 8-12. It amplifies the differential bridge voltage and gives a single-ended output voltage V_o. It also allows common-mode voltage $V_{\text{ref}}/2$ to be rejected by the instrumentation amplifier's CMRR. Example 8-11 illustrates the advantages of adding an IA.

EXAMPLE 8-11

The instrumentation amplifier of Fig. 8-12 has a differential gain of 500 and a $\text{CMRR}_{\text{dB}} = 100$ dB. A stress is applied to imbalance the bridge by $V_{\text{DIFF}} = 100\ \mu$V. Solve for (a) V_{OCM}; (b) V_o.

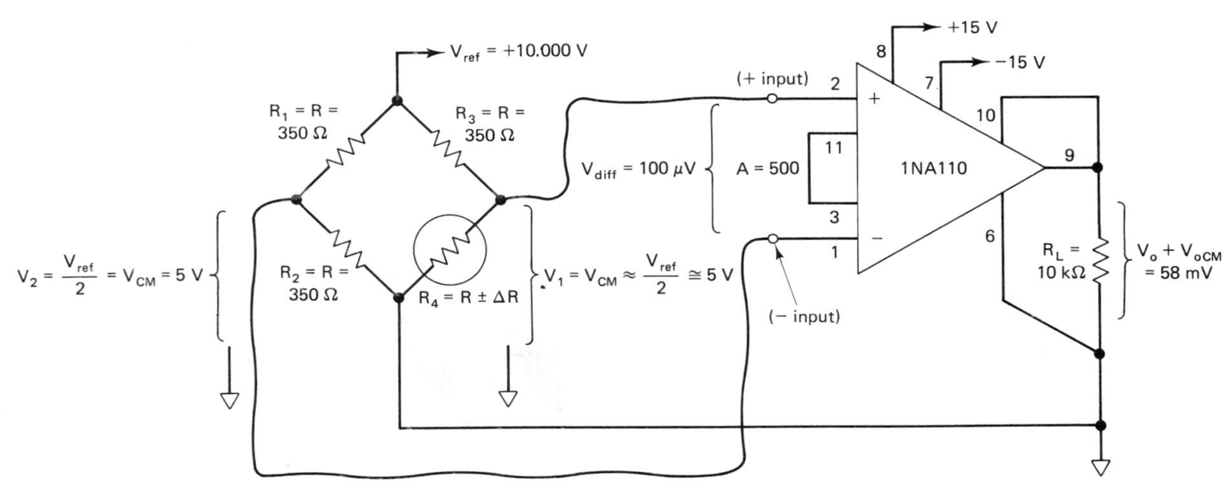

Figure 8-12 Addition of an instrumentation converts differential output voltage of the bridge V_1–V_2 into a single-ended output voltage V_o.

SOLUTION (a) From Eq. (8-6),

$$\text{CMRR} = 10^{\text{CMRR}_{dB}/20} = 10^{100/20} = 10^{5.5} = 316{,}228$$

From Eq. (8-4);

$$A_{\text{CM}} = \frac{A_{\text{DIFF}}}{\text{CMRR}} = \frac{500}{316{,}228} = 0.00158$$

The common-mode voltage that appears at the output is obtained from Eq. (8-3) as

$$V_{\text{OCM}} = A_{\text{CM}}\ V_{\text{CM}} = A_{\text{CM}}\ \frac{V_{\text{ref}}}{2}$$
$$= 0.00158(5\ \text{V}) = 7.9\ \text{mV} = 8\ \text{mV}$$

(b) The signal output voltage V_o is found from Eq. (8-2).

$$V_o = A_{\text{DIFF}}\ V_{\text{DIFF}} = 500(100\ \mu\text{V}) = 50\ \text{mV}$$

Example 8-11 demonstrates that the output voltage of the IA consists of a common-mode voltage of 8 mV superimposed on a strain gauge signal of 50 mV. The output voltage is then 58 mV. Care must be exercised in evaluating the output to exclude the 8 mV. In Section 8-5.2 we explain how to reduce this unwanted common-mode voltage.

8-5.2 Bipolar Bridge Excitation Voltage

One method of reducing V_{OCM} is by exciting the bridge with a bipolar ±5 V supply as in Fig. 8-13. Assume that the bridge is balanced. Voltage V_1 equals V_2 and both are at = O V with respect to ground. The IA will reduce this near-zero common-mode voltage to a very negligible value.

8-5.3 Temperature Effects on Bridge Balancing

Strain gauges are essentially wire and as such have *positive coefficients of temperature.* An increase in temperature increases the unstrained resistance R of the gauge. A decrease in temperature causes R to decrease slightly.

The R_2 resistor in all the bridges should have *exactly* the same positive temperature coefficient as the strain gauge. Otherwise, the bridge output will depend more on a ΔR due to temperature rather than ΔR due to stress.

This problem is overcome by replacing R_2 with a *dummy or temperature-compensating gauge* as in Fig. 8-14. This R_{temp} strain gauge should be placed immediately adjacent and at a right angle to the working strain gauge.

The temperature-compensating gauge must see the same thermal environment as the working gauge. As temperature changes, resistance of the compensating gauge will track that of the working gauge.

The active length of the dummy gauge is perpendicular to the sensing axis, so that its $\Delta R = 0\Omega$ due to stress. Thus $V_1 - V_2$ is due only to ΔR of the working gauge because its active length is along the sensing axis.

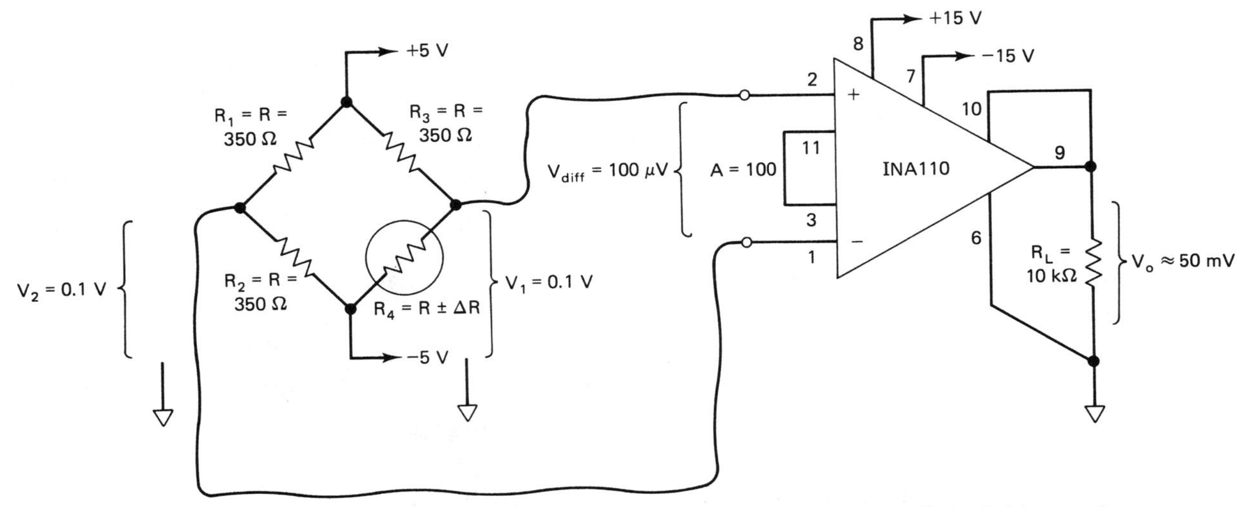

Figure 8-13 V_{OCM} can be reduced considerably if the bridge excitation voltage is a bi-polar supply.

8-5.4 Increasing the Bridge's Differential Output

The differential output voltage $V_1 - V_2$ of a bridge can be increased by the method shown in Fig. 8-15. Two working gauges R_1 and R_4, are placed in opposite arms of the bridge with their active lengths along the sensing axis. Two temperature-compensation gauges, R_2 and R_3, have active lengths perpendicular to the sensing axis.

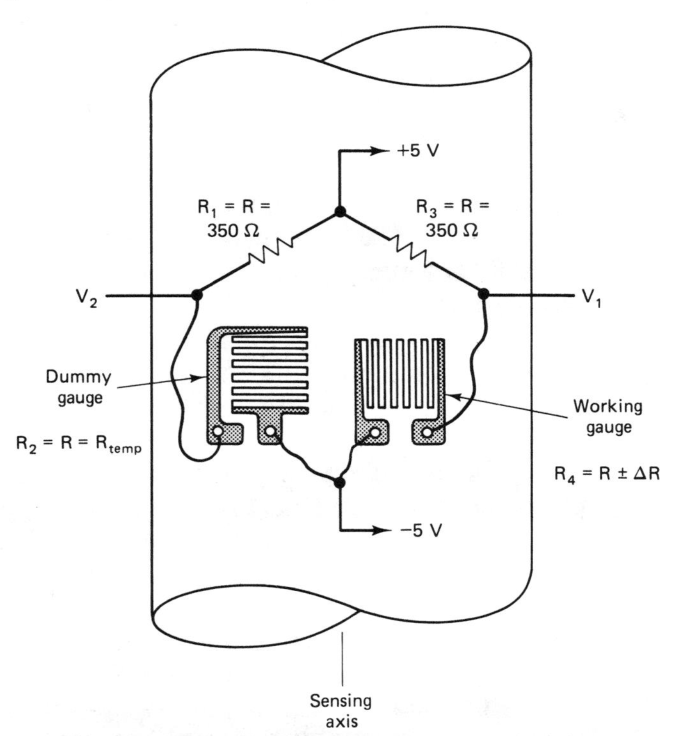

Figure 8-14 Proper addition of a dummy gauge to the bridge circuit compensates for the changes in temperature.

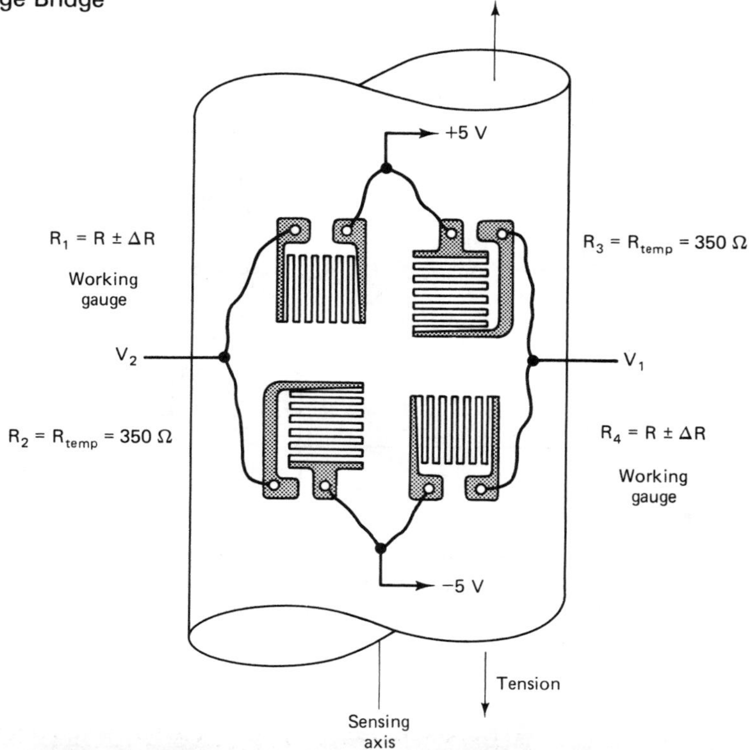

Figure 8-15 Two gauges are used for temperature compensation and two working gauges detect the strain. Output $V_1 - V_2$ is double that of a single working gauge.

Under tension, working gauges R_1 and R_4 increase to $R + \Delta R$. Dummy gauges R_2 and R_3 experience no resistance change. V_1 increased by the same amount that V_2 decreases. The differential output voltage $V_1 - V_2$ becomes *twice* that of a single working gauge, or

$$V_1 - V_2 = V_{\text{DIFF}} \simeq \left(\frac{\Delta R}{2R}\right) V_{\text{ref}} \tag{8-17}$$

EXAMPLE 8-12

Solve for the differential voltage $V_1 - V_2$ in Fig. 8-15 if ΔR equals 0.014 Ω.

SOLUTION Even though the bridge is excited by a ±5-V bipolar supply, V_{ref} is the total voltage, 10 V. From Eq. (8-17),

$$V_{\text{DIFF}} = V_1 - V_2 = \left(\frac{0.014\ \Omega}{2[350\ \Omega]}\right)(10\ \text{V}) = 200\ \mu\text{V}$$

8-5.5 Practical Bridge Balancing Technique

Suppose that the strain gauge setup of Fig. 8-15 uses gauges that have an unstrained resistance R equal to 350 Ω with a tolerance of + 0.1% or +350 mΩ. Suppose further that R_4 is at one end of the tolerance limit with a high value of 350.350 Ω and R_1 is at the other end with a low value 349.650 Ω. The bridge would be grossly imbalanced by 700 mΩ. It is not practical to improve the tolerance of the gauges to 0%. Some method must be used to balance the bridge prior to detecting a strain.

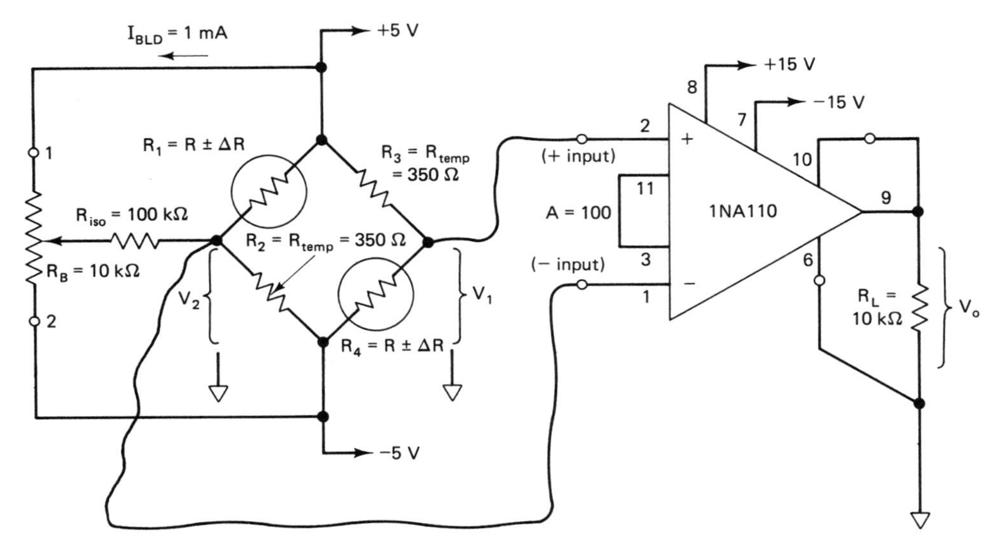

Figure 8-16 R_B and R_{iso} are added to provide a practical method of balancing or zeroing the strain-gauge bridge.

The strain gauge setup of Fig. 8-15 has been incorporated into Fig. 8-16 with the addition of potentiometer R_B and a balancing and isolating resistor R_{iso}. The value of R_B is chosen to be 10 kΩ. It is connected across the ±5-V supply to draw a standard 1-mA bleeder current. R_{iso} is selected to be 10 or more times the value of R_B. Therefore, R_{iso} does not load down the voltage-divider action of R_B. To balance or zero the strain gauge bridge in Fig. 8-16, proceed as follows:

1. Remove the load from the specimen to be tested so that the working gauges are under *no* strain.
2. Monitor V_o of the IA with an oscilloscope set on dc coupling.
3. Adjust R_B until V_o is as close to zero volts as possible.

End of Chapter Exercise

Name: ______________________

Date: __________ **Grade:** __

A. FILL-IN THE BLANKS

Fill in the blanks with the best answer.

1. The basic differential amplifier circuit has __________ resistors, and 1 op amp.
2. When differential gain is divided by common mode gain the result is called __________ .
3. The gain of a differential amplifier is changed by adjusting __________ resistor(s).
4. The gain of an instrumentation amplifier is changed by adjusting __________ resistor(s).
5. An output offset voltage of an instrumentation amplifier is accomplished by connecting an equal voltage to its __________ terminal.
6. Two typical values for strain gauge resistance are 120 Ω and __________ .
7. Temperature changes of a working or active gauge are compensated for by a companion __________ gauge.
8. A potentiometer and __________ resistor provide a practical way to balance a resistance bridge circuit.

B. TRUE/FALSE

Fill in **T** if the statement is true, and **F** if any part of the statement is false.

1. ______ A remote signal to be amplified has a noisy ground. You select an instrumentation amplifier.

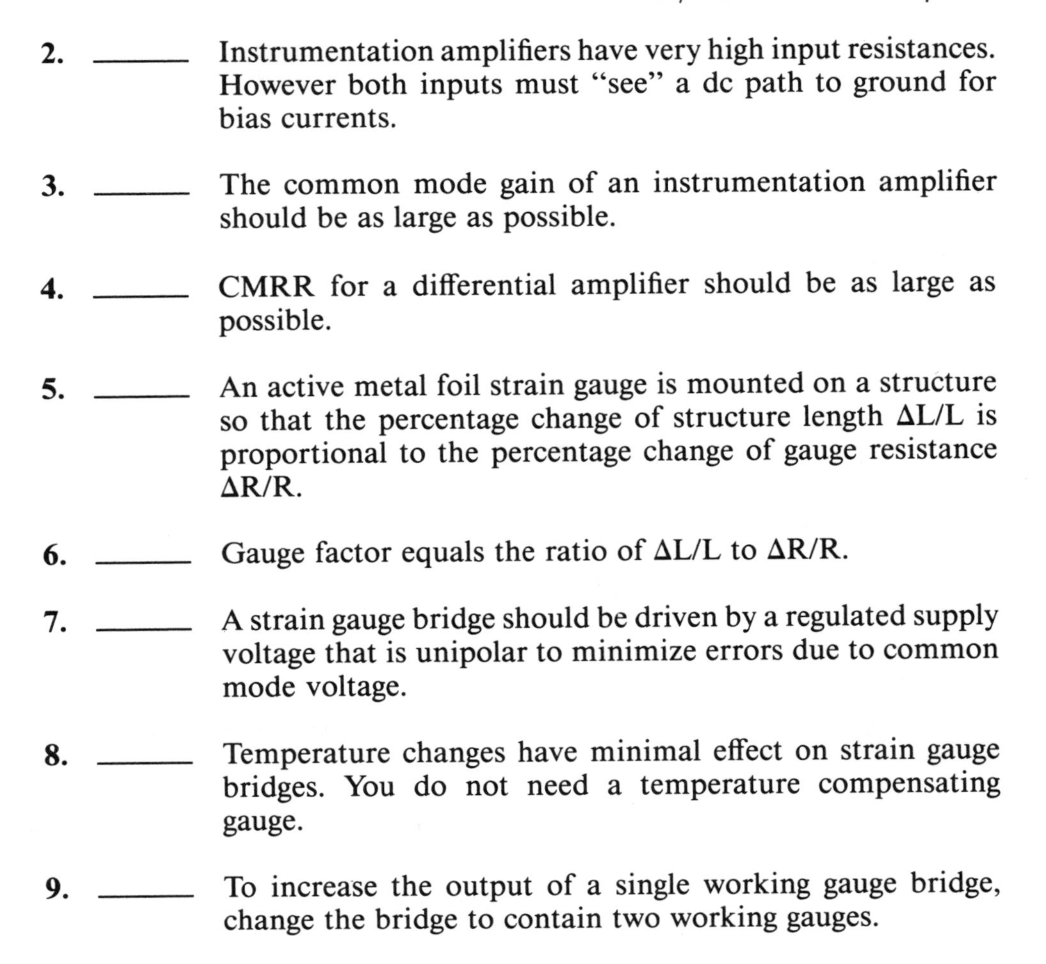

2. ______ Instrumentation amplifiers have very high input resistances. However both inputs must "see" a dc path to ground for bias currents.

3. ______ The common mode gain of an instrumentation amplifier should be as large as possible.

4. ______ CMRR for a differential amplifier should be as large as possible.

5. ______ An active metal foil strain gauge is mounted on a structure so that the percentage change of structure length $\Delta L/L$ is proportional to the percentage change of gauge resistance $\Delta R/R$.

6. ______ Gauge factor equals the ratio of $\Delta L/L$ to $\Delta R/R$.

7. ______ A strain gauge bridge should be driven by a regulated supply voltage that is unipolar to minimize errors due to common mode voltage.

8. ______ Temperature changes have minimal effect on strain gauge bridges. You do not need a temperature compensating gauge.

9. ______ To increase the output of a single working gauge bridge, change the bridge to contain two working gauges.

C. CIRCLE THE CORRECT ANSWER

Circle the correct answer for each statement.

1. A temperature compensation gauge is mounted (parallel, perpendicular) to its companion working gauge.
2. (Stress, strain) causes (stress, strain).
3. Compressing a member causes a change in length of ($+\Delta L$, $-\Delta L$).
4. Tension causes a change of (ΔR, $-\Delta R$) on the working gauge.
5. Tension causes a change of (ΔR, $-\Delta R$, negligible value) on the temperature compensating gauge.
6. The (reference, sense) terminal of an instrumentation amplifier is used to inject an offset voltage.
7. The output voltage of an instrumentation amplifier has an undesired component due to (differential, common mode) gain.
8. A strain gauge bridge is preferably activated by a (unipolar, bipolar) supply.
9. Gain of an instrumentation amplifier is adjusted by changing (one, two) resistors.

D. MATCHING

Match the name or symbol in column **A** with the statement that matches best in column **B**.

	COLUMN A		COLUMN B
1. ______	ε	**a.**	A_{DIFF}/A_{CM}
2. ______	CMRR	**b.**	Force/Area
3. ______	strain	**c.**	stress/strain
4. ______	stress	**d.**	$\Delta L/L$
5. ______	tension	**e.**	ΔR

PROBLEMS

8-1. Calculate the gain factor g of Fig. 8-1 if both input resistors, R, are changed to 15 kΩ.

8-2. Repeat Example 8-1 but with both input resistors R changed to 10 kΩ.

8-3. Revise the circuit of Fig. 8-1 for a differential gain of
(a) 10;
(b) 50;
(c) 100.
Leave the input resistors R at 15 kΩ.

8-4. Calculate the common-mode voltage gain A_{CM} of a differential amplifier that outputs V_{OCM} = 0.55 mV when V_{CM} = 1.00 V.

8-5. If A_{DIFF} = 100 and A_{cm} is obtained from Problem 8-4, calculate the CMRR.

8-6. Convert the following values of CMRR into their equivalent $CMRR_{dB}$ values:
(a) 1000;
(b) 10,000;
(c) 100,000;
(d) 1,000,000.

8-7. Convert the following values of $CMRR_{dB}$ into their equivalent CMRR values:
(a) 50 dB;
(b) 70 dB,
(c) 90 dB;
(d) 100 dB.

8-8. Repeat Example 8-5 with $CMRR_{dB}$ equal to 110 dB and both gR resistors changed to 2.2 MΩ.

8-9. Refer to Fig. 8-4 where $R = 20\ \text{k}\Omega, V_1 = 5.001$ V and $V_2 = 5.000$ V, and solve both A_{DIFF} and then V_o' if mR is
(a) 4.44 kΩ;
(b) 201 Ω;
(c) 80.2 Ω.

8-10. **(a)** Explain the function of the A_3 differential amplifier output stage of Fig. 8-4.
(b) What is the gain of this stage?

8-11. Draw a complete three-op-amp instrumentation amplifier with a differential gain of 100 and the ability to adjust its output by a dc offset of ±5 V.

8-12. Repeat Example 8-7 if A is
(a) 400;
(b) 700.

8-13. Repeat Example 8-8 for the INA110 set to a differential gain of 200 and a $\text{CMRR}_{\text{dB}} = 100$ dB. V_i is adjusted to a 1.5 mV (rms).

8-14. What are the two most common strain gauge resistances?

8-15. **(a)** What types of stress is applied to a strain gauge whose resistance becomes $R + \Delta R$?
(b) What is the stress if the gauge resistance becomes $R - \Delta R$?

8-16. Repeat Example 8-9 for the $V_{\text{ref}} = 5$V and $-\Delta R$ of 0.0015Ω.

8-17. Assume that the gauge factor GF in Problem 8-16 is 2.2. Solve for
(a) $-\Delta R/R$;
(b) the unit strain $-\Delta L/L$.

8-18. Repeat Example 8-11 for a $\text{CMRR}_{\text{dB}} = 110$ dB.

8-19. Explain how to increase the differential output voltage V_{DIFF} of the bridge circuit shown in Fig. 8-13.

8-20. Solve for V_{DIFF} in Fig. 8-15 if $V_{\text{ref}} = \pm 2.5$ V and ΔR is
(a) 0.8 mΩ; **(b)** 1.6 mΩ.

Laboratory Exercise 8

Name: ____________________

Date: __________ **Grade:** ___

INSTRUMENTATION AMPLIFIERS

OBJECTIVES: Upon completion of this laboratory exercise on instrumentation amplifiers, you will be able to (1) measure A_{DIFF} for a basic differential amplifier; (2) measure A_{CM} and calculate CMRR for a basic differential amplifier; (3) build and test a discrete three-op-amp instrumentation amplifier; (4) add an offset voltage to the reference terminal of an IA, and (5) test the characteristics of an INA110 instrumentation amplifier from Burr-Brown.

REFERENCES: Chapter 8 and INA110 datasheet

PARTS LIST

4	OP-07	1	100-Ω 1% resistor
1	INA110 instrumentation amplifier (see data sheets pp 285-89)	2	10-k Ω 1% resistors
1	10-Ω 5% resistor	7	20-k Ω 1% resistors
1	470-Ω 5% resistor	2	100-k Ω 1% resistors
1	15-k Ω 5% resistor	1	10-k Ω potentiometer
1	82-k Ω 5% resistor	1	50-k Ω potentiometer

Procedure A: Differential Voltage Gain

1. Build the test circuit of Fig. L8-1. With a DMM on the 2 Vdc range, measure both V_1 and V_2 with respect to ground and record the values on Fig. L8-1. Next, calculate the differential voltage across the 10-Ω resistor. $V_{DIFF} = V_1 - V_2 =$ __________.
2. Is the common-mode voltage, V_{CM}, in Fig. L8-1 V_1 or V_2?__________.

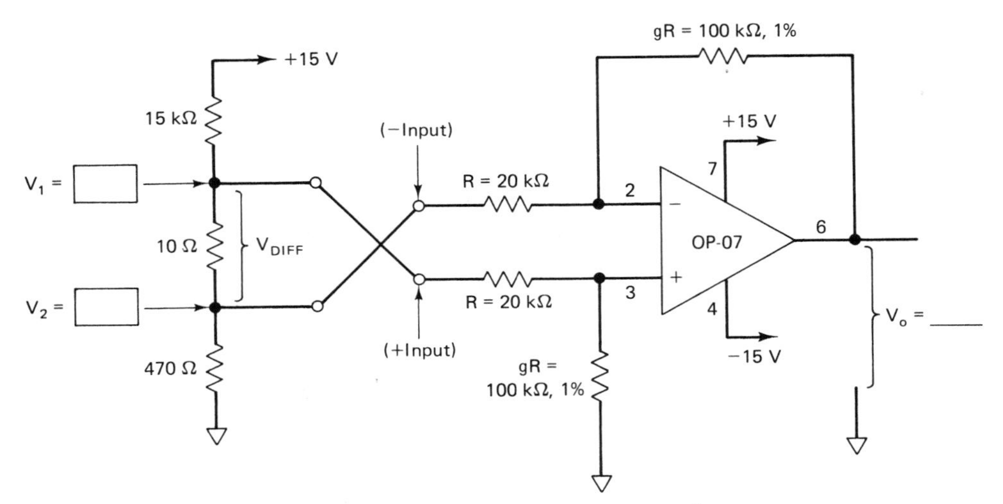

Figure L8-1 Test circuit to measure A_{DIFF}.

3. Calculate the differential voltage gain, A_{DIFF}, for the circuit shown and solve for the expected output voltage.

$$A_{DIFF} = \frac{V_o}{V_1 - V_2} = g = \frac{gR}{R}$$

A_{DIFF} = ________; V_o = ________.

4. Measure and record the value of V_o using a DMM on the most sensitive scale possible. Does the calculated value of V_o compare favorably with the measured value? ________.

5. What changes must be made to Fig. L8-1 to (a) increase A_{DIFF} to 100? ________________________________

____

(b) decrease A_{DIFF} to 1? ________________________________

____

Procedure B: Measuring Both A_{CM} and CMRR

6. Modify Fig. L8-1 to include a common-mode adjustment as shown in Fig. L8-2. Connect both inputs (+input and −input) together, and to V_2 the common-mode voltage, V_{CM}. Remeasure and record $V_2 = V_{CM}$. V_{CM} = ________.

7. Move the DMM to the output of the differential amplifier and adjust the 50-k Ω potentiometer for the *smallest* output voltage possible. Record this voltage as V_{OCM}. V_{OCM} = ________. (*Note*: V_{OCM} should be approximately 1 mV for this test.)

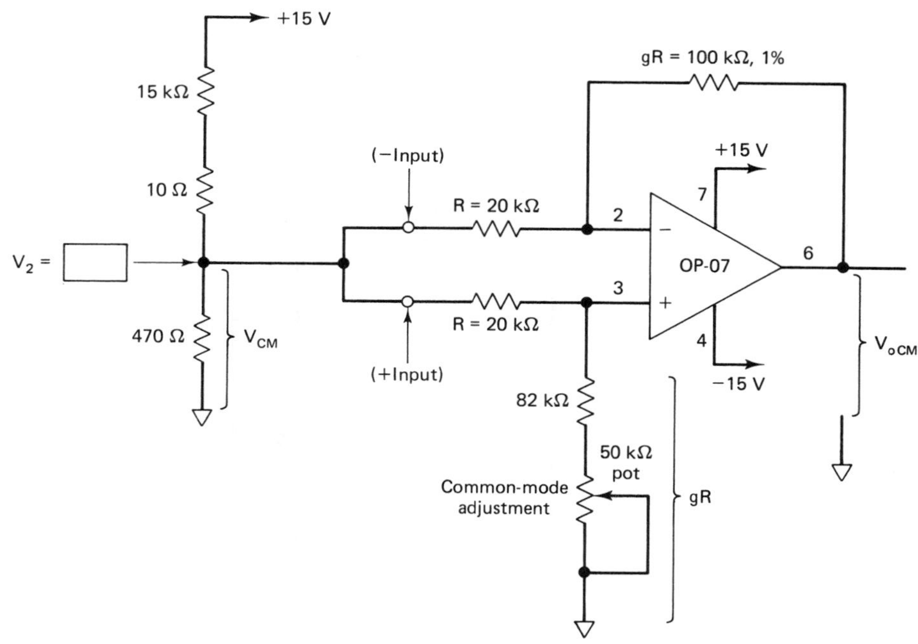

Figure L8-2 Test circuit to measure A_{CM} and CMRR.

8. Calculate the common-mode voltage gain. $A_{CM} = V_{OCM}/V_{CM}$. A_{CM} = __________. (*Note*: A_{CM} should be a very small value. Ideally, A_{CM} should equal zero.

9. Determine the common-mode rejection ratio (CMRR) using the equation: CMRR = A_{DIFF}/A_{CM}. CMRR = __________.

10. What change can be made to Fig. L8-2 to improve (increase) the value of CMRR determined in step 9? ________________

____

Procedure C: Three-Op-Amp Instrumentation Amplifier

11. The circuit shown in Fig. L8-3 is a three-op-amp instrumentation amplifier, with buffered inputs (A_1 and A_2) added to increase input impedance. Differential gain is adjusted by the *mR* resistor. The output is a single-ended voltage V_o and responds only to $V_{DIFF} = V_1 - V_2$. Build Fig. L8-3.

12. Install an *mR* resistor equal to 20 kΩ, 1%. Calculate the value of *m* from the equation $m = mR/R$. m = __________ .

$$A_{DIFF} = \frac{V_o'}{V_1 - V_2} = 1 + \frac{2}{m}$$

A_{DIFF} __________. Solve for the differential output voltage, V_o', of

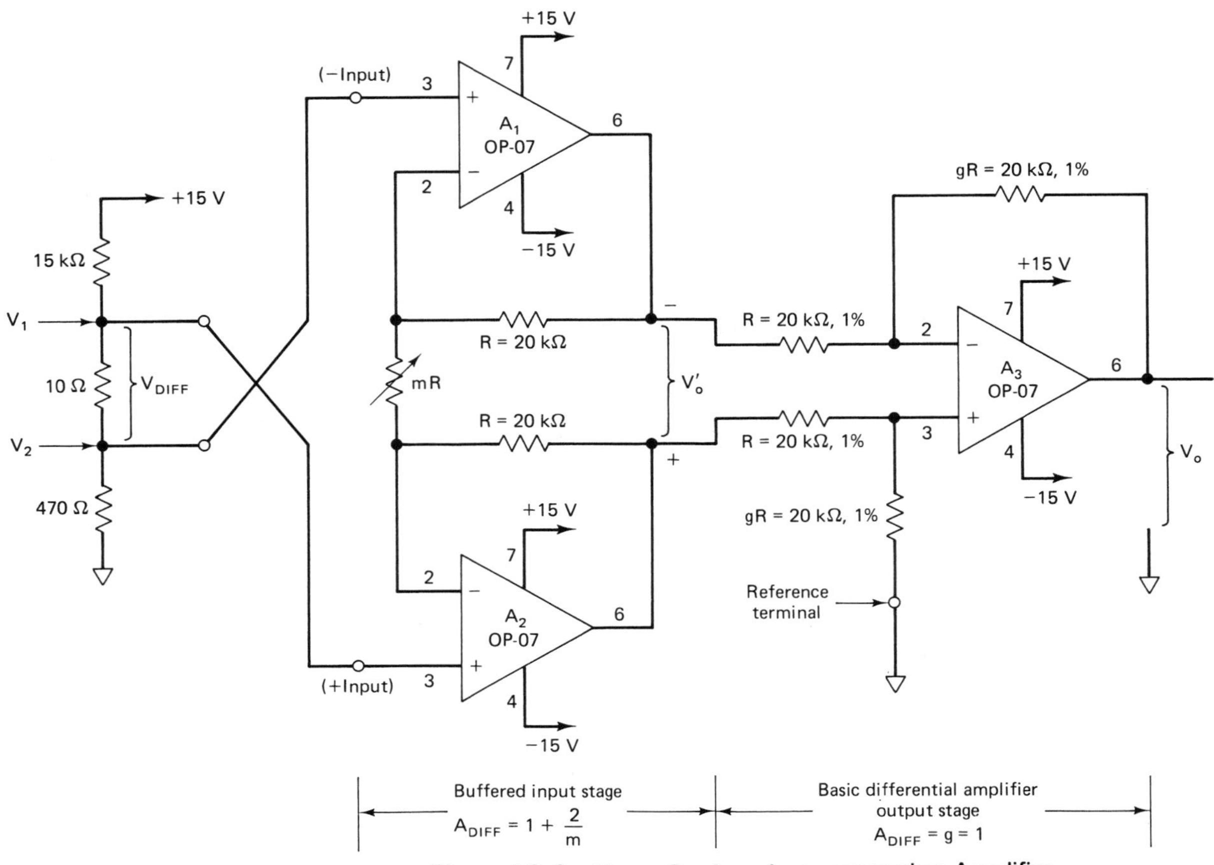

Figure L8-3 Three-Op-Amp Instrumentation Amplifier.

the input stage. $V_o' = A_{\text{DIFF}}\,(V_1 - V_2)$. $V_o' =$ __________ . (Since the differential gain of the output stage is $g = gR/R = 20\text{ k}\Omega/20\text{ k}\Omega = 1$, the overall circuit gain is

$$A_{\text{DIFF}} = \frac{V_o}{V_1 - V_2} = (1 + \frac{2}{m}) \cdot (1) = 1 + \frac{2}{m}$$

and $V_o = A_{\text{DIFF}}\,(V_1 - V_2)$. Determine V_o. $V_o =$ __________.

13. Use a DMM on dc volts to measure V_o'. $V_o' =$ __________. (*Note*: This measurement is across pins 6 of A_1 and A_2.) Next, measure V_o. V_o = __________. (*Note*: This measurement is from pin 6 of A_3 to ground.

14. Measure both V_1 and V_2 with respect to ground. V_1 = __________; $V_2 =$ __________. Calculate the circuit's differential gain from

$$A_{\text{DIFF}} = \frac{V_o}{V_1 - V_2}$$

A_{DIFF} = __________. Do the calculated values of V_o *and* A_{DIFF} compare favorably with the measured values? __________.

15. Change mR to a 100kΩ, 1% resistor to increase the IA's differential gain and calculate the following: m = __________; A_{DIFF} = __________; V_o' = __________; V_o = __________.
16. Use a DMM or dc volts to measure the following: V_o' = __________; V_o = __________; V_1 = __________; V_2 = __________. Calculate A_{DIFF} from these measured values. A_{DIFF} = __________. Do the calculated values of V_o and A_{DIFF} compare favorably with the measured values? __________.
17. Calculate the value of mR necessary to set the differential gain to 100.

$$A_{\text{DIFF}} = 1 + \frac{2}{m}$$

Therefore,

$$m = \frac{2}{A_{\text{DIFF}} - 1} \text{ and } mR = (m)\,\text{R}$$

mR = __________ .

Procedure D: Adding a Voltage to the Reference Terminal

18. Figure L8-4 illustrates how the reference terminal of the instrumentation amplifier shown in Fig. L8-3 can be used to provide an offset voltage at its single-ended output.
19. Build Fig. L8-4 using a fourth OP-07 op amp. Disconnect the reference terminal of Fig. L8-3 from ground and connect it to V_{ref} of Fig. L8-4.
20. Apply power and adjust the 10-kΩ potentiometer until V_{ref} equals 0 V.
21. Measure and record V_o (Fig. L8-3). V_o = __________. This value of V_o just measured should equal the value of V_o measured in Step 16. Proceed if your answer is yes.
22. Calculate the value of V_o if $V_{\text{ref}} = +3$ V.

$$V_o = A_{\text{DIFF}}\,(V_1 - V_2) + V_{\text{ref}}$$

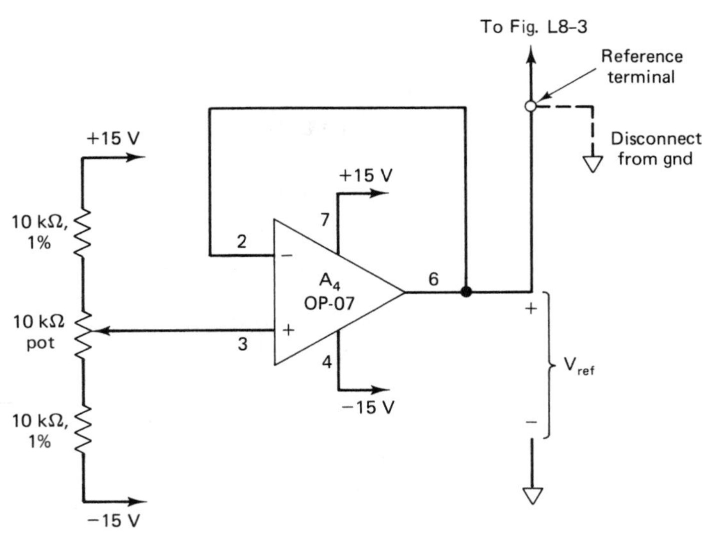

Figure L8-4 Adjustable Reference Voltage, V_{ref}.

V_o = __________. Adjust V_{ref} to + 3 V and measure V_o at the output of the IA. V_o = __________. Do the calculated and measured values for V_o compare favorably? __________.

23. Calculate the value of V_o if V_{ref} is set to −6.0 V. Use the equation given in step 22: V_o __________. Adjust V_{ref} to −6.0 V and measure V_o at the output of the IA. V_o = __________. Does the calculated and measured values for V_o compare favorably? __________.

24. Give one application of adding V_{ref} to an instrumentation amplifier. ____

____

Procedure E: INA110 Instrumentation Amplifier

25. Build the instrumentation amplifier shown in Fig. L8-5. Set the differential gain A_{DIFF} to 10 by connecting pin 3 (RG) to pin 13 (A = 10). Again measure V_1 and V_2 with respect to ground. V_1 = __________; V_2 = __________.

26. Predict the output voltage V_o from the equation, $V_o = A_{DIFF} (V_1 - V_2)$. V_o = __________. Measure V_o · V_o = __________.

27. For a differential gain of 100 remove the gain strap between pins 13

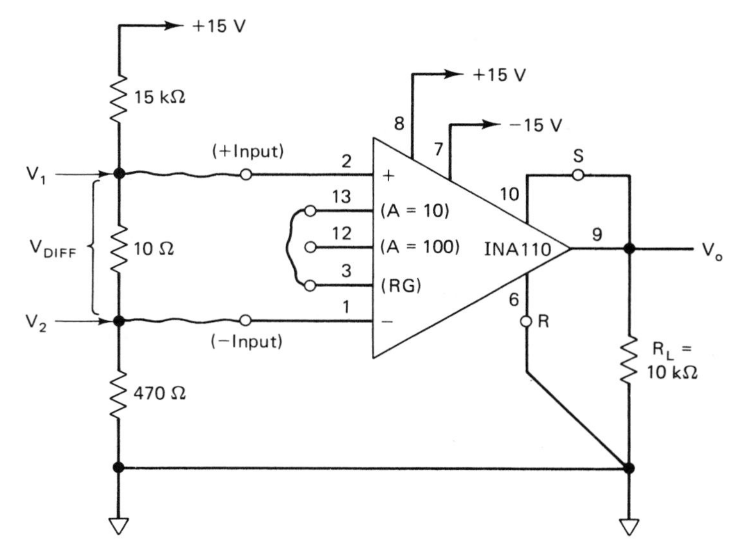

Figure L8-5 Instrumentation Amplifier.

to 3 and connect pins 12 to 3. Predict V_o. $V_o = 100\,(V_1 - V_2)$. V_o = __________. Measure V_o and compare values: V_o = __________.

28. To measure the common-mode voltage gain of the INA110 instrumentation amplifier, connect both inputs (pin 1 and pin 2) together and to V_2 (see Fig. L8-6) the common-mode voltage V_{CM}. Remeasure V_2. $V_2 = V_{CM}$ = __________.

29. Move the DMM to the output and measure V_{OCM} on the lowest sensitivity scale possible. V_o = __________.

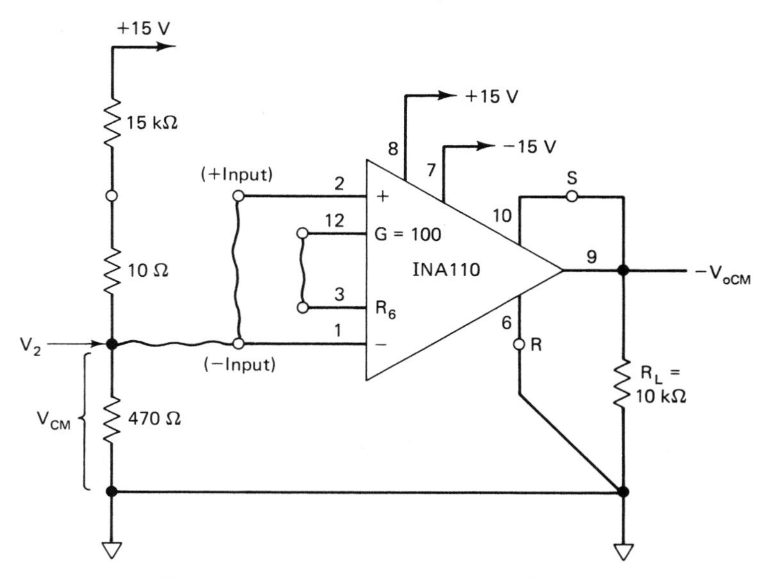

Figure L8-6 Measurement of CMRR.

30. Solve for the common-mode voltage gain. $A_{CM} = V_{OCM}/V_{CM}$. A_{CM} = __________.

31. Calculate the value of CMRR from the equation CMRR = A_{DIFF}/A_{CM}. CMRR = __________. Convert to a common-mode rejection ratio in dB units using the equation $CMRR_{(dB)} = 20 \log CMRR$.

    CMRR(db) = __________.

32. Compare this measured value of CMRR(dB) with the typical data sheet value. Do they agree? __________.

CONCLUSION:

INA110

Fast-Settling FET-Input Very High Accuracy INSTRUMENTATION AMPLIFIER

FEATURES

- LOW BIAS CURRENT: 50pA, max
- FAST SETTLING: 4μs to 0.01%
- HIGH CMR: 106dB, min; 90dB at 10kHz
- CONVENIENT INTERNAL GAINS: 1, 10, 100, 200, 500
- VERY-LOW GAIN DRIFT: 10 to 50ppm/°C
- LOW OFFSET DRIFT: 2μV/°C
- LOW COST
- PINOUT COMPATIBLE WITH AD524 AND AD624, allowing upgrading of many existing applications

APPLICATIONS

- Fast scanning rate multiplexed input data acquisition system amplifier
- Fast differential pulse amplifier
- High speed, low drift gain block
- Amplification of low level signals from high impedance sources and sensors
- Instrumentation amplifier with input low pass filtering using large series resistors
- Instrumentation amplifier with overvoltage input protection using large series resistors
- Amplification of signals from strain gauges, thermocouples, and RTDs

DESCRIPTION

The INA110 is a monolithic FET input instrumentation amplifier with a maximum bias current of 50pA. The circuit provides fast settling of 4μs to 0.01%. Laser trimming guarantees exceptionally good DC performance. Voltage noise is low, and current noise is virtually zero. Internal gain set resistors guarantee high gain accuracy and low gain drift. Gains of 1, 10, 100, 200, and 500 are provided.

The inputs are inherently protected by P-channel FETs on each input. Differential and common-mode voltages should be limited to $\pm V_{CC}$. When severe overvoltage exists, use diode clamps as shown in the application section.

The INA110 is ideally suited for applications requiring large input resistors for overvoltage protection or filtering. Input signals from high source impedances can easily be handled without degrading DC performance. Fast settling for rapid scanning data acquisition systems is now achievable with one component, the INA110.

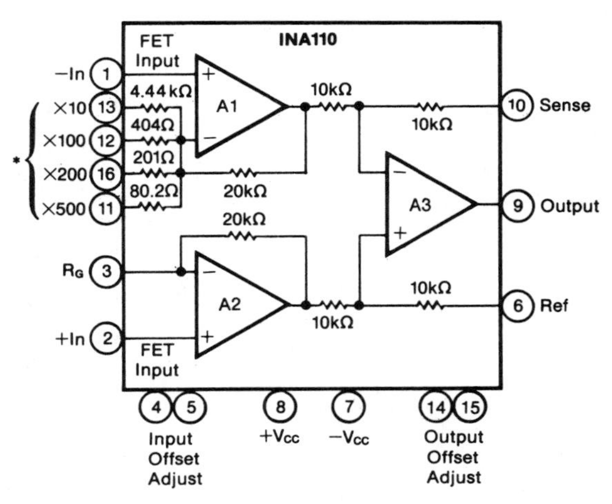

International Airport Industrial Park - P.O. Box 11400 - Tucson, Arizona 85734 - Tel. (602) 746-1111 - Twx: 910-952-1111 - Cable: BBRCORP - Telex: 66-6491

PDS-645

Printed In U.S.A. March, 1986

SPECIFICATIONS

ELECTRICAL

At +25°C, $\pm V_{CC}$ = 15VDC, R_L = 2kΩ unless otherwise noted.

PARAMETER	CONDITIONS	INA110AG			INA110BG			UNITS
		MIN	TYP	MAX	MIN	TYP	MAX	
GAIN								
Range of Gain		1		800	*		*	V/V
Gain Equation[1]				G = 1 + [40K/(R_G + 50Ω)]				V/V
Gain Error, DC: G = 1			0.002	0.04		*	0.02	%
G = 10			0.01	0.1		0.005	0.05	%
G = 100			0.02	0.2		0.01	0.1	%
G = 200			0.04	0.4		0.02	0.2	%
G = 500			0.1	1.0		0.05	0.5	%
Gain Temp. Coefficient: G = 1			±3	±20		*	±10	ppm/°C
G = 10			±4	±20		±2	±10	ppm/°C
G = 100			±6	±40		±3	±20	ppm/°C
G = 200			±10	±60		±5	±30	ppm/°C
G = 500			±25	±100		±10	±50	ppm/°C
Nonlinearity, DC: G = 1			±0.001	±0.01		±0.0005	±0.005	% of FS
G = 10			±0.002	±0.01		±0.001	±0.005	% of FS
G = 100			±0.004	±0.02		±0.002	±0.01	% of FS
G = 200			±0.006	±0.02		±0.003	±0.01	% of FS
G = 500			±0.01	±0.04		±0.005	±0.02	% of FS
OUTPUT								
Voltage, R_L = 2kΩ	Over temp	±10	±12.7		*	*		V
Current	Over temp	±5	±25		*	*		mA
Short-Circuit Current			±25			*		mA
Capacitive Load	Stability'		5000			*		pF
INPUT								
OFFSET VOLTAGE[2]								
Initial Offset			±(100+1000/G)	±(500+5000/G)		±(50+600/G)	±(250+3000/G)	μV
vs Temperature			±(2+20/G)	±(5+100/G)		±(1+10/G)	±(2+50/G)	μV/°C
vs Supply	V_{CC} = ±6V to ±18V		±(4+60/G)	±(30+300/G)		±(2+30/G)	±(10+180/G)	μV/V
BIAS CURRENT								
Initial Bias Current	Each input		20	100		10	50	pA
Initial Offset Current			2	50		1	25	pA
Impedance: Differential			$5\times10^{12}\|\|6$			*		Ω||pF
Common-Mode			$2\times10^{12}\|\|1$			*		Ω||pF
VOLTAGE RANGE	V_{IN} Diff. = 0V[3]							
Range, Linear Response		±10	±12					V
CMR with 1kΩ Source Imbalance:								
G = 1	DC	70	90		80	100		dB
G = 10	DC	87	104		96	112		dB
G = 100	DC	100	110		106	116		dB
G = 200	DC	100	110		106	116		dB
G = 500	DC	100	110		106	116		dB
NOISE, Input[4]								
Voltage, f_O = 10kHz			10			*		nV/√Hz
f_B = 0.1Hz to 10Hz			1			*		μVp-p
Current, f_O = 10kHz			1.8			*		fA/√Hz
NOISE, Output[4]								
Voltage, f_O = 10kHz			65			*		nV/√Hz
f_B = 0.1Hz to 10Hz			8			*		μVp-p
DYNAMIC RESPONSE								
Small Signal: G = 1	−3dB		2.5			*		MHz
G = 10			2.5			*		MHz
G = 100			470			*		kHz
G = 200			240			*		kHz
G = 500			100			*		kHz
Full Power	V_{OUT} = ±10V, R_L = 2kΩ	190	270		*	*		kHz
Slew Rate	G = 1 to 200	12	17		*	*		V/μs
Settling Time:								
0.1%, G = 1	V_O = 20V step		4			*		μs
G = 10			2			*		μs
G = 100			3			*		μs
G = 200			5			*		μs
G = 500			11			*		μs

ELECTRICAL (CONT)

PARAMETER	CONDITIONS	INA110AG			INA110BG			UNITS
		MIN	TYP	MAX	MIN	TYP	MAX	
Settling Time:								
0.01%, G = 1	V_O = 20V step		5	12.5		*	*	μs
G = 10			3	7.5		*	*	μs
G = 100			4	7.5		*	*	μs
G = 200			7	12.5		*	*	μs
G = 500			16	25		*	*	μs
Overload Recovery(5)	50% overdrive		1			*		μs
POWER SUPPLY								
Rated Voltage			±15			*		V
Voltage Range		±6		±18	*		*	V
Quiescent Current	V_O = 0V		±3.0	±4.5		*	*	mA
TEMPERATURE RANGE								
Specification		−25		+85	*		*	°C
Operation		−55		+125	*		*	°C
Storage		−65		+150	*		*	°C
θ_{JA}			100			*		°C/W

* Same as INA110AG.

NOTES: (1) Gains other than 1, 10, 100, 200, and 500 can be set by adding an external resistor, R_G, between pin 3 and pins 11, 12, and 16. Gain accuracy is a function of R_G and the internal resistors which have a ±20% tolerance with 20ppm/°C drift. (2) Adjustable to zero. (3) For differential input voltage other than zero, see Typical Performance Curves. (4) $V_{NOISE\ RTI} = \sqrt{V_{N\ INPUT}^2 + (V_{N\ OUTPUT}/Gain)^2}$ (5) Time required for output to return from saturation to linear operation following the removal of an input overdrive voltage.

MECHANICAL

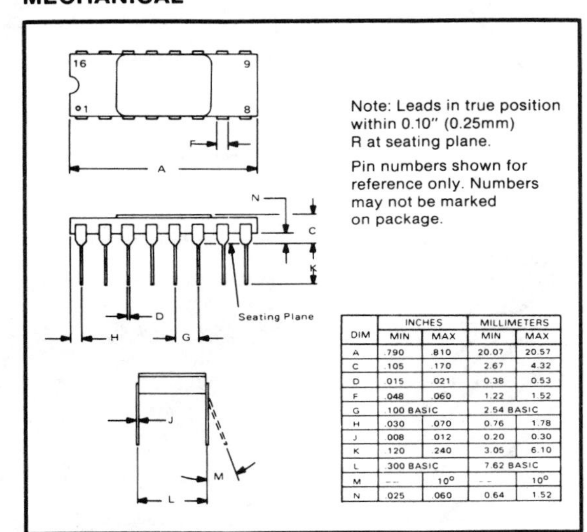

DIM	INCHES		MILLIMETERS	
	MIN	MAX	MIN	MAX
A	.790	.810	20.07	20.57
C	.105	.170	2.67	4.32
D	.015	.021	0.38	0.53
F	.048	.060	1.22	1.52
G	.100 BASIC		2.54 BASIC	
H	.030	.070	0.76	1.78
J	.008	.012	0.20	0.30
K	.120	.240	3.05	6.10
L	.300 BASIC		7.62 BASIC	
M	---	10°	--	10°
N	.025	.060	0.64	1.52

ORDERING INFORMATION

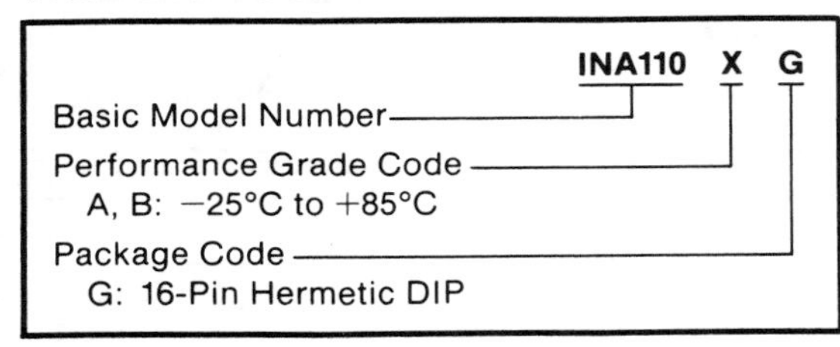

ABSOLUTE MAXIMUM RATINGS

Supply .... ±18V
Input Voltage Range .... $\pm V_{CC}$
Operating Temperature Range .... −55°C to +125°C
Storage Temperature Range .... −65°C to +150°C
Lead Temperature (soldering 10 seconds) .... +300°C
Output Short-Circuit Duration ... Continuous to Common

PIN CONFIGURATION

−In	1	16	×200
+In	2	15	Output Offset Adjust
R_G	3	14	Output Offset Adjust
Input Offset Adjust	4	13	×10
Input Offset Adjust	5	12	×100
Reference	6	11	×500
$-V_{CC}$	7	10	Output Sense
$+V_{CC}$	8	9	Output

PRICES

Quantity	INA110AG	INA110BG
1-24	$16.85	$23.60
25-99	12.65	17.75
100s	8.85	12.35

TYPICAL PERFORMANCE CURVES

$T_A = 25°C$, $\pm V_{CC} = 15VDC$ unless otherwise noted.

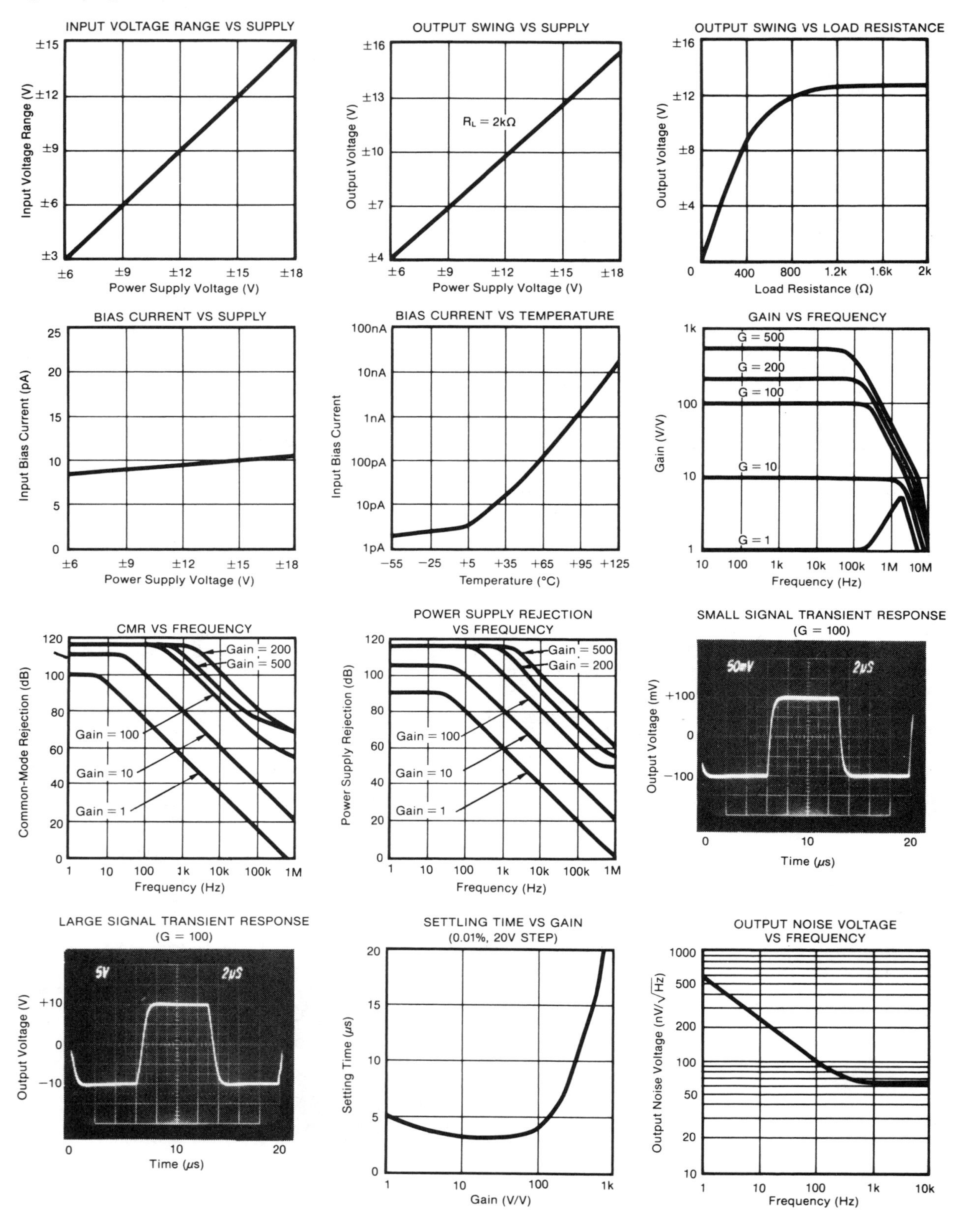

TYPICAL PERFORMANCE CURVES (CONT)

$T_A = 25°C$, $\pm V_{CC} = 15VDC$ unless otherwise noted.

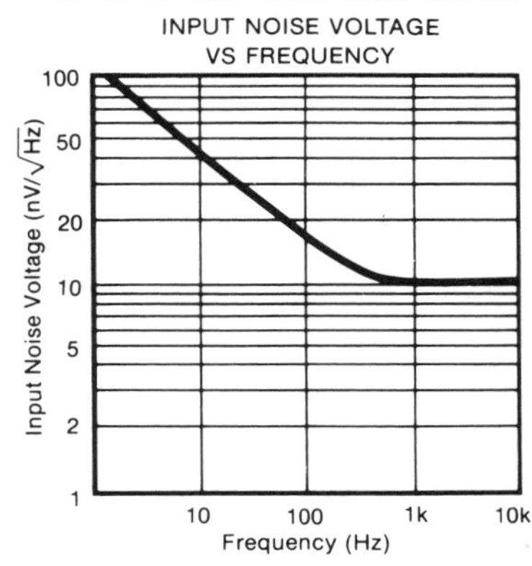

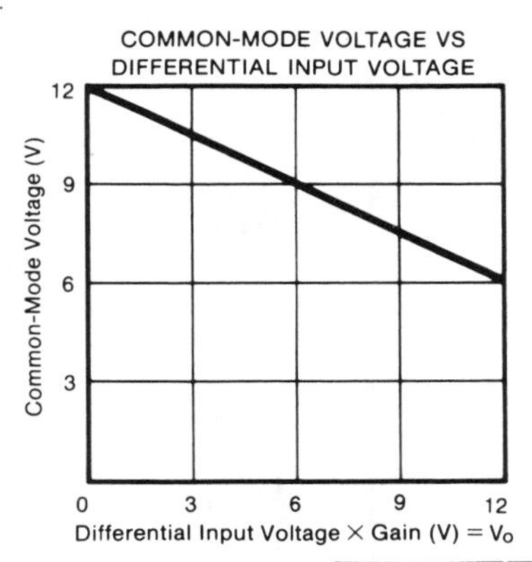

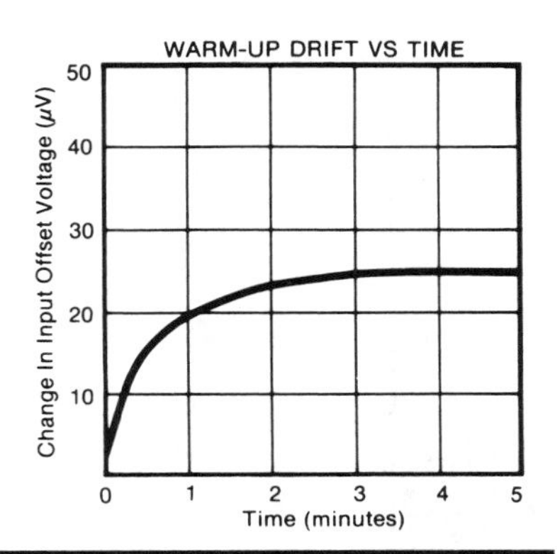

DISCUSSION OF PERFORMANCE

A simplified diagram of the INA110 is shown on the first page. The design consists of the classical three operational amplifier configuration with precision FET buffers on the input. The result is an instrumentation amplifier with premium performance not normally found in integrated circuits.

The input section (A_1 and A_2) incorporates high performance, low bias current, and low drift amplifier circuitry. The amplifiers are connected in the noninverting configuration to provide high input impedance ($10^{12}\Omega$). Laser-trimming is used to achieve low offset voltage. Input cascoding assures low bias current and high CMR. Thin-film resistors on the integrated circuit provide excellent gain accuracy and temperature stability.

The output section (A_3) is connected in a unity-gain difference amplifier configuration. Precision matching of the four $10k\Omega$ resistors, especially over temperature and time, assures high common-mode rejection.

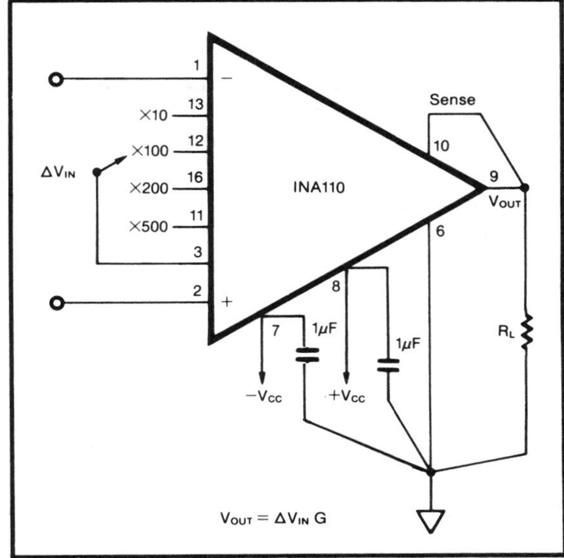

FIGURE 1. Basic Circuit Connection.

BASIC POWER SUPPLY AND SIGNAL CONNECTIONS

Figure 1 shows the proper connections for power supply and signal. Supplies should be decoupled with $1\mu F$ tantalum capacitors as close to the amplifier as possible. To avoid gain and CMR errors introduced by the external circuit, connect grounds as indicated, being sure to minimize ground resistance. Resistance in series with the reference (pin 6) will degrade CMR. Also to maintain stability, avoid capacitance from the output to the gain set, offset adjust, and input pins. The layout shown in Figure 2 is suggested for best performance.

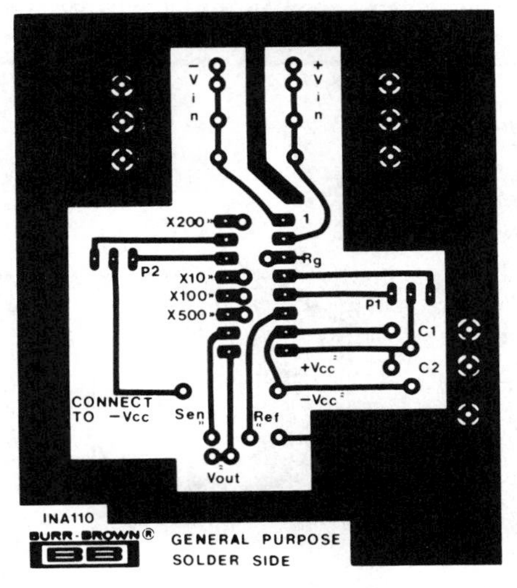

FIGURE 2. Recommended PC Board Layout for INA110.

OFFSET ADJUSTMENT

Figure 3 shows the offset adjustment circuit for the INA110. Both the offset of the input stage and output stage can be adjusted separately. Notice that the offset referred to the INA110's input (RTI) is the offset of the input stage plus the offset of the output stage divided by

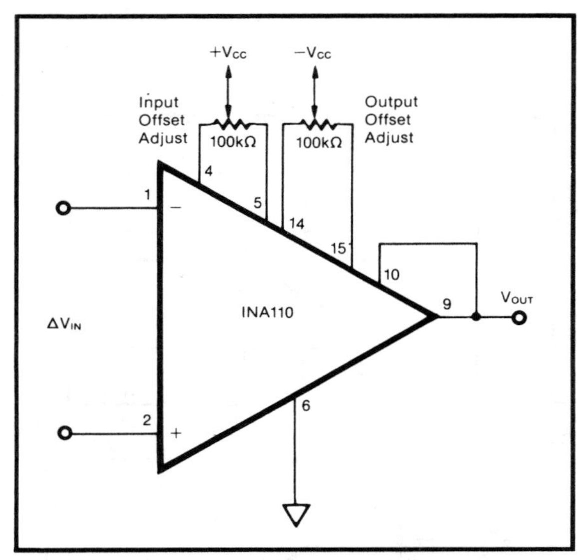

FIGURE 3. Offset Adjustment Circuit.

the gain of the input stage. This allows specification of offset independent of gain.

For systems using computer autozeroing techniques, neither offset nor offset drift are of concern. In many other applications the factory-trimmed offset gives excellent results. When greater accuracy is desired, one adjustment is usually sufficient. In high gains (>100) adjust only the input offset, and in low gains the output offset. For higher precision in all gains, both can be adjusted by first selecting high gain and adjusting input offset and then low gain and adjusting output offset. The offset adjustment will, however, add to the drift by approximately $0.33\mu V/°C$ per $100\mu V$ of input offset voltage that is adjusted. Therefore, care should be taken when considering use of adjustment.

Output offsetting can be accomplished as shown in Figure 4 by applying a voltage to the reference (pin 6)

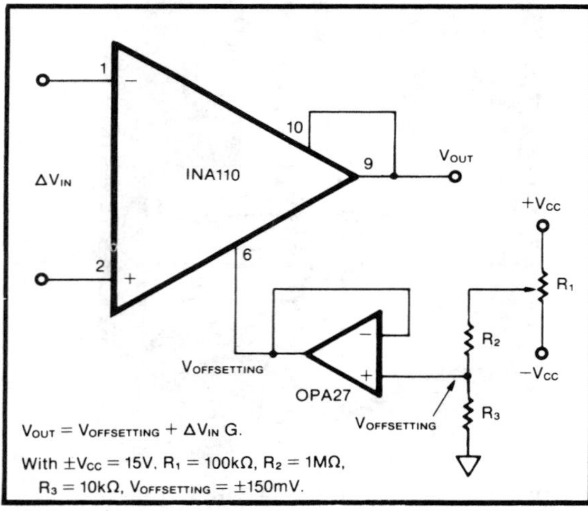

FIGURE 4. Output Offsetting.

through a buffer. This limits the resistance in series with pin 6 to minimize CMR error. Be certain to keep this resistance low. Note that the offset error can be adjusted at this reference point with no appreciable degradation in offset drift.

GAIN SELECTION

Gain selection is accomplished by strapping the appropriate pins together on the INA110. Table I shows possible gains from the internal resistors. Keep the connections as short as possible to maintain accuracy.

TABLE I. Internal Gain Connections.

Gain	Connect pin 3 to pin —	Gain Accuracy (%)	Gain Drift (ppm/°C)
The following gains have guaranteed accuracy:			
1	none	0.02	10
10	13	0.05	10
100	12	0.1	20
200	16	0.2	30
500	11	0.5	50
The following gains have typical accuracy as shown:			
300	12 & 16	0.25	10
600	11 & 12	0.25	40
700	11 & 16	2.0	40
800	11, 12, & 16	2.0	80

Gains other than 1, 10, 100, 200, and 500 can be set by adding an external resistor, R_G, between pin 3 and pins 12, 16, and 11. Gain accuracy is a function of R_G and the internal resistors which have a ±20% tolerance with 20ppm/°C drift. The equation for choosing R_G is shown below.

$$R_G = \frac{40k}{G-1} - 50\Omega$$

Gain can also be changed in the output stage by adding resistance to the feedback loop shown in Figure 5. This is useful for increasing the total gain or reducing the input stage gain to prevent saturation of input amplifiers.

The output gain can be changed as shown in Table II. Matching of R_1 and R_3 is required to maintain high CMR. R_2 sets the gain with no effect on CMR.

TABLE II. Output Stage Gain Control.

Output Stage Gain	R_1 and R_3	R_2
2	1.2kΩ	2.74kΩ
5	1kΩ	511Ω
10	1.5kΩ	340Ω

COMMON-MODE INPUT RANGE

It is important not to exceed the input amplifiers' dynamic range (see Typical Performance Curves). The differential input signal and its associated common-mode voltage should not cause the output of A_1 and A_2 (input amplifiers) to exceed approximately ±10V with ±15V supplies or nonlinear operation will result. Such large common-mode voltages, when the INA110 is in high gain, can cause saturation of the input stage even though the differential input is very small. This can be avoided by reducing the input stage gain and increasing the output stage gain with an H pad attenuator (see Figure 5).

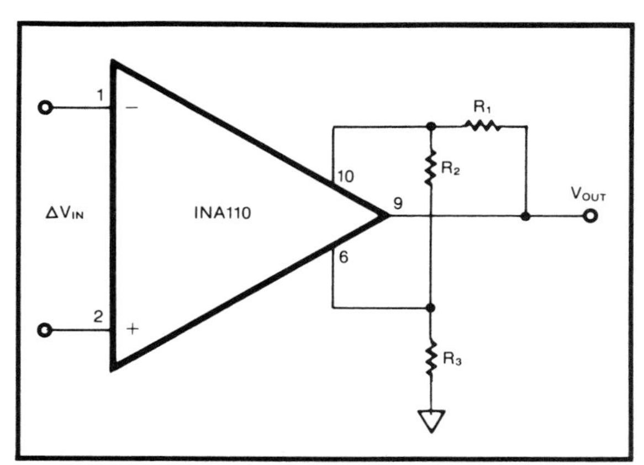

FIGURE 5. Gain Adjustment of Output Stage Using H Pad Attenuator.

OUTPUT SENSE

An output sense has been provided to allow greater accuracy in connecting the load. By attaching this feedback point to the load at the load site, IR drops due to load currents are eliminated since they are inside the feedback loop. Proper connection is shown in Figure 1. When more current is to be supplied, a power booster can be placed within the feedback loop as shown in Figure 6. Buffer errors are minimized by the loop gain of the output amplifier.

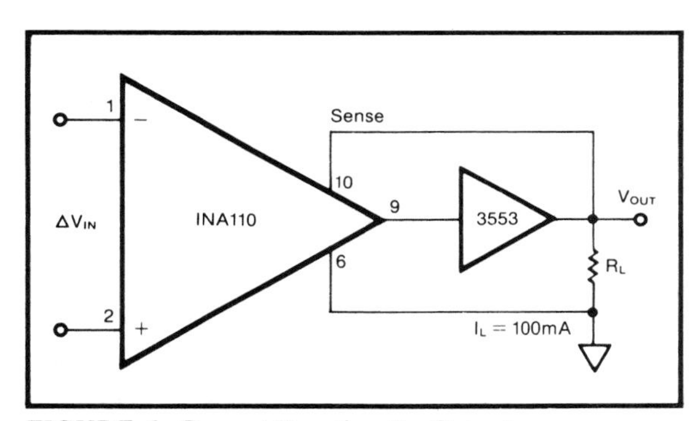

FIGURE 6. Current Boosting the Output.

LOW BIAS CURRENT OF FET INPUT ELIMINATES DC ERRORS

Because the INA110 has FET inputs, bias currents drawn through input source resistors have a negligible effect on DC accuracy. The picoamp levels produce no more than microvolts through megohm sources. Thus, input filtering and input series protection are readily achievable.

A return path for the input bias currents must always be provided to prevent charging of stray capacitance. Otherwise the output can wander and saturate. A 1MΩ to 10MΩ resistor from the input to common will return floating sources such as transformers, thermocouples, and AC-coupled inputs (see Applications section).

DYNAMIC PERFORMANCE

The INA110 is a fast-settling FET input instrumentation amplifier. Therefore, careful attention to minimize stray capacitance is necessary to achieve specified performance. High source resistance will interact with input capacitance to reduce the overall bandwidth. Also, to maintain stability, avoid capacitance from the output to the gain set, offset adjust, and input pins (see Figure 2 for PC board layout).

When using high source resistance (>20kΩ) in the positive input only, bypass it to ground with at least 50pF to maintain stability.

The INA110 is designed for fast settling with easy gain selection. It has especially excellent settling in high gain. It can also be used in fast-settling unity-gain applications. As with all such amplifiers, the INA110 does exhibit signficant gain peaking when set to a gain of 1. It is, however, unconditionally stable. The gain peaking can be cancelled by band-limiting the negative input to 400kHz with a simple external RC circuit for applications requiring flat response.

Another distinct advantage of the INA110 is the high frequency CMR response. High frequency noise and sharp common-mode transients will be rejected. To preserve AC CMR, be sure to minimize stray capacitance on the input lines. Matching the RCs in the two inputs will help to maintain high AC CMR.

Timers and Oscillators

9

LEARNING OBJECTIVES

Upon completion of this chapter on timers and oscillators, you will be able to:

- Describe the operating principles of an integrated-circuit 555 timer.
- Use a 555 timer IC to provide "delayed" start or "initialization" timing pulses, where required.
- Make inexpensive square-wave oscillators with a 555.
- Stretch or shrink pulses by wiring a 555 for one-shot operation.
- Make an inexpensive triangle-wave oscillator out of two op amps, three resistors, and a capacitor.
- Describe operating principles of an AD630 balanced modulator/demodulator when used as a switched gain amplifer.
- Use the AD630 to make *precision* triangle or square wave oscillators whose output voltage amplitude and frequency of operation can, easily and precisely, be adjusted *independently*.
- Wire a state-of-the-art IC AD639A trigonometric function generator to compute the sine of any angle between 0 and $\pm 500°$.
- Connect the AD639A and AD630 to make a *precision* sine-wave generator whose frequency is easy to adjust precisely.

9-0 INTRODUCTION

A timer IC such as the 555 produces precise output timing intervals in response to an input command. It can also function as an oscillator if its output is connected back to its own input. This versatile chip can be used in a wide variety of applications such as pulse or square-wave generators. It can also solve a variety of timer applications. For inexpensive triangle, ramp, or sawtooth generators, we will use two op amps, a capacitor, and a few resistors.

There are many techniques for generating sine waves. However, the state-of-the-art way to generate a *precision sine wave* and also a *precision triangle wave* is with Analog Devices' AD639 trigonometric function generator IC and AD630 balanced modulator/demodulator IC. We begin our study of timers and oscillators with the standard 555 IC.

9-1 TERMINALS OF THE 555

9-1.1 Packaging and Power Supply Terminals

The 555 has eight terminals, as shown in Fig. 9-1. Pin 1 is the ground terminal. Pin 8 is the supply voltage terminal V_{CC}. V_{CC} typically ranges from +5 V to +15 V and has a maximum rating of +18 V.

Three equal 5-kΩ resistors are internally connected in series between V_{CC} and ground. As shown in Fig. 9-1b, this resistor voltage divider sets the (−) input of comparator 1 at an *upper threshold voltage,* V_{UT}, where

$$V_{UT} = \tfrac{2}{3}V_{CC} \qquad \text{(9-1a)}$$

The same voltage divider sets the (+) input of comparator 2 at the lower threshold voltage, V_{LT}, where

$$V_{LT} = \tfrac{1}{3}V_{CC} \qquad \text{(9-1b)}$$

V_{UT} and V_{LT} will be associated in Section 9-1.6 with the controlling inputs: the pin 6 *threshold* and pin 2 *trigger* terminals.

9-1.2 Discharge Terminal

The discharge terminal, pin 7, allows the control logic to close a switch (NPN transistor) when output pin 3 is in the low state. Pin 7 is usually used to charge or discharge a timing capacitor in timer or oscillator applications. The process is explained in Section 9-3.1 and Fig. 9-4.

9-1.3 Control Voltage Terminal

The control voltage terminal, pin 5, allows the user to (within narrow limits) change the values of both V_{UT} and V_{LT} by wiring external resistors to ground or another control voltage. However, in most applications, pin 5 is connected via a 0.01- or 0.1-uF capacitor to ground. This bypasses any power supply ripple or noise voltages to ground and minimizes their effects on both V_{UT} and V_{LT}.

9-1.4 Output Terminal

The output terminal, pin 3, acts much like a transfer switch. When the control logic tells it to go high, a low-resistance connection is made to pin 8, V_{CC}. In this state, pin 3 can *source* current exceeding 20 mA (200 mA for pulse operation). When the control logic tells pin 3 to go low, a

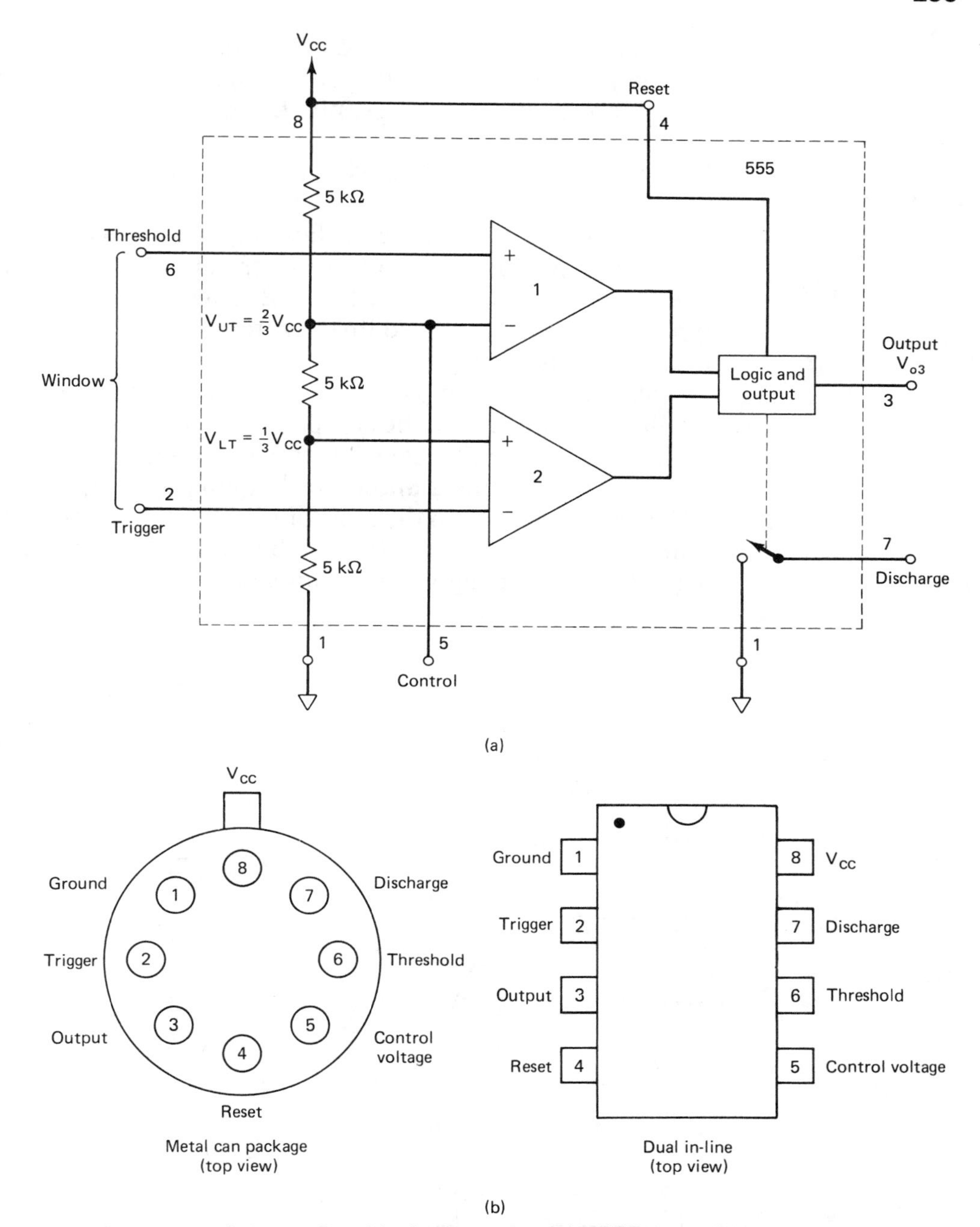

Figure 9-1 (a) Simplified schematic of the 555 timer; (b) pin connections and packages.

low-resistance connection is made to grounded pin 1. Now the output terminal can *sink* up to 200 mA.

This output voltage V_{o3} at pin 3 is either near V_{CC} potential (high) or ground potential (low).

9-1.5 Reset Terminal

Reset terminal, pin 4, is an input. Usually, it is wired to supply terminal V_{CC} at pin 8, to disable the *reset* feature. However, *Reset* may be used for instantaneous override of commands from the normally used input terminals, *Trigger* and *Control.* (Their functions are explained in the next section.) If *Reset* is grounded, either directly or via a transistor, the 555's *Output* pin 3, and *Discharge,* pin 7, are brought low.

9-1.6 Input Trigger and Threshold Terminals

Perhaps the best way to learn how input terminals *Trigger* (pin 2) and *Threshold* (pin 6) exert control over the 555 is to perform a simple operating test. As shown in Fig. 9-2, a test voltage V_i is applied to *both* input terminals and varied linearly from 0 to 15 V, and back to 0 V. V_i is plotted versus time *t.*

A CRO is used to plot V_{o3} versus V_i and clearly identifies a hysteresis graph that is much like the comparators with positive feedback discussed in Chapter 4. The 555 thus has three distinct operating states.

When *both* pins 2 and 6 are *below* V_{LT}, the V_{o3} output goes *high,* as indicated by state H on the plots of Fig. 9-2. If both pins are *above* V_{UT}, the output always goes *low,* indicated by state L. This indicates that the 555 acts like an inverting voltage comparator.

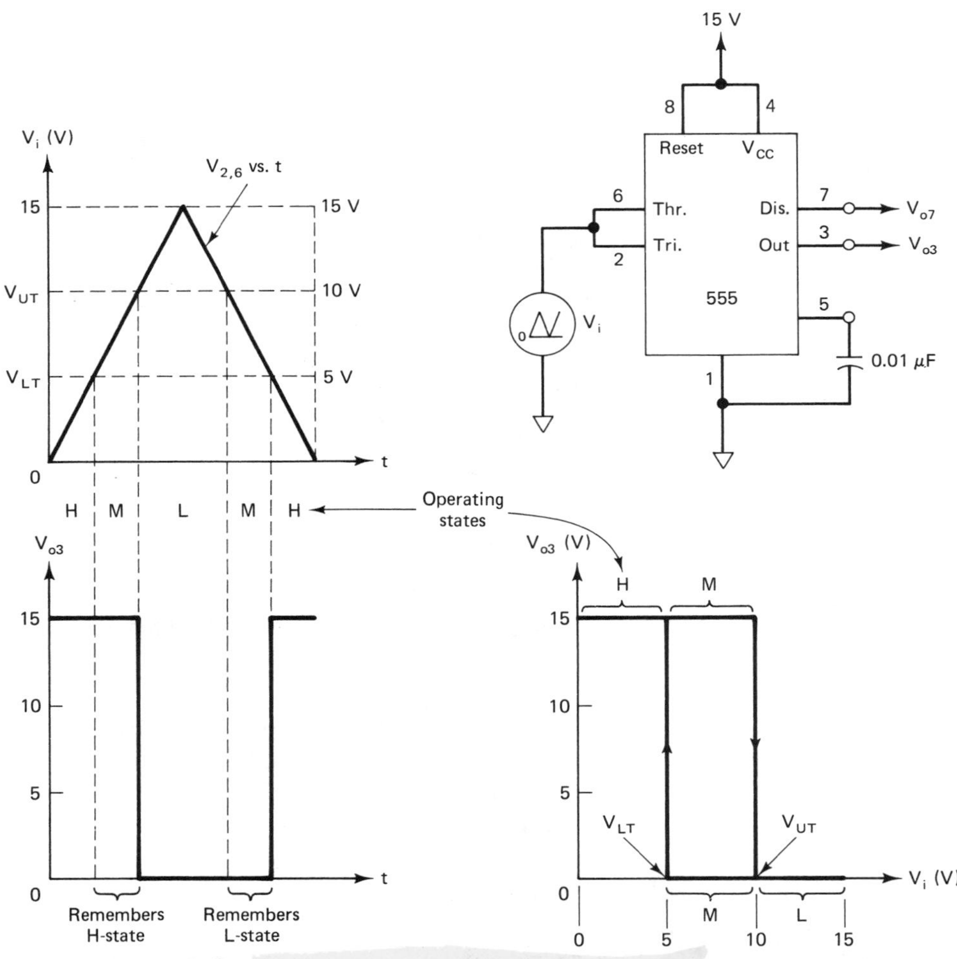

Figure 9-2 The four operating states of a 555 timer are portrayed by plotting V_i versus *t,* V_o versus *t* and V_o versus V_i on a CRO.

However, the M state is a *memory state.* If *both* trigger and threshold terminals lie between the threshold voltages, the 555 "remembers" its last high (H) or low (L) output state. There are only four possible combinations of input voltages because trigger (pin 2) is compared with V_{LT} and threshold (Pin 6) is compared with V_{UT} (Fig. 9-lb). These combinations are presented in Table 9-1.

TABLE 9-1

Operating States of the 555 Timer; V_{UT} = 2/3 V_{CC}, V_{LT} = 1/3 V_{CC}, high ≃ V_{CC}, low or ground ≃ 0 V

Operating state	*Trigger pin 2*	*Threshold pin 6*	*Output Terminal's State*	
			Output 3	*Discharge 7*
H	Below V_{LT}	Below V_{UT}	High	Open
H	Below V_{LT}	Above V_{UT}	High	Open
M	Above V_{LT}	Below V_{UT}	Remembers last state	
L	Above V_{LT}	Above V_{UT}	Low	Ground

9-2 555 DELAY APPLICATIONS

9-2.1 Delayed Start

We will use the operating test data of Table 9-1 and Fig. 9-2 to learn how the 555 timer performs in two timing applications. They are "delayed start" and "initializing."

When power is applied to a system, it is sometimes necessary to apply power immediately to one portion of the system and delay power application to another portion. For example, suppose that an alarm circuit sounded if a motor was not at the correct speed. If both motor and alarm circuit were powered up together, the alarm would immediately sound. We want to apply power to the motor and allow it to start and simultaneously delay power to the alarm circuit until the motor reaches operating speed.

When the switch is closed in Fig. 9-3a, power is applied to both the 555 timer and system. The uncharged capacitor initially acts as a short circuit. Therefore, pins 2 and 6 are at V_{CC} and tell the output to go low (state L). As the capacitor charges, the voltage at pins 2 and 6 decreases and passes through memory state M, where the output stays low. When $V_{2,6}$ drops just below V_{LT}, state H sends the output high. The delay interval, T, from switch closure to the time that V_o goes high is established by

$$T = 1.1\ R_A C \tag{9-2}$$

9-2.2 Initializing

When systems are powered up, it is often necessary to ensure that certain subsystems start up in a predetermined condition. For example, a counting system counts how many people enter a store each day, or how many cars roll out of an assembly line each day.

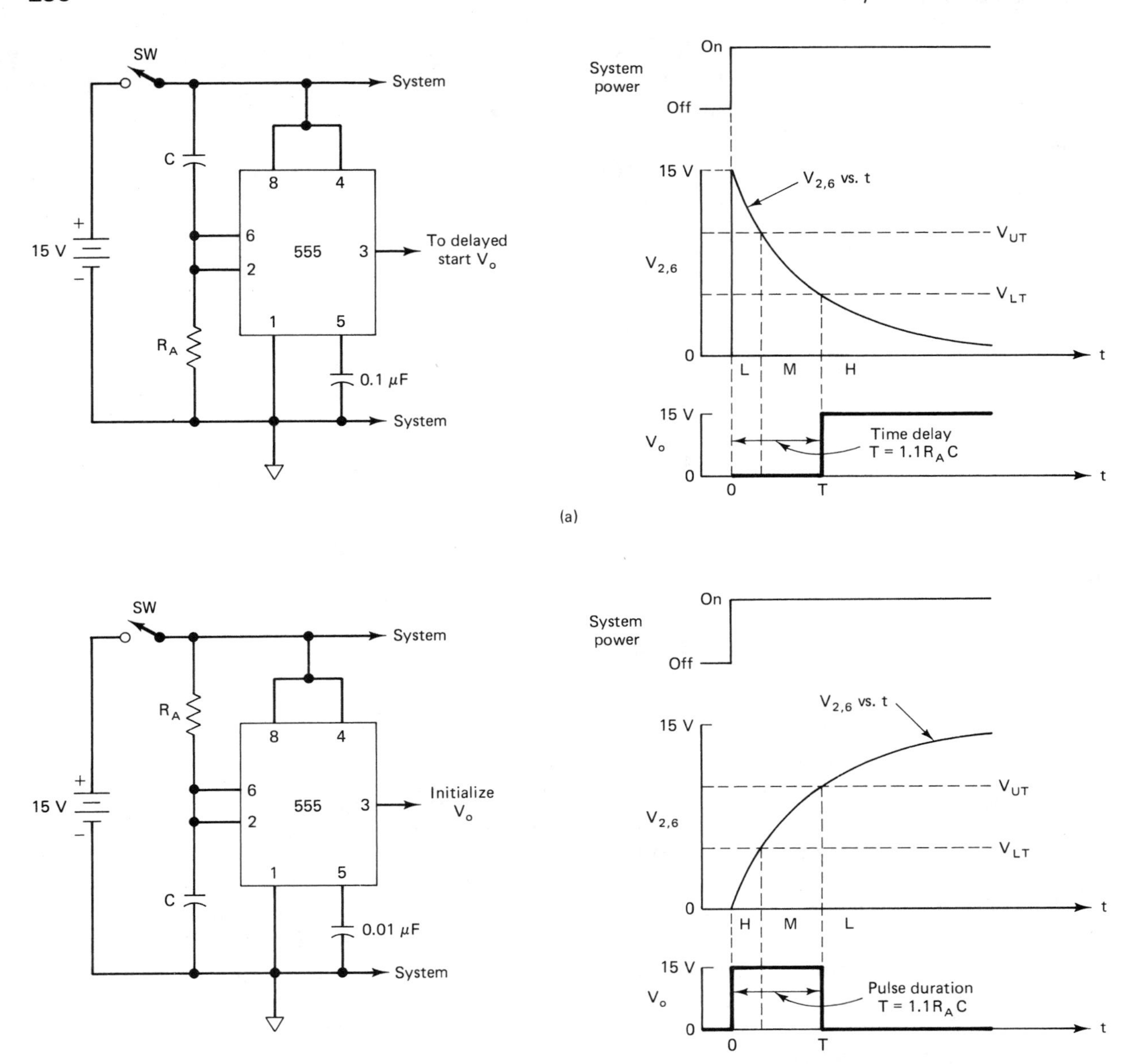

Figure 9-3 System power is applied when switch SW is closed. In (a) there is a time delay *T* before V_o applies power to a second system. In (b), output V_o applies power for time *T* to perform initialization. (a) Output V_o is delayed from going high, after switch SW is closed, by time interval *T*: (b) Output V_o goes high for time interval *T* after switch SW is closed.

These types of applications require that the counters all be automatically reset to 0000 when power is applied at opening of business. This type of problem is called *initializing* or *automatic reset.* It is solved nicely by the circuit of Fig. 9-3b.

When power is applied to the reset circuit of Fig. 9-3b, capacitor *C* initially acts as a short circuit at time zero. $V_{2,6}$ rises exponentially, crossing memory state M, where V_o remains high. When $V_{2,6}$ crosses V_{UT}, state L operation sends the output low. Operation of the counters is passed over to the counting circuitry. Duration, T, of the output pulse is found from Eq. (9-2).

EXAMPLE 9-1

In Figure 9-3a, $C = 1\ \mu F$ and $R_A = 1\ M\Omega$. Find (a) V_{UT} and V_{LT}; (b) time delay T.

SOLUTION (a) From Eqs. (9-1a) and (9-1b),

$$V_{UT} = \tfrac{2}{3}(15\ V) = 10\ V \text{ and } V_{LT} = \tfrac{1}{3}(15\ V) = 5\ V$$

(b) From Eq. 9-2),

$$T = 1.1\ R_A C = 1.1\ (1\ M\ \Omega)\ (1\ \mu F) = 1.1\ s$$

EXAMPLE 9-2

Find the pulse duration in Fig. 9-3b if $R_A = 10\ k\Omega$ and $C = 0.01\ \mu F$.

SOLUTION From Eq. (9-2),

$$T = 1.1 R_A C = 1.1 \times 10^4 \times 0.01 \times 10^{-6}\ F = 0.011\ s = 11\ ms$$

9-3 555 SIGNAL GENERATION

9-3.1 Circuit Operation

The 555 timer chip is wired to operate as an oscillator or astable multivibrator in Fig. 9-4a. Input pins 2 and 6 are connected to the junction of R_B and C. Discharge pin 7 is wired to the junction of R_A and R_B.

To understand how this circuit operates, refer to the waveforms in Fig. 9-4b. At time $t = 0$, capacitor voltage $V_C = V_{2,6}$ is very briefly below V_{LT}. From Table 9-1 we see that state H operation occurs. Therefore, output pin 3 or V_o goes high. Also, discharge terminal pin 7 acts as an open, as shown in Fig. 9-3c.

Capacitor C then charges via V_{CC}, R_A, and R_B. V_C increases exponentially from V_{LT} toward V_{UT}, as shown in Fig. 9-4b. Output V_o remains high during this charge time t_{high} (state M).

When V_C reaches and just barely exceeds V_{UT}, state (L) operation sends the output low and pin 7 grounds the junction of R_A and R_B. R_A must have a value greater than 1 kΩ to protect pin 7 from excessive current.

As shown in Fig. 9-4d, C discharges through R_B from V_{UT} to V_{LT} (state M) during time interval t_{low}. When V_C drops just below V_{LT}, another cycle begins. The time for one cycle T is given by

$$T = t_{high} + t_{low} \tag{9-3a}$$

and frequency of oscillation f is

$$f = \frac{1}{T} \tag{9-3b}$$

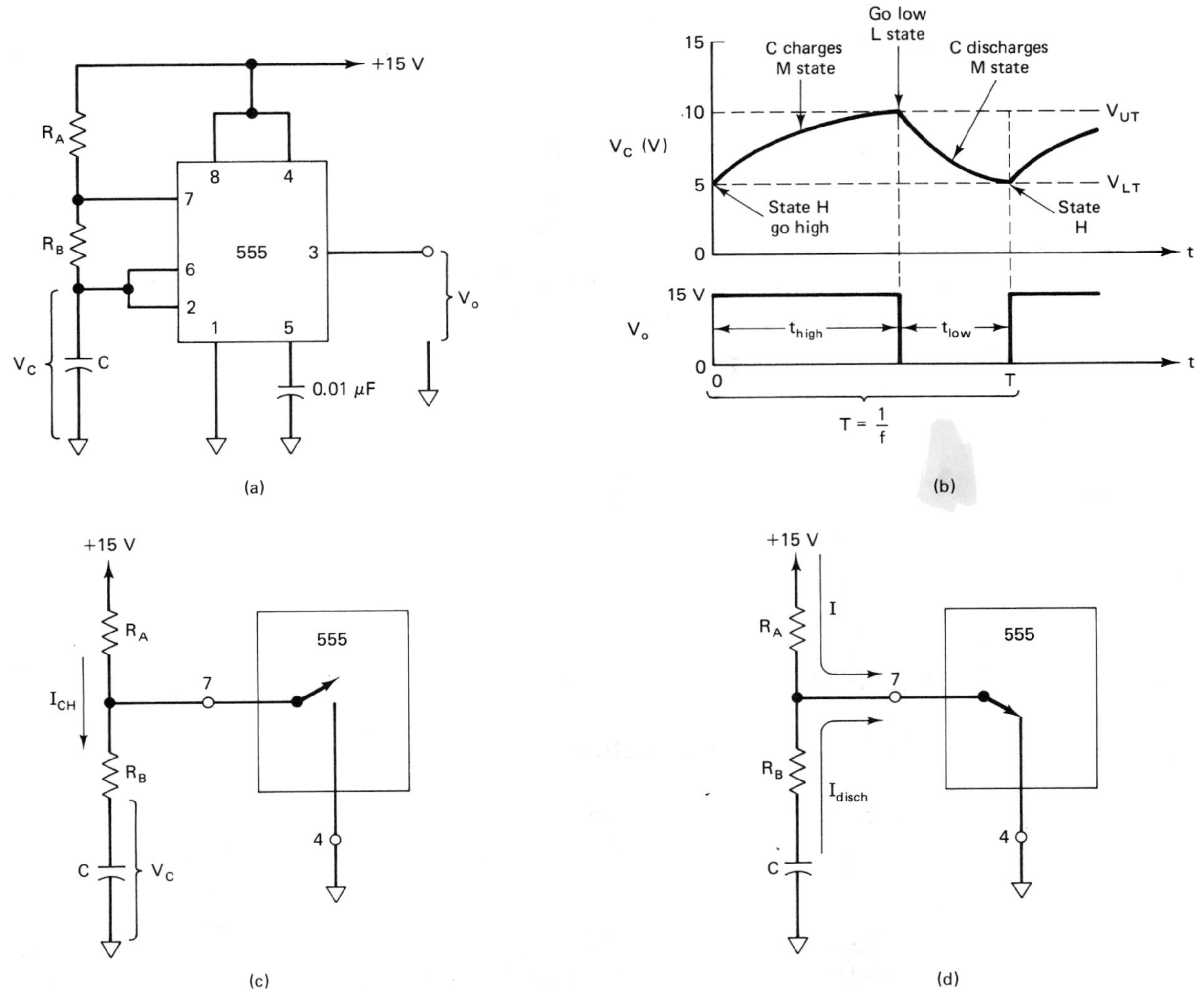

Figure 9-4 The 555 is connected to act as an oscillator or a stable multivibrator in (a). Output signal voltages V_C and V_O are shown in (b). Operation of discharge terminal 7 is shown in (c) and (d). (a) Oscillator circuit; (b) voltage waveforms; (c) C charges during t_{high}; (d) C discharges during t_{low}.

9-3.2 Frequency of Oscillation

Two output voltage waveshapes are available as shown in Fig. 9-4b. V_o is an unsymmetrical square wave and V_C is an unsymmetrical exponential wave. The period and frequency of oscillation are determined as follows: High and low times for each cycle are calculated from

$$t_{high} = 0.69(R_A + R_B)C \tag{9-4a}$$

and

$$t_{low} = 0.69\,(R_B)C \tag{9-4b}$$

Period T is then found from:

$$T = t_{high} + t_{low} = 0.69\,(R_A + 2R_B)C \tag{9-4c}$$

and frequency f from

$$f = \frac{1}{T} \tag{9-4d}$$

EXAMPLE 9-3

$R_A = R_B = 10\ \text{k}\Omega$ and $C = 0.01\ \mu\text{F}$ in Fig. 9-4a. Find the frequency of oscillation.

SOLUTION From Eq. (9-4),

$$t_{\text{high}} = 0.69(10\ \text{k}\Omega + 10\ \text{k}\Omega)\ (0.01 \times 10^{-6}\ \text{F}) = 0.138\ \text{ms}$$

$$t_{\text{low}} = 0.69(10\ \text{k}\Omega)\ (0.01 \times 10^{-6}\ \text{F}) = 0.069\ \text{ms}$$

$$T = (0.138 + 0.069)\ \text{ms} = 0.207\ \text{ms}$$

$$f = \frac{1}{0.207\ \text{ms}} = 4830\ \text{Hz}$$

EXAMPLE 9-4

Make an oscillator to output a frequency of 1 kHz.

SOLUTION Arbitrarily, choose $C = 0.01\ \mu\text{F}$. Let $R_A = R_B$ so that $t_{\text{high}} = 2t_{\text{low}}$, and $T = 3t_{\text{low}}$. See Eqs. (9-4a) and (9-4b). Find $t_{\text{low}} = T/3 = 1\ \text{ms}/3 = 0.33$ ms. From Eq. (9-4b);

$$R_B = \frac{t_{\text{low}}}{0.69C} = \frac{0.33 \times 10^{-3}\ \text{s}}{0.69 \times 0.01 \times 10^{-6}\ \mu\text{F}} = 47.8\ \text{k}\Omega$$

Thus $R_A = R_B = 47.8\ \text{k}\Omega$.

9-3.4 Tone-Burst Oscillator

Two 555s are interconnected in Fig. 9-5 to make a *tone-burst* or *chirp* oscillator. The B timer's oscillating frequency can be adjusted from approximately 1 kHz to 15 kHz (audio range) with the 0- to 10-k Ω potentiometer. The A 555 oscillates at a much lower frequency that can be reduced to about 1 Hz minimum with the 1-MΩ pot.

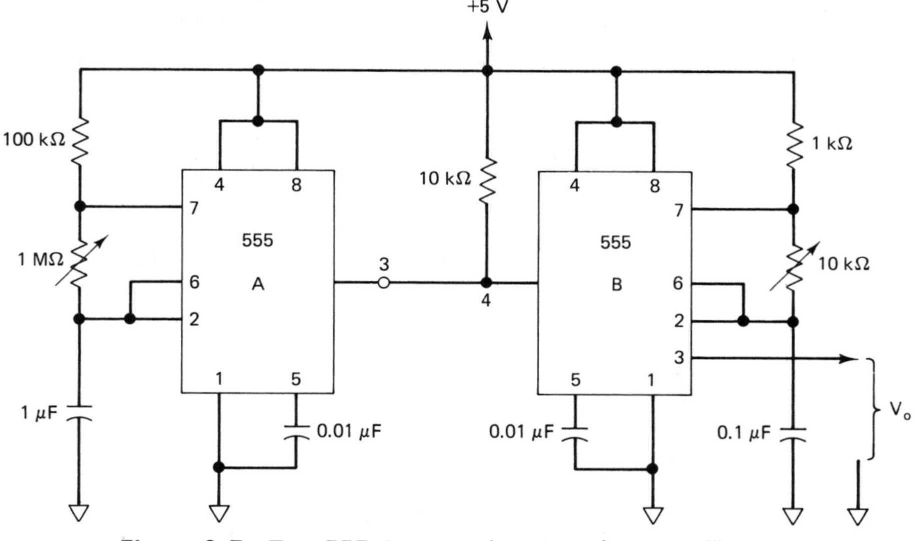

Figure 9-5 Two 555 timers make a tone-burst oscillator.

When output pin 3 of A goes high, it forces reset pin 4 of B to go high. This enables 555 B to oscillate. When pin 3 of A goes low, reset pin 4 of B goes low to stop oscillations. The result is a periodic burst of oscillation. If V_o is connected via a transistor to a speaker, its output sounds like a repetitive chirp.

9-4 ONE-SHOT MULTIVIBRATOR

9-4.1 Definition

A one-shot (or monostable) multivibrator issues an output pulse in response to an input command pulse. The duration of the output pulse is independent of the duration of the input pulse because only the *negative edge* of the input pulse triggers the output pulse. A single input negative edge triggers a single output pulse: hence the name "one-shot."

One-shot applications are programmable time delays. The output pulse is usually longer than the input pulse (pulse stretcher), although the output pulse can also be made shorter (pulse shrinker).

9-4.2 Circuit Operation

The one-shot multivibrator circuit is shown in Fig. 9-6a. In its idle state, input pin 2 is high (at V_{CC}) and pins 36 and 7 are low (state L) (Fig. 9-6b). Voltages across *both* capacitors C and C_i are zero.

Let V_i go low (negative transition) at time N. This pulls trigger pin 2 down to ground, where state H* sends V_o high and opens pin 7. V_o goes high to initiate the output pulse.

Capacitor C then charges toward V_{UT}. See $V_{6,7}$ in Fig. 9-6b. When $V_{6,7}$ reaches V_{UT} after an elapsed time of t_{high} or T, state L operation places a ground on pin 6-7 and V_o goes low. This terminates the output pulse. The 555 one-shot also returns to its idle state L.

Duration of the output pulse T is determined by

$$T = 1.1R_AC \tag{9-5}$$

The time constant for R_i and C_i should always be less than R_A and C. Diode D prevents pin 2 from being driven above V_{CC}. The time constant $R_i \times C_i$ should be about one-tenth of time T.

EXAMPLE 9-5

Calculate output pulse duration for the one-shot of Fig. 9-6.

SOLUTION From Eq. (9-5),

$$T = 1.1R_AC = 1.1(10\ \text{k}\Omega)\ (0.1\ \mu\text{F}) = 1.1\ \text{ms}$$

9-4.3 Missing-Pulse Detector

We use the results of Example 9-5 to study an unusual application for the one-shot. It is the *missing-pulse detector* of Fig. 9-7a.

Refer to time zero in Fig. 9-7b. V_i is low to hold pin 2 low and the

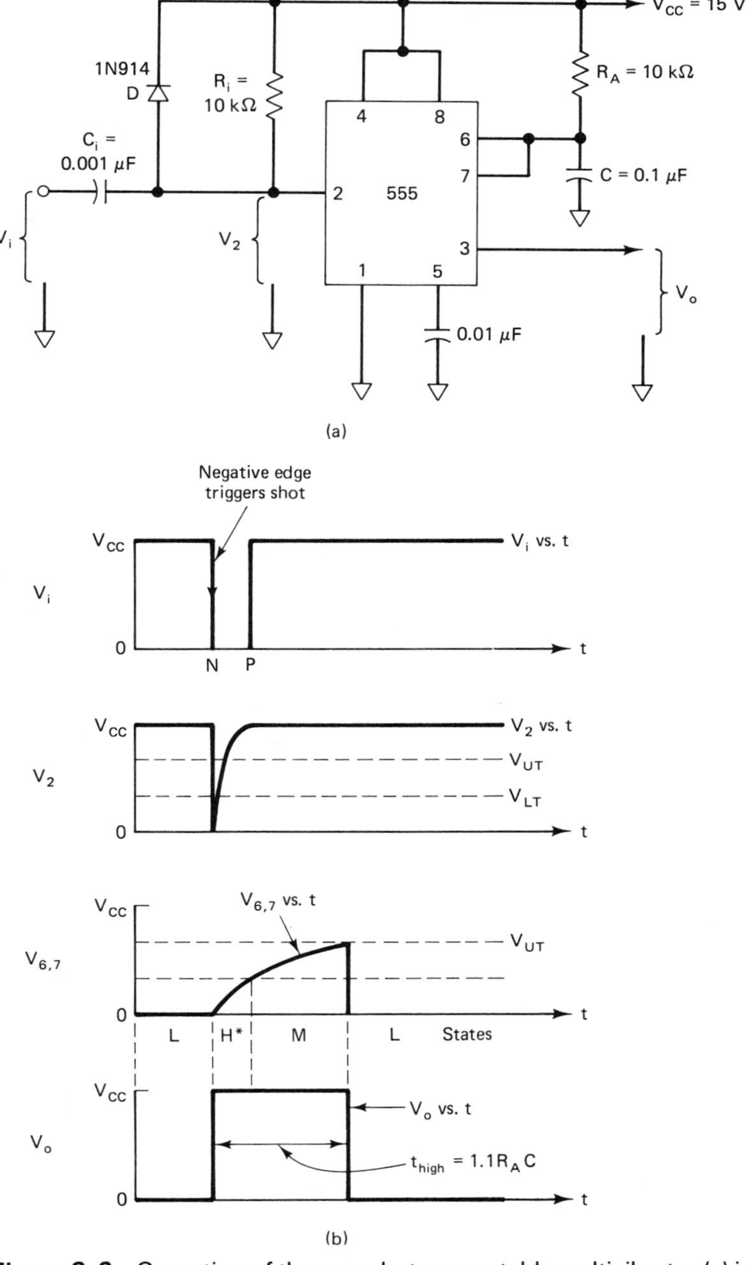

Figure 9-6 Operation of the one-shot monostable multivibrator (a) is explained by the operating waveshapes (b).

transistor on. The transistor shorts capacitor C to hold pin to 6 low. The 555 is in state H operation with pin 3 high and pin 7 open.

At time 0.2 ms, V_i goes high announcing a pulse beginning. The transistor cuts off and V_C charges toward V_{UT}. Before V_C reaches V_{UT}, the next pulse at 1 ms saturates the transistor and discharges V_C to 0 V. Every pulse has a high time of 0.8 ms and is repeated every 1.0 ms. When a missing pulse occurs, the capacitor voltage reaches V_{UT}, where state L sends the output low.

The next pulse at time 3.0 ms sends V_o high. Thus a low pulse is seen at V_o every time there is a missing pulse at input V_i. With modification, this circuit can detect missing heartbeats and drive a counter to give a total count of how many missed beats occur per hour.

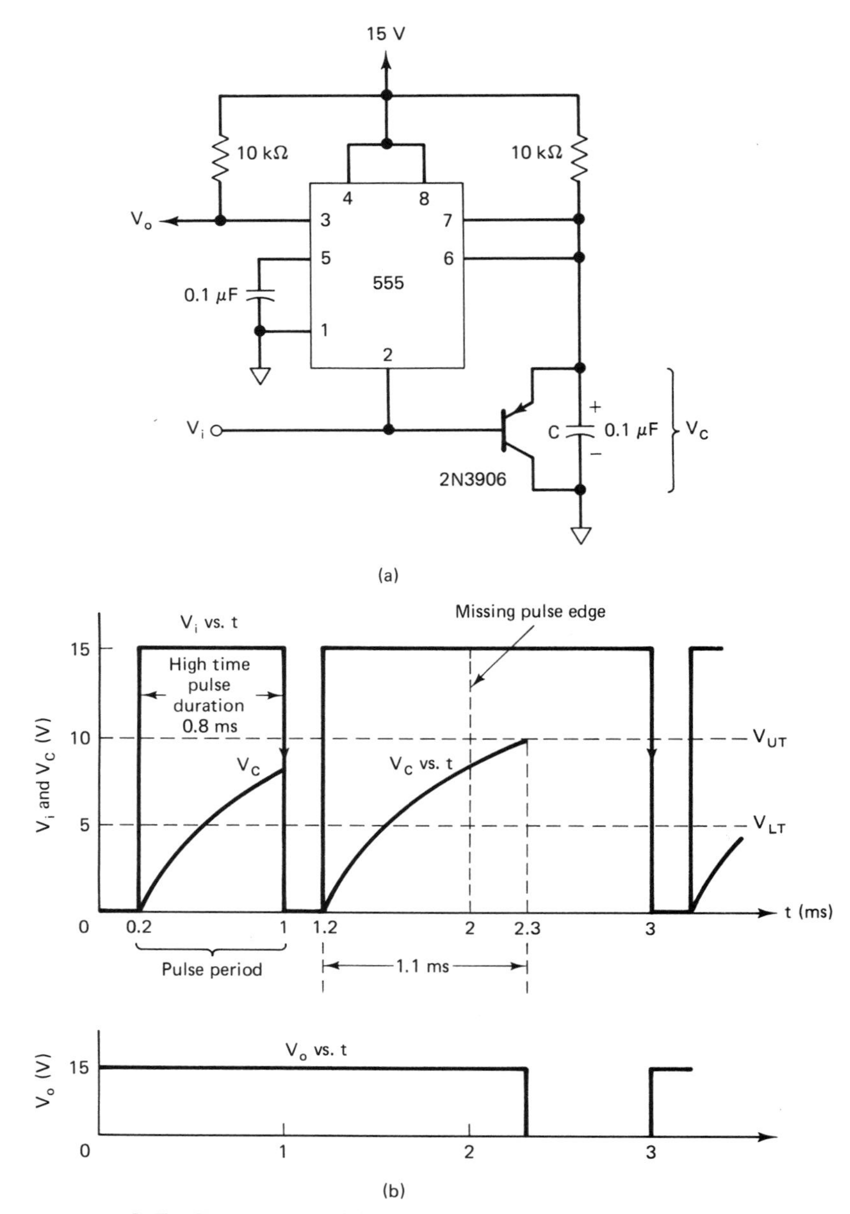

Figure 9-7 The circuit in (a) detects missing pulses as shown by the waveforms in (b).

9-5 TRIANGLE-WAVE GENERATORS

9-5.1 Theory of Operation

A basic bipolar triangle-wave generator circuit is presented in Fig. 9-8a. The triangle wave, V_A, is available at the output of the 741 integrator circuit. An additional square-wave signal, V_B, is available at the output of the 301 comparator.

To understand circuit operation, refer to time interval 0 to 1 ms in Fig. 9-8b. Assume that V_B is high at $+V_{sat}$. This forces a constant current (V_{sat}/R_i) through C (left to right) to drive V_A negative from V_{UT} to V_{LT}. When V_A reaches V_{LT}, pin 3 of the 301 goes negative and V_B snaps to $-V_{sat}$ at t = 1 ms.

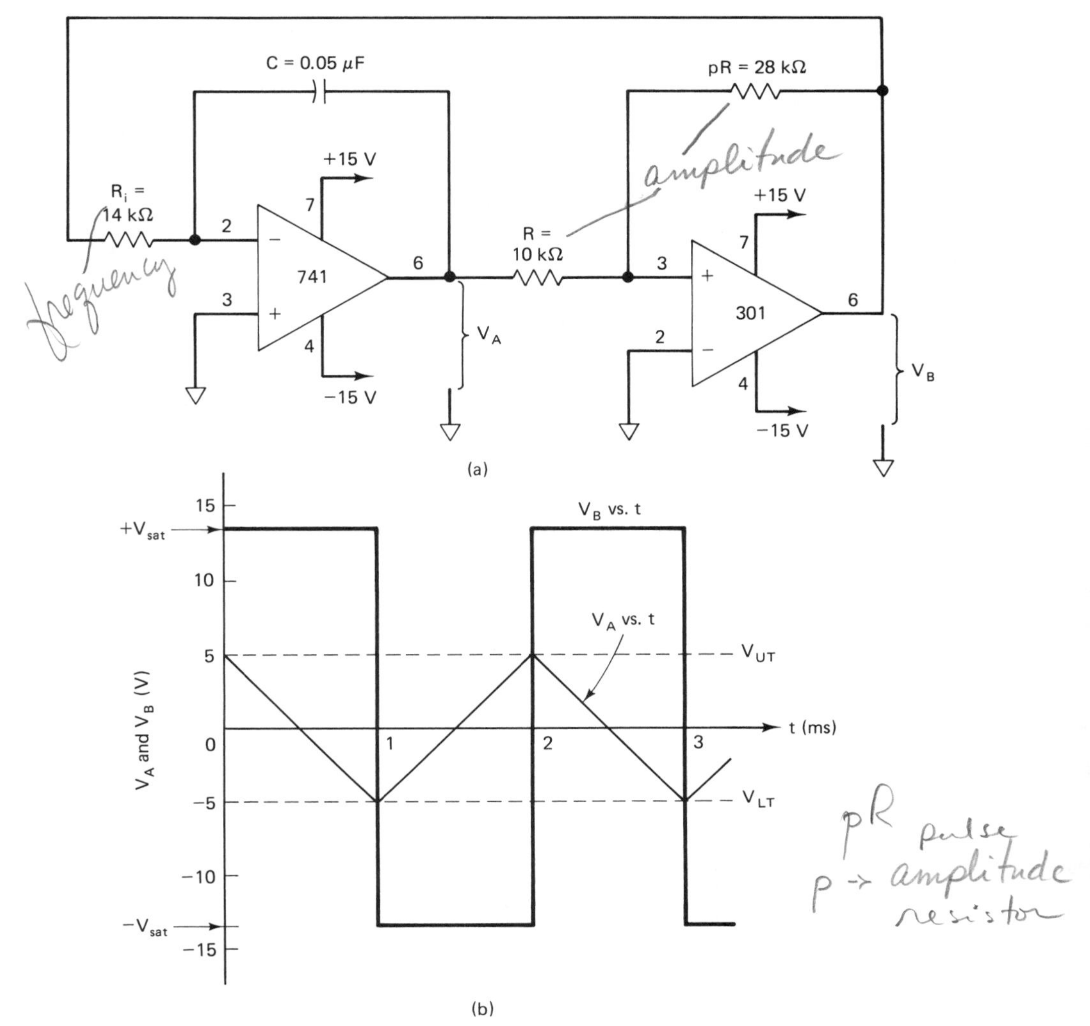

Figure 9-8 The bipolar triangle-wave generator circuit in (a) generates triangle-wave and square-wave oscillator signals as in (b). (a) Basic bipolar triangle-wave generator oscillating frequency is 1000 Hz; (b) output voltage waveshapes.

When V_B is at $-V_{sat}$, it forces a constant current (right to left) through C to drive V_A positive from V_{LT} toward V_{UT}. See time 1 to 2 ms. When V_A reaches V_{UT} at $t = 2$ ms, pin 3 of the 301 goes positive and V_B snaps to $+V_{sat}$. This initiates the next cycle of oscillation.

9-5.2 Frequency of Operation

The peak values of the triangular wave are established by the *ratio* of resistor pR by R and the saturation voltages. They are given by

$$V_{UT} = -\frac{-V_{sat}}{p} \tag{9-6a}$$

$$V_{LT} = -\frac{+V_{sat}}{p} \tag{9-6b}$$

where

$$p = \frac{pR}{R} \tag{9-6c}$$

If the saturation voltages are reasonably equal, the frequency of oscillation, f, is given by

$$f = \frac{p}{4R_iC} \tag{9-7}$$

EXAMPLE 9-6

A triangle-wave generator oscillates at a frequency of 1000 Hz with peak values of approximately +5 V. Calculate the required values for pR, R_i, and C in Fig. 9.8a.

SOLUTION *First* we work on the calculation for the comparator resistor ratio p that controls peak triangle wave output voltages, V_{UT} and V_{LT}. $+V_{sat}$ is practically + 14.2 V and $-V_{sat}$ is typically −13.8 V for a ±15 V supply. This observation points out one deficiency in our low-cost triangle-wave generator. It does *not* have *precisely* equal positive and negative peak outputs. (We will remedy this problerm, at a higher cost, in Section 9-10). From Eq. (9-6a), solve for p:

$$p = -\frac{-V_{sat}}{V_{UT}} = -\frac{-13.8\text{ V}}{5\text{ V}} = +2.76 \approx 2.8$$

Choose R = 10 kΩ. Then from Eq. (9-6c) we solve for pR as

$$pR = (p)R = 2.8\ (10\text{ k }\Omega) = 28\text{ k}\Omega$$

Next we select R_i and C. Begin by making a *trial* choice for C = 0.05 μF. Then calculate a value for R_i to see if R_i comes up greater than 10.0 kΩ. From Eq. (9-7),

$$R_i = \frac{p}{4fC} = \frac{2.8}{4\ (1000\text{ Hz})\ 0.05\ \mu\text{F})} = 14\text{ k}\Omega$$

In practice it would be prudent to make R_i up from a 12-kΩ resistor in series with a 0 −5-kΩ pot. The 5-kΩ pot would then be adjusted for an oscillation frequency of precisely 1.00 kHz.

9-5.3 Unipolar Triangle Wave Generator

The bipolar triangle-wave generator circuit of Fig. 9-8 can be changed to output a unipolar triangle wave. Simply add a diode in series with pR as shown in Fig. 9-9a. Circuit operation is studied by reference to the waveshapes in Fig. 9-9b.

When V_B is at $+V_{sat}$, the diode stops current flow through pR and sets V_{LT} at 0 V. When V_B is at $-V_{sat}$, the diode allows current flow through pR and sets V_{UT} at a value of

$$V_{UT} = -\frac{(-V_{sat} + 0.6\text{ V})}{p} \tag{9.8a}$$

Frequency of oscillation is then given approximately by

$$f \approx \frac{p}{2R_iC} \tag{9-8b}$$

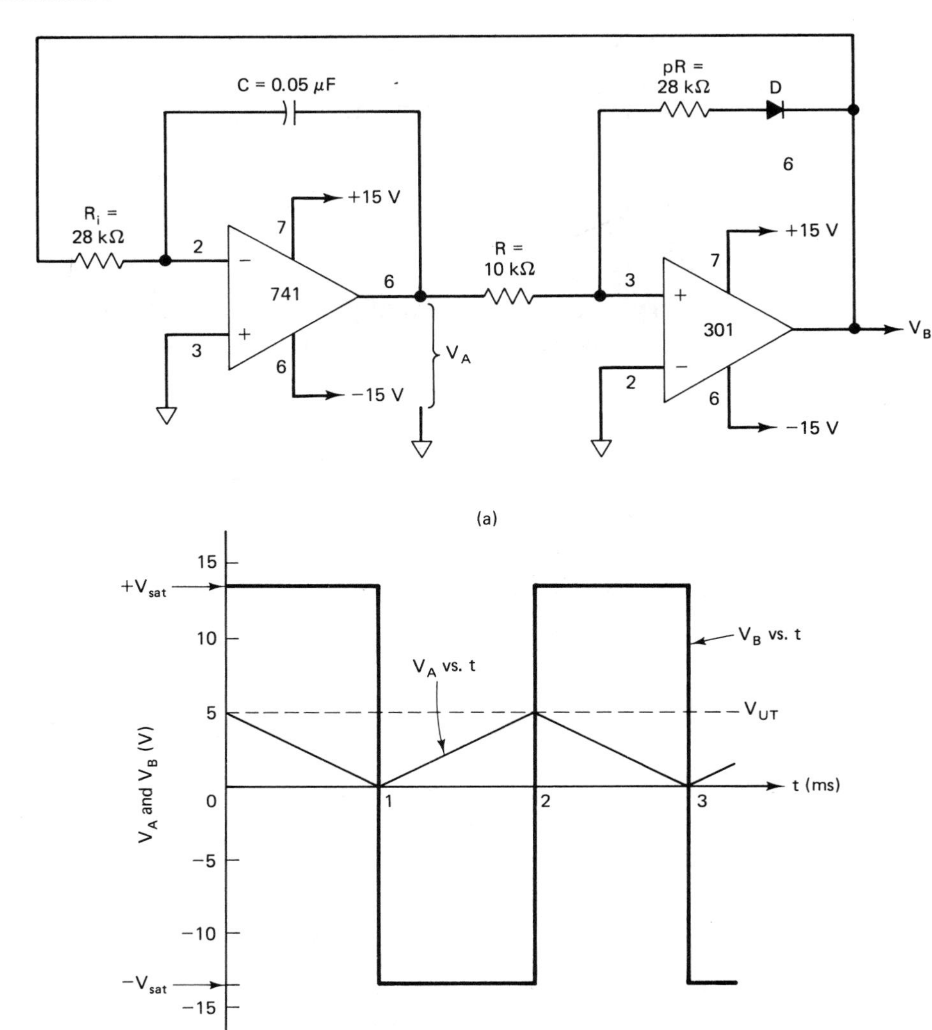

Figure 9-9 Diode *D* in (a) converts the bipolar triangle-wave generator into a unipolar triangle-wave generator. Waveshapes are shown in (b). (a) Basic unipolar triangle-wave generator, oscillating frequency is 1000 Hz; (b) output voltage waveshapes.

EXAMPLE 9-7

Find the approximate peak voltage and frequency for the unipolar triangle-wave generator in Fig. 9-9.

SOLUTION Calculate

$$p = \frac{pR}{R} = \frac{28\text{ k}\Omega}{10\text{ k}\Omega} = 2.8$$

Find the peak value of V_A from Eq. (9-8a):

$$V_{\text{UT}} = -\left(\frac{-V_{\text{sat}} + 0.6\text{ V})}{p}\right) = -\left(\frac{-13.8\text{ V} + 0.6\text{ V}}{2.8}\right) \simeq 4.7\text{ V}$$

From Eq. (9-8b),

$$f = \frac{p}{2R_iC} = \frac{2.8}{2(28\text{ k}\Omega)(0.05\ \mu\text{F})} = 1000\text{ Hz}$$

9-5.4 Disadvantages

The triangle-wave generator of Fig. 9-8 is inexpensive and reliable. However, it has two disadvantages. The rates of rise and fall of the triangle wave are unequal. Also, the peak values of both triangle-wave and square-wave outputs are unequal. This is because the magnitude of $+V_{sat}$ and $-V_{sat}$ are unequal.

In the next section we substitute an AD630 for the comparator. This will give the equivalent of precisely equal square-wave ± voltages that will *also* be equal to the ± peak values of triangle-wave voltage.

9.6 BALANCED MODULATOR/DEMODULATOR

The AD630 (see data sheets 333-38) is an advanced integrated circuit. It has 20 pins, which allow this versatile switched voltage gain IC to act as a modulator, demodulator, phase detector and multiplexer, as well as performing other signal conditioning tasks.

We connect the AD630, as in Fig. 9-10a, as a controlled switched gain (+1 or −1) amplifier. This particular application will be examined by discussing the role performed by the dominant terminals.

9-6.1 Input and Output Terminals

The input signal V_{ref} is connected to *modulation* pins 16 and 17 in Fig. 9-10a, and thus to the inputs of two amplifiers A and B. The gain of A is programmed for −1 and B and +1 by shorting (1) terminals 13 to 14, (2) 15 to 19 to 20, (3) 16 to 17, and (4) grounding pin 1.

The carrier input terminal, pin 9 (in this application), determines which amplifier A or B is connected to the output terminal. If pin 9 is *above* the voltage at pin 10 (ground), then B amplifier is selected. Voltage at output pin 13 then equals V_{ref} times (+1).

If pin 9 voltage is *below* ground (negative), the A amplifier is selected and output pin 13 equals V_{ref} times (−1). (Note that in communication circuits, V_{ref} is called the *analog data* or *signal voltage.* V_C is called a chopper or *carrier voltage.* V_o is the *modulated* output. That is, the *amplitude* of the low-frequency signal voltage is impressed upon the higher-frequency carrier wave—hence the names selected for the AD630's input and output terminals.)

9-6.2 Input–Output Waveforms

V_{ref} is a dc voltage of 5.0 V in Fig. 9-10b. V_C is 100-Hz square wave with peak amplitudes that must exceed ±1 mV. Output voltage V_o is shown to switch synchronously with V_C from $+V_{ref}$ to $-V_{ref}$ and vice versa in Fig. 9-10b. We are going to replace the unpredictable $\pm V_{sat}$ of the 301 comparator in Figs. 9-9 or 9-8 with precisely + or − V_{ref}.

Moreover, V_{ref} can be adjusted easily to any required value. As shown in the next section, V_{ref} will set the positive and negative peak values of both triangle-wave and square-wave generators.

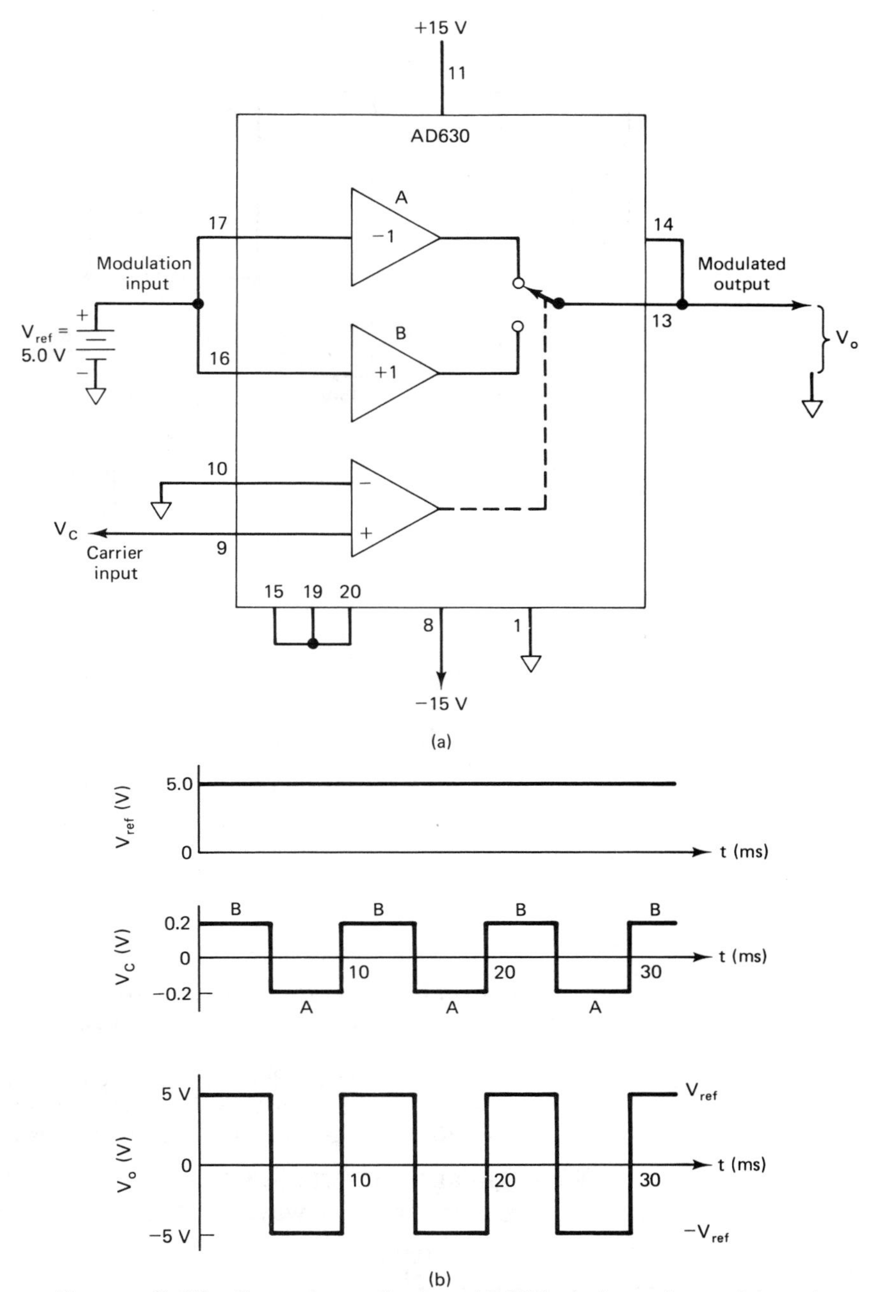

Figure 9-10 Operation of the AD630 balanced modulator/demodulator as a switched gain amplifier. (a) Wiring for switched gains of (+)1 or (−)1; (b) carrier V_C selects gains of +1 or −1 for input V_{ref}. Output V_O is equal to precisely either V_{ref} or $-V_{ref}$.

9-7 PRECISION TRIANGLE/SQUARE-WAVE GENERATOR

9-7.1 Circuit Operation

Only six parts plus voltage source, V_{ref}, make up a versatile triangle- and square-wave generator of Fig. 9-11a. Circuit operation is explained by reference to the waveshapes in Fig. 9-11b.

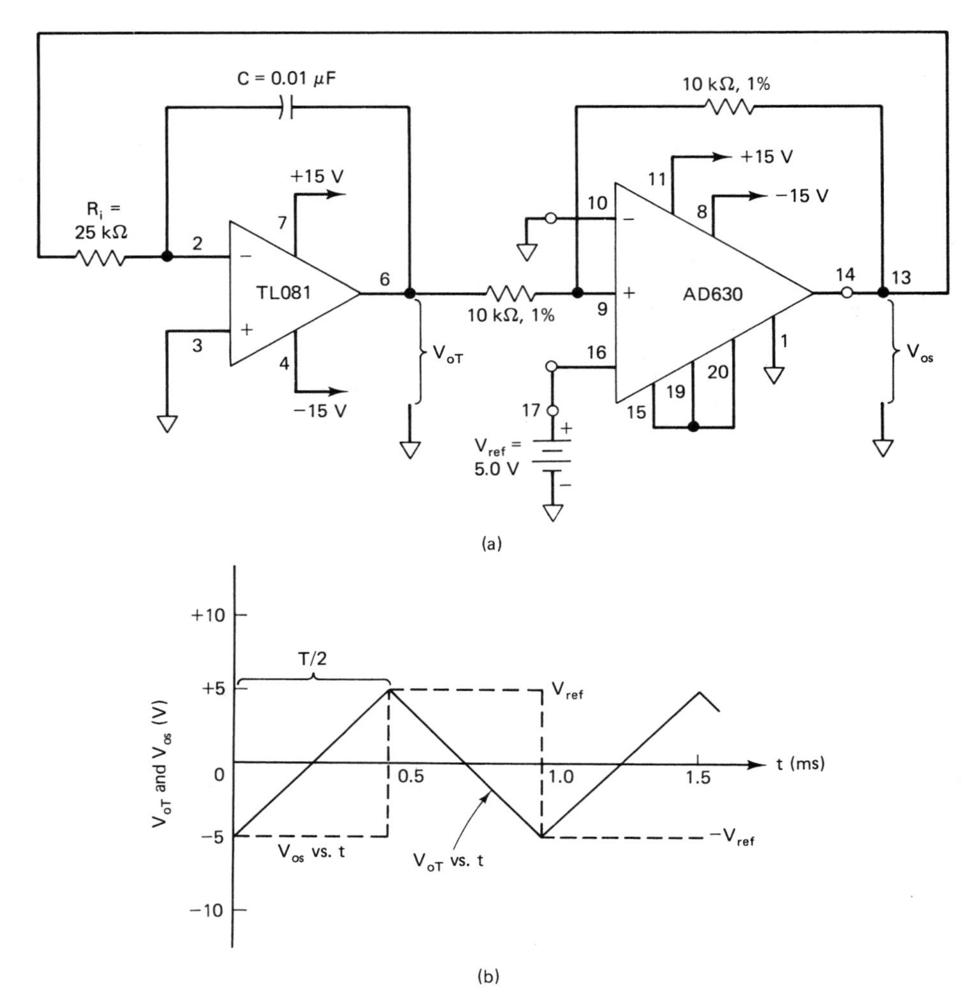

Figure 9-11 The precision triangle/square-wave oscillator in (a) has the output waveshapes in (b). V_{ref} should be buffered for a low-impedance source voltage. (a) Precision triangle/square-wave oscillator (compare with Fig. 9-8a). V_{ref} must be a low-impedance source. V_{ref} sets the ± peak values or V_{OT} and R_i adjusts the frequency. (see data sheets 330-32) (b) Square- and triangular-wave output waveshapes.

We begin at time zero. Square-wave output V_{os} begins at $-V_{ref}$ or -5 V. This forces the triangle-wave V_{oT} to go positive from a starting point of $-V_{ref} = -5$ V. During this time, pin 9 is below ground to select an AD630 gain of -1 and holds V_{os} at $-$ 5V.

At time $T/2 = 0.5$ ms, V_{oT} reaches $+5$ V, where pin 9 is driven slightly positive to select an AD630 gain of $+1$. This snaps V_{os} to $V_{ref} = +5$ V. V_{os} then drives V_{oT} negative. When V_{oT} reaches -5 V, pin 9 goes negative at $T = 1.0$ ms and snaps V_{os} negative to -5 V. This completes one cycle of oscillation and begins another.

9-7.2 Frequency of Oscillation

The easiest way to find the frequency of oscillation is to begin with the rate of rise of the triangle wave, V_{oT}/t, in volts per second. Rate of rise of the triangle wave, from 0 to 0.5 ms in Fig. 9-11b, is found from

$$\frac{V_{oT}}{t} = \frac{V_{ref}}{R_iC} \qquad (9\text{-}9a)$$

The time t for a half-cycle is $T/2$, and during this time, V_{oT} changes by $2V_{\text{ref}}$. We substitute for t and V_{oT} into Eq. (4-9a) to obtain

$$\frac{2V_{\text{ref}}}{T/2} = \frac{V_{\text{ref}}}{R_i C} \tag{9-9b}$$

and solve for both period T and frequency of oscillation f:

$$T = 4R_i C \quad \text{and} \quad f = \frac{1}{T} = \frac{1}{4R_i C} \tag{9-9c}$$

Note that V_{ref} cancels out in Eqs. (9-9b) and (9-9c). This is a very important advantage. The peak output voltages of both square- and triangle-wave signals are set by $+V_{\text{ref}}$. As V_{ref} is adjusted, the frequency of oscillation does *not* change.

EXAMPLE 9-8

Make a triangle/square-wave generator that has peak voltages of ± 5 V and oscillates at a frequency of 1.0 kHz.

SOLUTION Choose $V_{\text{ref}} = 5.0$ V. V_{ref} should be the output of an op amp for low impedance. Arbitrarily choose $C = 0.01\ \mu\text{F}$. From Eq. (9-9c),

$$R_i = \frac{1}{4fC} = \frac{1}{4(1000)(0.01\ \mu\text{F})} = 25.0\ \text{k}\Omega.$$

To refine the output frequency, make R_i up of a 22-kΩ resistor in series with a 5-kΩ or 10-kΩ variable resistor.

9-8 SINE-WAVE GENERATION SURVEY

Commercial function generators produce triangular, square, and sinusoidal signals whose frequency and amplitude can be changed by the user. To obtain a sine-wave output, the triangle wave is passed through a shaping network made of carefully selected resistors and diodes. The sine waves thus produced are reasonably good. However, there is inevitably some distortion, particularly at the peaks of the sine wave.

When an application requires a single-frequency sine wave, conventional oscillators use phase-shifting techniques that usually employ (1) two R-C tuning networks and (2) complex amplitude limiting circuitry. The limit circuit must usually be custom-adjusted for each oscillator to minimize distortion. Frequency of this oscillator is difficult to vary because two R-C networks must be varied and their values must track within $\pm 1\%$.

The recent invention of two state-of-the-art ICs has eliminated the disadvantages of difficult frequency adjustment and difficult amplitude control. They are the AD630 and AD639.

The AD630 has already been used to generate a precision triangle wave whose frequency and amplitudes are precise and easy to adjust. We will connect the triangle-wave output V_{oT} of Fig. 9-11a to an AD639 universal trigonometric function generator. The resulting circuit will have the best qualities of a *precision sine-wave generator whose frequency will be easily adjustable.*

9-9 *UNIVERSAL TRIGONOMETRIC FUNCTION GENERATOR*

The AD639 (see data sheets 339-43) is a state-of-the-art trigonometric function generator. It will perform all trigonometric functions *in real time*, including sin, cos, tan, cosec, sec and cotan. When a calculator performs a trig function, the operator punches in a number corresponding to the number of angular degrees and punches SIN. The calculator pauses, then displays a number indicating the sin of the angle. That is, a number for angle θ is entered and the calculator produces a number for sin θ.

The AD639 accepts an input voltage that represents the angle. It is called the *angle voltage*, V_{ang}. For the AD639, the angle voltage is found from

$$V_{ang} = \left(\frac{20\text{ mV}}{1°}\right)\theta = \left(\frac{1\text{ V}}{50°}\right)\theta \tag{9-10}$$

Four input terminals are available. We will look at only the single active input that generates sin functions. The output voltage will equal sin θ or 10 sin θ, depending how the internal gain control is pin programmed.

9-9.1 Sine Function Operation

The AD639 is wired to output $V_o = 1 \sin\theta$ in Fig. 9-12a. There are four input terminals: 1, 2, 7, and 8. Wired as shown, the chip performs a sin function. Pins 3, 4, and 10 control gain. Normally, 3 and 4 are grounded so that pin 10 can activate the internal gain control. A gain of 1 results when pin 10 is wired to $-V_s$ or pin 9. Wire pin 10 to $+V_s$ or pin 16 and obtain a gain of +10. *Then* $V_o = 10 \sin\theta$. Pin 6 is a precision 1.80-V reference voltage that corresponds to an angle voltage of 90° [see Eq. (9-10)]. We analyze sin function operation by an example.

EXAMPLE 9-9

Calculate the required input angle voltage and resultant output voltage for angles of (a) ±45°; (b) ±90°; (c) ±225°; (d) ±405°.

SOLUTION From Eq. (9-10) and Fig. 9-12):

(a) $V_{ang} = \left(\frac{20\text{ mV}}{1°}\right)(\pm 45°) = \pm 0.90$ V, $V_o = 1 \sin(\pm 45°) = \pm 0.707$ V.

(b) $V_{ang} = \left(\frac{20\text{ mV}}{1°}\right)(\pm 90°) = \pm 1.80$ V, $V_o = 1 \sin(\pm 90°) = \pm 1.0$ V.

(c) $V_{ang} = \left(\frac{20\text{ mV}}{1°}\right)(\pm 225°) = \pm 4.50$ V, $V_o = 1 \sin(\pm 225°) = \mp 0.707$ V.

(d) $V_{ang} = \left(\frac{20\text{ mV}}{1°}\right)(\pm 405°) = \pm 8.10$ V, $V_o = 1 \sin(\pm 405°) = \pm 0.707$ V.

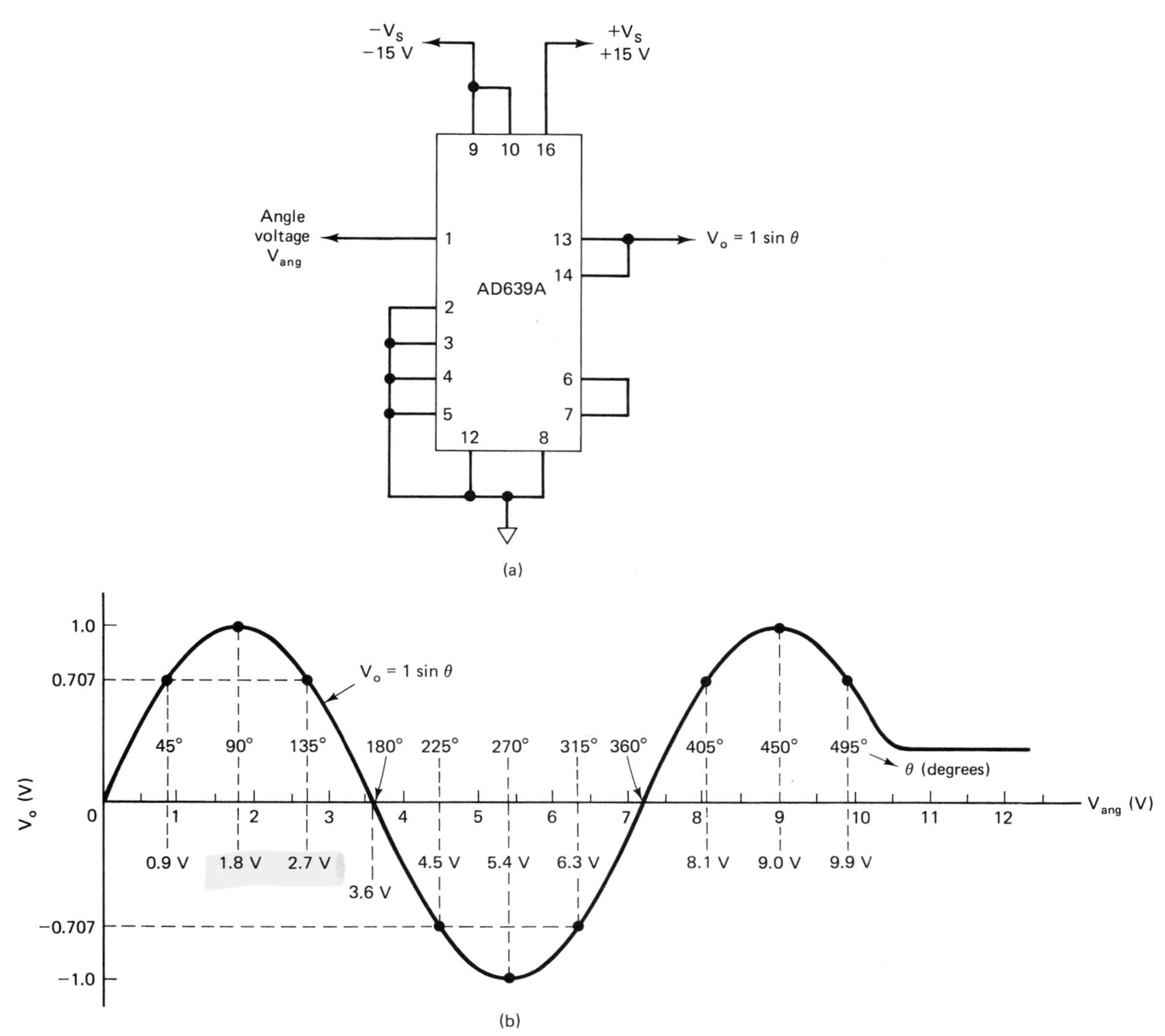

Figure 9-12 The AD639 is pin-programmed in (a) to act as a sine function generator. Each ±20 MV of input angle voltage corresponds to an input angle of $\theta = \pm 1^0$. Output V_0 equals $1 \times \sin\theta$. (a) The AD639A is pin-programmed to output the sine of the angle voltage; (b) output voltage V_0 equals the sine of θ if θ is represented by an angle voltage of 20 mV per angular degree.

Example 9-9 clearly illustrates that the AD639, remarkable as it is, cannot output the sin of, for example, 36,000°. This would require an angle voltage of 720 V. The normal ±15-V supply limits the guaranteed usable input angle to ±500° or +10.000 V. We extend the results of Example 9-9 to summarize briefly the performance of the sin function generator in Table 9-2 and Fig. 9-12.

In Fig. 9-12b, V_o is plotted against both V_{ang} and θ. A study of this figure shows that if V_{ang} *could be varied linearly* by a triangle wave, V_o *would vary sinusoidally.* Further, if frequency of the triangle wave could be varied easily, the sine-wave frequency could *easily* be tuned, adjusted, or varied. We pursue this observation in the next section.

TABLE 9-2
AD639 Sin Functions[a]

Input		*Output (V)*	
(angular degrees) θ	*Angle voltage,* V_{ang} *(V)*	$V_o = 1 \sin\theta$ *(wire pin 10 to 9*	$V_o = 10 \sin\theta$ *(wire pin 10 to 16*
0	0.00	0.000	0.000
±45	±.90	±0.707	±7.07
±90	±1.80	±1.000	±10.07
±135	±2.70	±0.707	±7.07
±180	±3.60	0.000	0.000
±225	±4.50	∓0.707	∓7.07
±270	±5.40	∓1.000	∓10.00
±315	±6.30	∓0.707	∓7.07
±360	±7.20	0.000	0.00
±405	±8.10	±0.707	±7.07
±450	±9.00	±1.000	±10.00
±495	±9.90	±0.707	±7.07
±500	±10.00	±0.643	±6.43

[a]Connect terminal 10 to 9 to pin program $V_o = \sin\theta$, or 11 to 16 to pin program $V_o = 10 \sin\theta$. Input angle voltage $V_{ang} = 20$ mV/1°C θ.

9-10 PRECISION SINE-WAVE GENERATOR

9-10.1 Circuit Operation

Connect the precision triangle-wave oscillator of Fig. 9-11a to the sin function generator of Fig 9-12a to construct the *precision* sine-wave generator in Fig. 9-13a. As a bonus, we also have precision triangle-wave and square-wave outputs. The 1.80-V reference voltage of the AD639 is connected to modulation inputs 16 and 17 of the AD630 modulator (Fig. 9-10). Circuit operation is now examined by reference to Fig. 9-13b.

Triangle-wave Rise Time, 0 to T/2 in Fig. 9-13b

1. AD630:
   (a) Pin 13 is at $-V_{ref} = -1.8$ V, causing:
   (b) Pin 9 to select gain $= -1$ to hold 13 at -1.8 V, and
   (c) Op-amp output voltage to ramp up.
2. *Op Amp:*
   (a) Pin 6 ramps from $-V_{ref} = -1.8$ V *toward* $+V_{ref} = 1.8$ V to:
   (b) Hold pin 9 of the AD630 negative, and
   (c) Drive input 1 of the AD639 with an angle voltage linearly from -1.8 V to 1.8 V.
3. *AD639:*
   (a) Pin 1 's input angle voltage corresponds to an input angle that varies linearly from $-90°$ to $+90°$.
   (b) Pin 13 outputs $V_o = 10 \sin\theta$ from -10 V to $+10$ V.

When op amp pin 6 reaches $+1.8$ V, pin 9 of the AD630 goes positive to select a gain of $+1$. Its output, in 13, snaps to $+1.8$ V. This begins the *fall time.*

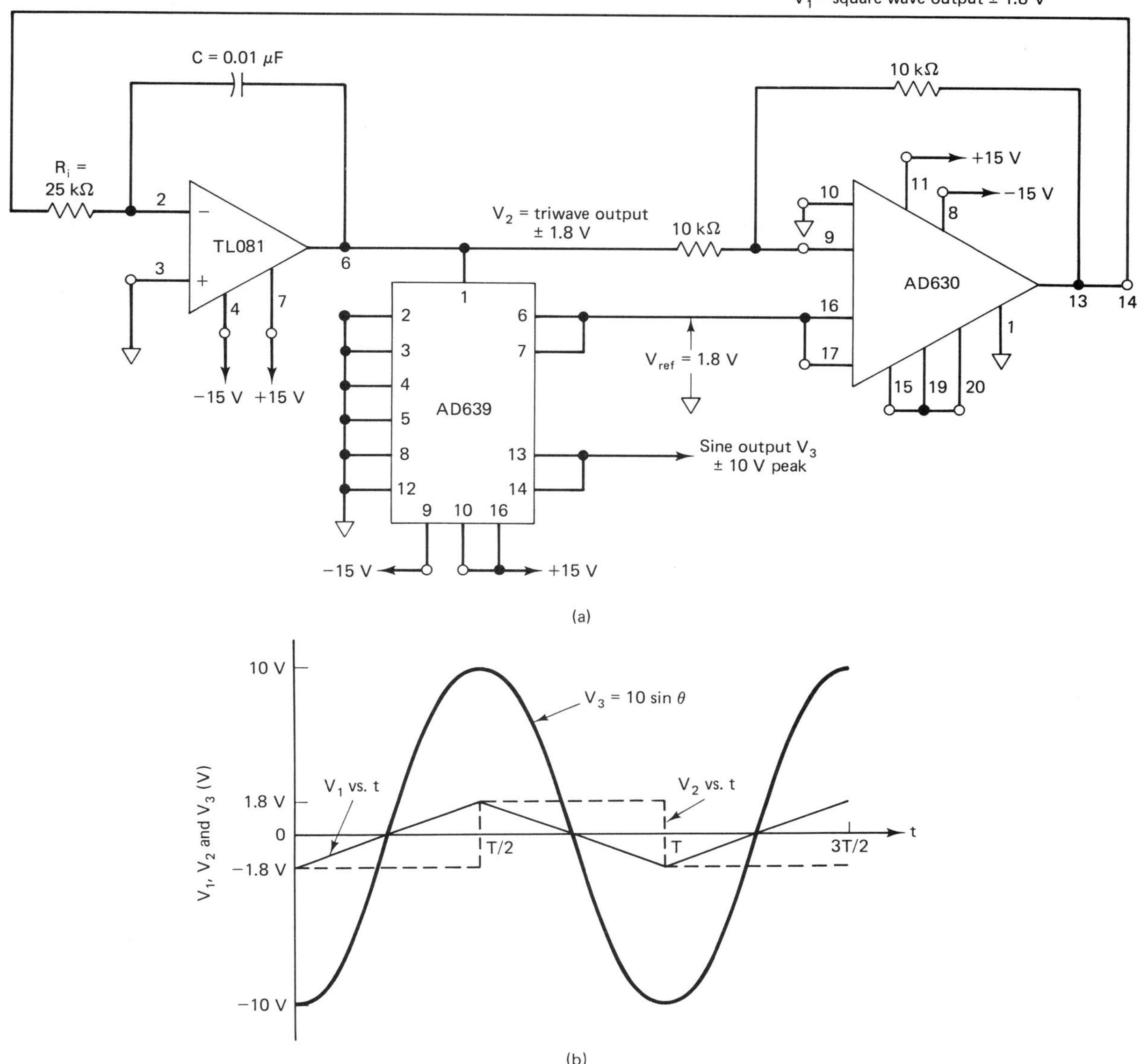

Figure 9-13 The signal generator in (a) outputs precision square, sine, or triangle waves. Frequency of oscillation can be set precisely by adjusting R_i. Frequency range can be changed by changing C. (a) Precision sine–triangle–square-wave generator; (b) output waveshapes. Peak amplitudes are equal and constant in value. Frequency can be varied by changing R_i or C.

Triangle-wave Fall Time, T/2 to T in Fig. 9-13b

1. *AD630:* Causes the triangle-wave to ramp down from +1.8 V to −1.8 V. At −1.8 V, gain is switched to −1 and a new cycle begins.
2. *Op Amp:* Applies an angle voltage to input pin 1 of the AD639 that varies linearly from +1.8 V to −1.8 V.
3. *AD639:* Its input angle voltage corresponds to an input angle of θ = +90° to −90°. Pin 13 outputs a sine wave that varies from +10 V to −10 V.

9-10.2 Frequency of Oscillation

The frequency of oscillation, f, is determined by R_i, C and the op amp in Fig 9-13a from

$$f = \frac{1}{4R_iC} \tag{9-11}$$

Peak amplitudes of the triangle wave and square wave are precisely equal to ±1.8 V. The sine wave has peak amplitudes of ±10 V and is synchronized to the triangle-wave (for ±1 V peaks change the AD639 pin 10 connection to $-V_S$).

EXAMPLE 9-10

Let $C = 0.025\ \mu F$ in Fig. 9-13a (two 0.05 μF in series). How does frequency change as R_i is changed from 10 kΩ to 100 kΩ?

SOLUTION From Eq. (9-11),

$$f = \frac{1}{4(10\ k\Omega)(0.025\ \mu F)} = 1\ k\ Hz \qquad f = \frac{1}{4(100\ k\Omega)(0.025\ \mu F)}$$
$$= 100\ Hz$$

Example 9-10 shows the overwhelming superiority of this sine-wave generator. Frequency is easily tuned and with precision.

End of Chapter Exercises

Name: ____________________

Date: __________ **Grade:** __

A. FILL IN THE BLANKS

Fill in the blanks with the best answer.

1. The controlling input terminals of a 555 timer are pin 6 __________ and pin 2 __________ terminals.
2. The output terminal of a 555 timer can either source or sink up to __________mA of current.
3. If both the trigger and threshold terminals of a 555 timer lie between V_{UT} and V_{LT} the output will __________.
4. The delayed start interval of Fig 9-3a can be increased by __________ the value of R_A.
5. Refer to Fig. 9-4. When V_C increases to V_{UT}, V_o switches to __________ volts. When V_C drops to V_{LT}, V_o switches to __________ volts.
6. The peak amplitude of the triangle wave generator of Fig 9-8 is controlled by which two resistors? __________ and __________.
7. Frequency of operation of Fig. 9-8 is controlled by which resistor? __________
8. Figure 9-9 can be changed from a positive unipolar triangle wave generator to a negative unipolar triangle wave generator by __________ the diode D.
9. The peak amplitude of the precision triangle wave generator shown in Fig. 9-11 can be increased by increasing the __________ voltage.

10. Figure 9-13 produces a precision sine wave with a frequency that can be changed by changing the value of ________.

B. TRUE/FALSE

Fill in **T** if the statement is true, and **F** if any part of the statement is false.

1. ______ The output terminal of a 555 timer IC acts like a transistor switch

2. ______ The 555 timer has only two distinct operating states

3. ______ When both pins 2 and 6 are below V_{LT} the output of a 555 timer IC goes low.

4. ______ Refer to Fig. 9-8. If the value of R_i is increased, the frequency of triangle wave output at pin 6 of the 741 op amp decreases

5. ______ A positive voltage V_C at the carrier input of Fig 9-10 will cause V_o to be at +5.0 V.

6. ______ The output voltage of Fig 9-12 has a peak amplitude of 10 V when pin 10 is wired to $+V_S$.

C. CIRCLE THE CORRECT ANSWER

Circle the correct answer for each statement.

1. The frequency of oscillation of Fig 9-4 can be increased by (increasing, decreasing) the value of C.
2. If V_B of the triangle wave generator shown in Fig 9-8 is at a voltage $+V_{sat}$, V_A will be forced (positive, negative).
3. Increasing R_i from 25 kΩ to 50 kΩ in Fig. 9-11 will (increase, decrease) the triangle wave frequency.
4. If the angle voltage at pin 1 of Fig 9-12 is changed from 1.8 V to 2.7 V, V_o will (increase, decrease) in value.
5. Increasing the value of capacitor in Fig 9-13 will (increase, decrease) the generated sine waves frequency.

D. MATCHING

Match the name or symbol in column **A** with the statement that matches best in column **B**.

	COLUMN A		COLUMN B
1.	______ trigger	a.	AD639
2.	______ threshold	b.	V_{UT}
3.	______ memory state	c.	pin 6
4.	______ V_{ang}	d.	pin 2
5.	______ $\frac{2}{3}V_{CC}$	e.	$V_{UT} < V_{2,6} > V_{LT}$

PROBLEMS

9-1. A 555 timer has a supply voltage of 5 V.
(a) Calculate the upper and lower threshold voltages, V_{UT} and V_{LT}. What output voltage is measured at pin 3 if
(b) reset pin 4 is grounded;
(c) reset is at +5 V with pins 2 and 6 grounded;
(d) reset is at +5 V with pins 2 and 6 at +5 V?

9-2. Draw the schematic for a "delayed start" circuit using a 555. Its output should go high 3.3s after power is applied.

9-3. Draw the schematic for a 555 initializing circuit. Its output pulse should remain high for 3.3 ms after power is applied.

9-4. Make a 555 oscillator that outputs a frequency of 10 kHz. Low time for the output square wave should equal one-third of the period of oscillation.

9-5. In the tone-burst oscillator of Fig. 9-5, the 1-MΩ variable resistor is set to 500 kΩ and the 10-kΩ variable resistor is set to 5kΩ. What frequency is the tone generated by 555B, and how many tone bursts per second are generated by 555A?

9-6. Capacitor C is reduced to 0.01 μF in the one-shot circuit of Fig. 9-6a.
(a) What type of edge is required at input V_i to initiate the one-shot?
(b) What is the duration of the output pulse?

9-7. Refer to the triangle-wave generator in Fig. 9-8. What is the resulting peak voltages and frequency if
(a) resistor R_i is doubled;
(b) R_i stays at 14 kΩ and resistor pR is doubled?

9-8. Diode D is reversed in Fig. 9-9. Draw the new waveshapes for V_A and V_B.

9-9. Reference and carrier voltage waveshapes are drawn in Fig. P9-9. They are applied to the AD630 in Fig. 9-10. Sketch output V_o.

9-10. Refer to the precision triangle-wave generator of Fig. 9-11. What are the new peak output voltages and frequency of oscillation if
(a) V_{ref} is reduced to 1.0 V;
(b) V_{ref} remains at 5.0 V and R_i is reduced to 10 kΩ?

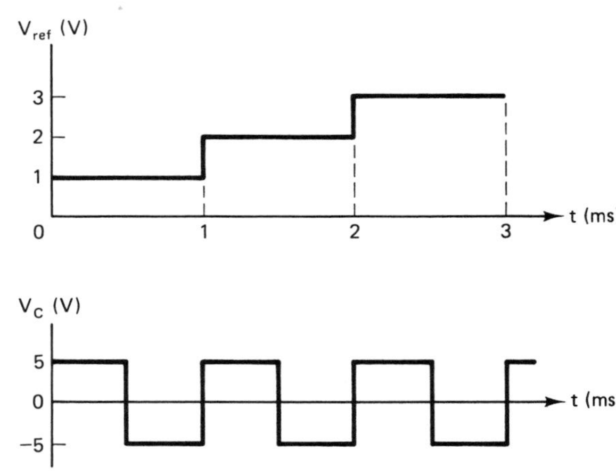

Figure P9-9 Waveshapes for Problem 9-9.

9-11. An angle voltage of 1.2 V is applied to the trigometric function generator of Fig. 9-12.

(a) What input angle θ is represented by the 1.2 V?

(b) What is the resulting output voltage V_o?

9-12. Suppose that the input angle voltage in Fig. 9-12 rose in a straight line from 0 to 1.8 V. Describe the resulting output voltage.

9-13. R_i is set to 10.0 kΩ and $C = 0.01\ \mu F$ in the precision sine-wave generator of Fig. 9-13. Find the frequency of oscillation.

Laboratory Exercise 9

Name: ____________________

Date: __________ **Grade:** ___

TIMERS AND OSCILLATORS

OBJECTIVES: Upon completion of this laboratory exercise on timers and oscillators, you will be able to (1) build an oscillator with a 555 timer IC; (2) make a low-cost triangle-wave generator; (3) operate an AD630 IC as a switched-gain amplifier; (4) use it to make a *precision* triangle-wave generator; (5) operate an AD639 as a sine function generator; and (6) connect the AD639 to your precision triangle-wave generator to make a precision sine-wave generator.

REFERENCE: Chapter 9

PARTS LIST

1	555 IC	1	1-kΩ resistor
2	741 op amps	1	5-kΩ resistor
1	301 op amp	3	10-kΩ 1% resistors
1	TL081 op amp (optional) (See data sheets pp 330-33)	1	15-kΩ resistor
		1	25-kΩ resistor
1	AD630 balanced modulator (See data sheets pp 333-38)	1	27-kΩ resistor
		1	0- to 10-kΩ pot
1	AD639 trigonometric function generator (See data sheets pp 339-43)	2	0.01-μF capacitors
		1	0.05-μF capacitor

Procedure A: Analyzing a 555 Oscillator Circuit

1. Build the circuit of Fig. L9-1 and connect a dual-trace CRO, as indicated. Zero the traces as in Fig. L9-2.

(a) Install $R_A = 10$ kΩ. Sketch both waveshapes for V_o and V_c as *dashed* lines on Fig. L9-2. Accurately measure t_{low}, t_{high}, and period T from V_o. Enter these data in Fig. L9-3. Calculate frequency from $f = 1/T$.

(b) Change R_A to 1 kΩ. Sketch the waveshapes for V_o and V_c as *solid* lines in Fig. L9-2. Enter your measured values for t_{low}, t_{high}, and T into Fig. L9-3.

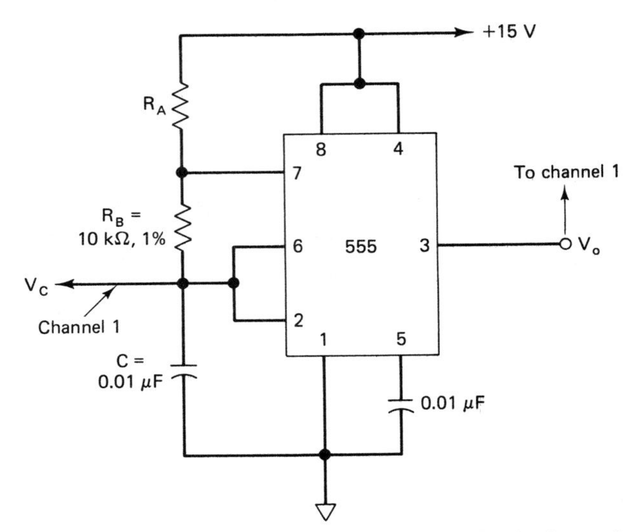

Figure L9-1 Using the 555 IC as an oscillator. Both channel vertical sensitivities set for 5 V/div dc-coupled. Time base is set for 50 μs/div. Zero the traces as in Fig. L9-2. Trigger on the negative edge of V_o.

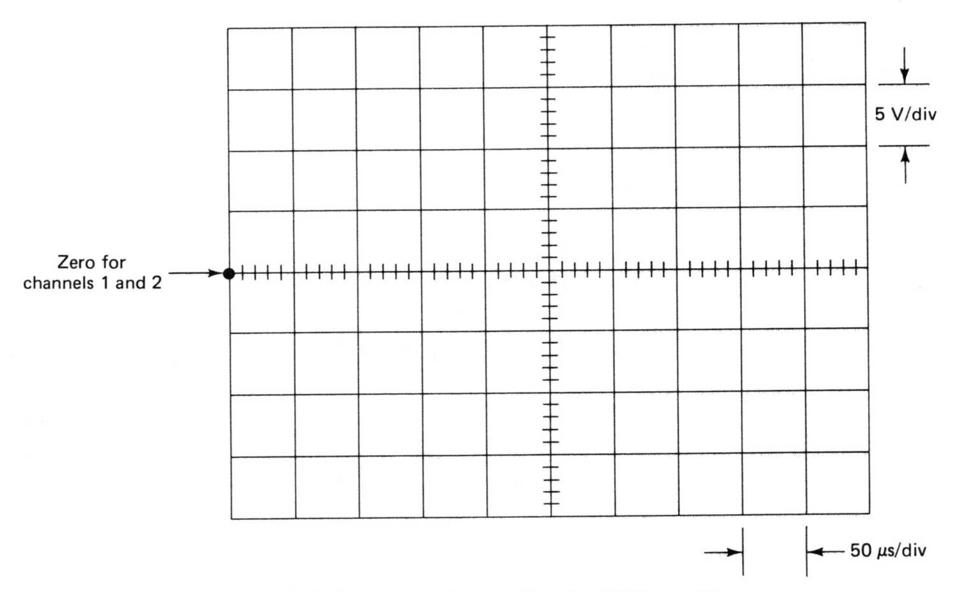

Figure L9-2 Waveshapes for the 555 oscillator.

R_A	T_{low}	T_{high}	T	f
10 kΩ				
1 kΩ				

Figure L9-3 Oscillator frequency, high and low times for the 555 oscillator of Fig. L9-2.

(c) When R_A was changed from 10 kΩ to 1 kΩ, did the following increase, decrease, or stay constant? T_{low}, ________; T_{high}, ________; T, ________; f, ________.

(d) Draw dashed lines through V_c to locate V_{UT} and V_{LT} on your sketch for V_c in Fig. L9-2. (*Hint:* See Fig. 9-4b.)

Procedure B: Building a Low-Cost Bipolar Triangle-Wave and Square-Wave Generator

2. Build the low-cost triangle-wave generator of Fig. L9-4 and connect a dual-trace CRO as shown. Sketch the waveshapes for V_A and V_B on Fig. L9-5.
3. Indicate the location of V_{UT} and V_{LT} on your sketch by dashed lines (see Fig. 9-8). Do the magnitudes of V_{UT} and V_{LT} differ? ________. Measure the frequency of oscillation. f = ________. Calculate the predicted frequency of oscillation from Eqs. (9-6c) and (9-7).

$$P = \frac{pR}{R} = \frac{27\ \text{k}\Omega}{10\ \text{k}\Omega} = \underline{\hspace{3cm}}$$

$$f = \frac{p}{4R_iC} = \underline{\hspace{3cm}}$$

What is the ratio you measured to predicted values of frequency? ________

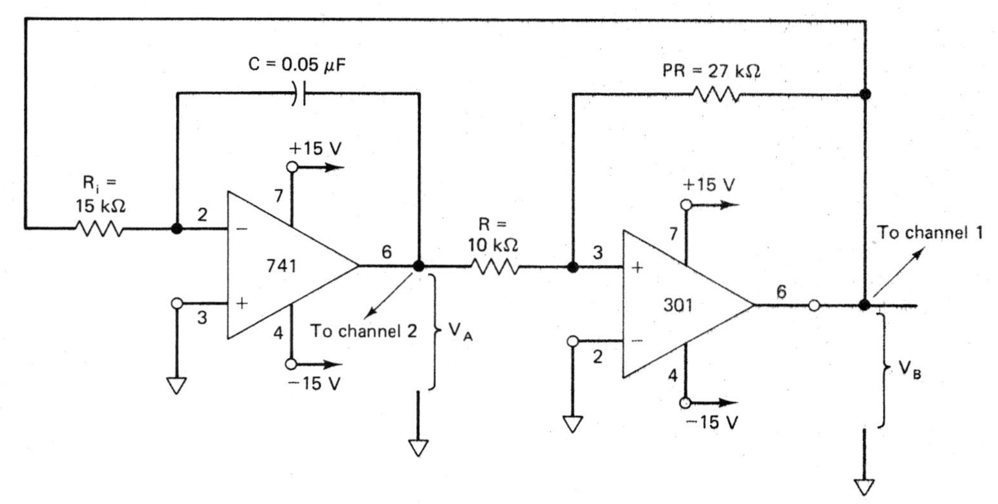

Figure L9-4 Demonstration circuit for a low-cost triangle-wave signal generator. Both CRO channels set for dc-coupling and sensitivities of 5 V/cm. Time base is set for 0.1 ms/div.

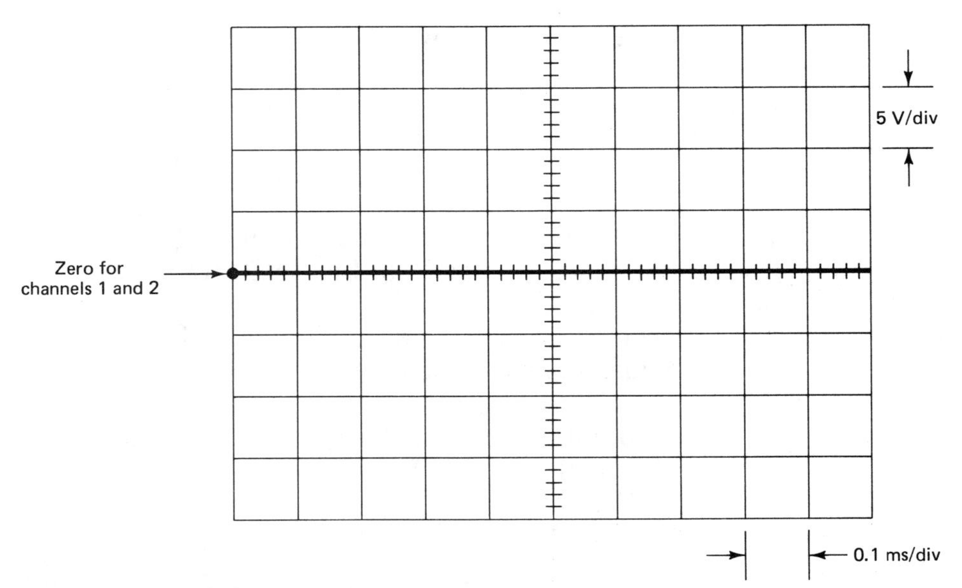

Figure L9-5 Waveshapes for the low-cost triangle-wave generator of Fig. L9-4.

Procedure C: Operating the AD630 Balanced Modulator/Demodulator

4. Build the circuit of Fig. L9-6 and adjust V_{ref} to equal 5.0 V. Jumper AD630 pin 9 to test point *a* so that the voltage on pin 9 of the AD630 is above pin 10. Measure V_{os} with a DMM. V_{os} = __________.

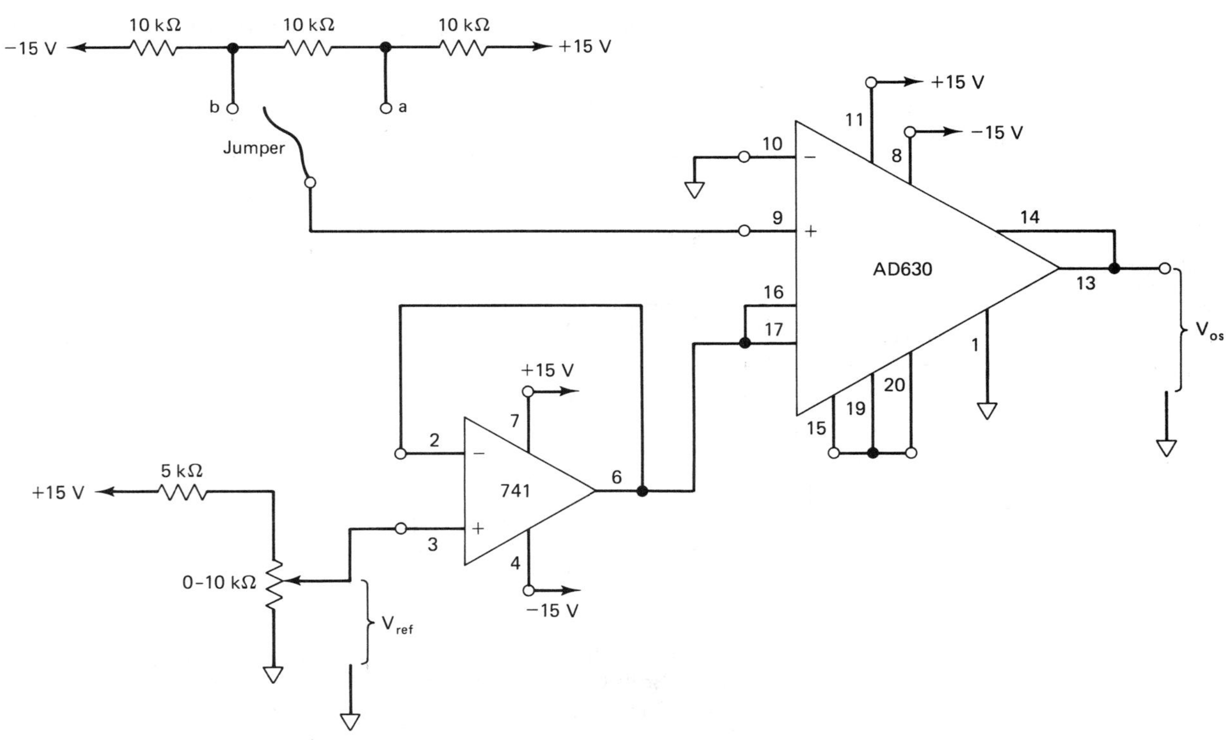

Figure L9-6 Circuit to test an AD630 balanced modulator/demodulator.

Does V_{os} equal $(+1) \times V_{ref}$ or $(-1) \times V_{ref}$? ________. Change the jumper to test point *b* so that pin 9 of the AD630 is below pin 10.

Measure $V_{os} \cdot V_{os} =$ ________. Does V_{os} equal $(V_{ref}) \times (+1)$ or $(V_{ref}) \times (-1)$? ________. Save the Figure L9-6 circuit for Procedure D but remove the 3 10-kΩ resistors. State how pin 9 of the AD630 switches the gain of the AD630 (that acts on V_{ref}) from (+) to (−). ____

____

Procedure D: Building a Precision Triangle-wave Generator

5. Wire the circuit of Fig. L9-7. Connect a dual-trace CRO as shown, and adjust V_{ref} for 5.0 V. Sketch the waveshapes for square-wave output V_{os} and triangle-wave output V_{oT} on Fig. L9-8. Are the magnitudes of peak voltages for the triangle-wave equal to those of the square wave? ________. Are all magnitudes equal in value but not in polarity to V_{ref}? ________. Measure the frequency of

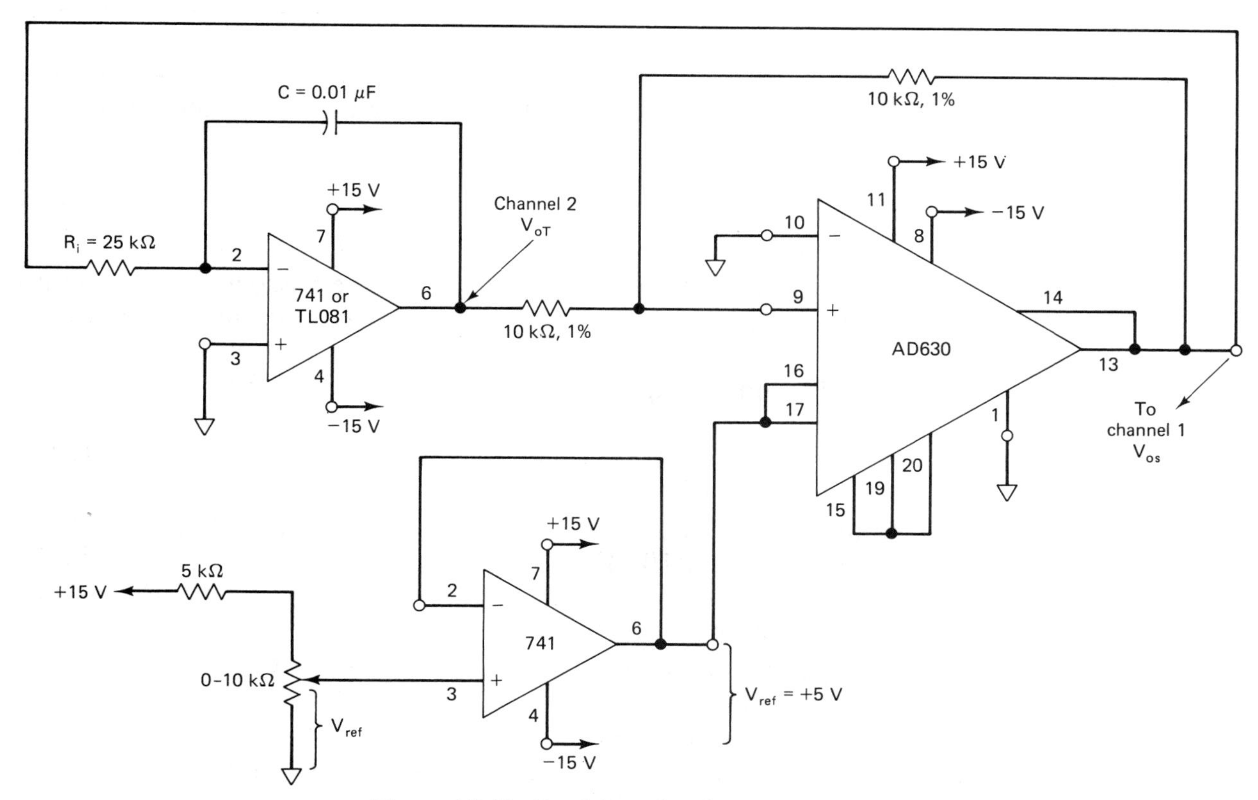

Figure L9-7 Precision triangle-wave generator. Both vertical sensitivities are set for dc-coupling 2 V/div. Time base = 0.2 ms/div. (See also Fig. L9-8.)

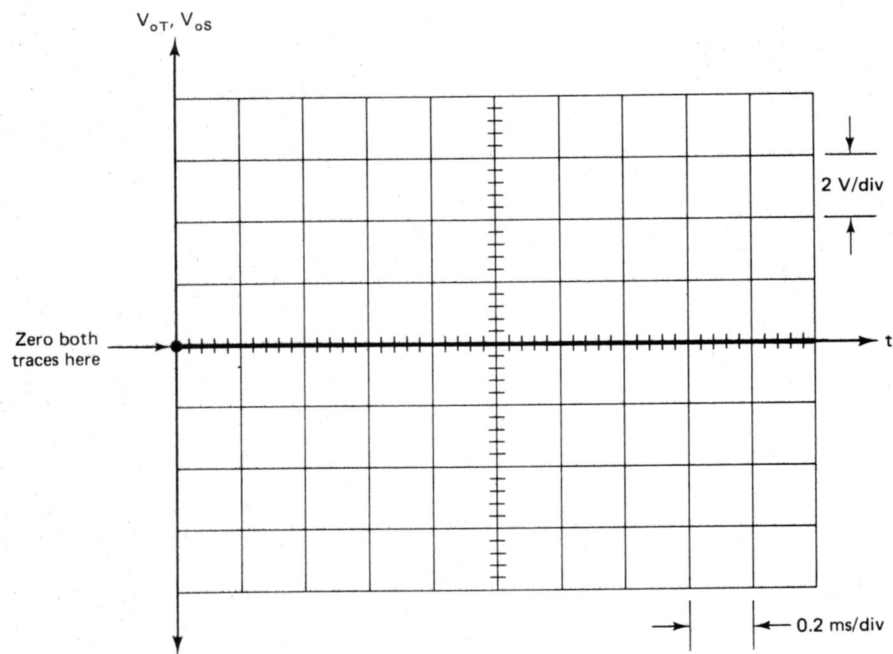

Figure L9-8 Waveshapes for the precision triangle-wave circuit of Fig. L9-7.

oscillation. f = __________. Tell how you would double the frequency by changing only one component. ________________

____

6. Decrease V_{ref} to +3 V. Do the following increase, decrease, or stay the same when V_{ref} was *decreased?* Peak value of V_{oT}, __________; frequency, __________; peak value of V_{os}, __________?
Remove the 5-kΩ resistor, 10-kΩ pot, and their associated 741 op amp from the circuit of Fig. L9-6. Save the remainder for Procedure F.

Procedure E: Generating a Sine Function with the AD639

7. Wire the sine function generator circuit of Fig. L9-9. Measure the voltage at pin 6 or 7 of the AD639.
8. Jumper point A to +15 V so that the angle voltages (V_{ang}) are positive. Use the 10-kΩ pot to adjust V_{ang} to the values shown in the second column of the data table given in Fig. L9-10. Record the corresponding values of V_o in the data table.
9. Change the jumper at point A to −15 V. Adjust V_{ang} to the negative values in the fifth column of Fig. L9-10. Record corresponding values of V_o in the data table.
10. Plot your measured values for V_o versus θ in Fig. L9-11. Remove the 10-kΩ pot and 5-kΩ resistor. Save the rest of the circuit for Procedure F.

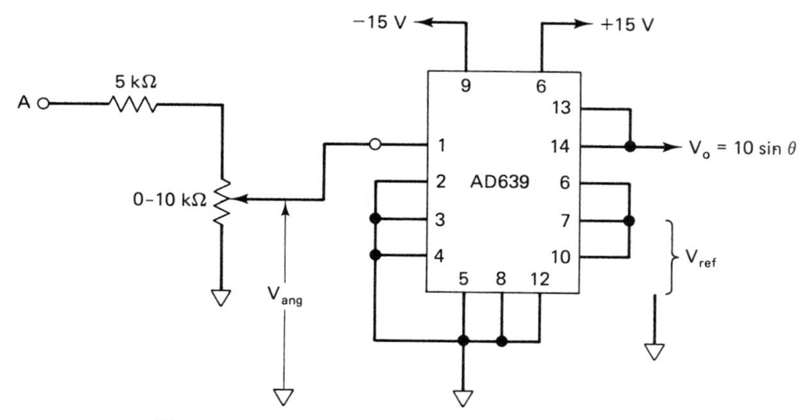

Figure L9-9 Sine function generator circuit.

Point A jumpered to +15 V			Point A jumpered to −15 V		
θ (deg)	V_{ang} (V)	V_0	θ (deg)	V_{ang} (V)	V_0
0	0		0	0	
45	0.90		−45	−0.9	
90	1.80		−90	−1.8	
135	2.70		−135	−2.7	
180	3.60		−180	−3.6	
225	4.50		−225	−4.5	
270	5.40		−270	−5.4	
315	6.30		−315	−6.3	
360	7.20		−360	−7.2	

Figure L9-10 Measured data for the sine function generator of Fig. L9-9.

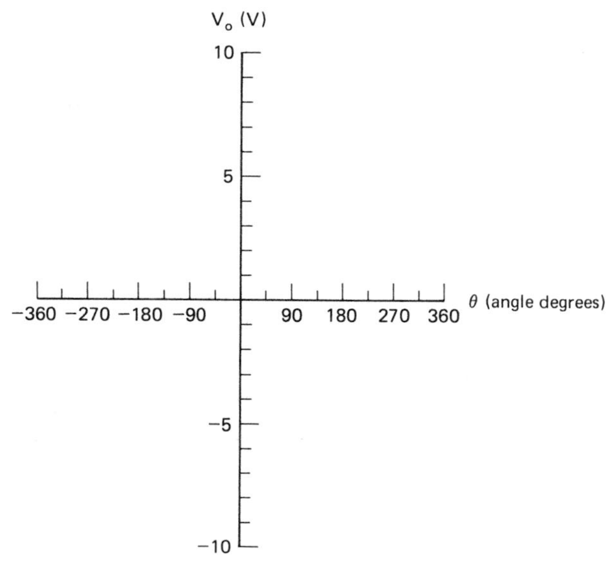

Figure L9-11 Graph of V_0 versus θ for the sine function generator of Fig. L9-9.

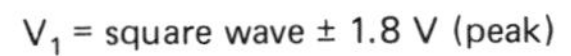

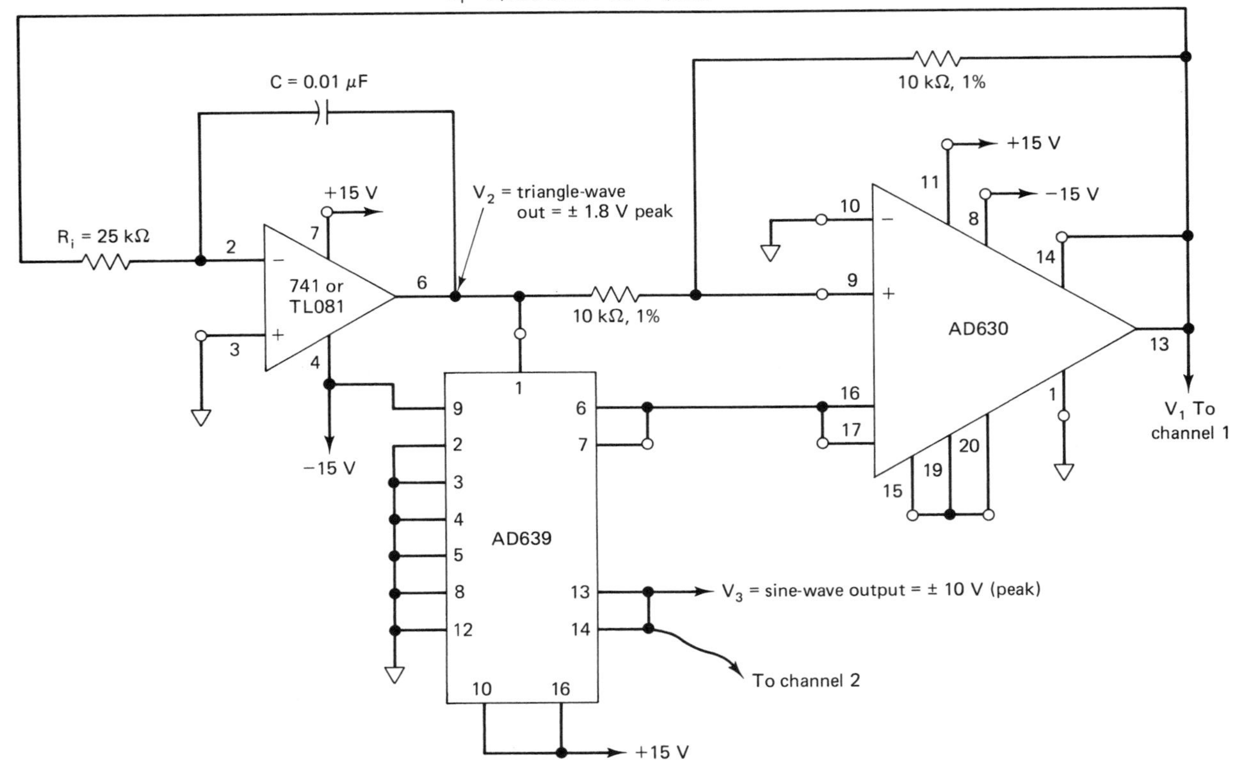

Figure L9-12 Precision sine-wave generator.

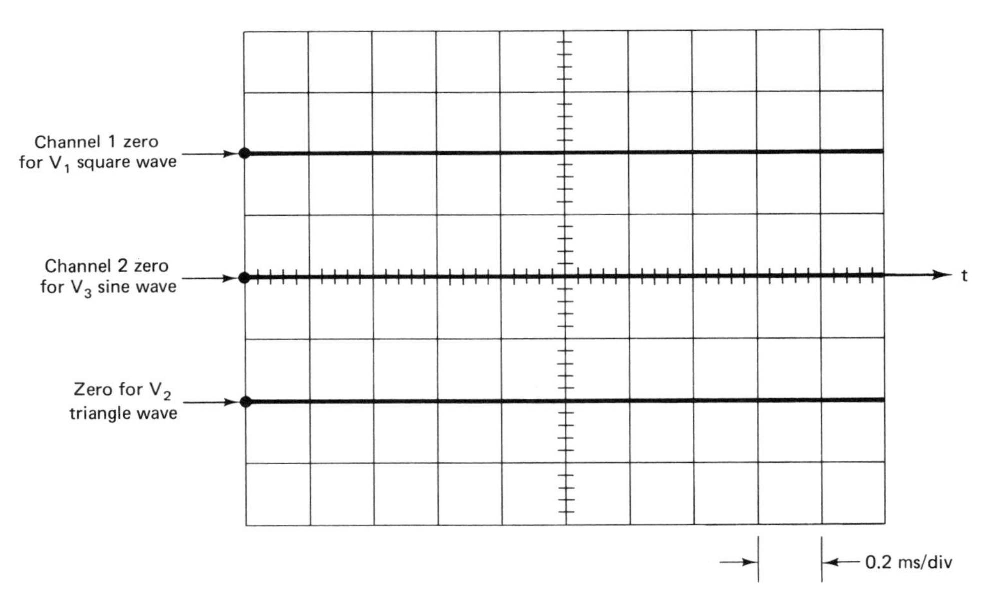

Figure L9-13 Waveshapes for the precision sine wave generator of Fig. L9-12.

Procedure F: Precision Sine-Wave Generator

11. Wire the precision triangle-wave generator (Fig. L9-7) to the AD639 (Fig. L9-9) to obtain the precision sine-wave generator in Fig. L9-12. Connect channel 1 of a dual-trace CRO to display the V_1 square wave. Vertical sensitivity is dc-coupled at 1 V/div. Connect the dc-coupled channel 2 to monitor sine wave V_3 at 5 V/div. Set the time base to 0.2 ms/div. Zero both traces as shown in Fig. L9-13. Sketch both V_1 and V_3 on Fig. L9-13.

12. Shift channel 1 of the CRO to display triangle-wave V_2. Sketch V_2 on Fig. L9-13.

13. From your sketches, measure the frequency of oscillation. f = ______________. Predict f from $f = 1/4R_iC$. Predicted f = ______________. Account for any significant differences in the "conclusion" section.

CONCLUSION:

TYPES TL080 THRU TL084, TL080A THRU TL084A, TL081B, TL082B, TL084B JFET-INPUT OPERATIONAL AMPLIFIERS

JULY 1977

26 DEVICES COVER COMMERCIAL, INDUSTRIAL, AND MILITARY TEMPERATURE RANGES

- Low Power Consumption
- Wide Common-Mode and Differential Voltage Ranges
- Low Input Bias and Offset Currents
- Output Short-Circuit Protection
- High Input Impedance . . . JFET-Input Stage
- Internal Frequency Compensation (Except TL080, TL080A)
- Latch-Up-Free Operation
- High Slew Rate . . . 13 V/μs Typ

description

The TL081 JFET-input operational amplifier family is designed to offer a wider selection than any previously developed operational amplifier family. Each of these JFET-input operational amplifiers incorporates well-matched, high-voltage JFET and bipolar transistors in a monolithic integrated circuit. The devices feature high slew rates, low input bias and offset currents, and low offset voltage temperature coefficient. Offset adjustment and external compensation options are available within the TL081 Family.

Device types with an "M" suffix are characterized for operation over the full military temperature range of −55°C to 125°C, those with an "I" suffix are characterized for operation from −25°C to 85°C, and those with a "C" suffix are characterized for operation from 0°C to 70°C.

TL080, TL080A

JG OR P DUAL-IN-LINE PACKAGE (TOP VIEW)

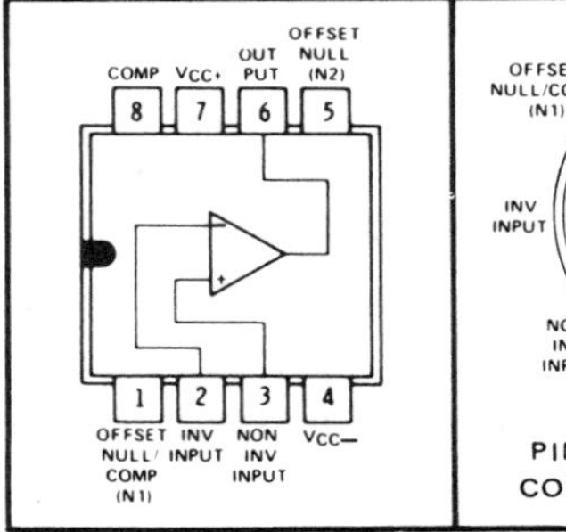

L PLUG-IN PACKAGE (TOP VIEW)

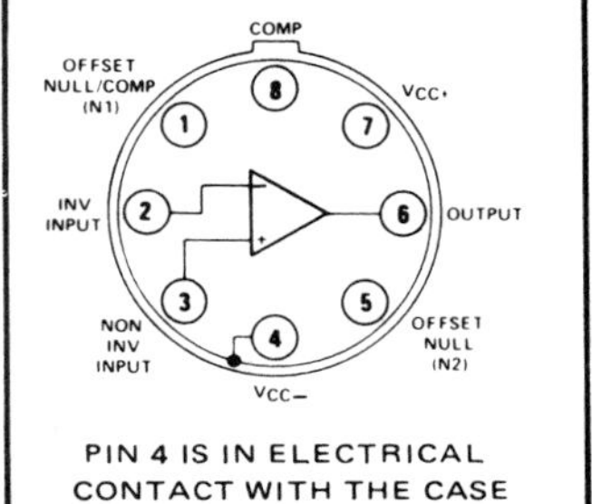

TL081, TL081A, TL081B

JG OR P DUAL-IN-LINE PACKAGE (TOP VIEW) L PLUG-IN PACKAGE (TOP VIEW)

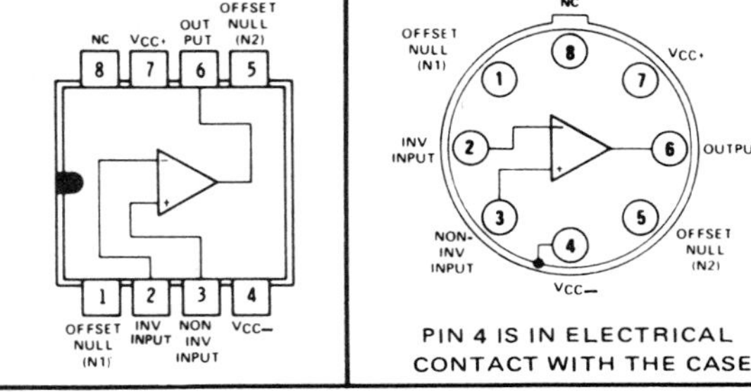

TL082, TL082A, TL082B

JG OR P DUAL-IN-LINE PACKAGE (TOP VIEW)

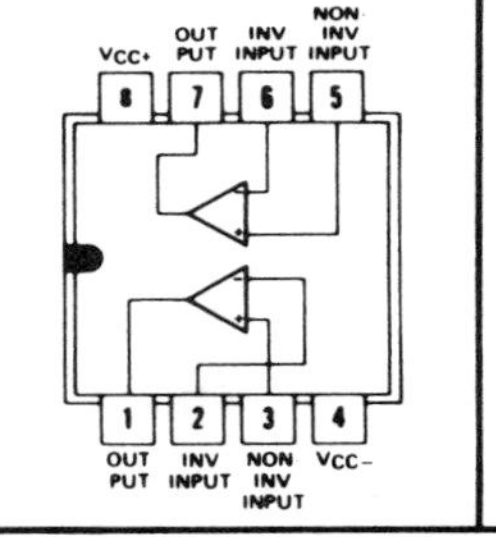

L PLUG-IN PACKAGE (TOP VIEW)

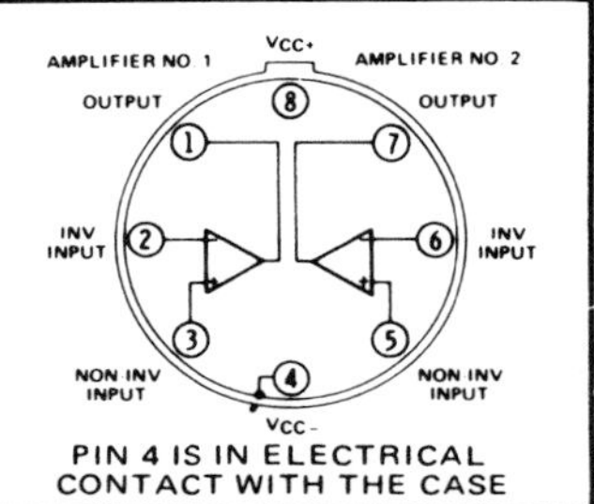

TL083, TL083A

J OR N DUAL-IN-LINE PACKAGE (TOP VIEW)

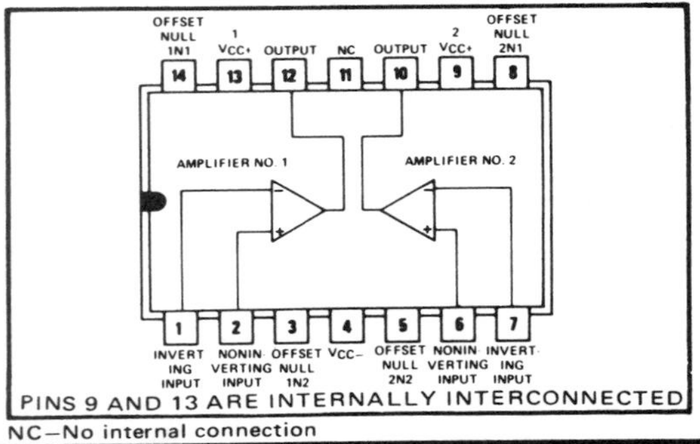

TL084, TL084A, TL084B

J OR N DUAL-IN-LINE PACKAGE (TOP VIEW)

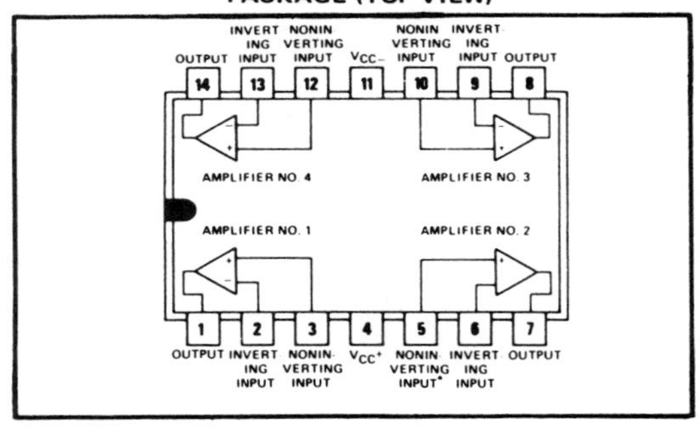

NC—No internal connection

*Courtesy of **Texas Instruments Inc.**

schematic (each amplifier)

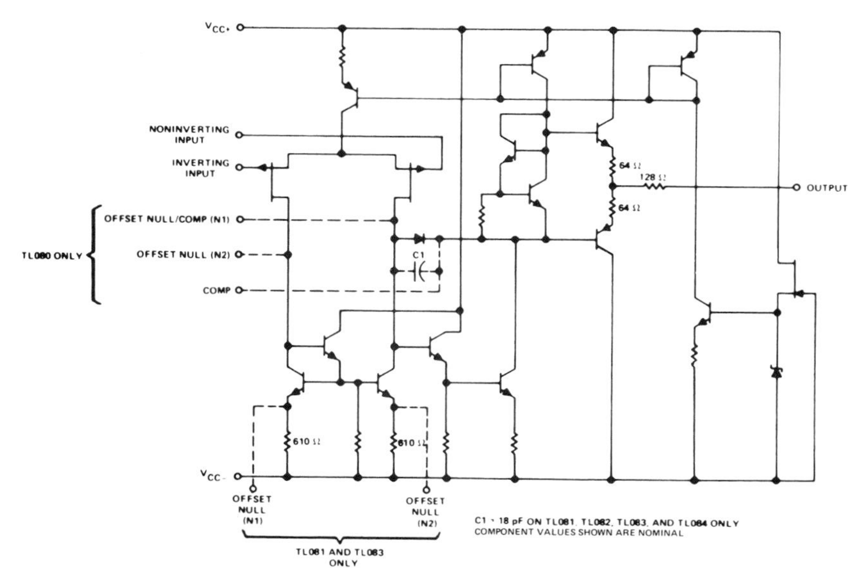

DISSIPATION DERATING TABLE

PACKAGE	POWER RATING	DERATING FACTOR	ABOVE T_A
J	680 mW	8.2 mW/°C	67°C
JG	680 mW	6.6 mW/°C	47°C
L	625 mW	5.0 mW/°C	25°C
N	680 mW	9.2 mW/°C	76°C
P	680 mW	8.0 mW/°C	65°C

absolute maximum ratings over operating free-air temperature range (unless otherwise noted)

		TL08_M TL08_AM	TL08_I	TL08_C TL08_AC TL08_BC	UNIT
Supply voltage, V_{CC+} (see Note 1)		18	18	18	V
Supply voltage, V_{CC-} (see Note 1)		−18	−18	−18	V
Differential input voltage (see Note 2)		±30	±30	±30	V
Input voltage (see Notes 1 and 3)		±15	±15	±15	V
Duration of output short circuit (see Note 4)		Unlimited	Unlimited	Unlimited	
Continuous total dissipation at (or below) 25°C free-air temperature (see Note 5)	J, JG, N, or P Package	680	680	680	mW
	L Package	625	625	625	
Operating free-air temperature range		−55 to 125	−25 to 85	0 to 70	°C
Storage temperature range		−65 to 150	−65 to 150	−65 to 150	°C
Lead temperature 1/16 inch from case for 60 seconds	J, JG, or L Package	300	300	300	°C
Lead temperature 1/16 inch from case for 10 seconds	N or P Package	260	260	260	°C

NOTES:
1. All voltage values, except differential voltages, are with respect to the zero reference level (ground) of the supply voltages where the zero reference level is the midpoint between V_{CC+} and V_{CC-}.
2. Differential voltages are at the noninverting input terminal with respect to the inverting input terminal.
3. The magnitude of the input voltage must never exceed the magnitude of the supply voltage or 15 volts, whichever is less.
4. The output may be shorted to ground or to either supply. Temperature and/or supply voltages must be limited to ensure that the dissipation rating is not exceeded.
5. For operation above 25°C free-air temperature, refer to Dissipation Derating Table.

operating characteristics, $V_{CC\pm} = \pm 15$ V, $T_A = 25°C$

	PARAMETER	TEST CONDITIONS		MIN	TYP	MAX	UNIT
SR	Slew rate at unity gain	$V_I = 10$ V, $C_L = 100$ pF,	$R_L = 2$ kΩ, $A_{VD} = 1$		13		V/μs
t_r	Rise time	$V_I = 20$ mV,	$R_L = 2$ kΩ,		0.1		μs
	Overshoot factor	$C_L = 100$ pF,	$A_{VD} = 1$		10%		
V_n	Equivalent input noise voltage	$R_S = 100\ \Omega$,	f = 1 kHz		47		nV/$\sqrt{\text{Hz}}$

electrical characteristics, $V_{CC\pm} = \pm 15$ V

PARAMETER		TEST CONDITIONS†		TL08_M TL08_AM MIN	TYP	MAX	TL08_I MIN	TYP	MAX	TL08_C TL08_AC TL08_BC MIN	TYP	MAX	UNIT
V_{IO}	Input offset voltage	R_S = 50 Ω, T_A = 25°C	TL08_		3	6		3	6		5	15	mV
			TL08_A‡		2	3					3	6	
			'81B,'82B,'84B								2	3	
		R_S = 50 Ω, T_A = full range	TL08_			9			9			20	
			TL08_A‡			5						7.5	
			'81B,'82B,'84B									5	
α_{VIO}	Temperature coefficient of input offset voltage	R_S = 50 Ω,	T_A = full range		10			10			10		μV/°C
I_{IO}	Input offset current§	T_A = 25°C	TL08_		5	100		5	100		5	200	pA
			TL08_A‡		5	100					5	100	
			'81B,'82B,'84B								5	100	
		T_A = full range	TL08_			20			10			5	nA
			TL08_A‡			20						3	
			'81B,'82B,'84B									3	
I_{IB}	Input bias current§	T_A = 25°C	TL08_		30	200		30	200		30	400	pA
			TL08_A‡		30	200					30	200	
			'81B,'82B,'84B								30	200	
		T_A = full range	TL08_			50			20			10	nA
			TL08_A‡			50						7	
			'81B,'82B,'84B									7	
V_{ICR}	Common-mode input voltage range	T_A = 25°C	TL08_	±12			±12			±10			V
			TL08_A‡	±12						±12			
			'81B,'82B,'84B							±12			
V_{OPP}	Maximum peak-to-peak output voltage swing	T_A = 25°C	R_L = 10 kΩ	24	27		24	27		24	27		V
		T_A = full range	R_L ⩾ 10 kΩ	24			24			24			
			R_L ⩾ 2 kΩ	20	24		20	24		20	24		
A_{VD}	Large-signal differential voltage amplification	R_L ⩾ 2 kΩ, V_O = ± 10 V, T_A = 25°C	TL08_	50	200		50	200		25	200		V/mV
			TL08_A‡	50	200					50	200		
			'81B,'82B,'84B							50	200		
		R_L ⩾ 2 kΩ, V_O = ± 10 V, T_A = full range	TL08_	25			25			15			
			TL08_A‡	25						25			
			'81B,'82B,'84B							25			
B_1	Unity-gain bandwidth	T_A = 25°C			3			3			3		MHz
r_i	Input resistance	T_A = 25°C			10^{12}			10^{12}			10^{12}		Ω
CMRR	Common-mode rejection ratio	R_S ⩽ 10 kΩ, T_A = 25°C	TL08_	80	86		80	86		70	76		dB
			TL08_A‡	80	86					80	86		
			'81B,'82B,'84B							80	86		
k_{SVR}	Supply voltage rejection ratio ($\Delta V_{CC\pm}/\Delta V_{IO}$)	R_S ⩽ 10 kΩ, T_A = 25°C	TL08_	80	86		80	86		70	76		dB
			TL08_A‡	80	86					80	86		
			'81B,'82B,'84B							80	86		
I_{CC}	Supply current (per amplifier)	No load, T_A = 25°C	No signal,		1.4	2.8		1.4	2.8		1.4	2.8	mA
V_{o1}/V_{o2}	Channel separation	A_{VD} = 100,	T_A = 25°C		120			120			120		dB

† All characteristics are specified under open-loop conditions unless otherwise noted. Full range for T_A is −55°C to 125°C for TL08_M and TL08_AM; −25°C to 85°C for TL08_I; and 0°C to 70°C for TL08_C, TL08_AC, and TL08_BC.

‡ Types TL080AM and TL083AM are not defined by this data sheet.

§ Input bias currents of a FET-input operational amplifier are normal junction reverse currents, which are temperature sensitive. Pulse techniques must be used that will maintain the junction temperature as close to the ambient temperature as is possible.

Balanced Modulator/Demodulator

AD630

FEATURES
Recovers Signal from +100dB Noise
2MHz Channel Bandwidth
45V/μs Slew Rate
−120dB Crosstalk @ 1kHz
Pin Programmable Closed Loop Gains of ±1 and ±2
0.05% Closed Loop Gain Accuracy and Match
100μV Channel Offset Voltage (AD630BD)
350kHz Full Power Bandwidth

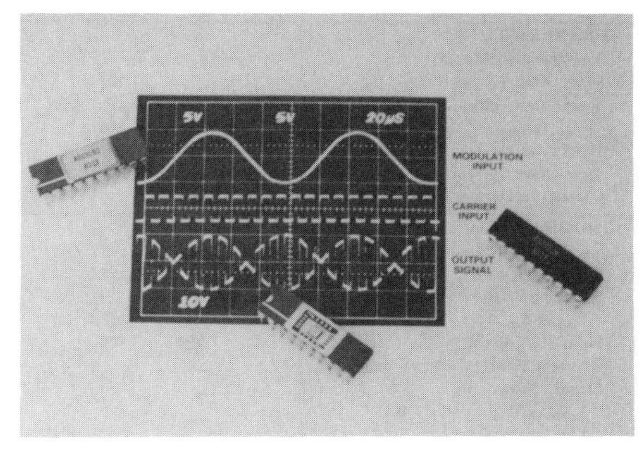

PRODUCT DESCRIPTION

The AD630 is a high precision balanced modulator which combines a flexible commutating architecture with the accuracy and temperature stability afforded by laser wafer trimmed thin film resistors. Its signal processing applications include balanced modulation and demodulation, synchronous detection, phase detection, quadrature detection, phase sensitive detection, lock-in amplification and square wave multiplication. A network of on-board applications resistors provides precision closed loop gains of ±1 and ±2 with 0.05% accuracy (AD630B). These resistors may also be used to accurately configure multiplexer gains of +1, +2, +3 or +4. Alternatively, external feedback may be employed allowing the designer to implement his own high gain or complex switched feedback topologies.

The AD630 may be thought of as a precision op amp with two independent differential input stages and a precision comparator which is used to select the active front end. The rapid response time of this comparator coupled with the high slew rate and fast settling of the linear amplifiers minimize switching distortion. In addition, the AD630 has extremely low crosstalk between channels of −100dB @ 10kHz.

The AD630 is intended for use in precision signal processing and instrumentation applications requiring wide dynamic range. When used as a synchronous demodulator in a lock-in amplifier configuration, it can recover a small signal from 100dB of interfering noise (see lock-in amplifier application). Although optimized for operation up to 1kHz, the circuit is useful at frequencies up to several hundred kilohertz.

Other features of the AD630 include pin programmable frequency compensation, optional input bias current compensation resistors, common mode and differential offset voltage adjustment, and a channel status output which indicates which of the two differential inputs is active.

PRODUCT HIGHLIGHTS

1. The configuration of the AD630 makes it ideal for signal processing applications such as: balanced modulation and demodulation, lock-in amplification, phase detection, and square wave multiplication.
2. The application flexibility of the AD630 makes it the best choice for many applications requiring precisely fixed gain, switched gain, multiplexing, integrating-switching functions, and high-speed precision amplification.
3. The 100dB dynamic range of the AD630 exceeds that of any hybrid or IC balanced modulator/demodulator and is comparable to that of costly signal processing instruments.
4. The op-amp format of the AD630 ensures easy implementation of high gain or complex switched feedback functions. The application resistors facilitate the implementation of most common applications with no additional parts.
5. The AD630 can be used as a two channel multiplexer with gains of +1, +2, +3 or +4. The channel separation of 100dB @ 10kHz approaches the limit which is achievable with an empty IC package.
6. The AD630 has pin-strappable frequency compensation (no external capacitor required) for stable operation at unity gain without sacrificing dynamic performance at higher gains.
7. Laser trimming of comparator and amplifying channel offsets eliminates the need for external nulling in most cases.

P.O. Box 280; Norwood, Massachusetts 02062 U.S.A.
Tel:617/329-4700 Twx: 710/394-6577
Telex: 924491 Cables: ANALOG NORWOODMASS

*Courtesy of **Analog Devices, Inc.**

SPECIFICATIONS (@ +25°C and $\pm V_S$ = ±15V unless otherwise specified)

Model	AD630J/A Min	Typ	Max	AD630K/B Min	Typ	Max	AD630S Min	Typ	Max	Units
GAIN										
Open Loop Gain	**90**	110		**100**	120		**90**	110		dB
±1, ±2 Closed Loop Gain Error		0.1				**0.05**		0.1		%
Closed Loop Gain Match		0.1				**0.05**		0.1		%
Closed Loop Gain Drift		2			2			2		ppm°C
CHANNEL INPUTS										
V_{IN} Operational Limit[1]		$(-V_S + 4V)$ to $(+V_S - 1V)$			$(-V_S + 4V)$ to $(+V_S - 1V)$			$(-V_S + 4V)$ to $(+V_S - 1V)$		Volts
Input Offset Voltage			**500**			**100**			**500**	μV
Input Offset Voltage T_{min} to T_{max}			**800**			**160**			**1000**	μV
Input Bias Current		100	**300**		100	**300**		100	**300**	nA
Input Offset Current		10	**50**		10	**50**		10	**50**	nA
Channel Separation @ 10kHz		100			100			100		dB
COMPARATOR										
V_{IN} Operational Limit[1]		$(-V_S + 3V)$ to $(+V_S - 1.5V)$			$(-V_S + 3V)$ to $(+V_S - 1.5V)$			$(-V_S + 3V)$ to $(+V_S - 1.3V)$		Volts
Switching Window			**±1.5**			**±1.5**			**±1.5**	mV
Switching Window T_{min} to T_{max}[2]			**±2.0**			**±2.0**			**±2.5**	mV
Input Bias Current		100	**300**		100	**300**		100	**300**	nA
Response Time (−5mV to +5mV step)		200			200			200		ns
Channel Status										
I_{SINK} @ $V_{OL} = -V_S + 0.4V$[3]	**1.6**			**1.6**			**1.6**			mA
Pull-Up Voltage			$(-V_S + 33V)$			$(-V_S + 33V)$			$(-V_S + 33V)$	Volts
DYNAMIC PERFORMANCE										
Unity Gain Bandwidth		2			2			2		MHz
Slew Rate[4]		45			45			45		V/μs
Settling Time to 0.1% (20V step)		3			3			3		μs
OPERATING CHARACTERISTICS										
Common-Mode Rejection	**85**	105		**90**	110		**90**	110		dB
Power Supply Rejection	**90**	110		**90**	110		**90**	110		dB
Supply Voltage Range	**±5**		±16.5	**±5**		±16.5	**±5**		±16.5	Volts
Supply Current		4	5		4	5		4	5	mA
OUTPUT VOLTAGE, @ $R_L = 2k\Omega$										
T_{min} to T_{max}	**±10**			**±10**			±10			Volts
Output Short Circuit Current		25			25			25		mA
TEMPERATURE RANGES										
Rated Performance – N Package	0		+70	0		+70		N/A		°C
D Package	−25		+85	−25		+85	−55		+125	°C

NOTES

[1] If one terminal of each differential channel or comparator input is kept within these limits the other terminal may be taken to the positive supply.
[2] This parameter guaranteed but not tested.
[3] I_{SINK} @ $V_{OL} = (-V_S + 1)$ volt is typically 4mA.
[4] Pin 12 Open. Slew rate with Pins 12 & 13 shorted is typically 35V/μs.

Specifications subject to change without notice.

Specifications shown in boldface are tested on all production units at final electrical test. Results from those tests are used to calculate outgoing quality levels. All min and max specifications are guaranteed, although only those shown in boldface are tested on all production units.

ABSOLUTE MAXIMUM RATINGS

Supply Voltage	±18V
Internal Power Dissipation	600mW
Output Short Circuit to Ground	Indefinite
Storage Temperature, Ceramic Package	−65°C to +150°C
Storage Temperature, Plastic Package	−55°C to +125°C
Lead Temperature, 10 sec. Soldering	+300°C
Max Junction Temperature	+150°C

ORDERING GUIDE

Model	Package
AD630JN	Plastic
AD630KN	Plastic
AD630AD	Ceramic
AD630BD	Ceramic
AD630SD	Ceramic

Typical Performance Characteristics

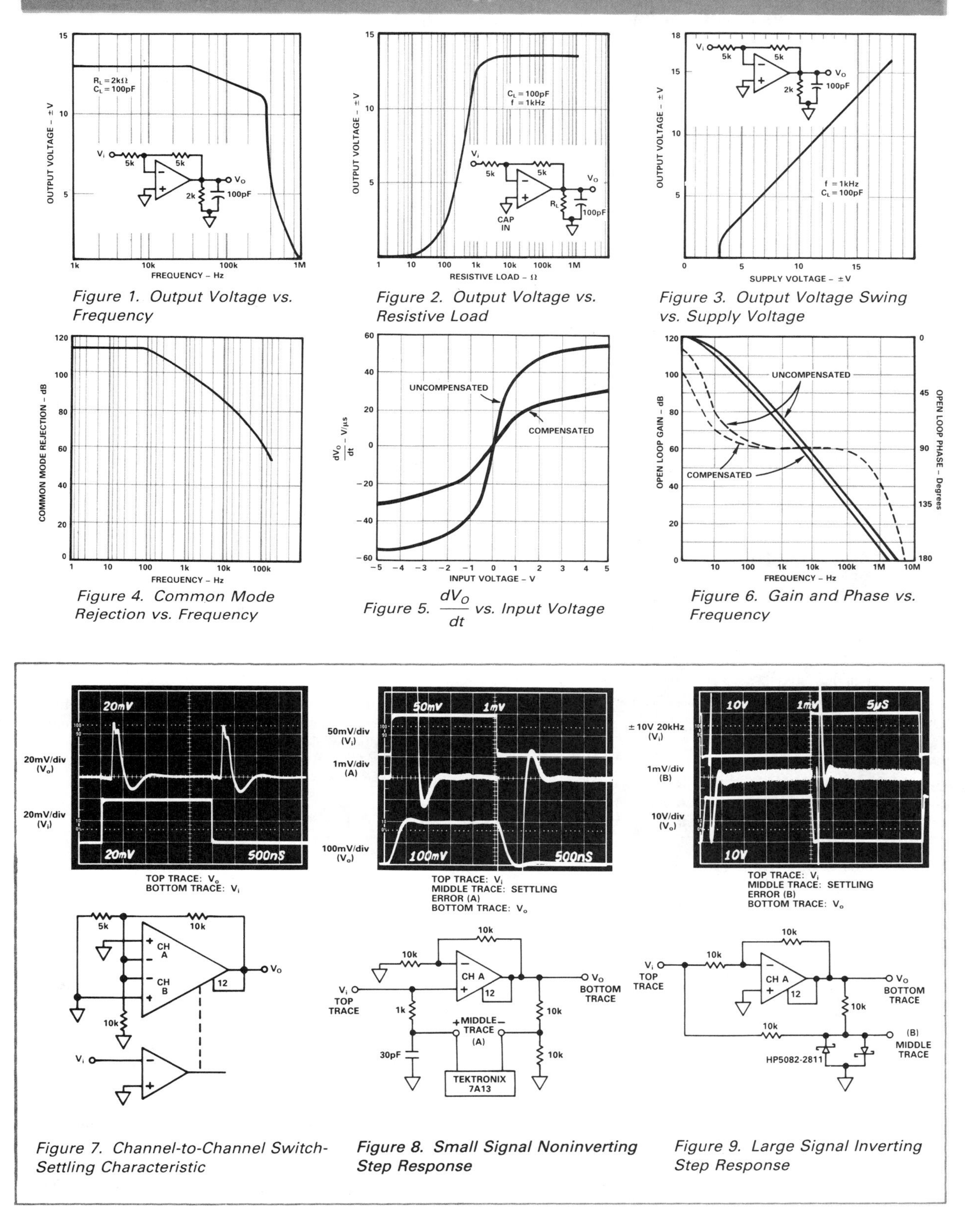

Figure 1. Output Voltage vs. Frequency

Figure 2. Output Voltage vs. Resistive Load

Figure 3. Output Voltage Swing vs. Supply Voltage

Figure 4. Common Mode Rejection vs. Frequency

Figure 5. $\frac{dV_O}{dt}$ vs. Input Voltage

Figure 6. Gain and Phase vs. Frequency

Figure 7. Channel-to-Channel Switch-Settling Characteristic

Figure 8. Small Signal Noninverting Step Response

Figure 9. Large Signal Inverting Step Response

OUTLINE DIMENSIONS

Dimensions shown in inches and (mm).

20-PIN CERAMIC DIP PACKAGE

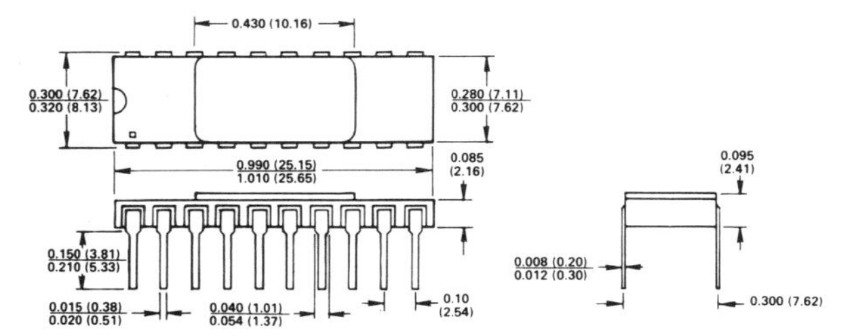

20-PIN PLASTIC DIP PACKAGE

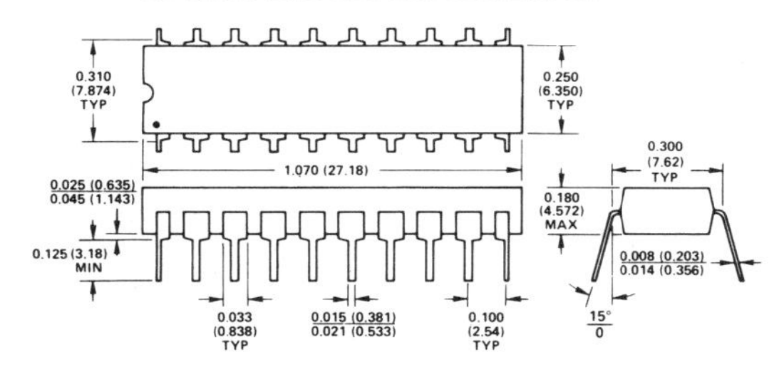

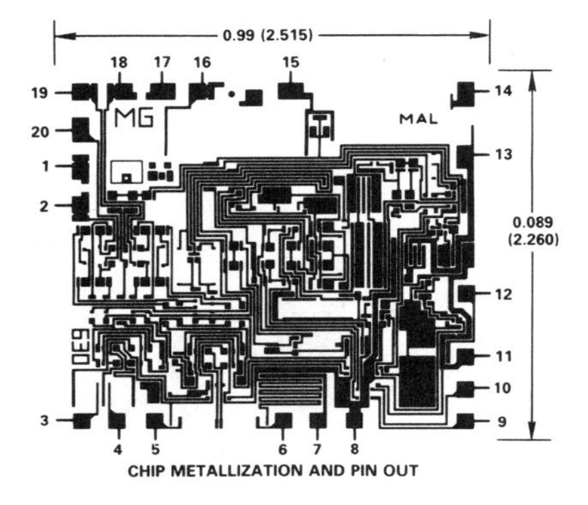

CHIP AVAILABILITY

The AD630 is available in laser trimmed, passivated chip form. The figure shows the AD630 metalization pattern, bonding pads and dimensions. AD630 chips are available; consult factory for details.

TWO WAYS TO LOOK AT THE AD630

Figure 10 is a functional block diagram of the AD630 which also shows the pin connections of the internal functions. An alternative architectural diagram is shown in Figure 11. In this diagram, the individual A and B channel pre-amps, the switch, and the integrator-output amplifier are combined in a single op amp. This amplifier has two differential input channels, only one of which is active at a time.

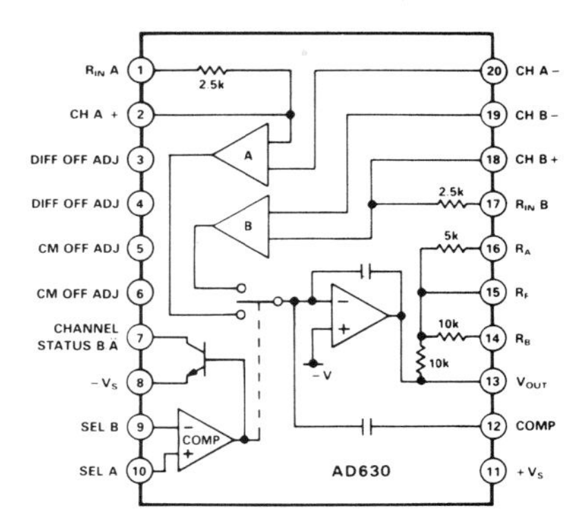

Figure 10. Functional Block Diagram

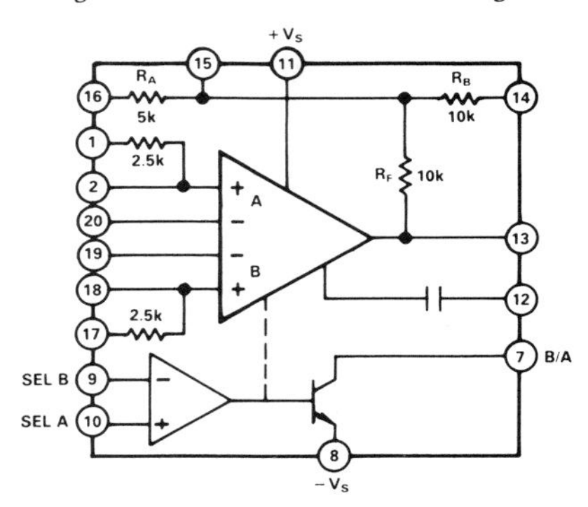

Figure 11. Architectural Block Diagram

HOW THE AD630 WORKS

The basic mode of operation of the AD630 may be more easy to recognize as two fixed gain stages which may be inserted into the signal path under the control of a sensitive voltage comparator. When the circuit is switched between inverting and noninverting gain, it provides the basic modulation/demodulation function. The AD630 is unique in that it includes Laser-Wafer-Trimmed thin-film feedback resistors on the monolithic chip. The configuration shown below yields a gain of ±2 and can be easily changed to ±1 by shifting R_B from its ground connection to the output.

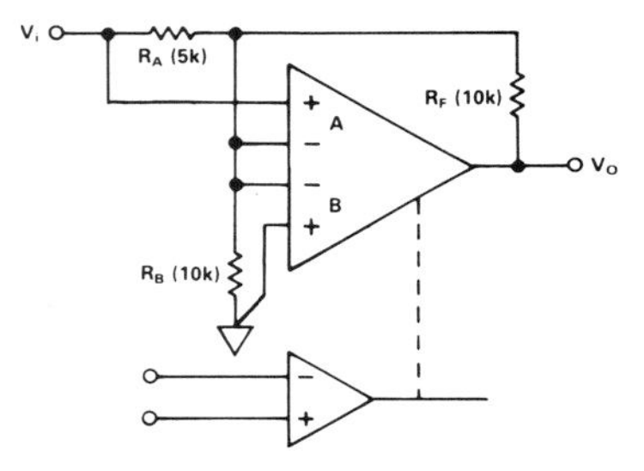

Figure 12. AD630 Symmetric Gain (±2)

The comparator selects one of the two input stages to complete an operational feedback connection around the AD630. The de-selected input is off and has negligible effect on the operation.

When channel B is selected, the resistors R_A and R_F are connected for inverting feedback as shown in the inverting gain configuration diagram in Figure 13. The amplifier has sufficient loop gain to minimize the loading effect of R_B at the virtual ground produced by the feedback connection. When the sign of the comparator input is reversed, input B will be de-selected and A will be

Universal Trigonometric Function Generator

AD639*

FEATURES
Complete, Fully-Calibrated Synthesis System
All Standard Functions: Sin, Cos, Tan, Cosec, Sec, Cot, Arcsin, Arccos, Arctan, etc.
Accurate Law Conformance (Sine to 0.02%)
Angular Range of ±500° (Sine Mode)
Function Programmable by Pin Strapping
1.5MHz Bandwidth (Sine Mode)
Multiplication via External Amplitude Input

APPLICATIONS
Continuous Wave Sine Generators
Synchro Sine/Cosine Multiplication
Coordinate Conversion and Vector Resolution
Imaging and Scanning Linearization Circuits
Quadrature and Variable Phase Oscillators

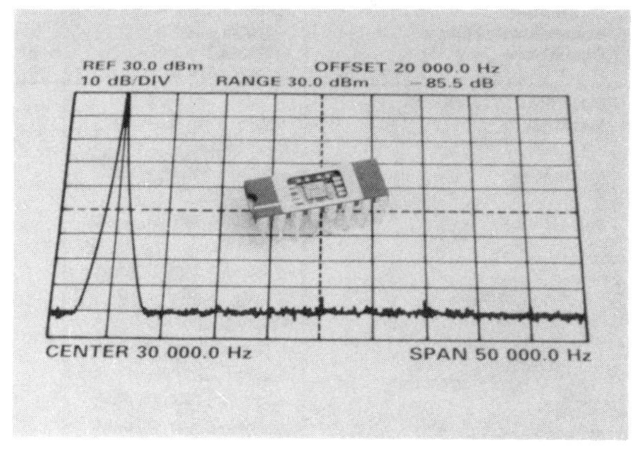

PRODUCT DESCRIPTION

The AD639 is a high accuracy monolithic function generator which provides all the standard trigonometric functions and their inverses via pin-strapping. Law conformance and total harmonic distortion surpass that previously attained using analog shaping techniques. Speed also exceeds that possible using ROM look-up tables and a DAC; in the sine mode, bandwidth is typically 1.5MHz. Unlike other function synthesis circuits, the AD639 provides a smooth and continuous sine conformance over a range of −500° to +500°. A unique sine generation technique results in 0.02% law conformance errors and distortion levels of −74dB in triwave to sinewave conversion.

The AD639 is available in three performance grades. The A and B are specified from −25°C to +85°C and the S is guaranteed over the extended temperature range of −55°C to +125°C. All versions are packaged in a hermetic TO-116, 16-pin ceramic DIP.

PRODUCT HIGHLIGHTS

The AD639 generates a basic function which is the ratio of a pair of independent sines:

$$W = U \frac{\sin(x_1 - x_2)}{\sin(y_1 - y_2)}$$

The differential angle arguments are proportional to the input voltages X and Y scaled by 50°/V. Using the 1.8V on-board reference any of the angular inputs can be preset to 90°. This provides the means to set up a fixed numerator or denominator ($\sin 90° = 1$) or to convert either sine function to a cosine ($\cos\theta = \sin(90° - \theta)$). Using the ratio of sines, all trigonometric functions can be generated (see Table I).

The amplitude of the function is proportional to a voltage U, which is the sum of an external differential voltage ($U_1 - U_2$) and an optional internal preset voltage (U_p). The control pin UP selects a 0V, 1V or 10V laser-trimmed preset amplitude which may be used alone ($U_1 - U_2 = 0$) or internally added to the $U_1 - U_2$ analog input. At the output, a further differential voltage Z can be added to the ratio of sines to obtain the offset trigonometric functions versine ($1 - \cos\theta$), coversine ($1 - \sin\theta$) and exsecant ($1 - \sec\theta$). A gating input is available which may be used to enable or disable the analog output. This pin also acts as an error flag output in situations where a combination of inputs will cause the output to saturate or to be undefined.

In the inverse modes, the argument can be the ratio of two input signals. This allows the user to compute the phase angle between the real and imaginary components of a signal using the arctangent mode.

*Protected by U.S. Patent Numbers 3,887,863; 4,475,169; 4,476,538.

Route 1 Industrial Park; P.O. Box 280; Norwood, Mass. 02062
Tel: 617/329-4700 **TWX: 710/394-6577**
West Coast 714/641-9391 **Mid-West** 312/653-5000 **Texas** 214/231-5094

selected. The new equivalent circuit will be the noninverting gain configuration shown below. In this case R_A will appear across the op-amp input terminals, but since the amplifier drives this difference voltage to zero the closed loop gain is unaffected.

The two closed loop gain magnitudes will be equal when $R_F/R_A = 1 + R_F/R_B$, which will result from making R_A equal to $R_FR_B/(R_F + R_B)$ the parallel equivalent resistance of R_F and R_B.

The 5k and the two 10k resistors on the AD630 chip can be used to make a gain of two as shown here. By paralleling the 10k resistors to make R_F equal 5k and omitting R_B the circuit can be programmed for a gain of ±1 (as shown in Figure 19a). These and other configurations using the on chip resistors present the inverting inputs with a 2.5k source impedance. The more complete AD630 diagrams show 2.5k resistors available at the noninverting inputs which can be conveniently used to minimize errors resulting from input bias currents.

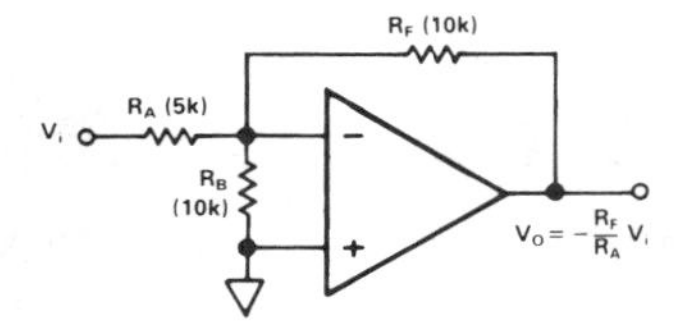

Figure 13. Inverting Gain Configuration

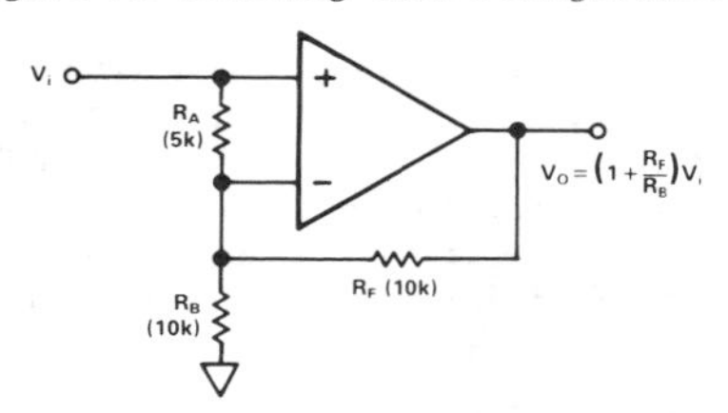

Figure 14. Noninverting Gain Configuration

CIRCUIT DESCRIPTION

The simplified schematic of the AD630 is shown in Figure 15. It has been subdivided into three major sections, the comparator, the two input stages and the output integrator. The comparator consists of a front end made up of Q52 and Q53, a flip-flop load formed by Q3 and Q4, and two current steering switching cells Q28, Q29 and Q30, Q31. This structure is designed so that a differential input voltage greater than 1.5mV in magnitude

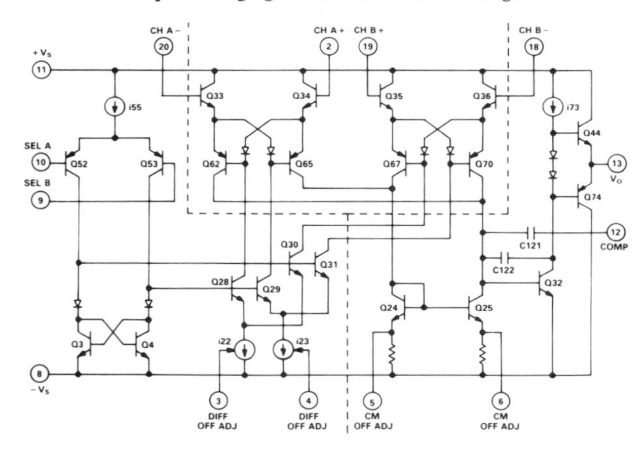

Figure 15. AD630 Simplified Schematic

applied to the comparator inputs will completely select one of the switching cells. The sign of this input voltage determines which of the two switching cells is selected.

The collectors of each switching cell connect to an input transconductance stage. The selected cell conveys bias currents i_{22} and i_{23} to the input stage it controls causing it to become active. The deselected cell blocks the bias to its input stage which, as a consequence, remains off.

The structure of the transconductance stages is such that they present a high impedance at their input terminals and draw no bias current when deselected. The deselected input does not interfere with the operation of the selected input insuring maximum channel separation.

Another feature of the input structure is that it enhances the slew rate of the circuit. The current output of the active stage follows a quasi-hyperbolic-sine relationship to the differential input voltage. This means that the greater the input voltage, the harder this stage will drive the output integrator, and hence, the faster the output signal will move. This feature helps insure rapid, symmetric settling when switching between inverting and noninverting closed loop configurations.

The output section of the AD630 includes a current mirror-load (Q24 and Q25), an integrator-voltage gain stage (Q32), and a complementary output buffer (Q44 and Q74). The outputs of both transconductance stages are connected in parallel to the current mirror. Since the deselected input stage produces no output current and presents a high impedance at its outputs, there is no conflict. The current mirror translates the differential output current from the active input transconductance amplifier into single ended form for the output integrator. The complementary output driver then buffers the integrator output to produce a low impedance output.

OTHER GAIN CONFIGURATIONS

Many applications require switched gains other than the ±1 and ±2 which the self-contained applications resistors provide. The AD630 can be readily programmed with 3 external resistors over a wide range of positive and negative gain by selecting R_B and R_F to give the noninverting gain $1 + R_F/R_B$ and subsequently R_A to give the desired inverting gain. Note that when the inverting magnitude equals the noninverting magnitude, the value of R_A is found to be $R_B\,R_F/(R_B + R_F)$. That is, R_A should equal the parallel combination of R_B and R_F to match positive and negative gain.

SPECIFICATIONS (typical @$T_A = 25°C$, $V_S = \pm 15V$, U or $U_p = 10V$ unless otherwise specified)

Parameter	Conditions	AD639A Min	AD639A Typ	AD639A Max	AD639B Min	AD639B Typ	AD639B Max	AD639S Min	AD639S Typ	AD639S Max	Units
SYSTEM PERFORMANCE											
SINE AND COSINE MODE ACCURACY											
Law Conformance[1]	−90° to +90°, U = 10V		0.02			0.02			0.02		%
Total Harmonic Distortion[2]	@10kHz, U = 10V		−74			−74			−74		dB
Mismatch of Six Peaks	−540° to +540°		0.05			0.05			0.05		%
Output Noise	@10kHz, U = 10V		2.8			2.8			2.8		μV/√Hz
	@10kHz, U = 1V		0.5			0.5			0.5		μV/√Hz
PEAK ABSOLUTE ERROR											
Sine Mode	−90° to +90°, U_p = 10V		0.4	0.8		0.2	0.4		0.4	0.8	%FS
	T_{min} to T_{max}		1.0			0.8	1.8		2.0	3.0	%FS
Cosine Mode	−90° to +90°, U_p = 10V		0.6	1.2		0.4	0.7		0.6	1.2	%FS
	T_{min} to T_{max}		1.5			1.2	2.0		2.0	3.3	%FS
Sine or Cosine	−180° to +180°, U_p = 10V		0.8	1.5		0.5	0.8		0.8	1.5	%FS
	T_{min} to T_{max}		1.7			1.3	2.5		2.8	4.0	%FS
	−360° to +360°, U_p = 10V		1.2			1.0			1.2		%FS
	−90° to +90°, U_p = 1V		1.3	2.5		1.0	1.7		1.3	2.5	%FS
	T_{min} to T_{max}		1.5			1.0	2.3		2.0	3.5	%FS
	−180° to +180°, U_p = 1V		1.5	3.0		1.2	2.0		1.5	3.0	%FS
	T_{min} to T_{max}		1.7			1.3	2.5		2.3	4.2	%FS
	−360° to +360°, U_p = 1V		2.0			1.8			2.0		%FS
vs Supply	−360° to +360°, U_p = 10V, V_s = ±15V ±1V		0.02			0.02			0.02		%FS/V
	−360° to +360°, U_p = 1V, V_s = ±15V ±1V		0.07			0.07			0.07		%FS/V
TANGENT MODE ACCURACY											
Peak Error[3]	−45° to +45°, U_p = 10V		2.0	3.5		1.0	2.0		2.0	3.5	%FS
	T_{min} to T_{max}		2.5			1.5	2.8		3.0		%FS
	−45° to +45°, U_p = 1V		3.0	5.0		1.5	3.0		3.0	5.0	%FS
	T_{min} to T_{max}		4.0			2.0	5.0				%FS
ARCTANGENT MODE ACCURACY											
Peak Angular Error											
Fixed Scale	U_p = 1V		1.5			1.5			1.5		Degrees
Variable Scale	U = 0.1V, −11V ≤ Z ≤ +11V		2.5			2.5			2.5		Degrees
	U = 10V, −11V ≤ Z ≤ +11V		0.5			0.5			0.5		Degrees
SECTIONAL SPECIFICATIONS											
ANGLE INPUTS (X1 & X2, Y1 & Y2)[4]											
Input Resistance to COM			3.6			3.6			3.6		kΩ
Nominal Scaling Factor			50			50			50		°/V
X1 & X2 Inputs											
Angular Range For Specified Error (X1 − X2)		−360		+360	−360		+360	−360		+360	Degrees
Scaling Error X1 or X2			0.2	0.65		0.2	0.65		0.2	0.65	%
Angular Offset X1 = X2 = 0			0.1	0.3		0.1	0.3		0.1	0.3	Degrees
Y1 & Y2 Inputs											
Angular Range For Specified Error (Y1 − Y2)		0		+180	0		+180	0		+180	Degrees
Scaling Error Y1 or Y2			1.0	2.0		1.0	2.0		1.5	2.0	%
Angular Offset Y1 = Y2 = 0			0.5	1.0		0.5	1.0		0.5	1.0	Degrees
AMPLITUDE INPUTS (U1 & U2)											
Input Resistance to COM			50			50			50		kΩ
Nominal Gain	X = Y = VR, W to Z1		1			1			1		V/V
Gain Error	U = 0.1 to 10V		0.01	0.5		0.01	0.5		0.01	0.5	%
	T_{min} to T_{max}		0.08			0.08			0.25		%
Voltage Offset	$U_1 = U_2$ = 0V		3.0	10		3.0	10		3.0	10	mV
	T_{min} to T_{max}		3.0			3.0			4.0		mV
Linearity Error	0 ≤ $U_1 - U_2$ ≤ 10V		0.1			0.1			0.1		%
AMPLITUDE PRESET (UP)											
1V Preset Enabled	UP tied to $-V_S$										
Amplitude Accuracy			0.4	2.0		0.4	2.0		0.4	2.0	%
	T_{min} to T_{max}		1.5			1.5			2.0		%
10V Preset Enabled	UP tied to $+V_S$										
Amplitude Accuracy			0.1	0.55		0.1	0.55		0.1	0.55	%
	T_{min} to T_{max}		1.0			1.0			1.5		%
INVERSE INPUTS (Z1 & Z2)											
Input Resistance to COM			50			50			50		kΩ

Parameter	Conditions	AD639A Min	AD639A Typ	AD639A Max	AD639B Min	AD639B Typ	AD639B Max	AD639S Min	AD639S Typ	AD639S Max	Units
SIGNAL OUTPUT (W)	$R_L \geq 2k\Omega$, $C_L \leq 100pF$										
Small Signal Bandwidth W to Z1											
	Cc = 0		1.5			1.5			**1.5**		MHz
	Cc = 200pF		30			30			**30**		kHz
Slew Rate	Cc = 0		30			30			**30**		V/μS
Output Voltage Swing		**±11**	±13		**±11**	±13		**±11**	**±13**		V
Short Circuit Current		**20**	30	**45**	**20**	30	**45**	**20**	**30**	**45**	mA
Output Offset	$Z_1 = Z_2 = 0$, $U_p = 10$		5	**30**		5	**30**		**5**	**30**	mV
	T_{min} to T_{max}		10			10					mV
	$Z_1 = Z_2 = 0$, $U_p = 1V$			**20**			**20**			**20**	mV
	T_{min} to T_{max}		7			7					mV
VOLTAGE REFERENCE (VR)											
$R_L \geq = 1.8K\Omega$											
Nominal Output			+1.8			+1.8			**+1.8**		V
Output Voltage Tolerance			0.05	**0.45**		0.05	**0.45**		**0.05**	**0.45**	%
	T_{min} to T_{max}		0.08			0.08	**0.5**		**0.2**	**0.7**	%
Supply Regulation	$+V_S = 5V$ to 18V		150			150			**150**		μV/V
Maximum Output Current			4			4			**4**		mA
GATE I/O (GT)											
Switching Threshold as an Input	Output Valid		+1.5			+1.5			**+1.5**		V
	Output Invalid		0.1			0.1			**0.1**		V
Voltage Output	Error, $R_L = 5k\Omega$		+2.25			+2.25			**+2.25**		V
	No Error, $R_L = 5k\Omega$		−0.25			−0.25			**−0.25**		V
POWER SUPPLIES											
Operating Range		**±4.75**		**±18**	**±4.75**		**±18**	**±4.75**		**±18**	V
$+V_S$ Quiescent Current	U = X = 0V, Y = Vr		7.5	**10**		7.5	**10**		7.5	**10**	mA
−V, Quiescent Current	U = X = 0V, Y = Vr		4.0	5		4.0	5		**4.0**	**5**	mA
TEMPERATURE RANGE											
Operating, Rated Performance		−25		+85	−25		+85	**−55**		**+125**	°C
Storage		−65		+150	−65		+150	**−65**		**+150**	°C
PRICES (Ceramic DIP-D)											
1 to 24			$17.30			$27.80					
25 to 99			$14.90			$24.00					
100 to 999			$11.95			$19.20					

NOTES
[1]Intrinsic accuracy measured at an amplitude of 10V using external adjustments to absorb residual errors in angular scaling, angular offset, amplitude scaling and output offset.
[2]Using a time and amplitude symmetric triangular wave of +3.6V peak-to-peak and external adjustments to absorb residual errors in angular scaling and offset.
[3]Full-scale is defined as the ideal output when the angle input is at either end of the limit specified.
[4]Specifications for the X inputs apply for range U = 1V to 10V, while the Y input errors are specifically given for U = 1V.

Specifications shown in boldface are tested on all production units at final electrical test. Results from those tests are used to calculate outgoing quality levels.
All min and max specifications are guaranteed, although only those shown in boldface are tested on all production units.
Shaded area denotes preliminary technical data. Contact the factory for details.

PIN CONFIGURATION & DIMENSIONS

Dimensions shown in inches and (mm).

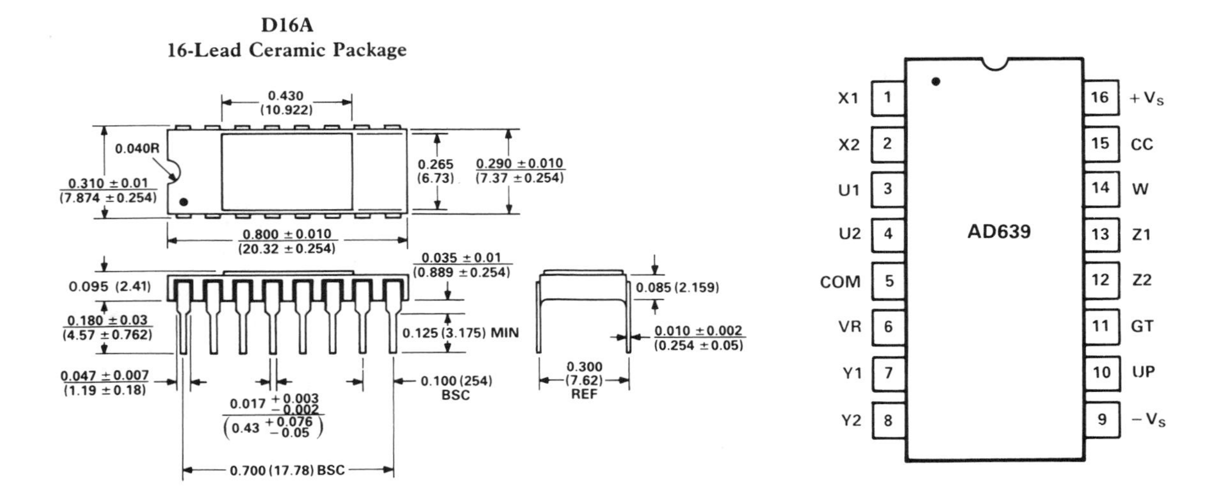

ABSOLUTE MAXIMUM RATINGS

	AD639A,B	AD639S
Supply Voltage	$\pm 18V$	*
Internal Power Dissipation	300mW	*
Output Short-Circuit to Ground	Indefinite	*
Input Voltages X_1, X_2, Y_1, Y_2[1]	$\pm 12V$	*
Input Voltages U_p, U_1, U_2, Z_1, Z_2[1]	$\pm 25V$	*
Operating Temperature Range	$-25°C$ to $+85°C$	$-55°C$ to $+125°C$
Storage Temperature Range	$-65°C$ to $+150°C$	*
Lead Temperature, Soldering	60sec, $+300°C$	*

NOTES
*Same as AD639A,B Specifications
[1]These inputs are purely resistive and the maximum inputs are determined by resistor dissipation limits, not the supply voltages.

$\sin(\theta) = \dfrac{\sin(\theta)}{1} = \dfrac{\sin(\theta - 0)}{\sin(90° - 0)}$		$\operatorname{cosec}(\theta) = \dfrac{1}{\sin(\theta)} = \dfrac{\sin(90° - 0)}{\sin(\theta - 0)}$	
$\cos(\theta) = \dfrac{\cos(\theta)}{1} = \dfrac{\sin(90° - \theta)}{\sin(90° - 0)}$		$\sec(\theta) = \dfrac{1}{\cos(\theta)} = \dfrac{\sin(90° - 0)}{\sin(90° - \theta)}$	
$\tan(\theta) = \dfrac{\sin(\theta)}{\cos(\theta)} = \dfrac{\sin(\theta - 0)}{\sin(90° - \theta)}$		$\cot(\theta) = \dfrac{\cos(\theta)}{\sin(\theta)} = \dfrac{\sin(90° - \theta)}{\sin(\theta - 0)}$	

Table I.

Principles Of Operation

Figure 1 is a functional equivalent of the AD639, intended to assist in understanding and utilizing the device: it is not a literal representation of the internal circuitry[1]. Two similar sine-shaping networks accept input voltages X_1, X_2, Y_1 and Y_2, proportional to the corresponding angles x_1, x_2, y_1 and y_2, with a scaling factor of 50°/V (20mV/°).

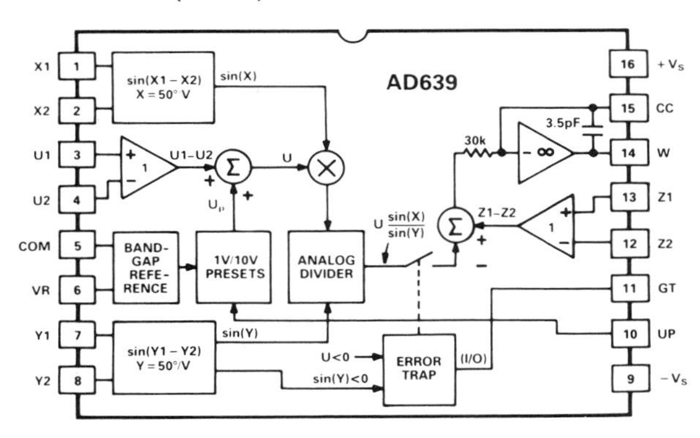

Figure 1. Equivalent Block Schematic of the AD639

The first of these networks generates an output proportional to the sine of $x = (x_1 - x_2)$ over a useful operating range in excess of $-500°$ to $+500°$ (see Figure 3). The accuracy of the function over the central $\pm 180°$ is excellent, a consequence of the optimized network design, further enhanced by precision laser wafer trimming during manufacture. The output of the X-network is multiplied by the amplitude-control voltage, U. This may be provided by applying inputs to U1 or U2, or pre-selected to be 1V or 10V by a control input to UP, or in combination; that is, the function amplitude is $U = (U_1 - U_2) + U_P$.

The second network generates an output proportional to the sine of $y = (y_1 - y_2)$. Although the X and Y networks are similar, other design considerations result in a smaller angular range for the Y-input. The principal range is from 0° to $+180°$; in the adjacent ranges ($+180°$ to $+360°$ and 0° to $-180°$) the error trap is activated.

The ratio of the two sines is generated by *implicit* division, rather than by use of a separate analog divider as indicated in Figure 1, and is summed with the voltage $Z = (Z_1 - Z_2)$. The difference is applied to the high-gain output op-amp. In the *normal* modes (see below) Z1 is connected to the output W, and Z2 is grounded. Under these conditions, the function is

$$W = U\,\frac{\sin(x_1 - x_2)}{\sin(y_1 - y_2)}$$

Either sine function can be converted to the cosine by applying the input to X2 or Y2 and introducing a $+90°$ offset, since

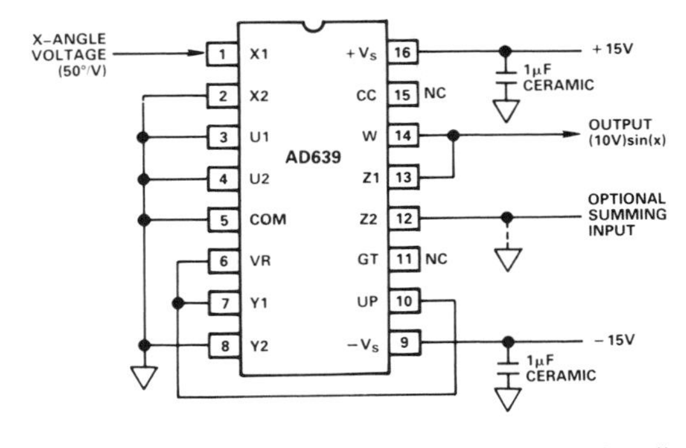

Figure 2. Connections for the Sine Mode with Amplitude Preset to 10V

[1]For details of the sine-network theory and design, see "A Monolithic Microsystem for Analog Synthesis of Trigonometric Functions and their Inverses," Barrie Gilbert, *IEEE Journal of Solid-State Circuits*, Vol. SC-17, No. 6, Dec. 1982, pp 1179-1191. Reprints available.

cos (θ) = sin(90° − θ). For example, by connecting the +1.8V reference output at pin 6 (VR) to X1 and the angle voltage, V_θ, to X2 the numerator becomes the cosine of angle θ. Alternatively, by connecting VR to either X1 or Y1 and grounding X2 or Y2, the numerator or denominator, respectively, becomes unity, since sin(90° − 0) = 1. By these means, the full set of *normal* functions shown in Table I can be generated. All functions can be sign-inverted by interchanging the X-inputs. The Z2 input can be used to sum another function to the output, W, with unity gain.

In addition to the *normal* modes providing sine, cosine, tangent, cosecant, secant and cotangent functions, the AD639 can generate the *offset* functions such as the versine, 1 − cos (θ), discussed below. The *inverse* functions such as arc-sine, arc-cosine and arc-tangent, are also supported by the AD639, by closing the feedback loop through the corresponding *normal* function. The output angle is limited to the principal range (for example, −90° to +90° for the arc-sine and arc-tangent, 0° to +180° or −180° to 0° for the arc-cosine).

TERMINOLOGY

When discussing a device having as many inputs and operating modes as the AD639, it is important to clarify the nomenclature and scaling conventions. In all cases *angles* are denoted by lower-case letters (x, y, θ) and have the dimension of angular degrees. Upper-case letters (A, V, U, W, X, Y and Z) refer to *voltages*; subscripts are used to refer to one or the other of a differential pair such as $X_1 - X_2$, or the preset value U_P. Numbered upper-case letters refer to the variable name or the package pin.

THE ANGLE INPUTS: X1, X2, Y1, Y2

The angles $x = (x_1 - x_2)$ and $y = (y_1 - y_2)$ are directly proportional to the differential voltages $X = (X_1 - X_2)$ and $Y = (Y_1 - Y_2)$ respectively, with a scaling factor of 50°/V. The X-inputs can be driven to ±12V pk, that is ±600°. The Y-input should be limited to 0 to +3.6V (0° to +180°) to satisfy certain internal requirements. The resistance at these inputs is nominally 3.6kΩ to COM.

The sine function exhibits odd-order symmetry: sin(−θ) = −sin(θ). By simply interchanging the X-inputs, the overall sign of any function can be inverted. The Y-inputs can also be interchanged to allow operation with a negative input voltage (0 to −3.6V) while maintaining the correct angular range.

It may occasionally be desirable to reduce the angular scaling factor. For example, to convert a triwave of ±10V amplitude into a continuous sinewave requires a scaling factor of 9°/V (since ±10V corresponds to ±90°). This can be achieved by using a resistor (in this case, about 16.4kΩ) in series with the X1 input; a resistor of equal value must be inserted in series with the X2 input to minimize angular offset error. Note that the on-chip thin-film resistors are not trimmed to absolute value, so a scaling adjustment is needed; however, once set, scaling will be stable.

THE AMPLITUDE-CONTROL INPUTS: U1, U2, UP

The amplitude of the function can be determined either by the application of an external voltage to the U1 and U2 inputs, or by enabling the internal preset voltage U_P by taking the control pin UP low or high, or via a combination of these modes. The net amplitude is $U = (U_1 - U_2) + U_P$. This sum must be greater than zero and less than $|-V_S|$; voltages beyond these limits activate the error trap.

In the external mode, the differential voltage $(U_1 - U_2)$ will generally be in the range 10mV to 10V. Positive inputs are applied to U1 while U2 is grounded; for negative inputs, interchange U1 and U2. The input resistance at U1 and U2 is nominally 50kΩ to analog common. A nominal bias current of −50μA is needed at the U-inputs; zero-valued inputs must therefore be connected to common to prevent offset error. The gain from the U-interface to the output is trimmed to be unity for sin(x)/sin(y) = 1. The effective gain can be lowered using a series resistor; to avoid offset an equal resistor must be used in the zero-valued input.

The UP control pin may be left unconnected (or grounded) to disable the internal amplitude preset, connected to $+V_s$ to set $U_P = 10V$, or to $-V_s$ to set $U_P = 1V$. An external resistor of 75kΩ ($\pm V_s = 5V$) to 360kΩ ($\pm V_s = 15V$) can be inserted in series with UP (which also has an input resistance of typically 50kΩ) to minimize power dissipation. Alternatively, V_r can be used to enable $U_P = 10V$. The UP input can be used to switch the output on or off under logic control, but requires a relatively long response time. The GT interface is more suitable for this purpose and it allows gating to any amplitude, U not just to the preset values of 1V or 10V.

THE REFERENCE OUTPUT: VR

The voltage V_r is laser-trimmed to +1.8V with respect to analog common. It can be used to fix the angle x or y to 90° and thus set sin(x) or sin(y) to unity. It can also provide a 90° offset to convert the numerator or denominator to a cosine function. Stable offsets less than 90° may be introduced using a voltage-dividing series-resistor (nominally 3.6kΩ for 45°). V_r can also be used as the amplitude input voltage $U_1 - U_2$, or as a convenient control input to set $U_P = 10V$. This output is short-circuit protected and can provide up to 4mA total load current.

THE ERROR-TRAP AND GATE: GT

In some applications it may be useful to know that the output is severely in error due to a dynamic combination of inputs. For example, the tangent, cotangent, secant and cosecant all exhibit regions where the function increases sharply for small angular changes, and the output may easily saturate. Consider the case where (10V) tan (θ) is being generated. W is 10V for θ = 45°, and the theoretical output of 17.3V at θ = 60° cannot be achieved using ±15V supplies. Likewise, the output is invalid whenever the angle y is outside of a valid range (principally 0 to +180°), or when $U < 0$ or $U > |-V_S|$. Under such conditions the AD639 generates a HIGH output at pin 11 and simultaneously clamps the analog output to zero (in fact, to the voltage Z_2). Grounding GT disables the error trap.

The GT pin may also be used as an *input* to gate the function output. This is achieved by raising pin 11 to a voltage above +1.5V. Response time is typically 500ns for a logic drive of 0 to +2V, and the ON/OFF ratio is greater than 83dB when used as a continuous-wave sine converter with a single-sided ±1.8V triwave drive at frequencies up to 10kHz, or 63dB at 100kHz; the feedthrough is entirely capacitive, and is equivalent to 5pF between X1 or X2 and the op-amp summing node. Feedthrough can be minimized by using a balanced drive to X1 and X2.

Operation In Normal Modes

In *normal* modes, the Z-input establishes a feedback path around the output op-amp, by connecting Z1 to the output, W, and Z2 to the ground associated with the load circuit. For the highest accuracy Z1 can be used to sense the output at the load terminals. Similarly, zero-valued angle inputs and the angle common (pin 5) should be connected to the ground associated with the source circuitry.

SINE MODE

The AD639 can generate either (1) a low-distortion continuous sinewave from a repetitive triwave input or (2) a high-accuracy sine function for use in computational applications. In most cases, the choice of preset or externally-controlled amplitude will make little difference to distortion or accuracy, and both methods are used in this section. In all of the *normal* modes, the Z2 input can be used either to sum a further signal to the output (or introduce an optional output offset trim). The denominator is set to unity by making $y=90°$, using the $+1.8V$ output. Figure 2 shows typical connections. The 10V preset is selected, using V_r as a control input to UP, and the ideal output is (10V)sin(x). In practice, five basic types of error arise:

1. *X-angle scaling error*: The amount by which the angle generated for each volt of X-input differs from 50°. In triwave-to-sinewave (CW) applications this introduces odd-order harmonic distortion, and is indistinguishable from an incorrect triwave amplitude.
2. *X-angle offset error*: The actual angle generated when $X=(X_1-X_2)=0$. In CW applications this introduces even-order harmonic distortion, as a non-zero mean in the triwave would.
3. *Amplitude scaling error*: The amount by which the peak-to-peak amplitude of the sinewave differs from the ideal value, U/sin(y). This error is usually critical only in computational applications. Errors associated with the Y-network also affect the amplitude in the sine mode.
4. *Output offset error*: The amount by which the *mean value* of the sinewave differs from zero (strictly, the voltage on Z2). This error is only important in computational applications. Note that the output may also be non-zero due to angular offset on the X-input. For example, the typical specified X-angle offset of 0.1° introduces an output error of 17.45mV when U/sin(y) = 10V, more than three times the specified mean offset component of 5mV.
5. *Law-conformance error*: The residual deviation between the output function and the ideal function when all of the above errors have been removed by trimming during manufacture or further external trimming, limiting the ultimate accuracy of the function.

Figure 3 shows the function when driven well beyond the specified angular range, using a differential X-input of ±18V peak. This also shows the AD639's ability to drive ±15V into a 600Ω load, with supplies of ±18V. Using an accurate data-acquisition system the output can be compared to a computer-generated sine function. When the first four types of errors are trimmed out, the peak error over the full input range is typically less than 0.5%. Over the central −90° to +90°, the peak law-conformance error is typically only 0.02%. Figure 4 shows the law conformance for four typical samples of AD639. The differential signal interfaces simplify the inclusion of optional offset correction to any of the variables.

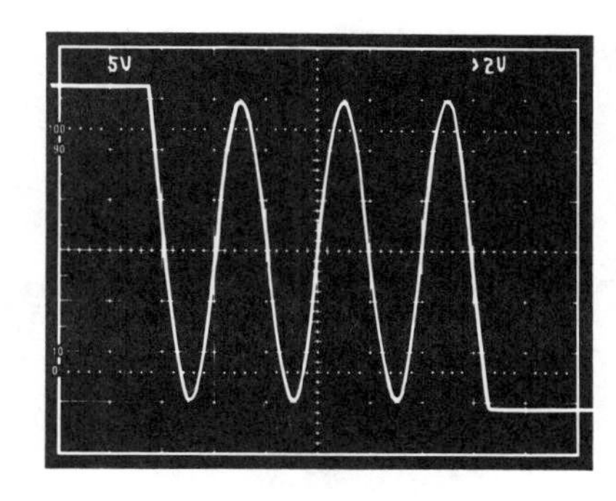

Figure 3. Output Function for Peak X-Input of ±18V, with $U=15V$, $R_L=600\Omega$ ($\pm V_S=18V$)

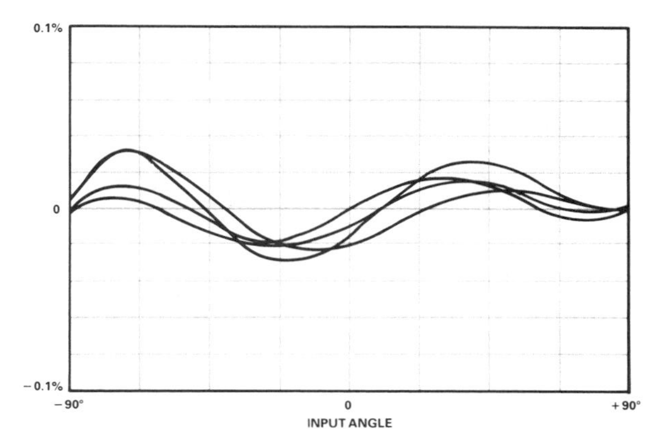

Figure 4. Residual Error Over Central 180° Using External Trimming

HARMONIC DISTORTION

The AD639 can generate continuous sinewaves of very low distortion using a linear, highly-symmetric triangle-wave of ±1.8V amplitude. Imperfections in the triwave will cause the following errors:

1. *Incorrect amplitude*: This causes odd-order distortion. Each 1% error (either too large or too small) generates 0.25% of HD3, 0.0833% of HD5 and a total harmonic distortion (THD) of 0.27% (−51.42dBs).
2. *Baseline offset*: This causes even-order distortion. Each millivolt of offset in a 1.8V triwave generates 0.037% of HD2, 0.0074% of HD4 and a THD of 0.038%, as well as a DC offset of 0.055% of the output amplitude.
3. *Time-asymmetry*: The run-up time, t_1, and run-down time, t_2, of the triwave may be unequal. This causes both odd- and even-order harmonics. Let the asymmetry in percent be $p=100(t_1-t_2)/(t_1+t_2)$. The even-order terms are proportional to p; the odd-order terms increase as p^2. A 1% time-asymmetry generates 0.57% of HD2, 0.00625% of HD3, 0.043% of HD4 and 0.00167% of HD5, and a THD of −44dBs. There is no DC term.
4. *Amplitude-nonlinearity*: This can take on many forms, such as an exponential nonlinearity in the triwave, amplitude compression, and so on. Distortion can be calculated for various special cases. Fortunately, it is fairly easy to avoid these types of imperfections in the triwave generator using appropriate design methods.

Active Filter Circuits

10

LEARNING OBJECTIVES

Upon completion of this chapter on active filter circuits, you will be able to:

- Name the four types of active filters and compare their performances with respect to frequency response, cutoff frequency, and gain.
- Distinguish between first-order or second-order low- and high-pass filters.
- Analyze performance of low- and high-pass filters.
- Make low- or high-pass filters for any required cutoff frequency.
- Cascade a low- and high-pass filter to make a wideband bandpass filter.
- Make a narrowband bandpass filter with one op amp, three resistors, and two capacitors.
- Make or analyze notch filter circuits.
- Fine-tune filter circuits to obtain the specified performance.

10-0 INTRODUCTION

An electronic filter is a circuit that ideally passes desired frequencies and does not pass unwanted frequencies. Filters can either be passive or active. *Passive filters* are made of passive components such as resistors, inductors, or capacitors. When a load is connected to a passive filter, its original filtering characteristics are changed.

With an op amp added to a passive filter, the resulting circuit is an *active filter*. The op amp does not load the passive components and does not change the original filtering characteristics. The op amp also brings the advantages of easier design and fine tuning, plus lower cost and a lower parts count, for improved filter characteristics.

Inexpensive resistors, capacitors, and op amps are used in this book to make active filters for frequency selectors in the range 1 Hz to 100 kHz.

Inductors are essentially useless in this frequency range because they are too large, expensive, have too many losses, and influence adjacent circuits.

10-1 TYPES OF ACTIVE FILTERS

Filters are first classified according to the band of frequencies that they pass. For example, a low-pass filter transmits low frequencies and eliminates high frequencies. Its performance is measured by the circuit of Fig. 10-1a, where the frequency of sine wave V_i is varied. Amplitude of V_i is held constant. V_o is measured at convenient frequencies and V_o is plotted against frequency.

The result is called an amplitude versus frequency response curve. *Low-pass* and *high-pass* filter performance is shown in Fig. 10-1b and c. Figure 10-1d shows that a bandpass filter passes a selected band of

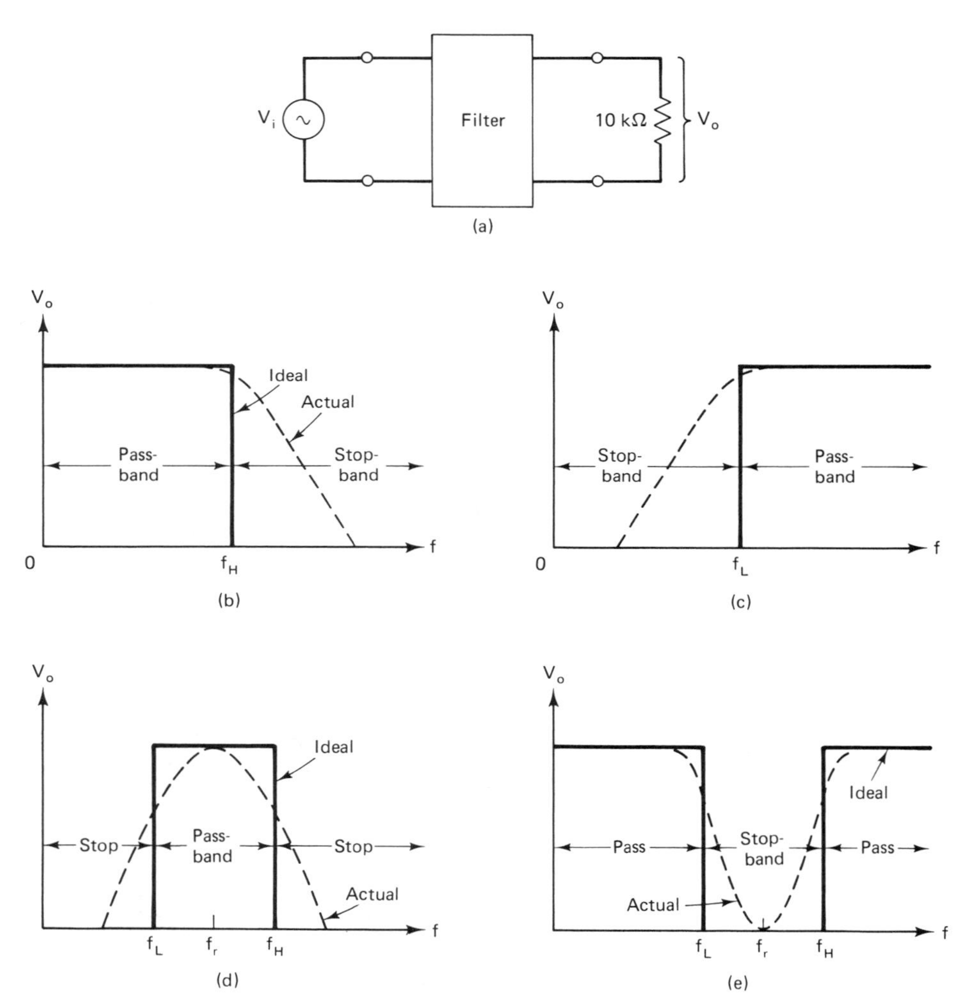

Figure 10-1 The magnitude of V_i is held constant and its frequency is varied in (a). V_O is measured at convenient frequencies and plotted to form the amplitude versus frequency response curves for each category of filter. (a) Basic filter test circuit; (b) low-pass filter; (c) high-pass filter; (d) band-pass filter; (e) band-elimination or notch filter.

frequencies. A *notch* or *band-elimination* filter eliminates a selected band of frequencies, as in Fig. 10-1e. Figure 10-1 shows the plots for ideal filters as solid lines. The dashed lines show the characteristics of practical filters.

10-2 BASIC LOW-PASS FILTER

10-2.1 Circuit Operation

A single-pole (single capacitor) low-pass filter circuit is shown in Fig. 10-2a. At low frequencies, capacitor C acts as an open circuit. The op-amp circuit acts as a voltage follower with a voltage gain of 1 or 0 dB. At high frequencies, the reactance of C becomes very low. Since V_i divides between R and C, the signal voltage at the (+) input, and consequently V_o

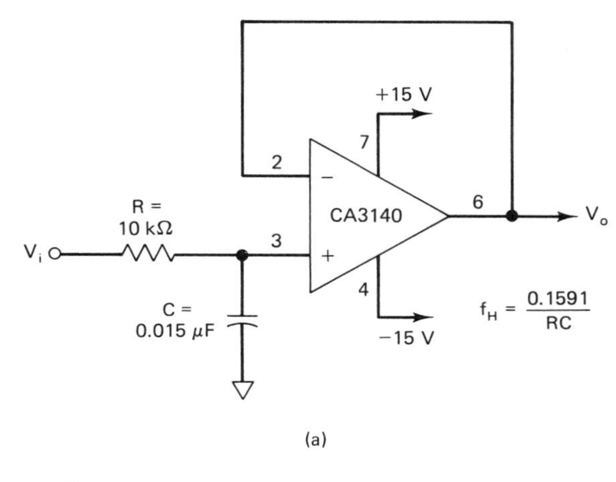

(a)

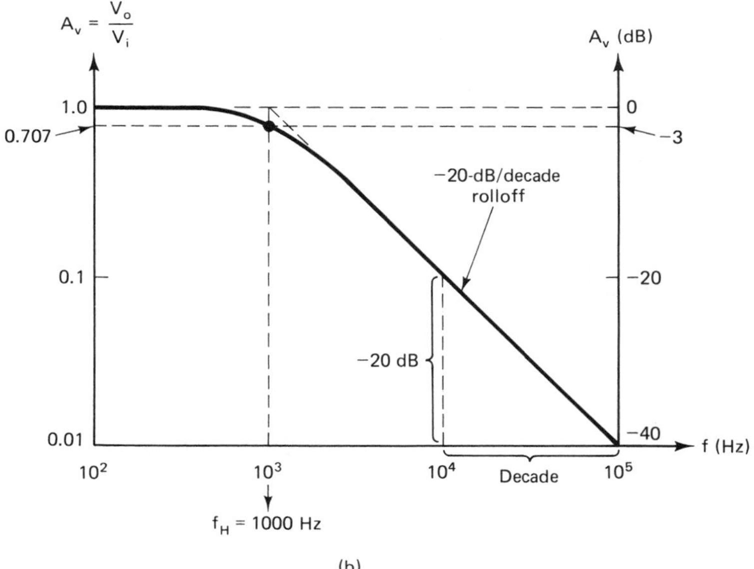

(b)

Figure 10-2 The single-pole low-pass filter in (a) transmits all frequencies *below* cutoff frequency f_H. Frequencies above f_H are attenuated as in (b). (a) Basic −20-dB/decade low-pass filter circuit; (b) frequency response for a single-pole, −20-dB/decade low-pass filter.

will approach zero. Thus low frequencies are passed from V_i to V_o and high frequencies will be attenuated.

10-2.2 Frequency Response

Our low-pass filter circuit has the frequency response shown in Fig. 10-2b. As the frequency of V_i is increased from zero, amplitude of V_o will remain constant in the passband and then begin to drop. When V_o drops to *exactly 0.707* times its low-frequency value, read the frequency of V_i. This frequency is called the *cut off* frequency, f_H. It is the highest *useful* frequency passed by the low-pass filter. Other names used to describe a cutoff frequency are: *corner,* 0.707, minus 3 dB, and half-power frequencies.

In Fig. 10-2 the frequency response curves are plotted on log-log paper for simplicity. Above cutoff frequency f_H, the voltage gain *decreases* by one-tenth (or −20 dB) when frequency is *increased* by a factor of 10. A frequency change of ×2 is called an *octave.* Gain will roll off at −6 dB per octave. A frequency change of X10 is called a *decade.* Gain rolls off at −20 dB per decade.

10-2.3 Calculating the Cutoff Frequency

For the low-pass filter of Fig. 10-2, the cutoff frequency f_H is expressed as

$$f_H = \frac{0.1591}{RC} \tag{10-1}$$

where R is in ohms and C is in farads.

EXAMPLE 10-1

(a) Calculate the cutoff frequency for the low-pass filter in Fig. 10-2a. (b) If V_i = 200 mV, find the magnitude of V_o at f_H.

SOLUTION (a) From Eq. (10-1),

$$f_H = \frac{0.1591}{RC} = \frac{0.1591}{(1 \times 10^4\ \Omega)(0.015 \times 10^{-6}\ \text{F})} = 1061\ \text{Hz} \approx 1\ \text{kHz}$$

(b) V_o will equal V_i or 200 mV at low frequencies such as 100 Hz. At f_H = 1 kHz, V_o = (0.707)(200 mV), or 141 mV. At 10 kHz ($10f_H$ = one decade), $V_o \simeq$ 20 mV.

10-2.4 Making a Low-Pass Filter

Equation (10-1) shows how to calculate cutoff frequency if you know the values of R and C in Fig. 10-2. To obtain a required low-pass cut off frequency, rewrite Eq. (10-1) as

$$R = \frac{0.1591}{f_H C} \tag{10-2}$$

The procedure to make a filter is illustrated by an example.

EXAMPLE 10-2

You need a filter that passes all frequencies up to 3000 Hz. Your company uses only 0.0033-, 0.033-, and 0.33-μF capacitors.

SOLUTION Arbitrarily pick $C = 0.33\ \mu$F. From Eq. (10-2),

$$R = \frac{0.1591}{f_H C} = \frac{0.1591}{(3000)(0.33 \times 10^{-6}\text{ F})} = 160\ \Omega$$

To avoid drawing excessive current from V_i, the value of R should exceed 2 kΩ. Therefore, make a better choice for C of 0.0033 μF. R will then equal 16 kΩ.

10-2.5 Practical Advice

Buy stable, non-polarized filter capacitors made from Mylar, polystyrene, or polypropylene dielectric material. However, capacitance value can vary by $\pm 20\%$ from the labeled value. If you need to refine f_H to *exactly* 3000 Hz, make up resistor R from a series circuit of 10 kΩ in series with a 0- to 10-kΩ variable resistor, as in Fig. 10-3. Adjust the signal generator's volume control to (1) an output of $V_i = 1.00$ V and (2) a frequency of $f_H =$ 3 kHz. Adjust R_i until V_o equals 0.707 V. The low-pass filter is now calibrated exactly for $f_H = 3000$ Hz.

10-3 BASIC HIGH-PASS FILTER

10-3.1 Circuit Operation

You make a basic high-pass filter simply by interchanging R and C in Fig. 10-2 or 10-3. The new basic high-pass filter circuit is shown in Fig. 10-4a. V_i divides between C and R. At low frequencies, the capacitive reactance is very large. Most of V_i is developed across C and very little across R. At high frequencies reactance is small and V_i is developed across R. The

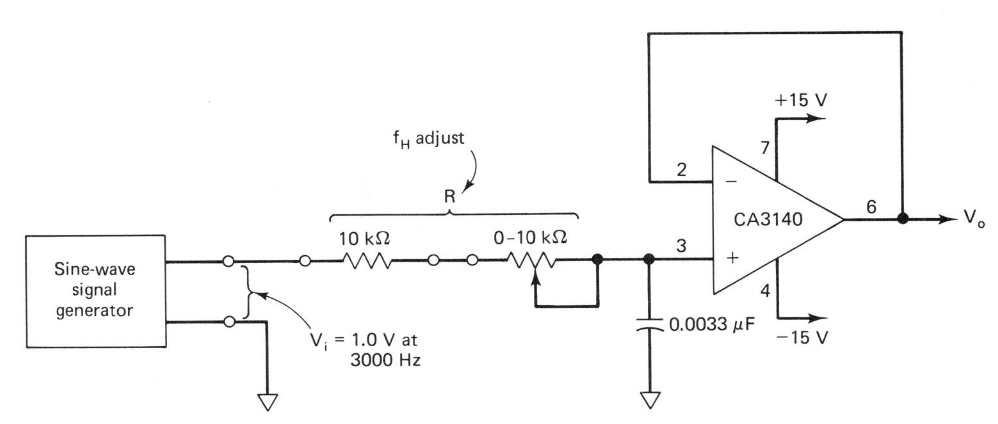

Figure 10-3 Practical solution to Example 10-2. With V_i set at 3000 Hz, adjust R until $V_o = 0.707$ V. Cutoff frequency f_H will then equal exactly 3000 Hz.

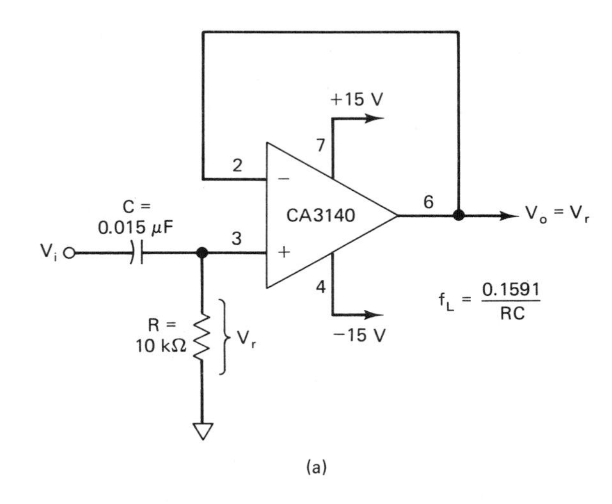

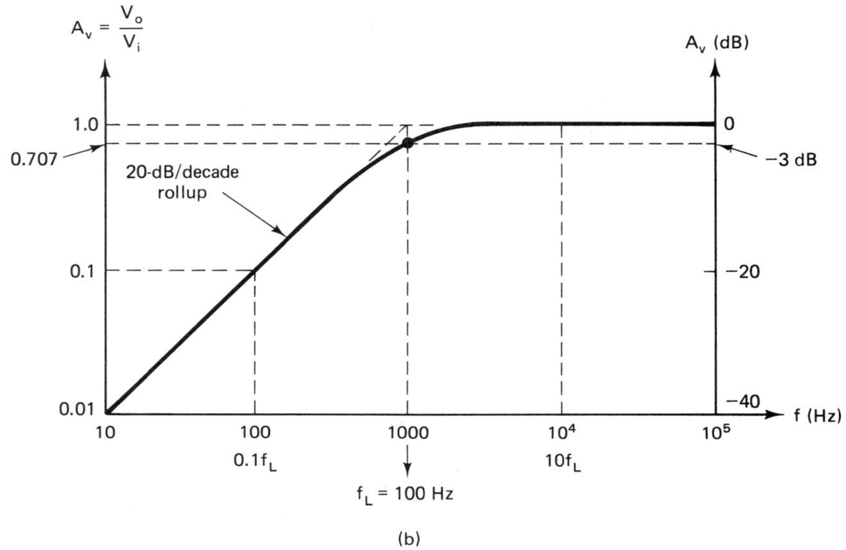

Figure 10-4 The single-pole high-pass filter in (a) transmits all frequencies *above* cutoff frequency f_L. Frequencies below f_L are attenuated as in (b). (a) Basic 20-dB/decade high-pass filter; (b) frequency response for a single-pole, 20-dB/decade high-pass filter.

CA3140 voltage follower forces V_o to equal V_r. Therefore, at high frequencies, $V_o = V_r \simeq V_i$.

10-3.2 Frequency Response

Frequency response of a single-pole (one capacitor) high-pass filter is shown in Fig. 10-4b. As frequency is increased by 10 from very low values (less than $0.01f_H$), the gain increases by a factor of 10 until the frequency exceeds the lower cutoff frequency f_L. That is, gain rolls up by 20 dB/decade or 6 dB/octave.

10-3.3 Calculating the Cutoff Frequency

Cutoff frequency f_L is calculated from

$$f_L = \frac{0.1591}{RC} \tag{10-3}$$

where f_L is the lowest *useful* frequency passed by the high-pass filter. For the filter of Fig. 10-4a, the lower cutoff frequency is

$$f_L = \frac{0.1591}{(1 \times 10^4 \ \Omega)(0.015 \times 10^{-6} \ \text{F})} \simeq 1000 \ \text{Hz}$$

We use an example to show how to make a high-pass filter.

EXAMPLE 10-3

You have available only three sizes of capacitors: 0.0033, 0.033, and 0.33 μF. Make a filter to pass all frequencies above 300 Hz.

SOLUTION From the problem statement, $f_L = 300$ Hz. Make a trial choice for C at 0.033 μF. Then rewrite Eq. (10-3):

$$R = \frac{0.1591}{f_L C} \tag{10-4}$$

$$R = \frac{0.1591}{(300)(0.033 \times 10^{-6} \ \text{F})} = 16 \ \text{k}\Omega$$

EXAMPLE 10-4

For the filter of Example 10-3, find the voltage gains at (a) f_L; (b) a frequency of $10f_L$ or 3000 Hz; (c) a frequency of $0.1f_L$ or 30 Hz.

SOLUTION (a) As shown in Fig. 10-4, the gain equals 0.707 at f_L (regardless of the value of f_L).

(b) For frequencies above f_L the gain approaches 1.

(c) Gain is reduced by 10 to 0.1 when frequency is reduced by one decade from f_L (Fig. 10-4b).

10-4 INTRODUCTION TO BANDPASS FILTERS

10-4.1 Frequency Response

A bandpass filter is a frequency selector. It allows one to select or pass only one particular band of frequencies from all other frequencies that may be present in a circuit. Its normalized frequency response is shown in Fig. 10-5.

This type of filter has a maximum gain at a resonant frequency f_r. In this chapter all bandpass filters will have a gain of 1 or 0 dB at f_r. There is one frequency below f_r where the gain falls to 0.707. It is the *lower cutoff frequency*, f_L. At *higher cutoff frequency*, f_H, the gain also equals 0.707, as in Fig. 10-5.

10-4.2 Bandwidth

The range of frequencies between f_L and f_H is called *bandwidth B*, or

$$B = f_H - f_L \tag{10-5}$$

The bandwidth is not exactly centered on the resonant frequency. (It is for this reason that we use the historical name "resonant frequency" rather than "center frequency" to describe f_r.)

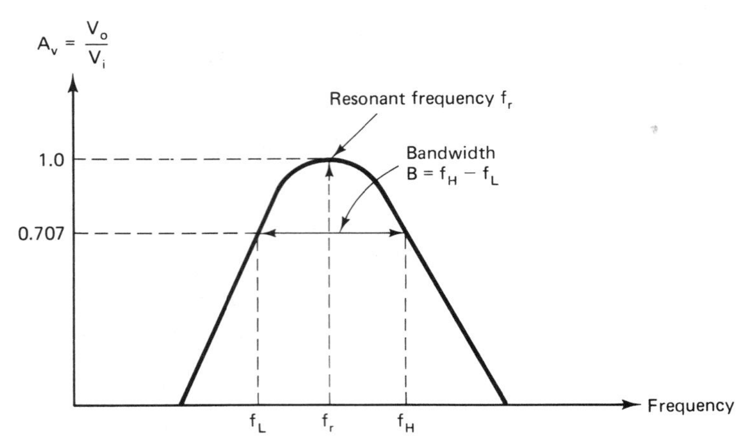

Figure 10-5 A bandpass filter has a maximum gain at resonant frequency f_r. The band of frequencies transmitted lies between f_L and f_H.

If you know the values for f_L and f_H, the resonant frequency can be found from

$$f_r = \sqrt{f_L f_H} \tag{10-6}$$

If you know the resonant frequency, f_r, and bandwidth, B, cutoff frequencies can be found from

$$f_L = \sqrt{\frac{B^2}{4} + f_r^2} - \left(\frac{B}{2}\right) \tag{10-7a}$$

$$f_H = f_L + B \tag{10-7b}$$

EXAMPLE 10-5

A bandpass voice filter has upper and lower cutoff frequencies of 300 and 3000 Hz. Find (a) the bandwidth; (b) the resonant frequency.

SOLUTION (a) From Eq. (10-5),

$$B = f_H - f_L = (3000 - 300) = 2700 \text{ Hz}$$

(b) From Eq. (10-6),

$$f_r = \sqrt{f_L f_H} = \sqrt{(300)(3000)} = 948.7 \text{ Hz}$$

[Note that f_r is always *below* the center frequency of (3000 + 300)/2 = 1650 Hz.]

EXAMPLE 10-6

A bandpass filter has a resonant frequency of 950 Hz and a bandwidth of 2700 Hz. Find its lower and upper cutoff frequencies.

SOLUTION From Eq. (10-7a),

$$f_L = \sqrt{\frac{B^2}{4} + f_r^2} - \left(\frac{B}{2}\right) = \sqrt{\frac{(2700)^2}{4} + (950)^2} - \frac{2700}{2}$$

$$= 1650 - 1350 = 300 \text{ Hz}$$

From Eq. (10-7b), $f_H = 300 + 2700 = 3000$ Hz.

10-4.3 Quality Factor

The *quality factor* Q is defined as the ratio of resonant frequency to bandwidth, or

$$Q = \frac{f_r}{B} \tag{10-8}$$

Q is a measure of the bandpass filter's *selectivity.* A high Q indicates that a filter selects a smaller band of frequencies (more selective).

10-4.4 Narrowband and Wideband Filters

A *wideband* filter has a bandwidth that is two or more times the resonant frequency. That is, $Q \leq 0.5$ for wideband filters. In general, wideband filters are made by cascading a low-pass filter circuit with a high-pass filter circuit. This topic is covered in the next section. A narrowband filter ($Q >$ 0.5) can usually be made with a single stage. This type of filter is presented in Section 10-7.

EXAMPLE 10-7

Find the quality factor of a voice filter that has a bandwidth of 2700 Hz and a resonant frequency of 950 Hz (see Examples 10-5 and 10-6).

SOLUTION From Eq. (10-8),

$$Q = \frac{f_r}{B} = \frac{950}{2700} = 0.35$$

This filter is classified as wideband because $Q < 0.5$.

10-5 BASIC WIDEBAND FILTER

10-5.1 Cascading

When the output of one circuit is connected in series with the input of a second circuit, the process is called *cascading* gain stages. In Fig. 10-6, the first stage is a 3000-Hz low-pass filter (see Example 10-2). Its output is connected to the input of a 300-Hz high-pass filter (see Example 10-3). The cascaded pair of active filters now form a bandpass filter from input V_i to output V_o. Note that it makes no difference if the high-pass is connected to the low-pass, or vice versa.

10-5.2 Wideband Filter Circuit

In general, a wideband filter ($Q < 0.5$) is made by cascading a low- and a high-pass filter (see Fig. 10-6). Cutoff frequencies of the low- and high-pass sections *must not overlap,* and each must have the same pass-band gain. Furthermore, the low-pass filter's cutoff frequency must be 10 or more times the high-pass filter's cutoff frequency.

For cascaded low- and high-pass filters, the resulting wideband filter will have the following characteristics:

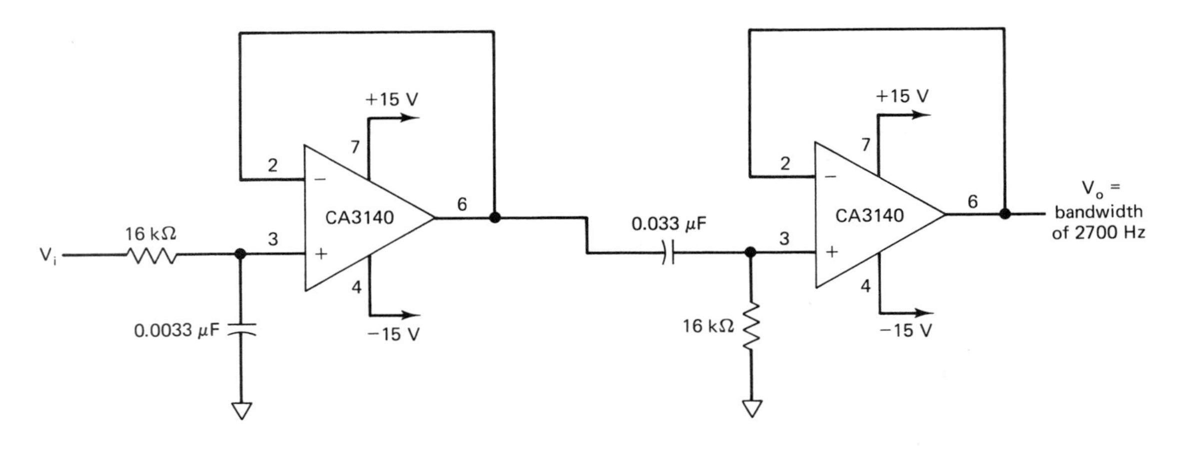

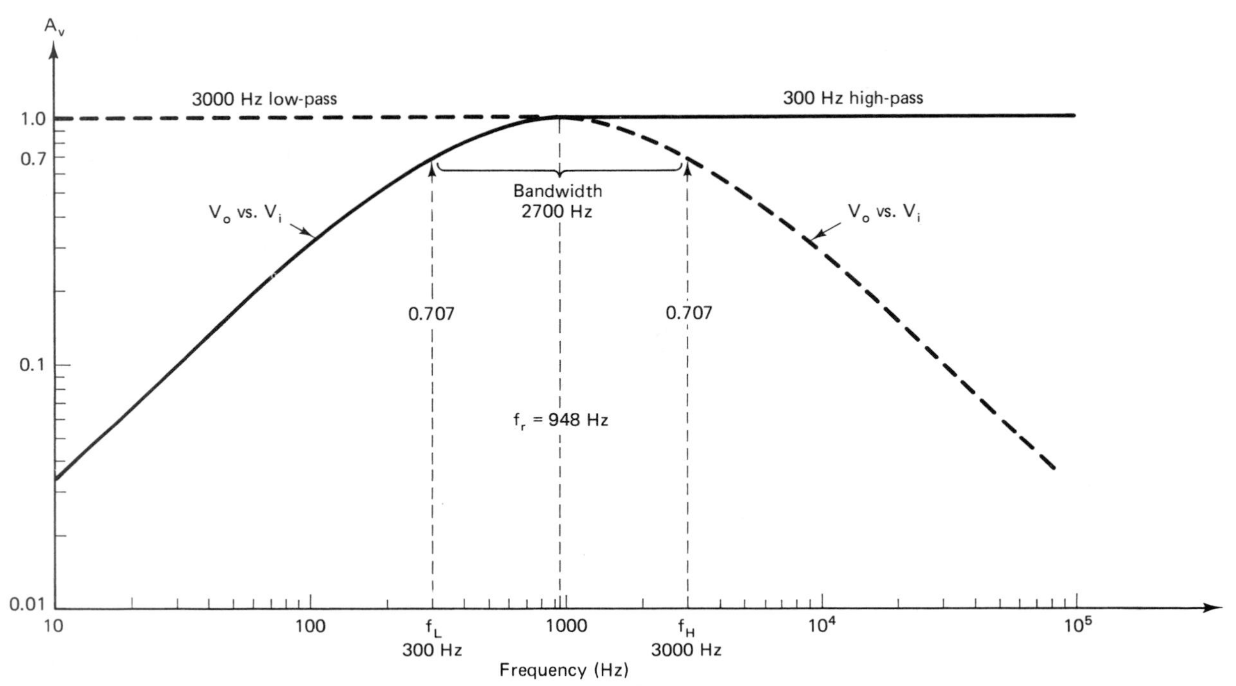

Figure 10-6 A wideband filter is made by cascading a low-pass and a high-pass filter. The high-pass filter sets f_L and the low-pass filter f_H.

1. The lower cutoff frequency, f_L, will be determined only by the high-pass filter.
2. The high cutoff frequency, f_H, will be set only by the low-pass filter.
3. Gain will be maximum at resonant frequency f_r and equal to the passband of either filter.

These principles are illustrated next.

10-5.3 Frequency Response

In Fig. 10-6, the frequency response of a basic 3000-Hz low-pass filter is plotted as a dashed line. The frequency response of 300-Hz high-pass filter is plotted as a solid line. The +20-dB/decade roll-up of the high-pass filter is seen to determine f_L. the -20-dB/decade roll-off of the low-pass sets f_H. Both roll-off curves make up the frequency response of the bandpass filter, V_o versus f. Observe that the resonant, low and high cutoff frequencies plus bandwidth agree exactly with the values calculated in Examples 10-5 and 10-6. Improved bandpass filters will be introduced in Sections 10-6.4 and 10-7. Discussion of notch filters is deferred until Sections 10-8 and 10-9.

10-6 SECOND-ORDER FILTERS

10-6.1 −40-dB/Decade Low-Pass Filter

The low-pass filter of Fig. 10-2 has one capacitor, rolls off at -20 dB/decade, and is called a first-order or single-pole filter. This filter can be improved by adding a second capacitor as in Fig. 10-7a. The two-capacitor circuit is called a second-order low-pass filter, and the roll-off is much sharper at -40 dB/decade or -12 dB/octave.

See Fig. 10-7b. When frequency is increased by 10 (above f_H), the gain is divided by 10×10 or 100. In general, each time a capacitor is properly added to a low-pass filter, the roll-off rate is increased by -20 dB/decade. Thus a three-capacitor (three-pole) filter would have a roll-off of -60 dB ($\div 1000$) per decade.

10-6.2 Low-Pass Filter Cutoff Frequency

The highest frequency passed by the second-order filter of Fig. 10-7 is called f_H and is found from

$$f_H = \frac{0.1125}{RC} \tag{10-9}$$

Voltage gain A_v equals 1 in the low passband and equals 0.707 at f_H. At a frequency of $10f_H$, gain has fallen by -40 dB to 0.01.

EXAMPLE 10-8

Find cutoff frequency f_H for the low-pass filter of Fig. 10-7.

SOLUTION From Eq. (10-9),

$$f_H = \frac{0.1125}{RC} = \frac{0.1125}{(7500\ \Omega)(0.015 \times 10^{-6}\ \text{F})} \simeq 1000\ \text{Hz}$$

EXAMPLE 10-9

Make a second-order low-pass filter with a cutoff frequency of 3000 Hz. Choose a standard capacitor for $C = 0.005\ \mu\text{F}$.

SOLUTION Rewrite Eq. (10-9) to solve for R:

$$R = \frac{0.1125}{f_H C} = \frac{0.1125}{(3000)(0.005 \times 10^{-6}\ \text{F})} = 7500\ \Omega$$

Then capacitor $2C = 0.01\ \mu\text{F}$.

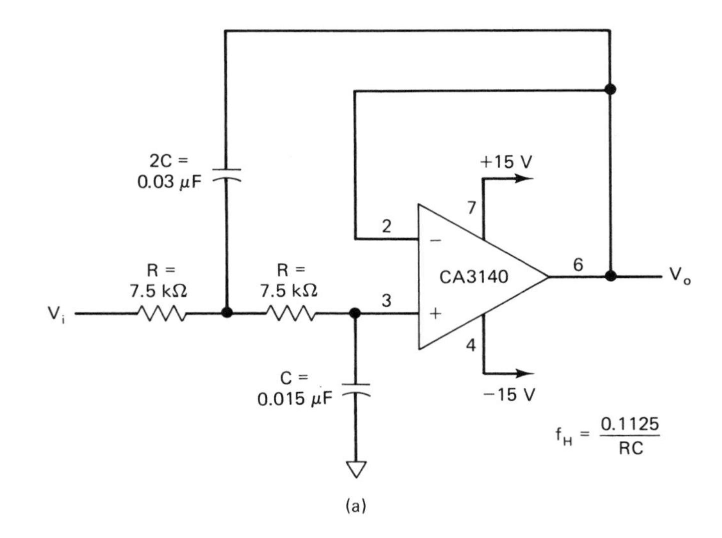

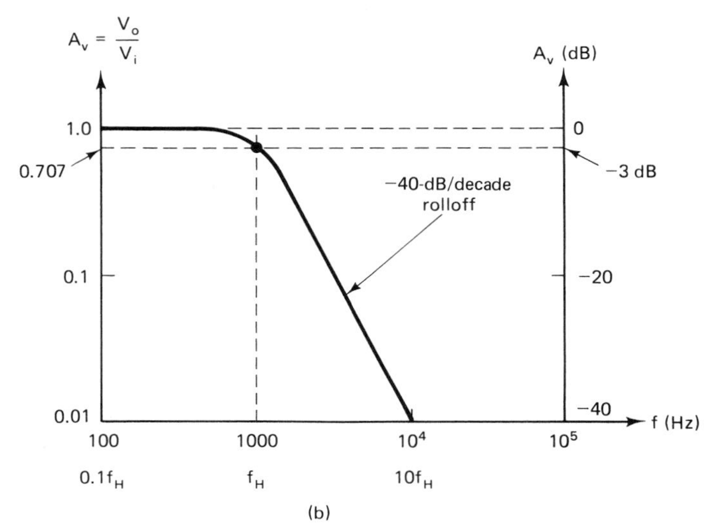

Figure 10-7 The second-order (two-capacitor) low-pass filter circuit in (a) has the frequency response curve shown in (b). For C = 0.015 μF, $2C$ = 0.03 μF, and R = 7.5 kΩ, f_H = l kHz. (a) Second-order low-pass filter; (b) frequency response of a second-order low-pass filter.

10-6.3 40-dB/Decade High-Pass Filter

The second-order high-pass filter of Fig. 10-8a has a much steeper roll-off than the basic high-pass in Fig. 10-4. As shown in Fig. 10-8b, the filter transmits all high frequencies above a lower limit f_L, where

$$f_L = \frac{0.1125}{RC} \tag{10-10}$$

At frequencies below f_L, gain is rolled off from $A_v = 1$ at a rate of 40 dB/decade. At cutoff frequency f_L, gain is 0.707. When frequency of V_i is reduced by a decade from f_L to $0.1f_L$, gain is reduced by 100 to 0.01 or −40 dB.

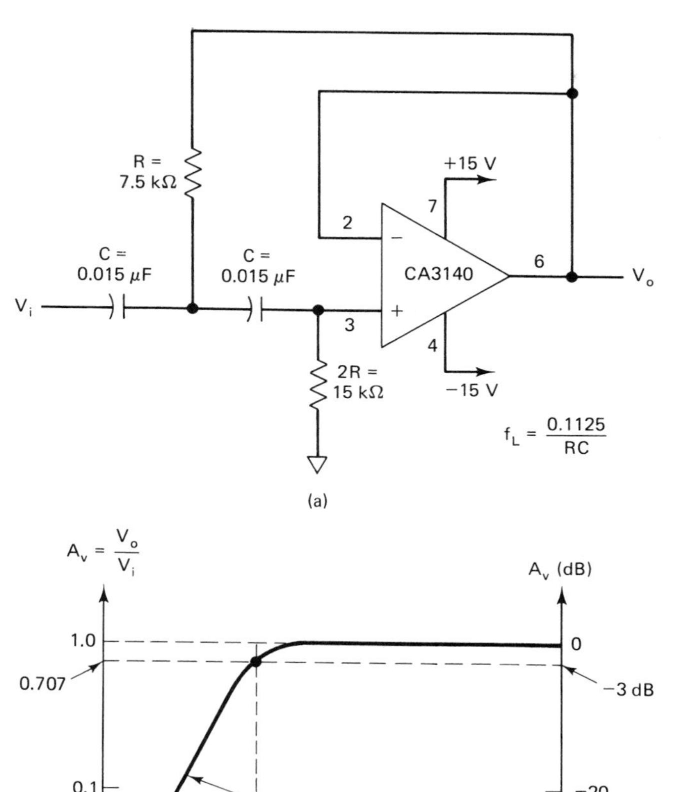

Figure 10-8 The second-order high-pass filter in (a) has the frequency response shown in (b). For $C = 0.015\ \mu\text{F}$, $R = 7.5\ \text{k}\Omega$, and $2R = 15.0\ \text{k}\Omega$, $f_L = 1\ \text{kHz}$. (a) Second-order high-pass filter; (b) frequency response of a second-order high-pass filter.

EXAMPLE 10-10

Find the lower cutoff frequency, f_L, for the high-pass filter of Fig. 10-8.

SOLUTION From Eq. (10-10),

$$f_L = \frac{0.1125}{RC} = \frac{0.1125}{(7500\ \Omega)(0.015 \times 10^{-6}\ \text{F})} \simeq 1\ \text{kHz}$$

EXAMPLE 10-11

Make a second-order filter with a low cutoff frequency of $f_L = 300$ Hz. Choose $C = 0.05\ \mu\text{F}$.

SOLUTION Rewrite Eq. (10-10) to solve for R:

$$R = \frac{0.1125}{f_L C} = \frac{0.1125}{(300)(0.05 \times 10^{-6}\ \text{F})} \simeq 7500\ \Omega$$

Then resistor $2R = 15\ \text{k}\Omega$.

10-6.4 Wideband Bandpass Filter

We cascade the second-order low- and high-pass filters of Examples 10-9 and 10-11 to obtain a good voice filter that passes frequencies in the range 300 to 3000 Hz. The complete circuit and frequency response is shown in Fig. 10-9.

The high-pass filter establishes the low cutoff frequency at $f_L = 300$ Hz and up off at 40 dB/decade. The low-pass filter sets the high cutoff frequency at $f_H = 3000$ Hz and rolls off at -40 dB/decade. This cascaded pair forms a bandpass filter that passes all frequencies between 300 and 3000 Hz. This passband is shown by the curves labeled V_o versus f in Fig. 10-9.

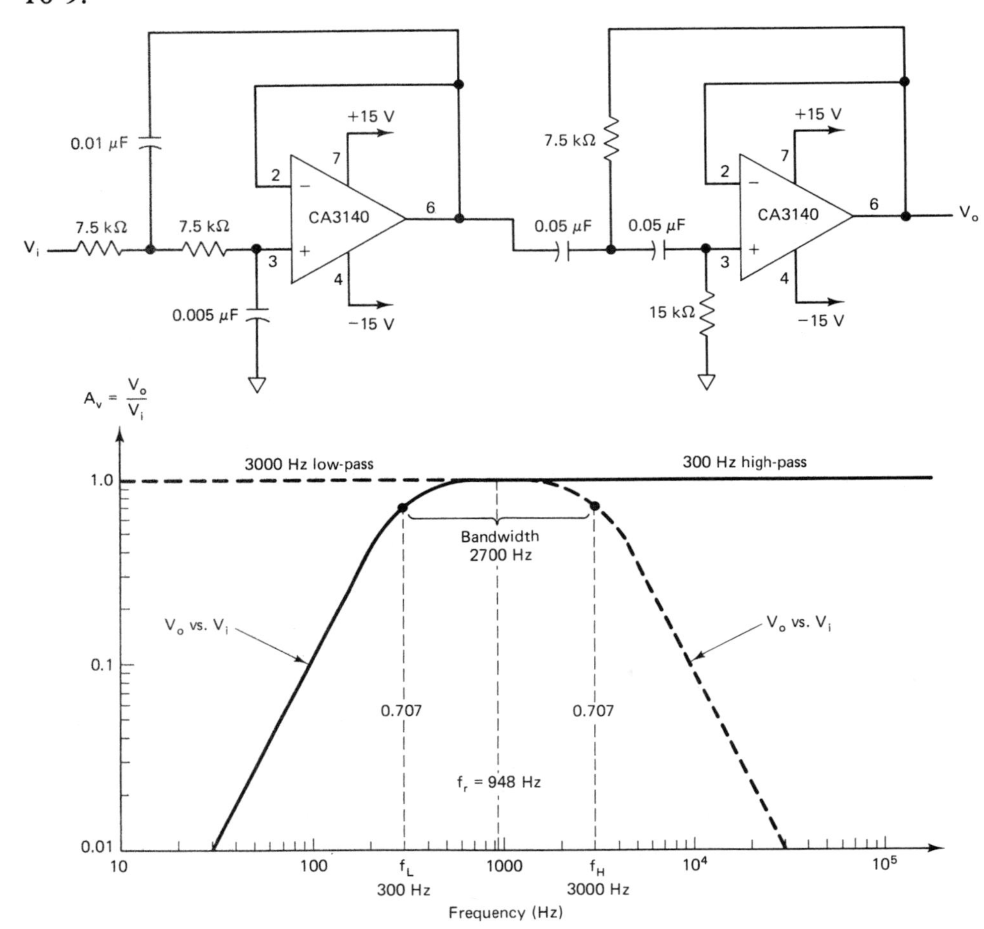

Figure 10-9 A 3000-Hz second-order low-pass filter is cascaded with a 300 Hz high-pass filter to form a 300- to 3000-Hz bandpass voice filter.

10-7 NARROWBAND BANDPASS FILTERS

Narrowband filters exhibit the typical frequency response shown in Fig. 10-10a. The analysis and construction of narrowband filters is considerably simplified if we stipulate that the narrowband filter will have a maximum gain of 1 or 0 dB at the resonant frequency f_r. Equations (10-5) to (10-7) and bandpass terms were presented in Section 10-4.1. They gave

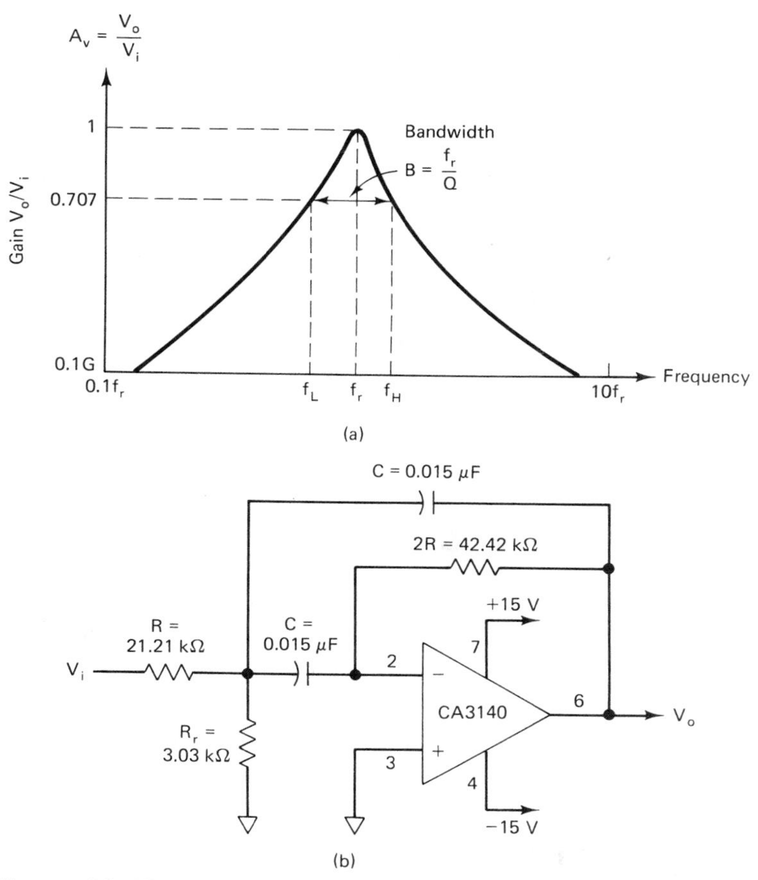

Figure 10-10 Narrow bandpass filter circuit and its frequency response for the component values shown; $f_r = 1000$ Hz, $B = 500$ Hz, $Q = 2$, $f_L = 780$ Hz, and $f_H = 1280$ Hz. (a) Typical frequency response of a bandpass filter; (b) narrow bandpass filter circuit.

an introduction to (cascaded pair) wideband filters. These equations and terms also apply to the narrowband filters that follow.

10-7.1 Narrowband Filter Circuit

A narrowband filter circuit uses only one op amp, as shown in Fig. 10-10b. (Compare with the two op-amp wideband filters in Figs. 10-6 and 10-9.) The filter's input resistance is established approximately by resistor R. If the feedback resistor ($2R$) is made two times the input resistor R, the filter's maximum gain will be 1 or 0 dB at resonant frequency f_r. By adjusting R_R, one can change (or exactly trim) the resonant frequency *without changing bandwidth or gain.*

10-7.2 Performance

Performance of the *unity*-gain narrowband filter in Fig. 10-10 is determined by only a few simple equations. Bandwidth B in hertz is determined by resistor R and the two (matched) capacitors C by

$$B = \frac{0.1591}{RC} \qquad (10\text{-}11a)$$

where

$$B = \frac{f_r}{Q} \tag{10-11b}$$

Gain is a maximum of 1 at f_r provided that feedback resistor $2R$ is twice the value of input resistor R.

Resonant frequency f_r is determined by resistor R_r according to

$$R_r = \frac{R}{2Q^2 - 1} \tag{10-11c}$$

If you are given component values for the circuit, its resonant frequency can be calculated from

$$f_r = \frac{0.1125}{RC}\sqrt{1 + \frac{R}{R_r}} \tag{10-11d}$$

10-7.3 Stereo-Equalizer Octave Filter

A stereo equalizer has 10 bandpass filters per channel. They separate the audio spectrum from approximately 30 Hz to 16 kHz into 10 separate octaves of frequency. Each octave can then be cut or boosted with respect to the other to achieve special sound effects, equalize room response, or equalize an automotive compartment to make the radio sound like it is playing in a large hall. The construction of one such equalizer will be analyzed by an example.

EXAMPLE 10-12

Octave equalizers have resonant frequencies at approximately 32, 64, 128, 250, 500, 1000, 2000, 4000, 8000, and 16,000 Hz. Q of each filter is chosen to have values between 1.4 and 2. Let's make a unity-gain narrowband filter to select the sixth octave. Specifically, make a filter with $f_r = 1000$ Hz and $Q = 2$.

SOLUTION From Eq. (10-11b),

$$B = \frac{f_r}{Q} = \frac{1000}{2} = 500 \text{ Hz}$$

[*Note:* From Eqs. (10-7), $f_L = 80$ and $f_H = 1280$ Hz.] Choose $C = 0.015$ μF. Find R from Eq. (10-11a).

$$R = \frac{0.1591}{BC} = \frac{0.1591}{(500)(0.015 \times 10^{-6}\text{ F})} = 21.21 \text{ k}\Omega$$

The feedback resistor will be $2R = 42.42$ kΩ. Find R_r from Eq. (10-11c).

$$R_r = \frac{R}{2Q^2 - 1} = \frac{21.21 \text{ k}\Omega}{2(2)^2 - 1} = \frac{21.21 \text{ k}\Omega}{7} = 3.03 \text{ k}\Omega$$

EXAMPLE 10-13

Given a bandpass filter circuit with the component values in Fig. 10-10b. Find (a) the resonant frequency; (b) the bandwidth.

SOLUTION (a) From Eq. (10-11d),

$$f_r = \frac{0.1125}{RC}\sqrt{1 + \frac{R}{R_r}} = \frac{0.1125}{(21.21 \times 10^3)(0.015 \times 10^{-6})}\sqrt{1 + \frac{21.21 \text{ k}\Omega}{3.03 \text{ k}\Omega}}$$

$$= (353.6 \text{ Hz})\sqrt{1 + 7} = 353.6 \text{ Hz} \times 2.83 \simeq 1000 \text{ Hz}$$

(b) From Eq. (10-11a),

$$B = \frac{0.1591}{RC} = \frac{0.1591}{(21.21 \times 10^3)(0.015 \times 10^{-6})} = 500 \text{ Hz}$$

10-8 NOTCH FILTERS

The notch or band-reject filter is named for the characteristic shape of its frequency response curve in Fig. 10-11. Unwanted frequencies are attenuated in the stopband B. The desired frequencies are transmitted in the passband that lies on either side of the notch.

Notch filters usually have a passband gain of unity or 0 dB. The equations for Q, B, f_L, f_H, and f_r are identical to those of its associated bandpass filter. The reasons for this last statement will be presented next.

10-8.1 Notch Filter Theory

As shown in Fig. 10-12, a notch filter is made by subtracting the output of a bandpass filter from the original signal. For frequencies in the notch filter's passband, the output of the bandpass filter section approaches zero. Therefore, input V_i is transmitted via adder input resistor R_1 to drive V_o to a value equal to $-V_i$. Thus $V_o = -V_i$ in both lower and upper passbands of the notch filter.

Suppose that the frequency of V_i is adjusted to resonant frequency f_r of the narrow bandpass filter component. (*Note:* f_r of the bandpass sets the notch frequency.) V_i will exit from the bandpass as $-V_i$ and then is inverted by R_1 and R to drive V_o to $+V_i$. However, V_i is transmitted via R_2 to drive V_o to $-V_i$. Thus V_o responds to both inputs of the adder and becomes $V_o = V_i - V_i = 0$ v at f_r.

In practice, V_o only approaches zero at f_r. The depth of the notch depends on how closely the resistors and capacitors are matched in the bandpass filter and judicious fine adjustment of resistor R_1 at the inverting adder's output. This procedure is explained in Section 10-9.4

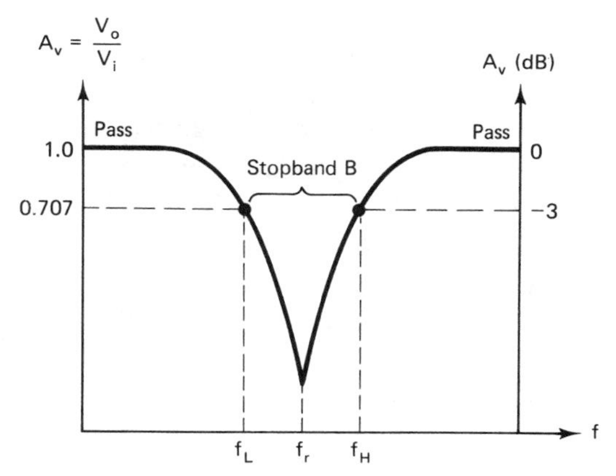

Figure 10-11 A notch filter transmits frequencies in the passband and rejects undesired frequencies in the stopband.

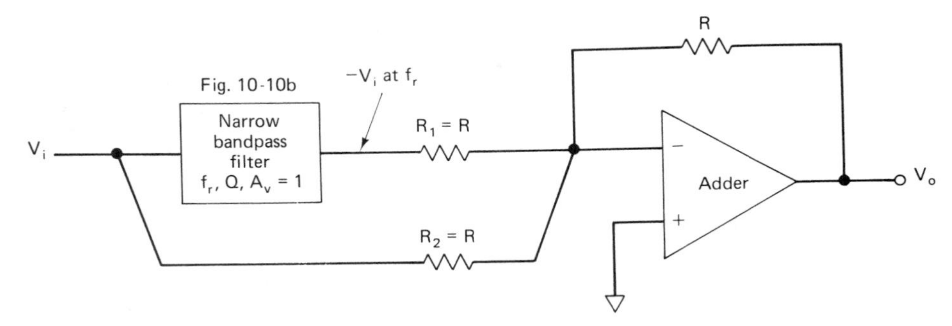

Figure 10-12 A notch filter is made by a circuit that subtracts the output of a bandpass filter from the original signal.

10-9 120-HZ NOTCH FILTER

10-9.1 Need for a Notch Filter

In applications where low-level signals must be amplified, there may be present one or more of an assortment of unwanted noise signals. Examples are 50-, 60-, or 400-Hz frequencies from power lines, 120-Hz, ripple from full-wave rectifiers, or even higher frequencies from regulated switching-type power supplies or clock oscillators.

If both signals and a single-frequency noise component are passed through a notch filter, only the desired signals will exit from the filter. The noise frequency is "notched out." As an example, let us make a notch filter to eliminate 120-Hz hum.

10-9.2 Statement of the Problem

The problem is to make a notch filter with a notch (resonant) frequency of $f_r = 120$ Hz. Let us select a stopband of $B = 12$ Hz. Gain of the notch filter in the passband will be unity (0 dB), so that the desired signals will be transmitted without attenuations.

We use Eq. (10-11b) to determine a value for Q that is required by the notch filter:

$$Q = \frac{f_r}{B} = \frac{120}{12} = 10$$

This high value of Q means that (1) the notch and component bandpass filter will have narrow bands with very sharp frequency response curves, and (2) the bandwidth is essentially centered on the resonant frequency. Accordingly, this filter will transmit all frequencies from 0 to $(120 - 6) = 114$ Hz, and also all frequencies above $(120 + 6) = 126$ Hz. The notch filter will stop all frequencies between 114 and 126 Hz.

10-9.3 Procedure to Make a Notch Filter

The procedure to make a notch filter is performed in two steps:

1. Make a bandpass filter that has the same resonant frequency, bandwidth, and consequently Q as the notch filter.

2. Connect the inverting adder of Fig. 10-12 by selecting equal resistors for R. Usually, R = 10 kΩ. (A practical fine tuning procedure is presented in the next section.)

10-9.4 Bandpass Filter Component

The first step in making a 120-Hz notch filter is illustrated by an example (see Fig. 10-13).

EXAMPLE 10-14

Design a bandpass filter with a resonant frequency of f_r = 120 Hz and a bandwidth of 12 Hz so that Q = 10. Thus gain of the bandpass section will be 1 at f_r and approach zero at the output of the notch labeled V_o.

SOLUTION Choose C = 0.33 μF. From Eq. (10-11),

$$R = \frac{0.1591}{BC} = \frac{0.1591}{(12)(0.33 \times 10^{-6})} = 40.2\ \text{k}\Omega$$

Then the bandpass feedback resistor will be $2R$ equals 80.4 kΩ.

$$R_r = \frac{R}{2Q^2 - 1} = \frac{40.2\ \text{k}\Omega}{2(10)^2 - 1} = \frac{40.2\ \text{k}\Omega}{199} = 201\ \Omega$$

This bandpass filter component is built first and f_r is fine-tuned by adjusting R_r (see Section 10-7.2).

10-9.5 Final Assembly

Please refer to Fig. 10-13. Simply connect an inverting adder (CA3140B or TL081) with equal 1% input and feedback 10-kΩ resistors as shown. The resultant notch filter (from V_i to V_o) exhibits a respectable performance that is an acceptable solution to the problem. The notch depth can be increased by fine trimming R_1 or R_2.

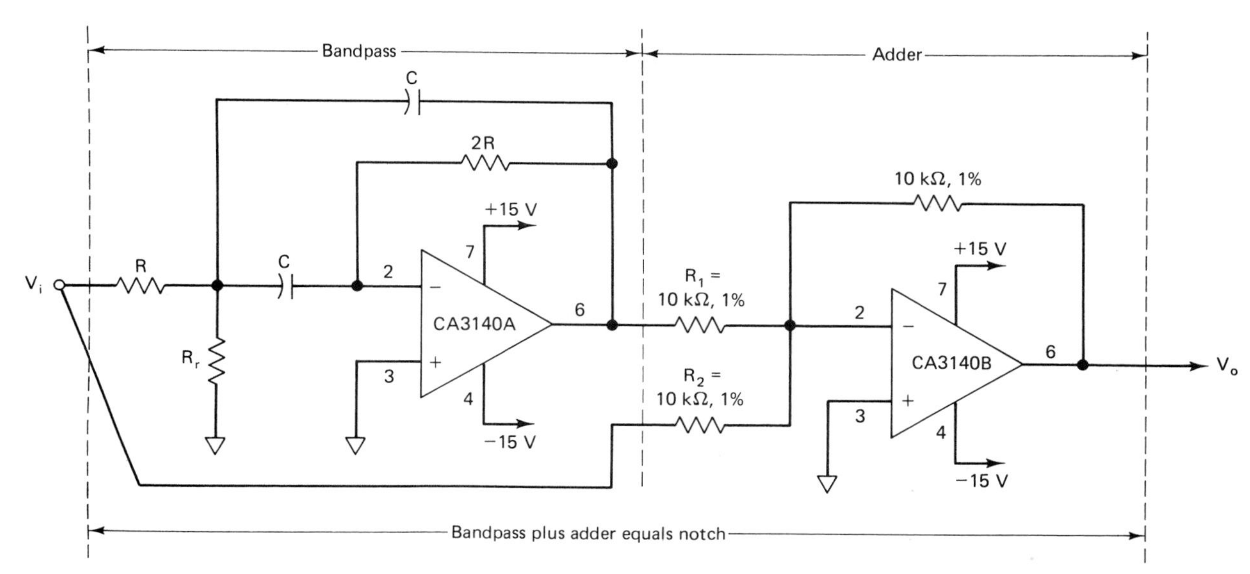

Figure 10-13 This two-op-amp notch filter is made from a bandpass filter plus an inverting adder. If C = 0.33 μF, R = 40.2 kΩ, and R_r = 201 Ω, the notch frequency will be 120 Hz and reject a bandwith of 12 Hz.

End of Chapter Exercises

Name: ____________________

Date: __________ Grade: ___

A. FILL IN THE BLANKS

Fill in the blanks with the best answer.

1. Another name for cutoff frequency is __________.
2. A single-pole low-pass filter will roll-off at the rate of __________ per decade above f_H.
3. A single-pole high-pass filter rolls-up at the rate of __________ per octave.
4. Figure 10-5 illustrates a bandpass filter. The range of frequencies between f_L and f_H is called the filter's __________.
5. Q is a measure of bandpass filter's selectivity. Q is defined as the ratio of resonant frequency to __________.
6. Second-order high-pass and low-pass filters have much steeper roll-offs than single-pole filters. Second-order filters require __________ capacitors.
7. Bandpass filter with a $Q > 0.5$ are called __________ filters while those with a $Q < 0.5$ are called __________ filters.
8. The maximum gain of a wideband filter will occur at the __________ frequency.
9. Refer to Fig. 10-10. The resistor R_r can be changed to adjust the __________ frequency.
10. Notch filters, such as shown in Fig. 10-11 are used to __________ unwanted frequencies.

B. TRUE/FALSE

Fill in **T** if the statement is true, and **F** if any part of the statement is false.

1. ______ A single-pole filter has one capacitor.

2. ______ Low-pass filters reject frequencies above f_H.

3. ______ The bandwidth of a bandpass filter is exactly centered on the resonant frequency, f_r.

4. ______ Wideband filters, made from cascading a low-pass and a high-pass filter, must have cutoff frequencies at least a decade (factor of ten) apart.

5. ______ If in Fig. 10-10 R_r is changed both the bandwidth and resonant frequency will be changed.

C. CIRCLE THE CORRECT ANSWER

Circle the correct answer for each statement.

1. Filters are classified according to the band of frequencies that they (pass, reject).
2. Double-pole low-pass filters roll-off at the rate of (−20 dB, −40 dB) per decade.
3. High-pass filters have a cutoff frequency labeled as (f_L, f_H).
4. A narrow band filter, usually made with a single op amp should have a Q (greater, less) than 0.5.
5. If R in narrow bandpass filter of Fig. 10-10 is increased from 21.21 kΩ to 42.42 kΩ the bandwidth will (increase, decrease).

D. MATCHING

Match the name or symbol in column **A** with the statement that matches best in column **B**.

	COLUMN A		COLUMN B
1. ______	sensitivity	**a.**	f_H-f_L
2. ______	resonant frequency	**b.**	−40 dB/decade
3. ______	bandwidth	**c.**	f_r/B
4. ______	roll-off	**d.**	f_r
5. ______	roll-up	**e.**	+20 dB/decade

PROBLEMS

10-1. Name the four categories of active filters.

10.2. A low-pass filter has a cutoff frequency of $f_H = 1000$ Hz. What range of frequencies are
- **(a)** transmitted in the passband;
- **(b)** attenuated in the stopband?

10-3. A lowpass single-capacitor active filter has a gain of 1 in the passband, a cutoff frequency of 1000 Hz, and a roll-off of −20 dB/decade. If the input signal is a sine wave with a value of 1.0 V (rms), find the output voltage at
- **(a)** 100 Hz;
- **(b)** 1000 Hz;
- **(c)** 10,000 Hz.

10-4. A high-pass second-order filter has a gain of 1 in the passband, a cutoff frequency of 1000 Hz, and an input voltage of 1.0 V rms.
- **(a)** What is its roll-off in dB/decade? Find the output voltage at
- **(b)** 10,000 Hz;
- **(c)** 1000 Hz;
- **(d)** 100 Hz.

10-5. Refer to the circuit of Fig. P10-5.
- **(a)** Is the op amp *A* circuit, a low-pass filter, or a high-pass filter?
- **(b)** Is the op-amp *B* circuit a low-pass filter or a high-pass filter?
- **(c)** What is the order of each filter and the roll-offs?
- **(d)** When op amp *A* is cascaded with op amp *B,* what type of filter results?

10-6. What is the passband gain and cutoff frequency for
- **(a)** op amp *A* in Fig. P10-5;
- **(b)** op amp *B?*
- **(c)** What are their gains at f_L and f_H?

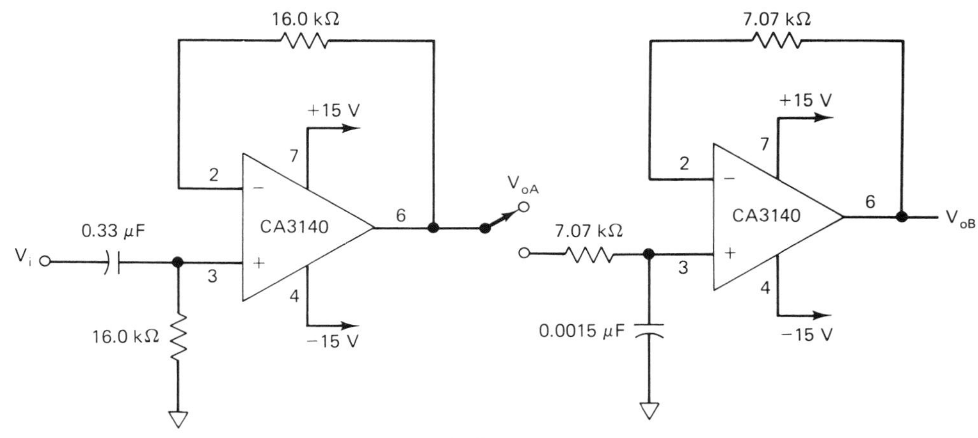

Figure P10-5

10-7. If op amps A and B are cascaded in Fig. P10-5, identify the
- **(a)** passband of frequencies;
- **(b)** stopbands of frequencies.
- **(c)** Find the resonant frequency f_r.
- **(d)** Find the value of Q.

10-8. **(a)** Identify the circuit of Fig. P10-8. Calculate its
- **(b)** cutoff frequency;
- **(c)** roll-off;
- **(d)** passband of frequences;
- **(e)** passband gain.

10-9. For the circuit of Fig. P10-9, find
- **(a)** the cutoff frequency;
- **(b)** passband of frequencies:
- **(c)** passband gain;
- **(d)** gain at the cutoff frequency

10-10. Refer to the bandpass filter of Fig. 10-10 plus Examples 10-12 and 10-13. If capacitors C are doubled, what changes occur in
- **(a)** resonant frequency;
- **(b)** bandwidth;
- **(c)** Q?

10-11. Again refer to the bandpass filter of Fig. 10-10 and Example 10-13. If resistor R_r is doubled, find the effect on
- **(a)** resonant frequency;
- **(b)** bandwidth;
- **(c)** Q.

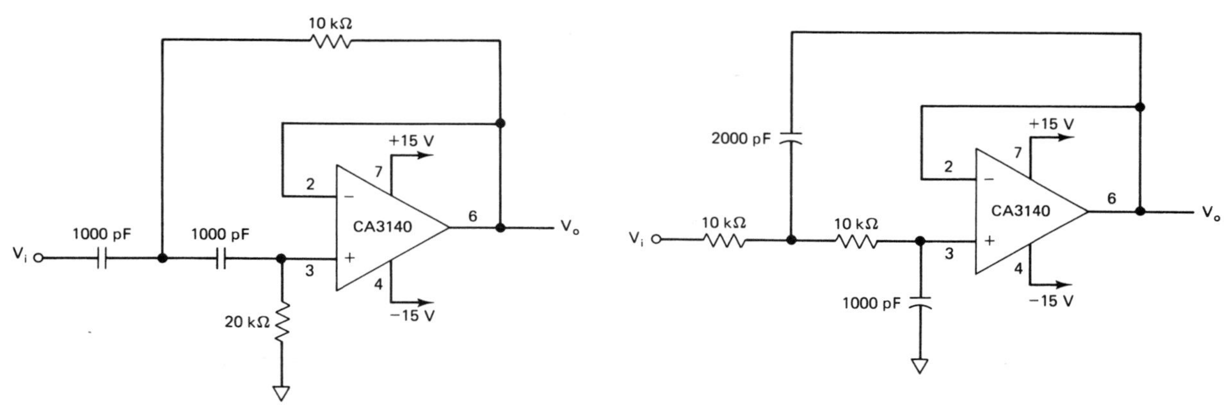

Figure P10-8

Figure P10-9

Laboratory Exercise 10

Name: ____________________

Date: __________ **Grade:** ___

ACTIVE FILTER CIRCUITS

OBJECTIVES: Upon completion of this laboratory exercise on active filters, you will be able to (1) make a single-pole low or high-pass filter and measure its frequency response; (2) make a second-order low- or high-pass filter and measure its frequency response; (3) build a band pass filter and (4) add to it an inverting adder to make a notch filter.

REFERENCE: Chapter 9

PARTS LIST

2	CA3140 (or TL081) op amps	1	0.005-μF capacitor
1	56 Ω resistor	1	0.01-μF capacitor
2	7.5-kΩ1% resistors	2	0.05-μF capacitors
6	10-kΩ1% resistors	1	0.015-μF capacitor
1	15-kΩ1% resistor	2	0.15-μF capacitors (matched)

Procedure A: Measuring Cutoff Frequency and Frequency Response of a −20-dB/Decade Low-Pass Filter

1. Before you build the low-pass filter circuit of Fig. L10-1, measure the resistance and capacitance values for R and C. R = __________; C = __________. *Throughout this laboratory exercise,* the signal V_i should be a sine wave from a low-impedance (50-Ω) function or sine-wave generator. The amplitude of V_i is to be kept at 1.0 V (rms). All voltages should be measured with an analog voltmeter. DMMS are *essentially useless* when you are looking for a null or a peak. Also, their frequency response may be inadequate.
2. Build the circuit of Fig. L10-1. Adjust the frequency of V_i to 50 Hz. Measure the value of V_o = __________. Record this value in Fig. L10-1. Multiply your measured value of V_o by 0.707 to obtain V_{oH}.

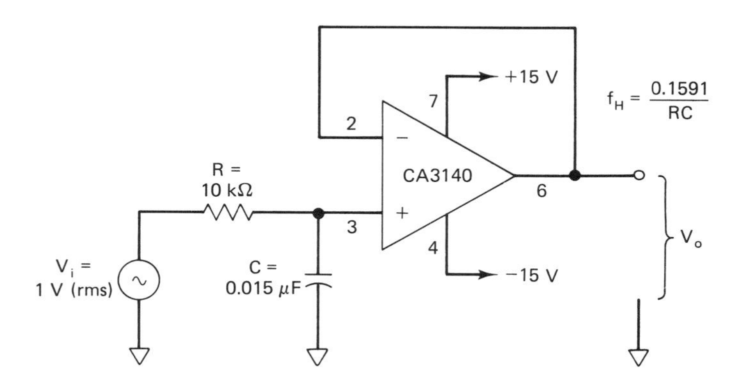

Frequency (Hz)	50	100	200	500	1 k	2 k	5 k	10 k	100 k
V_o									

Figure L10-1 Circuit and data table for a 1000-Hz −20-dB/decade low-pass filter.

V_{oH} = __________. Increase the frequency of V_i until $V_o = V_{oH}$. Measure the frequency of V_i and call it cutoff frequency. f_H = __________. Use a CRO or more conveniently a frequency counter to measure frequency. Does f_H equal within ±5% the predicted value obtained from the equation in Fig. L10-1?

__________.

3. Measure V_o at each frequency listed in Fig. L10-1 and record each value in Fig. L10-1. Plot V_o versus frequency f as a solid line on Fig. L10-2. [*Note:* Since V_i = 1.0 V rms and gain = V_o/V_i, the V_o plot is also a plot of gain versus frequency (Bode plot).]

Procedure B: Measuring Cutoff Frequency and Frequency Response of a 20-dB/Decade High-Pass Filter

4. Build the high-pass filter circuit of Fig. L10-3 with the *same R* and *C* used in step 2. Set V_i to 1.0 V (rms) at 5000 Hz. Measure V_o. V_o = __________. Multiple 0.707 × V_o to obtain V_{oL}. V_{oL} = __________. Reduce the frequency of V_i until $V_o = V_{oL}$. Record this value of frequency. f_L = __________. Is this value of f_L reasonably equal to the value of f_H measured in step 2?

   __________.

5. Calculate the predicted values of f_H and f_L from $f_H = f_L = \frac{0.1591}{RC}$. f_H = __________; f_L = __________. Account for any significant differences between measured and predicted values for f_L and f_H.

   ____

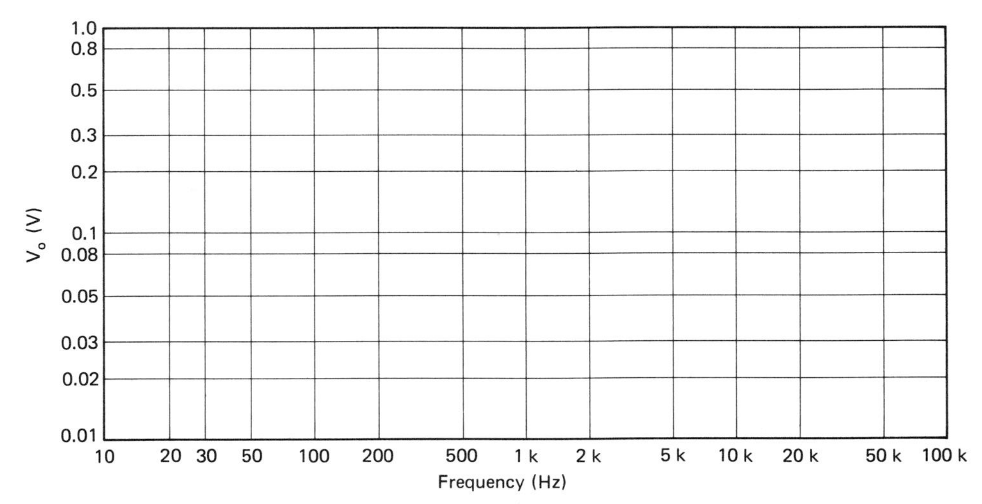

Figure L10-2 Frequency response curves for the low-pass filter of Fig. L10-1 and the high-pass filter of Fig. L10-3.

6. Adjust the frequency of V_i to the values given in Fig. L10-3 and record the corresponding values of V_o. Plot V_o versus f as a dashed line on Fig. L10-2.

Procedure C: Measuring the Frequency Responses of Second-Order Low-Pass and High-Pass Filters

7. Build the low-pass filter of Fig. L10-4. Adjust V_i for 1.0 V (rms) at 100 Hz. Measure V_o. V_o = ______. Calculate $V_{oH} = 0.707 \times V_o$. V_{oH} = ______. Increase frequency until $V_{oH} = V_{oH}$ and measure upper cutoff frequency f_H. f_H = ______. Adjust the frequency of V_i to each value in Fig. L10-4 and record corresponding

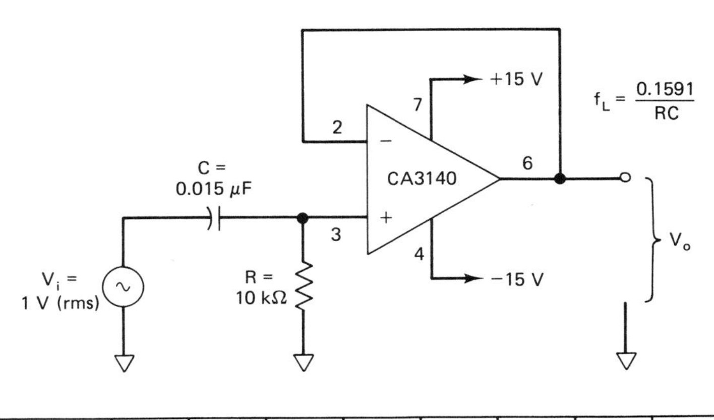

Frequency (Hz)	50	100	200	500	1 k	2 k	5 k	10 k	100 k
V_o									

Figure L10-3 Circuit and data table for a 1000-Hz 20-dB/decade high-pass filter.

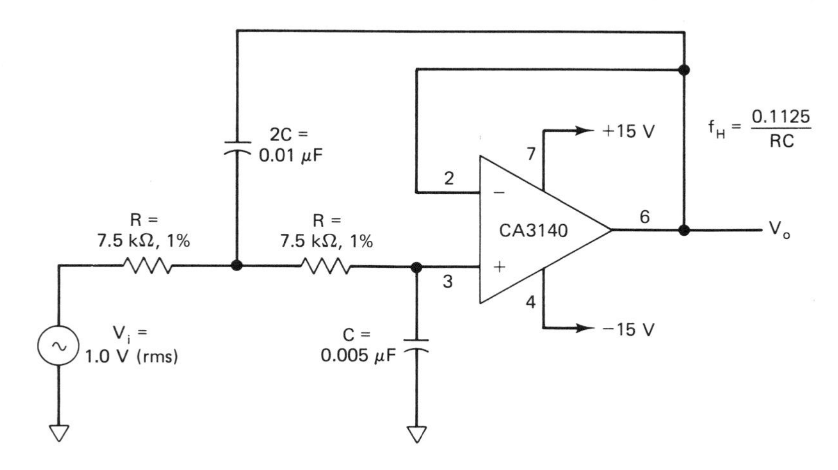

Frequency (Hz)	100	1 k	2 k	3 k	5 k	10 k	20 k	30 k
V_o								

Figure L10-4 Second-order low-pass filter. Resistor and capacitor values must equal those shown to within ±1% to obtain agreement with the equation for f_H.

values of V_o. Plot these points to construct the frequency response curve as a *dashed* line in Fig. L10-6.

8. Build the high-pass filter of Fig. L10-5. Adjust V_i to 1.0 V rms at 10 kHz. Measure V_o. V_o = __________. Calculate $V_{oL} = 0.707 \times V_o$. V_{oL} = __________. Reduce frequency until $V_o = V_{oL}$ and record the lower cutoff frequency. f_L = __________. Take data to complete the table in Fig. L10-5 and plot the frequency response curve as a solid line in Fig. L10-6.

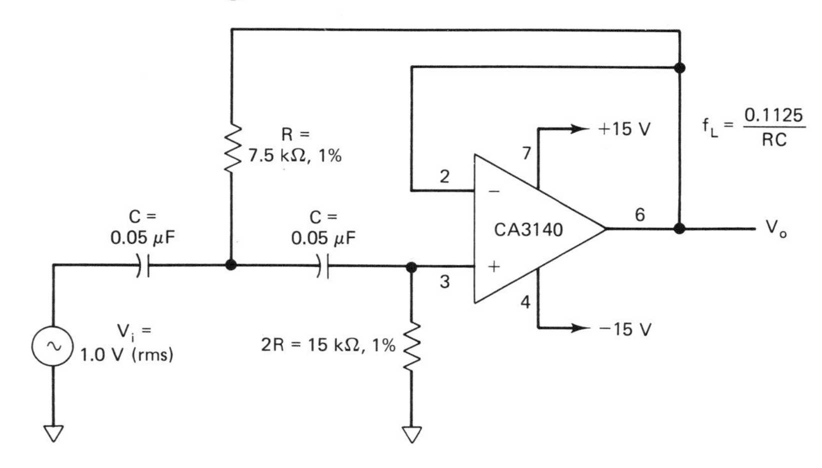

Frequency (Hz)	50	100	200	300	1 k	3 k	10 k	30 k
V_o								

Figure L10-5 Second-order high-pass filter *R* and *C* values must equal those shown within ±1% for agreement with the equation for f_L.

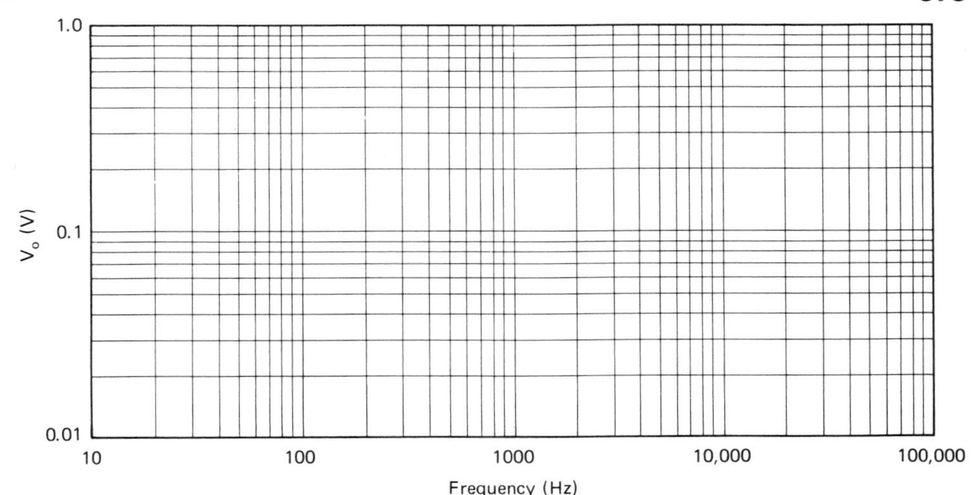

Figure L10-6 Frequency response curves for the second-order low- and high-pass filters of Fig. L10-4 and L10-5.

9. Calculate predicted values for f_L and f_H in Fig. L10-4 and L10-5 from the equation $f_L = f_H = 0.1125/RC$. f_L = __________; f_H = __________. Account for any significant differences between measured and predicted values for f_L and f_H. ________________

____

Procedure D; Measuring Frequency Response of a BandPass Filter

10. Refer to Fig. L10-7. Measure and record the *average* value for the R resistors and the *average* value of the C capacitors. R = __________; C = __________. Measure the value of R_r.

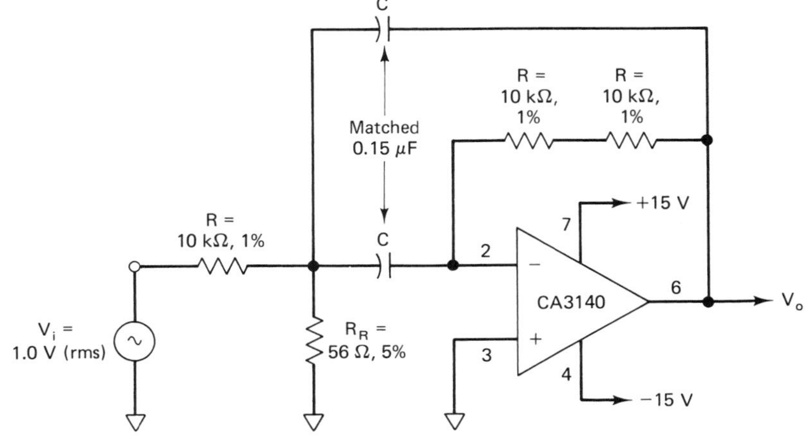

Figure L10-7 Narrow bandpass filter. The R resistors must be 1% tolerance. Capacitors C should be selected to match within ±1%.

R_r = __________. Using these measured values, predict the bandpass filters performance as follows, using measured values for R, C, and R_r: (a) Predict bandwith B from Eq. (10-12a):

$$B = \frac{0.1591}{RC} \qquad B = \_\_\_\_\_\_\_\_\_\_$$

(b) Predict resonant frequency f_r from Eq. (10-12):

$$f_r = \frac{0.1125}{RC}\sqrt{1 + \frac{R}{R_r}} \qquad f_r = \_\_\_\_\_\_\_\_\_\_$$

(c) Find the lower cutoff frequency f_L from Eq. (10-7a):

$$f_L = \sqrt{\frac{B^2}{4} + f_r^2} - \frac{B}{2} \qquad f_L = \_\_\_\_\_\_\_\_\_\_$$

(d) Find upper cutoff frequency f_H from Eq. (10-7b):

$$f_H = f_L + B \qquad f_H = \_\_\_\_\_\_\_\_\_\_$$

(e) Finally, find quality factor Q from Eq. (10-8):

$$Q = \frac{f_r}{B} \qquad Q = \_\_\_\_\_\_\_\_\_\_$$

To check your calculations, find f_r from $f_r = \sqrt{f_L f_H}$. f_r = __________. If this value equals that in step (b), your work is correct.

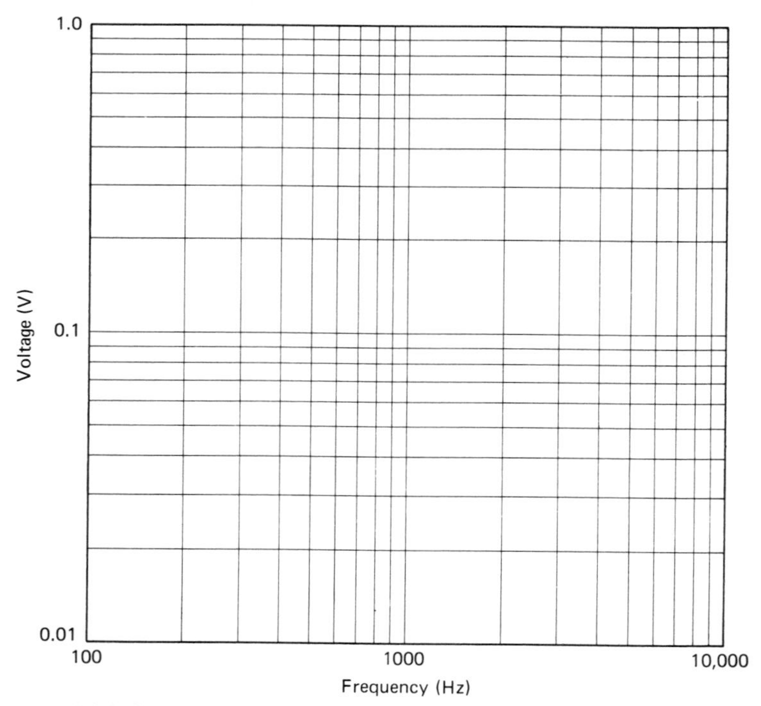

Figure L10-8 Frequency response curves for the bandpass filter of Fig. L10-7 and the notch filter of Fig. L10-9.

11. Build the bandpass filter circuit of Fig. L10-7. Set V_i to 1.0 V (rms) and frequency to 1000 Hz. Monitor V_o with an analog ac voltmeter. Carefully vary the frequency until V_o is at its maximum value.

Record this value of V_o as V_r = __________. Measure the corresponding value of frequency. (Use a frequency counter or CRO.)

f_r = __________.

12. Predict the values of V_o at both frequencies from $V_{oL} = V_{oH} = 0.707 \times V_{or}$. $V_{oL} = V_{oH}$ = __________. Adjust frequency to either side of f_r until $V_o = V_{oL}$ or V_{oH}. Measure the corresponding cutoff frequencies. f_L = __________; f_H = __________. Calculate the actual bandwidth B and Q from your measured values. $B = f_H - f_L$ = __________; $Q = f_r/B$ = __________.

13. Compare your predicted with measured values for f_r, f_L, and f_H and account for any significant differences. ____________________

____

Plot the three data points of V_o versus f (f_L, fr, and f_H) on Fig. L10-8. Adjust frequency to plot other convenient points to draw a smooth response curve over the range 100 to 10,000 Hz. Save this circuit for Procedure E.

Procedure E: Measuring Frequency Response of a Notch Filter

14. Add an inverting adder circuit to the bandpass filter of Fig. L10-7 to obtain the notch filter of Fig. L10-9. Adjust V_i to 1.0 V (rms) at a frequency of 50 Hz. Measure V_o. V_o = __________. Multiply V_o by 0.707 to obtain $V_{oL} = V_{oH}$ = __________.

15. Monitor V_o with an analog ac voltmeter or CRO. Vary the frequency of V_i until V_o is at a minimum. Measure this notch frequency (preferably with a frequency counter). f_r = __________. Carefully vary the frequency on either side of the notch until $V_o = V_{oL}$ or V_{oM}. Measure and record the corresponding cutoff frequencies. V_{oL} = __________; V_{oH} = __________.

Increase frequency to 10 kHz and measure V_o. V_o = __________.

16. Plot the five data points on Fig. L10-8 for f = 50 Hz, 10,000 Hz, f_L, f_r, and f_H. Change frequency and take several more data points to construct a smooth curve on Fig. L10-8.

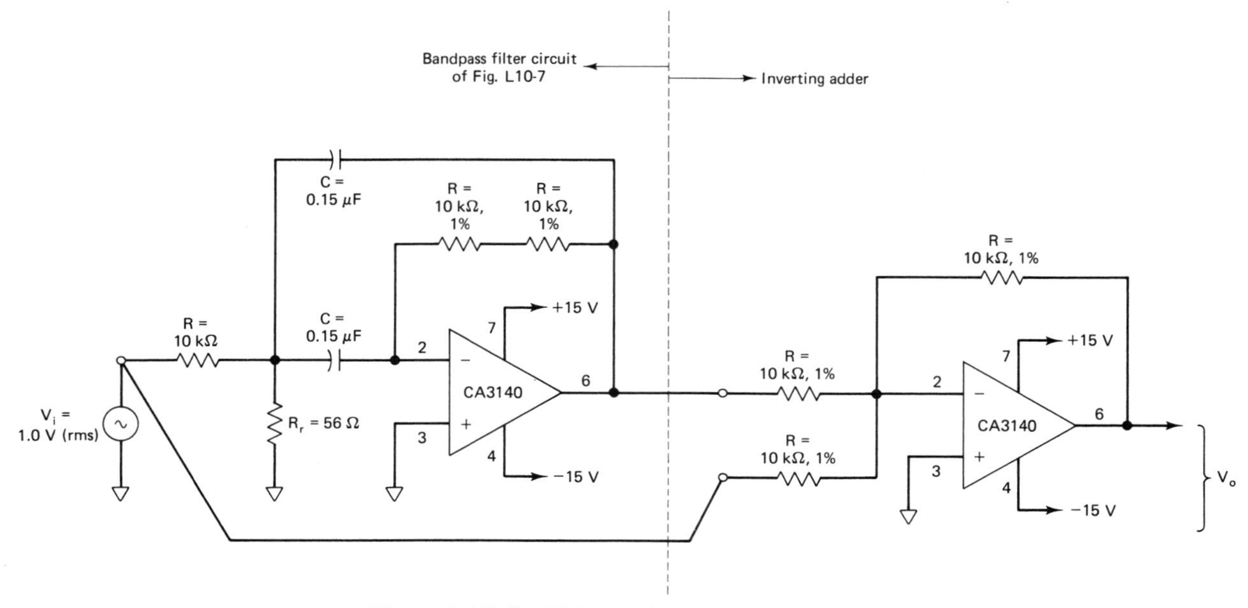

Figure L10-9 This notch or band-reject filter is made by connecting an inverting adder to the bandpass filter of Fig. L10-7. All R resistors are 10 kΩ, 1%.

17. Calculate the stopband from $B = f_H - f_L$. $B =$ __________. Does the stopband and notch frequency of the notch filter equal (within +5%) the bandwidth and resonant frequency of the bandpass filter? __________.

CONCLUSION:

Digital-to-Analog Converters

11

LEARNING OBJECTIVES

Upon completion of this chapter on digital-to-analog converters, you will be able to:

- Explain the need for digital-to-analog converters.
- Plot the output–input characteristics of a DAC.
- Draw an R-$2R$ ladder and calculate all branch currents.
- Draw a basic 4-bit or 8-bit DAC with voltage output, and find its analog output voltage for any digital input word.
- Distinguish between a DAC and a multiplying DAC.
- Show how a DAC-08 can be wired to give unipolar or bipolar analog output voltages.
- Write the output–input equations for discrete component or integrated-circuit DACs.
- Define resolution of a DAC.
- Explain how a microprocessor selects one DAC out of many other peripheral devices.
- Explain the operation of a microprocessor-compatible DAC, the AD558.
- Perform either static or dynamic tests on an 8-bit DAC.

11-0 INTRODUCTION TO CONVERTERS

Converters interface between the analog and digital worlds. Real-world physical processes are analog in nature. That is, they vary continuously. Transducers convert physical processes such as temperature or pressure into electrical quantities such as voltage, current, or resistance.

It is not easy to store, manipulate, compare, perform accurate calculations, or retrieve data in analog form. However, computers can store huge masses of data as digital codes, where they can be stored, processed, and manipulated. The results of such data processing can be retrieved quickly to provide information or control.

Thus developed the need for converters. *Analog-to-digital* converters (ADCs) are required so that analog voltage representations of physical processes can be transmitted to a computer. Digital-to-analog converters (DACs) allow the computer to send data or control signals back to the real world. Throughout this chapter and the next, we assume that the reader knows how to convert straight binary numbers to decimal integer numbers, and vice versa. For example, binary 101 has a decimal value of 5.

We choose to present D/A converters before A/D converters for one compelling reason. The widely used "successive approximation" A/D converter is made from a D/A converter plus additional hardware.

11-1 INTRODUCTION TO D/A CONVERTERS

11-1.1 Ideal Output–Input Characteristics

The analog output voltage of an ideal 3-bit digital-to-analog converter (hereinafter called a DAC) is plotted versus its digital input in Fig. 11-1. This particular presentation is a de facto standard in the converter literature.

Often digital inputs are represented by fractional binary numbers (0.101 represents $\frac{5}{8}$). Analog outputs are then represented as fractions of full-scale output voltage V_{FS}. This standard presentation is confusing

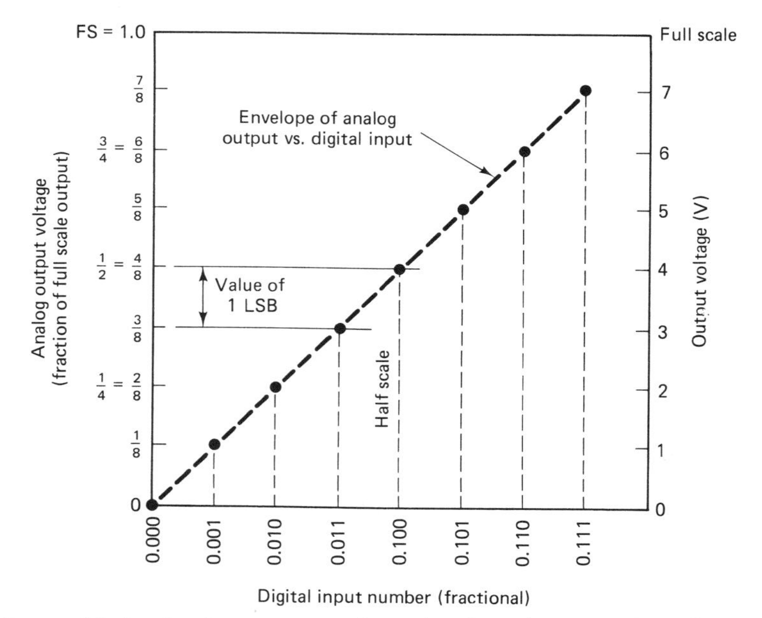

Figure 11-1 Analog output voltage is plotted as a series of dots versus a digital input fractional number for an ideal 3-bit DAC. Analog output voltage shown as either a fraction of full-scale voltage or simply as output voltage.

because V_{FS} is *never measured* at the DAC's output. For this reason let us seek another path.

Let us define *full-scale output* of a DAC as that *output voltage measured* when the *digital input is all 1's.* Suppose that we had a 3-bit DAC and every least significant bit was worth 1 V. A digital input of 001 causes an output of 1 V, as shown in Fig. 11-2. A full-scale digital input of 111 (integer decimal value = 7) would cause a full-scale output of 7 V. To obtain a more realistic definition of full-scale output, redraw the output–input DAC characteristics, as in Fig. 11-2. It will be used to define *resolution* of a DAC.

11-1.2 Resolution

Refer to Fig. 11-2, which represents the output versus input characteristic of an ideal 3-bit converter. There are only eight possible input combinations (2^3) for a 3-bit digital input. Count them: 000 to 111. A zero to full-scale analog output results from these combinations; *therefore, the analog output can be divided into 8 or 2^3 separate values.*

We have just discovered *one* definition for *resolution* of a DAC. For a 3-bit DAC, resolution equals 8 = 2^3. For an *n*-bit DAC;

$$\text{resolution} = 2^n \tag{11-1a}$$

There is another definition of resolution: *Resolution equals the value of 1 LSB:*

$$\text{resolution} = \text{value of 1 LSB} \tag{11-1b}$$

This definition points the way to the performance equation for a DAC.

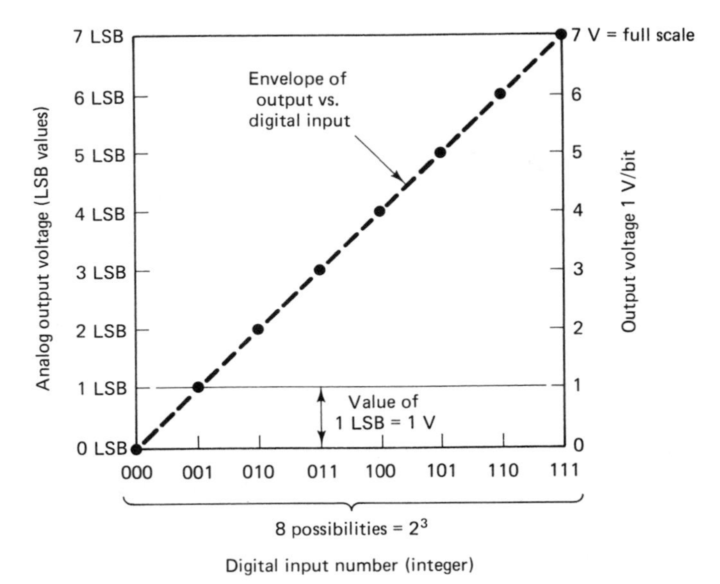

Figure 11-2 Analog output voltage versus digital input integer number for a 3-bit DAC.

11-1.3 DAC Output–Input Equation

Refer to Eq. (11-1b) and Fig. 11-2. Suppose that the value of one *least significant bit* (LSB) equals 1 V. That is, for a digital input word of 001, analog output voltage equals 1 V.

It follows that the characteristic output–input equation of this DAC is

$$V_o = (\text{value of 1 LSB})D \tag{11-2}$$

where V_o is the analog output voltage, and D is the *decimal* value of the digital input word. The digital input word is an *integer* number, *not* fractional.

EXAMPLE 11-1

Calculate the analog output voltage for digital inputs of (a) 001 and (b) 111 if a DAC's resolution is 10 mV/bit.

SOLUTION From Eq. (11-2);

$$V_o = (\text{value of 1 LSB})D$$

A value of 1 LSB equals the 10 mV/bit resolution [see Eq. (11-1b)].

(a) The decimal value of $D = 001$ is 1.

$$V_o = \left(\frac{10 \text{ mV}}{\text{bit}}\right)(1 \text{ bit}) = 10 \text{ mV}$$

(b) The decimal value of $D = 111$ is 7.

$$V_o = \left(\frac{10 \text{ mV})}{\text{bit}}\right)(7) = 70 \text{ mV}$$

Full-scale output results when the digital input code is a maximum of all 1's. For an n-bit DAC, with an input of all 1's, the decimal value of D in Eq. (11-2) equals $2^n - 1$. (See Example 11-1b, where $D = 2^3 - 1 = 7$ for an $n = 3$ bit DAC.) Measured full-scale output, FS, is found from

$$\text{full scale output} = (\text{value of LSB})(2^n - 1) \tag{11-3}$$

EXAMPLE 11-2

An 8-bit DAC has a resolution of 10 mV/bit. What is the full-scale output voltage?

SOLUTION From Eq. (11-3);

$$\text{full-scale output} = (10 \text{ mV})(2^8 - 1) = (10 \text{ mV})255 = 2.550 \text{ V}$$

11-1.4 D/A Conversion Process

The block diagram for a basic DAC is drawn in Fig. 11-3. Reference voltage V_{ref} is connected to a resistance network. A digital input code, via control circuitry, flips switches (one for each bit), connected to the resistance network. The output of the resistance network is in the form of a current. This current may then be converted to a voltage. Both current and voltage outputs are analog representations of the digital input code.

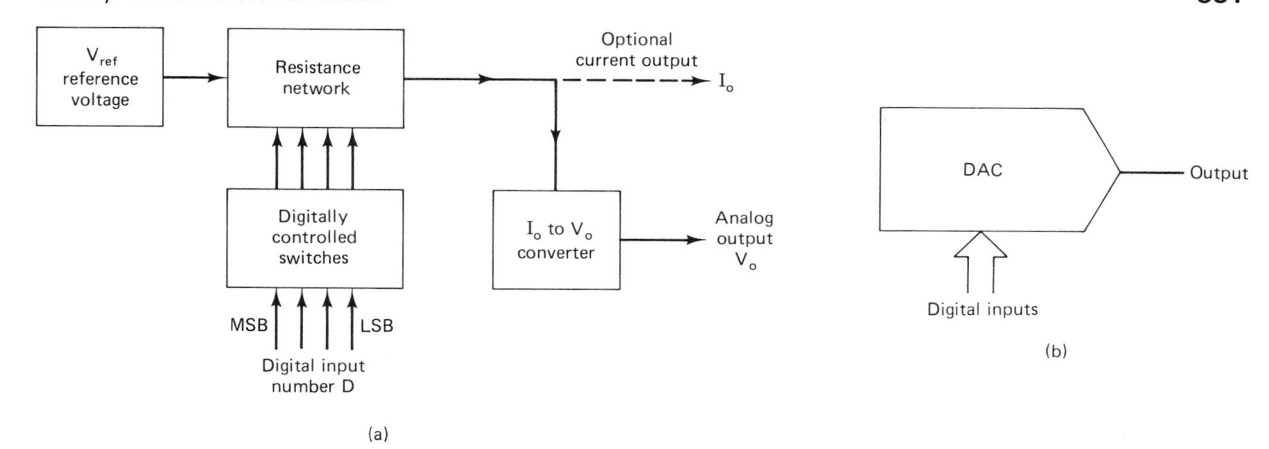

Figure 11-3 (a) Principles of D/A conversion are shown in this block diagram of a basic DAC; (b) schematic symbol.

The actual D/A conversion takes place within the resistance network. Accordingly, we begin our study of DAC circuitry by looking next at the (industry) standard resistance network. It is called an *R-2R ladder.*

11-2 R-2R RESISTANCE LADDER

11-2.1 Resistance Pattern

A 4-bit R-$2R$ ladder network is made from three equal R resistors and five equal $2R$ resistors in Fig. 11-4. Typically, $R = 10\ \text{k}\Omega$ and $2R = 20\ \text{k}\Omega$. Four current switches are activated to the "0" or "1" position depending on the digital input number. Switch positions are shown that correspond to a digital input of 1111. The wiper of each switch is *always* at ground potential *regardless* of switch position. Thus wiper current is constant except when the switch is in transition.

Refer to node 0. Note that R_o sees $2R$ via D_o in parallel with the $2R$ terminating resistor, or $2R \parallel 2R = R$. Looking to the right into node 1, R_1 sees $2R$ to ground via D_1, in parallel with $2R$ from $R_o + R = 2R$. Thus if we look into *any* node *from left to right,* we always see a net resistance to ground equal to R. If we look into the ladder at node 3, the entire ladder looks like a single R resistor.

11-2.2 Ladder Currents

In Section 11-2.1 it is clear from Fig. 11-4 that we see a resistance of R to ground looking (right) into any node. Any current entering this node sees two paths of equal resistance. One path is $2R$, via the bit switch; the other path is also $2R$, via one of the three horizontal R resistors in series with an R resistance looking right into the next node.

The current pattern of the R-$2R$ network in Fig. 11-4 is found as follows:

1. Input current to the ladder is set by

$$I = \frac{V_{\text{ref}}}{R} \tag{11-4}$$

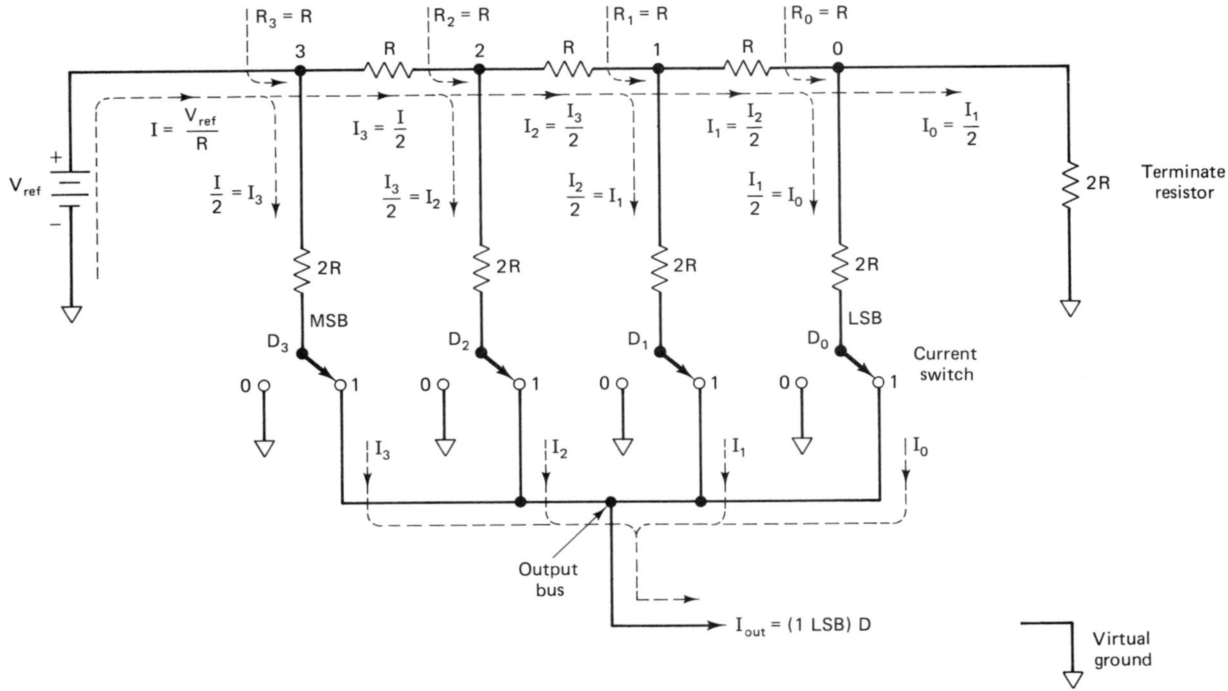

Figure 11-4 This teaching model of a 4-bit current output DAC is made from an *R*-2*R* ladder network, reference voltage, and current switches. The digital input word changes the current switcher to either the "0" or "1" position. Switches are shown positioned for a digital input of 1111.

2. Current I splits equally into two parts at node 3. One part (vertical) is bit 3 output current $I_3 = I/2$. The second part (horizontal) flows toward node 2, where it also will divide into two equal parts of $I_2 = I_3/2 = I/4$.

Thus the ladder output currents are *constant* and found from

$$I_3 = \frac{I}{2} \qquad I_2 = \frac{I_3}{2} = \frac{I}{4}$$
$$I_1 = \frac{I_2}{2} = \frac{I}{8} \qquad I_o = \frac{I_1}{2} = \frac{I}{16} = \text{LSB} \tag{11-5}$$

11-2.3 Output Current

The ladder's output current I_{out} is the *sum* of each ladder current that *reaches* the output bus. When the D_0 switch is closed, a current value of 1 LSB enters the output bus. If D_1 is "1," a current of 2 LSB enters the output bus. D_2 can output 4 LSB and D_3 can output 8 LSB. Output current is then found in terms of the digital input by

$$I_{out} = (\text{current value of 1 LSB}) \times \text{D} \tag{11-6a}$$

where

$$\text{current value of 1 LSB} = I_o \tag{11-6b}$$

but Eq. (11-4) shows that the user determines I_o by choosing V_{ref} and R. Substituting Eq. (11-4) into (11-6b) gives us

$$\text{current value of 1 LSB} = \left(\frac{V_{ref}}{R}\right)\left(\frac{1}{2^n}\right) = I_o \qquad (11\text{-}6c)$$

EXAMPLE 11-3

Assume that $V_{ref} = 10$ V and the 4-bit R-$2R$ ladder of Fig. 11-4 is made from 10- and 20-kΩ resistors. The ladder's input resistance is $R = 10$ kΩ. Find (a) the value of 1 LSB (resolution); (b) I_{out} when the digital input is 1111.

SOLUTION (a) From Eq. (11-6c), with $n = 4$,

$$\text{current value of 1 LSB} = \left(\frac{10\text{ V}}{10\text{ k}\Omega}\right)\left(\frac{1}{2^4}\right) = (1\text{ mA})\left(\frac{1}{16}\right) = 0.0625\text{ mA}$$

Resolution is 0.0625 mA per bit.

(b) From Eq. (11-6a), $D = 15$ for an input of 1111:

$$I_{out} = (0.0625\text{ mA})(15) = 0.9375\text{ mA}$$

11-3 MULTIPLYING DACs

11-3.1 Voltage Output DAC

An op amp can be added to the very basic DAC of Fig. 11-4. The result is the voltage output DAC shown in Fig. 11-5. Output current is converted to a voltage V_o by feedback resistor R_F and the op amp.

$$V_o = -I_{out}R_F = -\begin{pmatrix}\text{current value}\\ \text{of 1 LSB}\end{pmatrix} \times R_F D \qquad (11\text{-}7a)$$

Multiply Eq. (11-6c) by R_F to obtain the resolution of our voltage output DAC:

$$\text{resolution} = \begin{pmatrix}\text{voltage value}\\ \text{of 1 LSB}\end{pmatrix} = \left(\frac{V_{ref}}{R}\right)\left(\frac{1}{2^n}\right) R_F \qquad (11\text{-}7b)$$

Combine Eqs. (11-7a) and (11-7b) to obtain V_o in terms of the digital input and circuit components.

$$V_o = -(\text{resolution})D \qquad (11\text{-}7c)$$

EXAMPLE 11-4

Refer to the 4-bit DAC of Fig. 11-5. Find (a) its voltage resolution and (b) the output voltage when the digital input is 1111.

SOLUTION (a) From Eq. (11-7b),

$$\text{resolution} = \left(\frac{10\text{ V}}{10\text{ k}\Omega}\right)\left(\frac{1}{2^4}\right)(10\text{ k}\Omega) = \frac{10\text{ V}}{16} = 0.625\text{ V/bit}$$

(b) The value of D is 15 for an input of 1111. From Eq. (11-7c), $V_o = -(0.625\text{ V})(15) = -9.375$ V.

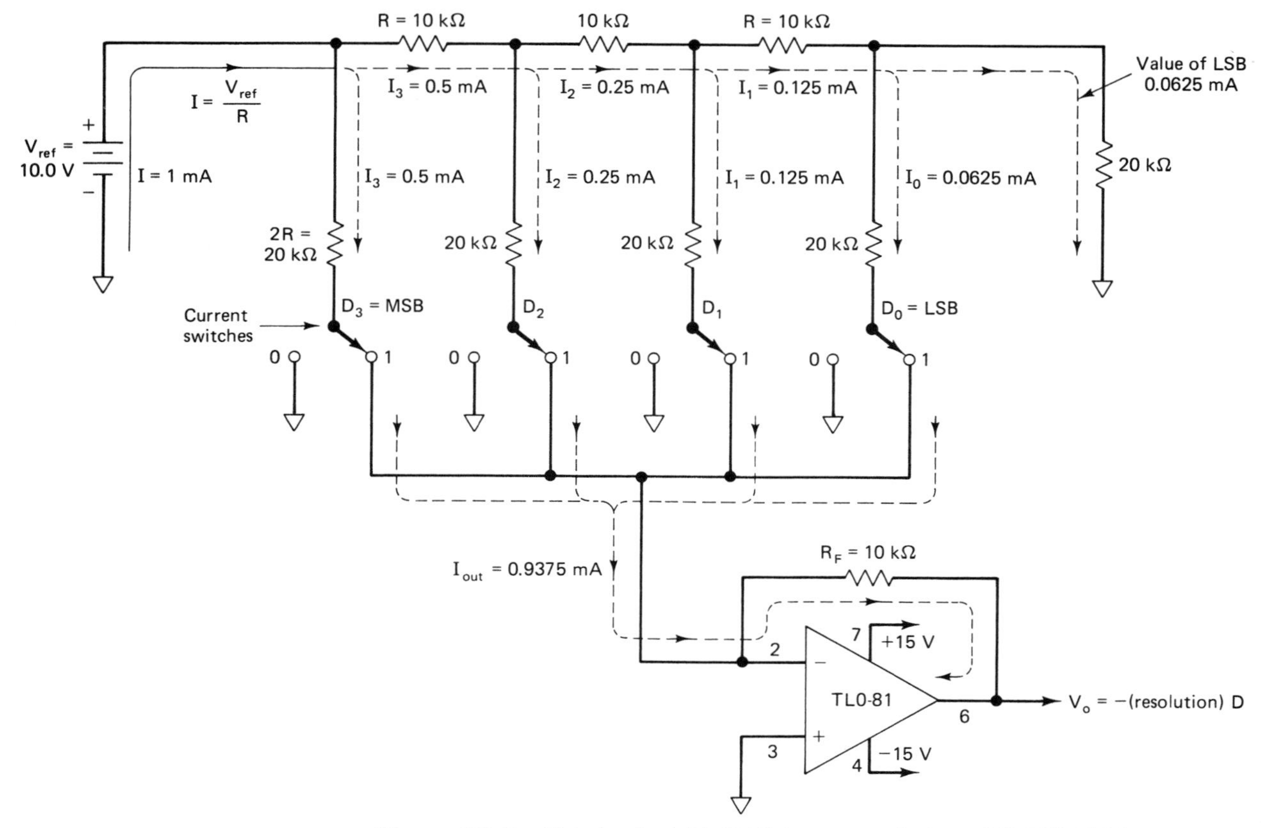

Figure 11-5 This basic 4-bit DAC has a resolution of 0.625 V/bit. I_{out} of the ladder network is converted to an analog output voltage V_O by the op amp and R_F.

11-3.2 Multiplying DAC

Equations (11-7b) and (11-7c) may be combined to express V_o as

$$V_o = -(\text{constant})V_{\text{ref}}D \tag{11-8}$$

where

$$\text{constant} = -\frac{R_F}{R} \times \frac{1}{2^n}$$

Equation (11-8) indicates V_{ref} is multiplied by D (and also a constant) to determine V_o. This observation describes a *multiplying* DAC converter or MDAC. One application for an MDAC allows a microprocessor to control volume of audio signals, as discussed next.

Suppose that V_{ref} is an audio signal with peak values of 10 V. From Example 11-4a, 1 LSB at the audio output of Fig. 11-4 equals 0.625 V. For digital inputs of 0000, 0001, 1000, or 1111, output V_o would be an audio signal with peak values increasing from 0 V, to + 0.625 V × 8 = 5 V to finally 0.625 V × 15 = 9.375 V. Thus the circuit acts like a volume control that is adjusted by a digitally controlled attenuator.

11-3.3 Toward Practical DACs

In practice, higher resolutions are required than the tutorial 3-bit and 4-bit DACs. DACs with at least 8-bit resolution are needed for most applications. The industry standards are 8-, 10-, 12-, 14-, and 16-bit. Higher-resolution DACs are available at substantial cost. It is best for a nonspecialist to purchase DACs from one of the expert manufacturers rather than to build one's own. In the remainder of this chapter, we present an 8-bit MDAC and an 8-bit microprocessor-compatible DAC that are readily available and low in cost.

11-4 BASIC 8-BIT D/A CONVERTER

A discrete-component 8-bit DAC is shown in Fig. 11-6. One R resistor and one $2R$ resistor are added to the ladder network for each additional bit. One obvious disadvantage of this circuit is that the digital input word must be loaded into the DAC by manually operating the current switches. We really want the switches to be operated by a TTL, CMOS, or other logic circuitry in a practical DAC.

The current output bus of Figs. 11-4 and 11-5 has been changed to two *complementary* output buses in Fig. 11-6. This allows greater flexibility. An op-amp can be connected to convert these current outputs to (1) positive voltage output, (2) negative voltage output, or (3) bipolar output.

Rather than build our own discrete-component 8-bit DAC, it is far less costly to buy one of the many superb IC DACs made by experts. For our first IC DAC, we selected an extremely versatile, low-cost industry standard, the DAC-08. It is a vast improvement over the DAC in Fig. 11-6.

11-5 EIGHT-BIT D/A CONVERTER

The DAC-08 is a low-cost multiplying, fast D/A converter. This versatile IC is housed in a 16-pin dual-in-line package. Its operating principles are introduced by reference to the tasks performed by each terminal as shown in Fig. 11-7.

11-5.1 Power Supply Terminals

Pins 13 and 3 are the positive and negative supply terminals, respectively, and can have any value from ± 4.5 to ± 18 V. They should be bypassed with 0.1-μF capacitors, as shown in Fig. 11-7.

11-5.2 Reference (Multiplying) Terminal

Flexibility of the DAC-08 is enhanced by having *two* rather than one reference input. Pins 14 and 15 allow positive or negative reference voltages, respectively. A positive reference voltage input is shown in Fig. 11-7.

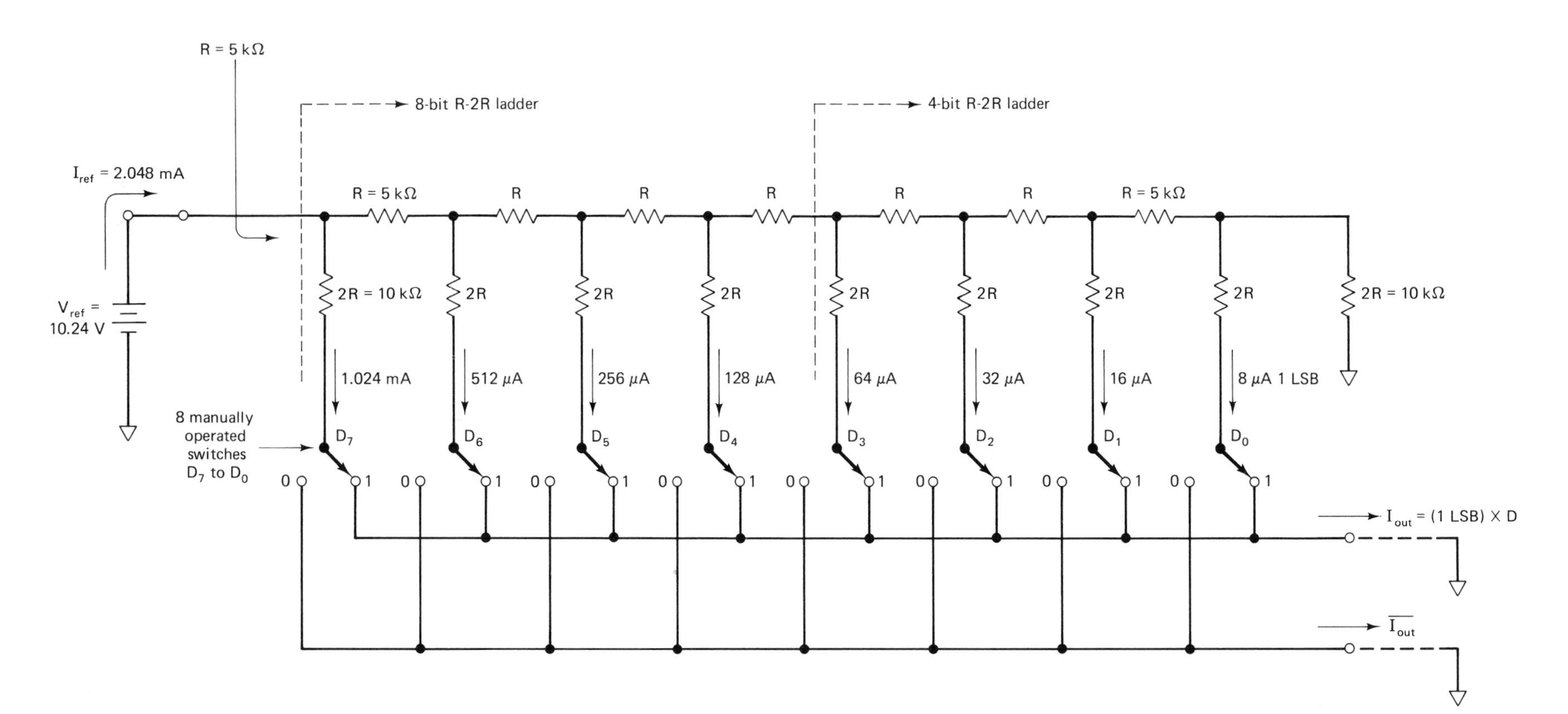

Figure 11-6 Rudimentary 8-bit discrete component DAC. $I_{out} = 8\ \mu A/bit$. Ladder input resistance is $R = 5\ k\Omega$. Reference current $I_{ref} = V_{ref}/R$.

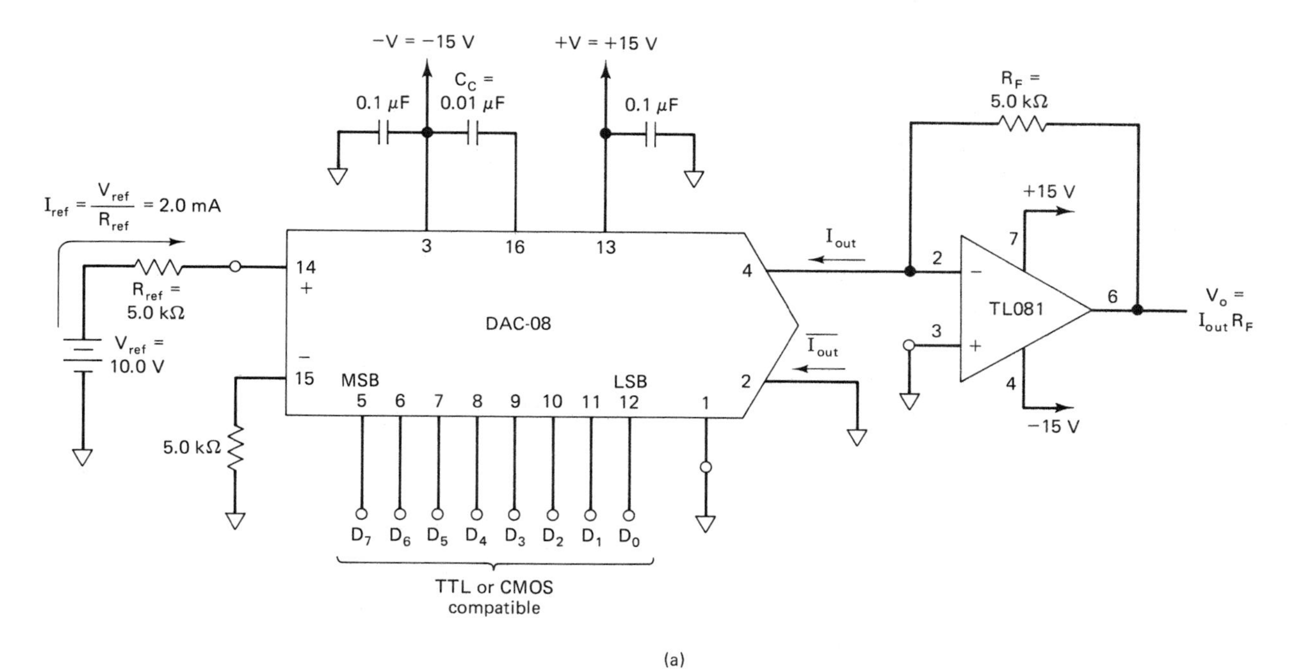

	Digital inputs								Analog output	
	D_7	D_6	D_5	D_4	D_3	D_2	D_1	D_0	I_{out}	V_o
LSB	0	0	0	0	0	0	0	1	7.812 μA	39 mV
Half-scale	1	0	0	0	0	0	0	0	1.000 mA	5.0 V
Full-scale	1	1	1	1	1	1	1	1	1.992 mA	9.96 V

(b)

Figure 11-7 The 8-bit multiply DAC-08 shown in (a) is wired for unipolar positive output voltage. Analog output current and voltage are shown in (b) for three digital input words. (a) DAC-08 wired for positive reference voltage V_{ref} and unipolar positive output voltage; (b) key digital inputs and analog outputs.

The user can adjust the DAC-08's input ladder current I_{ref} quite easily from

$$I_{ref} = \frac{V_{ref}}{R_{ref}} \tag{11-9}$$

I_{ref} is typically equal to 2 mA but can be set to any value from 4 μA to 4 mA.

11-5.3 Digital Input Terminals

Pins 5 through 12 identify the digital input terminals. Pin 5 is the most significant bit (MSB), D_7. Pin 12 is the LSB, D_0. The terminals are TTL or CMOS compatible. Logic input "0" is 0.8 V or less. Logic "1" is 2.0 V or more, regardless of the power supply voltages. Usually, pin 1, V_{LC}, is grounded. However, it can be used to adjust the *logic input threshold* voltage V_{TH} according to $V_{TH} = V_{LC} + 1.4$ V. These digital inputs control eight internal current switches.

11-5.4 Analog Output Currents

Two current output terminals are provided in Fig. 11-7a to increase the DAC-08's versatility. Pin 4 conducts output current I_{out} and pin 2 conducts its complement, $\overline{I_{out}}$. (Notice that their direction is *opposite* to their counterparts in Fig. 11-6.) If an internal switch is positioned to "1," its ladder rung current flows in the I_{out} bus. If positioned to "0," ladder rung current flows in the $\overline{I_{out}}$ bus.

The current value of 1 LSB (resolution) is found by modifying Eq. (11-6c):

$$\text{resolution} = (\text{value of 1 LSB}) = \left(\frac{V_{ref}}{R_{ref}}\right)\left(\frac{1}{2^n}\right) \qquad (11\text{-}10a)$$

I_{out} is calculated from

$$I_{out} = (\text{value of 1 LSB}) \times D \qquad (11\text{-}10b)$$

where D is the decimal value of the digital input work. The full-scale output current in the pin 4 output bus occurs when the digital input is 11111111, so that $D = 255$. Let's define this current as I_{FS}, where

$$I_{FS} = (\text{value of 1 LSB}) \times 255 \qquad (11\text{-}10c)$$

The sum of all ladder rung currents in the DAC-08 equals I_{FS}. Since this sum always divides between the I_{out} and $\overline{I_{out}}$, the value of $\overline{I_{out}}$ is given by

$$\overline{I_{out}} = I_{FS} - I_{out} \qquad (11\text{-}10d)$$

EXAMPLE 11-5

Calculate (a) the ladder input current I_{ref} of the DAC-08 in Fig. 11-7a; (b) the current value of 1 LSB.

SOLUTION (a) From Eq. (11-9);

$$I_{ref} = \frac{10\text{ V}}{5\text{ k}\Omega} = 2\text{ mA}$$

(b) From Eq. (11-10a);

$$\text{current value of 1 LSB or resolution} = \frac{10\text{ V}}{5\text{ k}\Omega} \times \frac{1}{2^8} = 7.812\ \mu\text{A}$$

EXAMPLE 11-6

For the DAC-08 circuit in Fig. 11-7, find the values of I_{out} and $\overline{I_{out}}$ when the digital input words are (a) 00000001; (b) 10000000; (c) 11111111.

SOLUTION Example 11-5a showed that current output resolution is 7.812 μA/bit. From Eq. (11-10c), evaluate I_{FS}.

$$I_{FS} = (\text{resolution})255 = 7.812\ \mu\text{A} \times 255 = 1.992\text{ mA}$$

The value of D is 1 for (a), 128 for (b), and 255 for (c). I_{out} can now be found from Eq. (11-10b):

(a) $I_{out} = 7.812\ \mu\text{A} \times 1 = 7.812\ \mu\text{A}$ for 00000001 input
(b) $I_{out} = 7.812\ \mu\text{A} \times 128 = 1.000\text{ mA}$ for 1000000 input
(c) $I_{out} = 7.812\ \mu\text{A} \times 255 = 1.992\text{ mA}$ for 11111111 input

From Eq. (11-10d),

(a) $\overline{I_{out}} = 1.992 \text{ mA} - 7.812\ \mu\text{A} = 1.984 \text{ mA}$

(b) $\overline{I_{out}} = 1.992 \text{ mA} - 1.0 \text{ mA} = 0.992 \text{ mA}$

(c) $\overline{I_{out}} = 1.992 \text{ mA} - 1.992 \text{ mA} = 0$

The results of Examples 11-5 and 11-6 are tabulated in Fig. 11-7b.

11-5.5 Unipolar Output Voltage

In Fig. 11-7a, the DAC-08's current output I_{out} is converted to an output voltage V_o by an external op amp and resistor R_F. This voltage output has a resolution of

$$\text{resolution} = \frac{V_{ref}}{R_{ref}} \times R_F \times \frac{1}{2^n} \tag{11-11a}$$

and V_o is given by

$$V_o = \text{resolution} \times D = I_{out}R_F \tag{11-11b}$$

EXAMPLE 11-7

For the DAC-08 circuit of Fig. 11-7a, find V_o for digital inputs of (a) 00000001; (b) 11111111.

SOLUTION From Eq. (11-11a),

$$\text{resolution} = (10 \text{ V}) \left(\frac{5 \text{ k}\Omega}{5 \text{ k}\Omega}\right) \times \left(\frac{1}{256}\right) = 39.0 \text{ mV/bit}$$

(a) From Eq. (11-11b) with the value of $D = 1$,

$$V_o = 39.0 \text{ mV} \times 1 = 39.0 \text{ mV for 00000001 input}$$

(b) The value of $D = 255$. From Eq. (11-11b),

$$V_o = 39.0 \text{ mV} \times 255 = 9.961 \text{ V for 11111111 input}$$

11-5.6 Bipolar Analog Output Voltage

Versatility of the DAC-08 is shown by wiring it to give a bipolar analog output voltage in response to a digital input word (Fig. 11-8a). The op amp and two resistors convert the *difference* between I_{out} and $\overline{I_{out}}$ into a voltage V_o:

$$V_o = (I_{out} - \overline{I_{out}})R_F \tag{11-12}$$

I_{out} drives V_o positive and $\overline{I_{out}}$ drives V_o negative. If the digital input word increases by 1 bit, I_{out} increases by 1 LSB. However, $\overline{I_{out}}$ must therefore *decrease* by 1 LSB. Therefore, the differential output current changes by 2 LSB; thus we would expect the bipolar output voltage span to be twice that of a unipolar output (Section 11-5.5).

V_{ref} has been increased slightly in Fig. 11-8a so that I_{ref} increases to 2.048 mA [Eq. (11-9)]. This increases the (current value of 1 LSB to an even 8 μA [Eq. (11-10a)]. We show how the output voltage responds to digital inputs by an example.

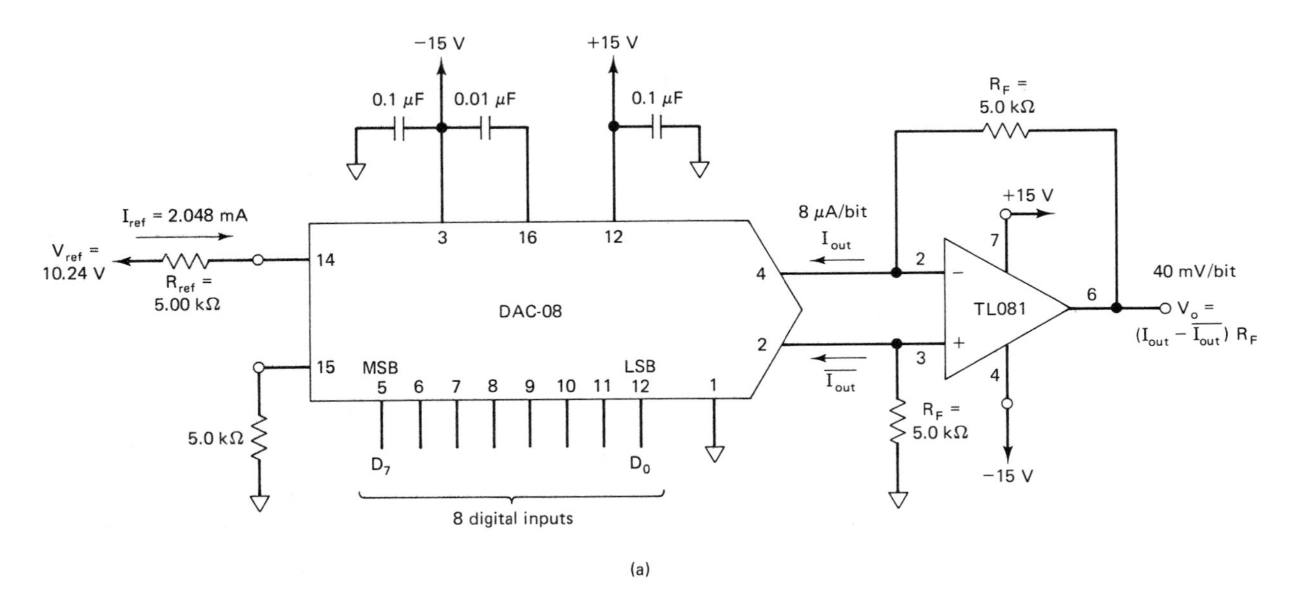

	Digital inputs								Analog outputs		
	D_7	D_6	D_5	D_4	D_3	D_2	D_1	D_0	I_{out} (mA)	$\overline{I_{out}}$ (mA)	V_o (V)
Negative full scale	0	0	0	0	0	0	0	0	0	2.040	−10.20
Negative zero	0	1	1	1	1	1	1	1	1.016	1.024	−0.040
Positive zero	1	0	0	0	0	0	0	0	1.024	1.016	0.040
Positive full scale	1	1	1	1	1	1	1	1	2.040	0	10.20

(b)

Figure 11-8 V_{ref} and R_{ref} give a resolution of 8 μA/bit. The outputs and op amp are connected to give a bipolar analog output voltage in response to a digital input. Resolution is 40 mV/bit. (a) DAC-08 wired for bipolar output voltage; (b) key digital inputs and analog outputs.

EXAMPLE 11-8

For the circuit of Fig. 11-8, calculate V_o for digital inputs of (a) 00000000; (b) 01111111; (c) 10000000; (d) 11111111.

SOLUTION The current value of 1 LSB equals 8 μA. From Eq. (11-10c), $I_{FS} = (8\ \mu A)255 = 2.040$ mA.

(a) From Eq. (11-10b), $I_{out} = (8\ \mu A) \times 0 = 0$. Then from Eq. (11-10d), $\overline{I_{out}} = 2.040\text{ mA} - 0 = 2.04$ mA. Find V_o from Eq. (11-12):

$$V_o = (0 - 2.04\text{ mA})(5\text{ k}\Omega) = -10.20\text{ V}$$

Values of I_{out}, $\overline{I_{out}}$, and V_o are calculated for (b), (c), and (d) and are summarized in Fig. 11-8b.

Note that full-scale negative input voltage of −10.20 V occurs for an all-zero digital input. All ones give a positive full-scale output of plus 10.20 V. Note also that V_o never goes to precisely zero volts. When I_{out} is less than $\overline{I_{out}}$ by 8 μA, (01111111), V_o equals −40 mV. Since this is the closest that V_o approaches to 0 V from negative full-scale, $V_o = -40$ mV is called negative zero.

Many DACs must operate under control of a microprocessor or computer. We therefore present a microprocessor-compatible DAC next.

11-6 MICROPROCESSOR COMPATIBILITY

11-6.1 Microprocessor-to-DAC Interfacing Principles

The programmer views a DAC, or any other peripheral device, as a memory address. A DAC is thought of as a "write only" address in memory because the microprocessor only *sends* data *to* the DAC. An A/D converter (Chapter 12) is thought of as a "read only" address in memory. For an ADC, the microprocessor reads the contents of a memory buffer register located in the converter.

Usually, addresses of the same type of peripheral device such as DACs are stored in their own block of addresses within the microprocessor. For example, Fig. 11-9 shows the map of 64k memory. The DAC locations are assigned 8XX0 to 8XXF.

The address for a particular DAC is placed on the address bus by the microprocessor. Only one of the decoder outputs goes low at a time to select the particular DAC, via a select line. This select line terminates at a DAC terminal called *chip select* ($\overline{CS}$), as shown in Fig. 11-10. The DAC is not yet completely selected.

The microprocessor has a terminal called MWR, MEMW, or R/$\overline{W}$ that is connected via a write line to *all* peripherals, including DACS. It provides a timing signal to complete the DAC selection process. The write line in Fig. 11-10 terminates on the *chip enable* terminal ($\overline{CE}$) of each DAC. When the microprocessor outputs data on the data bus, it is latched into the selected DAC, where the digital word is converted into an analog output voltage. Thus $\overline{CE}$ tells the DAC *when* to convert *if* the DAC is selected by a low on $\overline{CS}$.

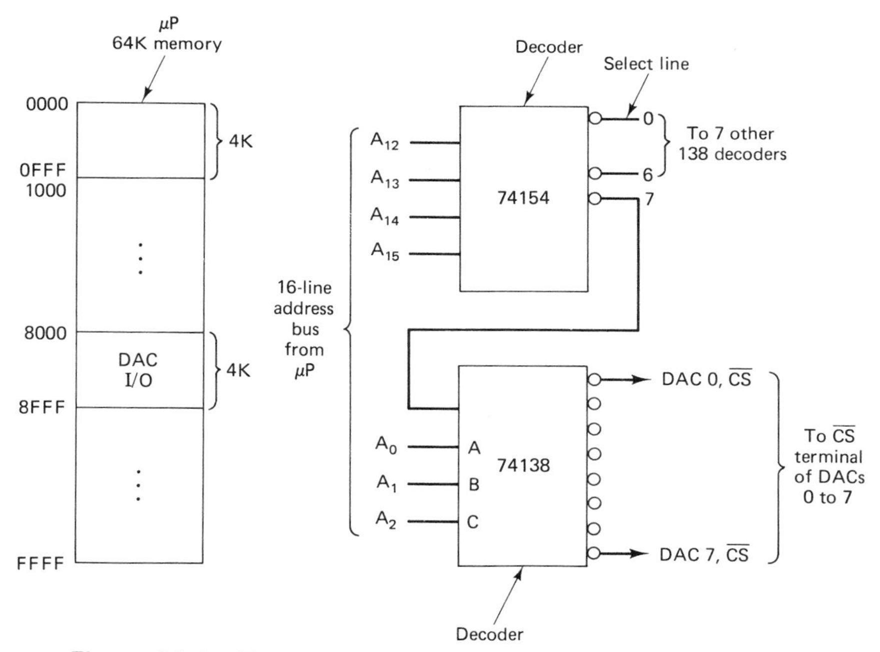

Figure 11-9 The computer programmer treats a DAC as if it is a "write only" address in memory. The address of one particular DAC is placed on the microprocessor's address line. A decoder circuit decodes the address and applies a low signal to the chip select $\overline{CS}$ terminal of a selected DAC.

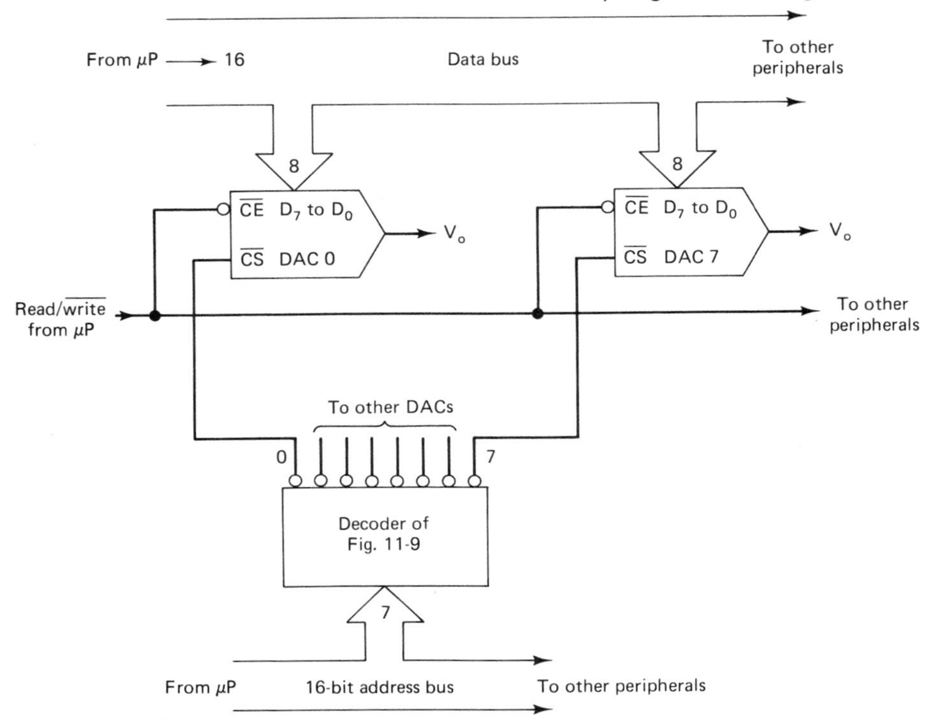

Figure 11-10 Simplified interface between a microprocessor and DAC peripherals.

11-6.2 Summary and Restatement

A programmer sends data to a DAC for conversion to an analog voltage (see Fig. 11-10). The DAC is treated as memory location, just as if the programmer would load or write digital data into any memory address.

The programmer partially selects one DAC via the *address* bus and decoders. Data from the microprocessor is then placed on the *data* bus. The programmer then issues a write command on the write line to all peripherals. Only the selected DAC responds to the digital word by latching data into a register and then performs the digital-to-analog conversion. The analog output stays until another word is sent to the DAC.

11-6.3 DAC Compatibility

The basic DAC will not work with a microprocessor. A microprocessor-compatibile DAC must have on-board logic that allows it to be selected. This logic is controlled by the $\overline{CS}$ and $\overline{CE}$ terminals in Figs. 11-10 and 11-11.

The DAC must also have on-board circuitry that:

1. Can write data from the data bus into a memory register (transparent) when selected
2. Can disconnect its digital inputs from the data bus when unselected and latch (remember) the last digital word

All this is required in addition to the basic reference voltage, resistance ladder, op amp, and digitally controlled switches. Fortunately,

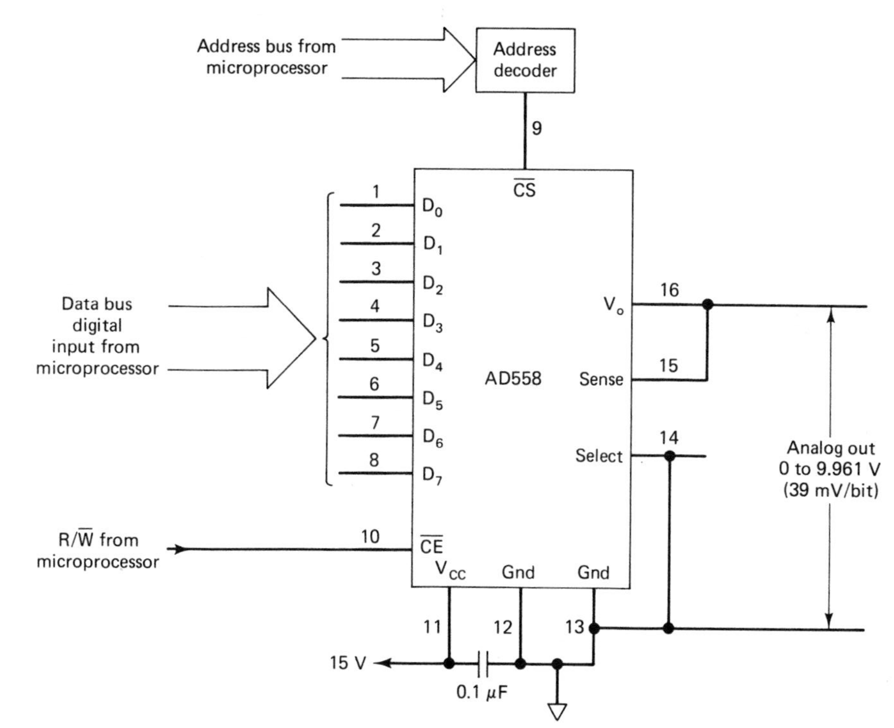

Figure 11-11 The AD558 is an 8-bit microprocessor-compatible D/A converter. It contains a precision voltage reference precision *R*-2*R* ladder, voltage output op amp, and digital input latches. On-board logic allows operation with or without a microprocessor data bus.

many DACS are available with all these features. We selected the AD558 as an excellent example.

11-7 EIGHT-BIT MICROPROCESSOR-COMPATIBLE DAC

An example of a complete 8-bit microprocessor-compatible D/A converter is introduced in Fig. 11-11. The AD558 (See data sheets pp. 408–14) can operate either continuously or it can be controlled by a microprocessor. It is complete with an on-board precision reference voltage, latching digital inputs, and select terminals. It also contains an op amp to give an analog output voltage that is pin-programmable for output ranges of 0 to 2.56 V or 0 to 10.0 V. Operation of the AD558 is studied by analyzing those tasks performed by its terminals.

11-7.1 Power Supply

Pin 11 is the power supply terminal V_{CC} in Fig. 11-11. It requires a minimum of +4.5 V and has a maximum rating of +16.5 V. Pins 12 and 13 are the digital and analog grounds, respectively. This allows the user to maintain *separate* analog and digital grounds throughout a system, joining them at *only* one point. Usually, pins 12 and 13 are wired together and a 0.1-μF bypass capacitor *must* be connected between V_{CC} and pin 12 or 13.

11-7.2 Digital Inputs

Pins 1 through 8 are the digital inputs D_0 to D_7 as in Fig. 11-11. D_0 is the LSB and D_7 is the MSB. They are compatible with standard TTL or *low*-voltage CMOS. Logic 1 is 2.0 V minimum for a "1" bit. Logic 0 is 0.8 V maximum for a "0" bit.

The digital input pins connect the data bus to the AD558's internal *memory latching register,* when the AD558 is selected. This condition is called *transparent.* When unselected, the latching register is essentially disconnected from the data bus and remembers the last word written into the latching register. This condition is called "latching."

11-7.3 Logic Circuitry

The microprocessor executes a *write* command over the address bus via an address decoder and write line to the AD558's logic control pins 9 and 10 (see Fig. 11-11). They are called chip select ($\overline{CS}$) and chip enable ($\overline{CE}$), respectively.

If a "1" is present on either $\overline{CS}$ or $\overline{CE}$, the digital inputs are in the "latching" mode. They are disconnected from the data bus. The input latches remember the last word written by the microprocessor over the data bus. If both $\overline{CS}$ and $\overline{CE}$ are "0," the AD558's inputs are "transparent" and connect the input memory latch register to the data bus. The microprocessor can now write data into the DAC. Digital-to-analog conversion takes place immediately and is completed in about 200 ns.

11-7.4 Analog Output

As shown in Fig. 11-11 and 11-12, analog output voltage appears between pins 16 (V_o) and pin 13 (analog ground). Pin 14 is called "select" (V_o gain). It is wired to ground to set the output voltage range at 0 to 2.56 V, as in Fig. 11-12a. The actual analog output range is 0 to 2.55 V or 10 mV/bit for a digital input of 00000000 to 11111111. A 0- to 10-V output range connection is shown in Fig. 11-12b. Actual range is 0 to 9.961 V or 38.9 mV/bit (power supply voltage must *exceed* maximum V_o by 2 V minimum). Sense terminal 15 allows remote load-voltage sensing to eliminate effects of *IR* drops in long leads to the load. It can also be used for current boost as in Fig. 11-12c.

11-8 TESTING AN 8-BIT DAC

11-8.1 Static Test

The simplest way to perform a static test on an 8-bit DAC is to use one jumper wire per digital input terminal. Connect each jumper wire to a "1" or "0" logic level, depending on the desired input code. Then measure V_o to see if it equals the digital input word times the value of one LSB. For example, if resolution is 10 mV/bit, a digital input of 11111111 should result in an analog output of 255 × 10 mV = 2.55 V.

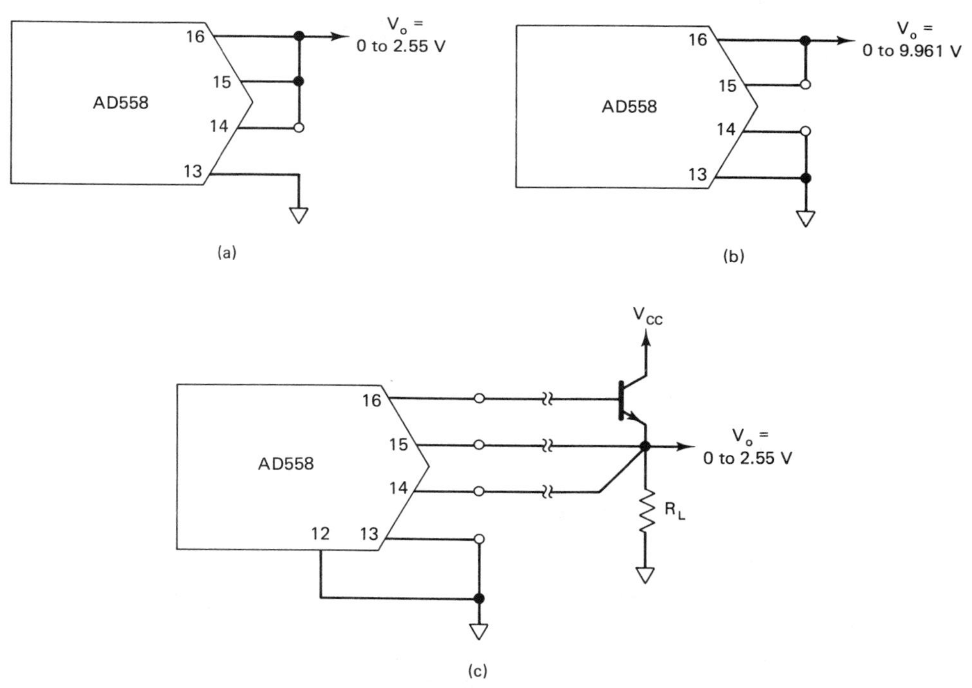

Figure 11-12 The AD558 may be pin-programmed for 0 to 2.55 V output as in (a) or 0 to 9.961 V output as in (b). Sense terminal 15 and gain select terminal 14 are extended to the load for current boost as in (c). (a) V_O = 0 to 2.55 V (V_{CC} = 4.5 to 16.5 V); (b) V_O = 0 to 9.961 V (V_{CC} = 11.4 to 16.5 V); (c) sense terminal 15 is extended to the load to eliminate V_{BE} drop in the current boost transistor.

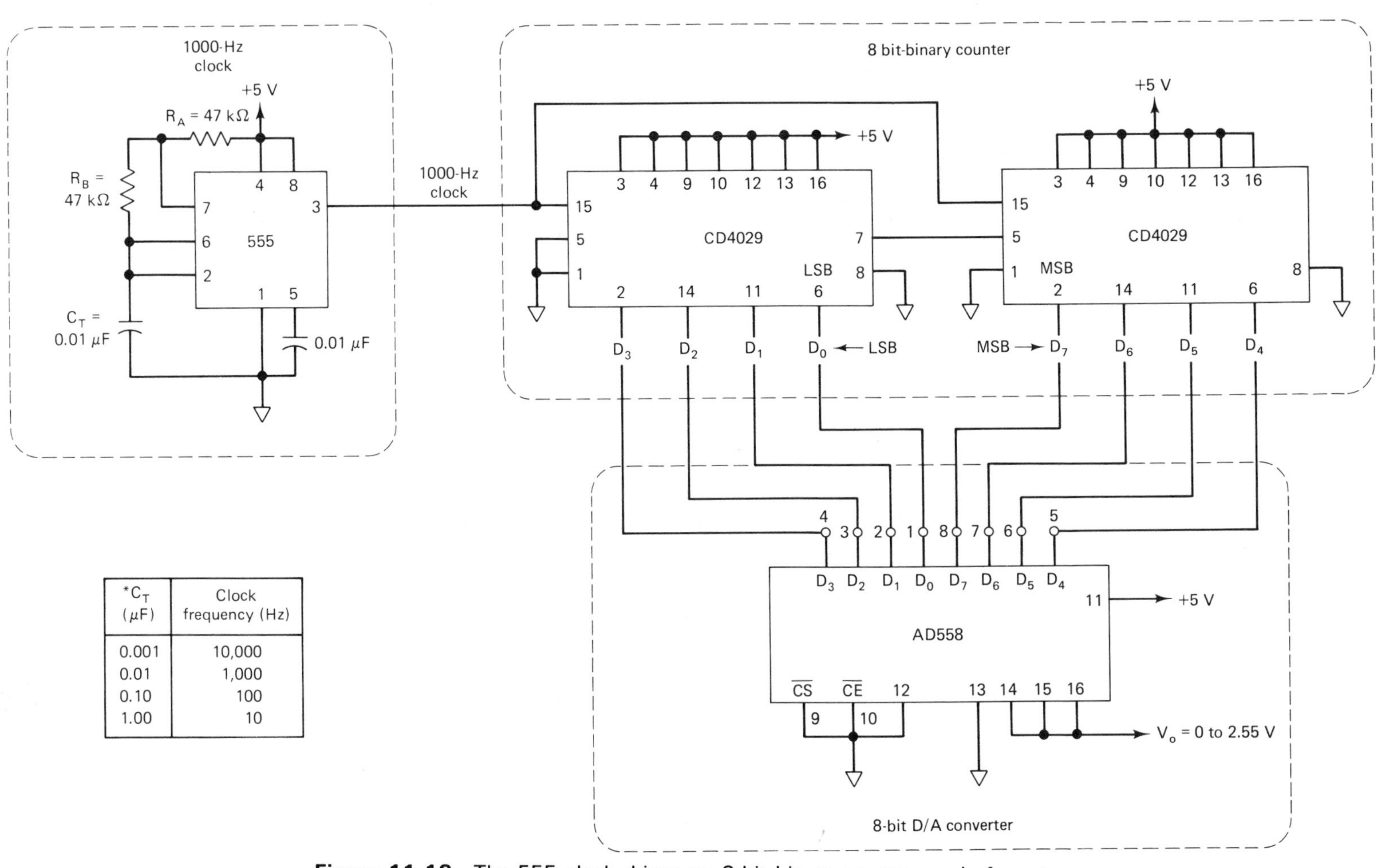

*C_T (μF)	Clock frequency (Hz)
0.001	10,000
0.01	1,000
0.10	100
1.00	10

Figure 11-13 The 555 clock drives an 8-bit binary counter made from two CD4049 ICs. Their outputs count in binary from 00000000 to 11111111 and then repeat. The digital count is converted by the DAC into an analog voltage that resembles a staircase.

11-8.2 Dynamic Test Circuit

The preceding manual test will not reveal the presence of glitches. A glitch can occur if the ladder switches or digital inputs do not change swiftly enough or at the same time. The glitch is a spike that appears on V_o, particularly at the major transitions 11111111 to 00000000 or 01111111 to 10000000.

A single AD558 can be tested dynamically *without* a microprocessor by the stand-alone, low-parts-count test circuit of Fig. 11-13. Pins 9 and 10 of the AD558 are grounded. This connects the AD558's input register (transparent) to an 8-bit synchronous counter that simulates a data bus.

The test circuit consists of three ICs. One 555 timer is wired as a 1-kHz clock. It steps an 8-bit synchronous binary counter made from two CD4029's, as in Fig. 11-13.

The counter's outputs are wired to the digital inputs of the DAC. A CRO is connected (dc-coupled) to display V_o. It should be externally triggered from the negative edge of the MSB (pin 8 of the AD558 or pin 2 of the right CD4029 in Fig. 11-13). The analog voltage waveshape appearing at V_o will resemble a staircase.

Each clock pulse steps the counter up by one count and increases V_o by 10 mV. Thus the risers will equal 10 mV and the staircase will have 256 treads from 0 to 2.55 V. The tread on each step will occupy about 1 ms. Thus one staircase waveshape is generated every 256 ms. Any glitch, nonlinearity, or other abnormality will be quite apparent. The capacitor C_T can be changed to give faster or slower clock frequencies, as indicated in Fig. 11-13. Any visible glitches can be minimized but *not* eliminated from V_o. However, a *sample-and-hold* or *follow-and-hold amplifier* can be connected to V_o. It waits until the glitch settles down, samples V_o, and holds this correct value. This principle of "deglitching" is presented in Section 12-7.

End of Chapter Exercises

Name: ____________________

Date: __________ **Grade:** __

A. FILL-IN THE BLANKS

Fill in the blanks with the best answer.

1. The resolution of a DAC equals the value of __________ LSB(s).
2. An R-2R ladder network is shown in Fig. 11-4. The resistance looking into any node is the same and equal to __________.
3. The resolution of the basic 4-bit DAC shown in Fig. 11-4 can be increased by increasing ______________________________ ______________________________.
4. Refer to Fig. 11-7. The reference current, I_{ref} of the DAC-08 is controlled by V_{ref} and the __________ resistor.
5. The full scale output voltage of Fig. 11-7 is __________ when $V_{ref} = 10$ V and $R_{ref} = 5$ kΩ.
6. Figure 11-8 shows the DAC-08 connected for a bipolar output. When I_{out} is less than $\overline{I_{out}}$ by 8μA, V_o equals −40 mV. $V_o = -40$ mV is called __________.
7. The programmer of a microprocessor based DAC can select a particular DAC via the __________ bus.
8. The AD558 DAC is __________ when the digital input pins connect the data bus to its internal memory latching register.
9. The select pin (14) of the AD558 is used to control the __________.
10. The time to complete one conversion for the AD558 DAC is about __________ ns.

B. TRUE/FALSE

Fill in **T** if the statement is true, and **F** if any part of the statement is false.

1. ______ The purpose of the op amp in Fig. 11-5 is to convert I_{out} into a voltage V_o.
2. ______ Industry standards for DACs are those with 8, 10, 12, 14, and 16 bits.
3. ______ Refer to Fig. 11-7. Output currents I_{out} and $\overline{I_{out}}$ are converted into an output voltage V_o by the DAC-08.
4. ______ The circuit of Fig. 11-8 is a bipolar DAC that can have an output voltage that is precisely zero.
5. ______ A programmer would view the AD558 DAC as a memory address he or she could write to.

C. CIRCLE THE CORRECT ANSWER

Circle the correct answer for each statement.

1. Resolution of a DAC equals (2^n, 1MSB).
2. The maximum output voltage of a DAC will (always, never) equal the applied reference voltage V_{ref}.
3. Refer to Fig. 11-7. If V_{ref} is increased from 10.0 V to 10.24 V the value of 1 LSB would change from 39 mV to (38 mV, 40 mV).
4. Microprocessor based DAC's are thought of as a (read only, write only) address in memory.
5. Microprocessor based DAC's are selected by the programmer via the (data, address) bus.

D. MATCHING

Match the name or symbol in column **A** with the statement that matches best in column **B**.

	COLUMN A		COLUMN B
1. ______	I_{ref}	**a.**	AD558
2. ______	resolution	**b.**	Ladder
3. ______	μP DAC	**c.**	V_{ref}/R_{ref}
4. ______	$\overline{CS}$	**d.**	1 LSB
5. ______	R-2R	**e.**	Terminal

PROBLEMS

11-1. What is the resolution of a 4-bit DAC? The value of 1 LSB is not known.

11-2. A 4-bit DAC has a full-scale output voltage of 150 mV with a digital input of 1111.
(a) What is its resolution in mV/bit?
(b) If the digital input word is 0101, find V_o.

11-3. An 8-bit DAC has a resolution of 10 mV/bit. What is the output voltage for inputs of
(a) 10000000;
(b) 11111111?

11-4. Resistor values are doubled to $R = 20\ k\Omega$ and $2R = 40\ k\Omega$ in the discrete component DAC of Fig. 11-6. What effect does this have on
(a) I_{ref};
(b) the value of the LSB current at I_{out};
(c) the value of I_{out} for an input word of 11111111;
(d) the value of $\overline{I_{out}}$ for an input of 11111111?

11-5. V_{ref} is changed to 10.24 V in the DAC-08 circuit of Fig. 11-7.
(a) Find the output voltage resolution. Calculate analog output V_o for inputs of
(b) 00000001;
(c) 10000000;
(d) full-scale of 11111111.

11-6. Refer to the DAC-08 bipolar output voltage circuit of Fig. 11-8. If the digital input word increases by one bit, what is the resulting change in **(a)** current I_{out};
(b) current $\overline{I_{out}}$;
(c) V_o?

11-7. What is the difference between a DAC and an MDAC?

11-8. A D/A converter is to be connected to a microprocessor.
(a) Does the programmer think of the DAC as a "read only" or "write only" address? Which bus, data, or address is used by the microprocessor to
(b) send information to the DAC;
(c) select the particular DAC?

11-9. Two terminals ($\overline{CE}$ and $\overline{CS}$) on the DAC are used to tell it
(a) *if* it should accept data from the microprocessor;
(b) *when* to accept data. Which does what?

11-10. A basic DAC contains an R−$2R$ ladder, external or internal reference voltage, and logic to position the ladder switches in response to digital input signals. Name two additional features needed so that the DAC can be used with a microprocessor.

11-11. The AD558 is a typical microprocessor-compatible DAC. When its logic control terminals $\overline{CS}$ and $\overline{CE}$ are high, describe the condition of
(a) the AD558 input register;
(b) the output terminal.

11-12. An AD558, 8-bit DAC is pin-programmed for an output of 0 to 2.55 V.

(a) What is the value of 1 LSB?

(b) What is V_o for a half-scale input of 10000000?

(c) What logic levels should be present on $\overline{CE}$ and $\overline{CS}$ to make the input latches transparent and allow the DAC to make a conversion?

Laboratory Exercise 11

Name: ______________________

Date: __________ **Grade:** __

OPERATING THE AD558 MICROPROCESSOR-COMPATIBLE DAC

OBJECTIVES: Upon completion of this laboratory exercise on a microprocessor-compatible DAC, you will be able to (1) measure its resolution; (2) measure the voltage on each digital input to see if it is receiving a high or low signal from the data bus, and also to tell when a digital input sees an open circuit; (3) observe how the control inputs cause the digital input register to assume either a transparent or latching condition; (4) connect a 555 and two CD4029 4-bit counters to make a continuously running 8-bit counter; and (5) connect the 8-bit counter to the AD558 to make a staircase generator to dynamically test the DAC.

REFERENCE: Chapter 11, AD558 Data Sheet

PARTS LIST

1	AD558 converter (see data sheets, pp. 408–14)	1	2.2-kΩ resistor
2	CD4029 binary counters	2	47-kΩ resistors
1	555 timer	1	0.01-μF capacitor

Procedure A: Measuring the Digital Input Voltages

1. Wire the circuit of Fig. L11-1. Measure the open circuit voltages at each digital input. Pin 1, __________; pin 2, __________; pin 3, __________; pin 4, __________; pin 5, __________; pin 6, __________; pin 7, __________; pin 8, __________. They should all be reasonably equal and typical for an open input of standard TTL logic.

2. Measure analog output V_o. V_o = __________.

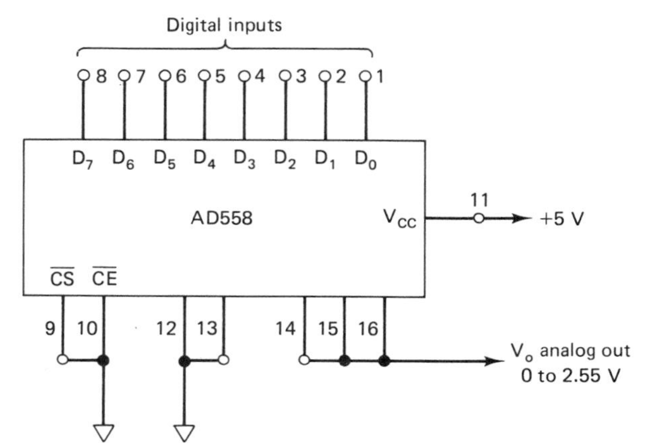

Figure L11-1 Circuit to measure input voltage levels.

3. Ground pins 1 through 8 to apply "zero" logic levels to the digital inputs. Measure V_o. V_o = ______.
4. Remove the grounds from pins 1 through 8. Connect +5 V to pins 1 through 8 to simulate "1" logic levels. Measure V_o. V_o = ______.
5. Does the AD558 interpret an open digital input lead as a "1" or a "0"? ______.

Procedure B: Measuring Resolution

6. Begin with the AD558 wired as in Fig. L11-1.
   (a) Jumper pins 8 and 1 to +5 V. Ground pins 2 through 7. You now have a digital input word, 10000001. Measure V_o. V_o = ______ (DVM).
   (b) Reduce the digital input by 1 LSB to the word 10000000. That is, remove the +5 V from pin 1 and substitute a ground. Measure V_o. V_o = ______ (DVM).
   (c) Subtract the value of V_o in step (b) from the value of V_o in step (a) to obtain ΔV_o. ΔV_o = ______. (ΔV_o is the output voltage change caused by an input code change of 1 LSB.)
   (d) Calculate resolution in mV/bit from resolution = ΔV_o/1 LSB. = Resolution = ______.
   (e) Alternate resolution measurement. Record the value of V_o obtained in step 4. V_o = ______. This corresponds to an input word 11111111 whose decimal value D equals 255. Calculate resolution = V_o/255. Resolution = ______.

(f) Explain any significant difference between the resolution value in steps (d) and (e). *Note:* Your DAC's performance equation is

$$\text{analog } V_o = \text{resolution [step (e)]} \times D$$

where D is the decimal value of the digital input word.)

Procedure C: Control Input Operation

7. Build the circuit of Fig. L11-2. Ground control logic pins 9 and 10. Monitor V_o with a DVM. The jumper wire from pin 8 will be used to change the digital input word. If the jumper is grounded, the input is 00000000. With the jumper open, the input is 10000000. Tapping the jumper to ground will change V_o if the AD558 is "transparent" but not if the AD558 is latching. Complete the table in Fig. L11-2 by observing V_o and applying an intermittent ground to pin 8 for each combination of control voltages on $\overline{\text{CS}}$ and $\overline{\text{CE}}$.

Procedure D: Building an 8-Bit Digital Counter

8. Refer to Fig. L11-3. Wire only the 555 clock circuit. Connect a CRO to pin 3 to measure the clock frequency. $f =$ __________
9. Add the A-CD4029 4-bit counter. Measure the frequencies

Logic pin 9	Control pin 10	Transparent or latching
GND	GND	
GND	+5 V	
+5 V	GND	
+5 V	+5 V	
Open	Open	

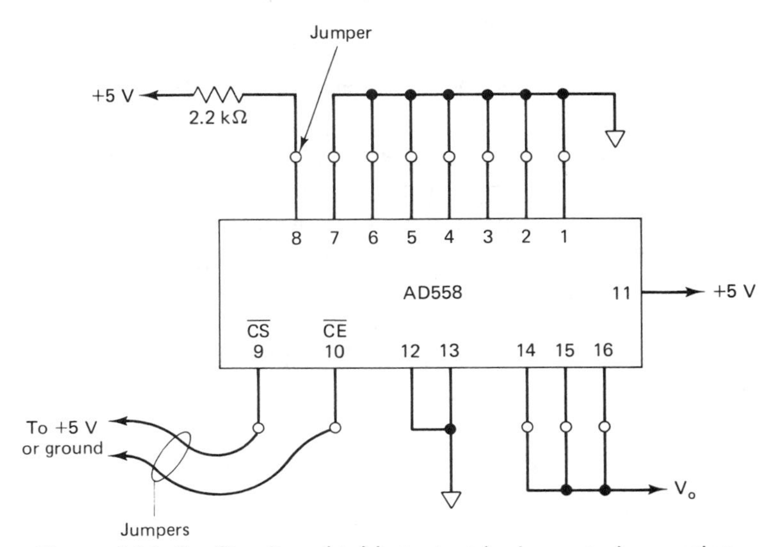

Figure L11-2 Circuit and table to test logic control operation.

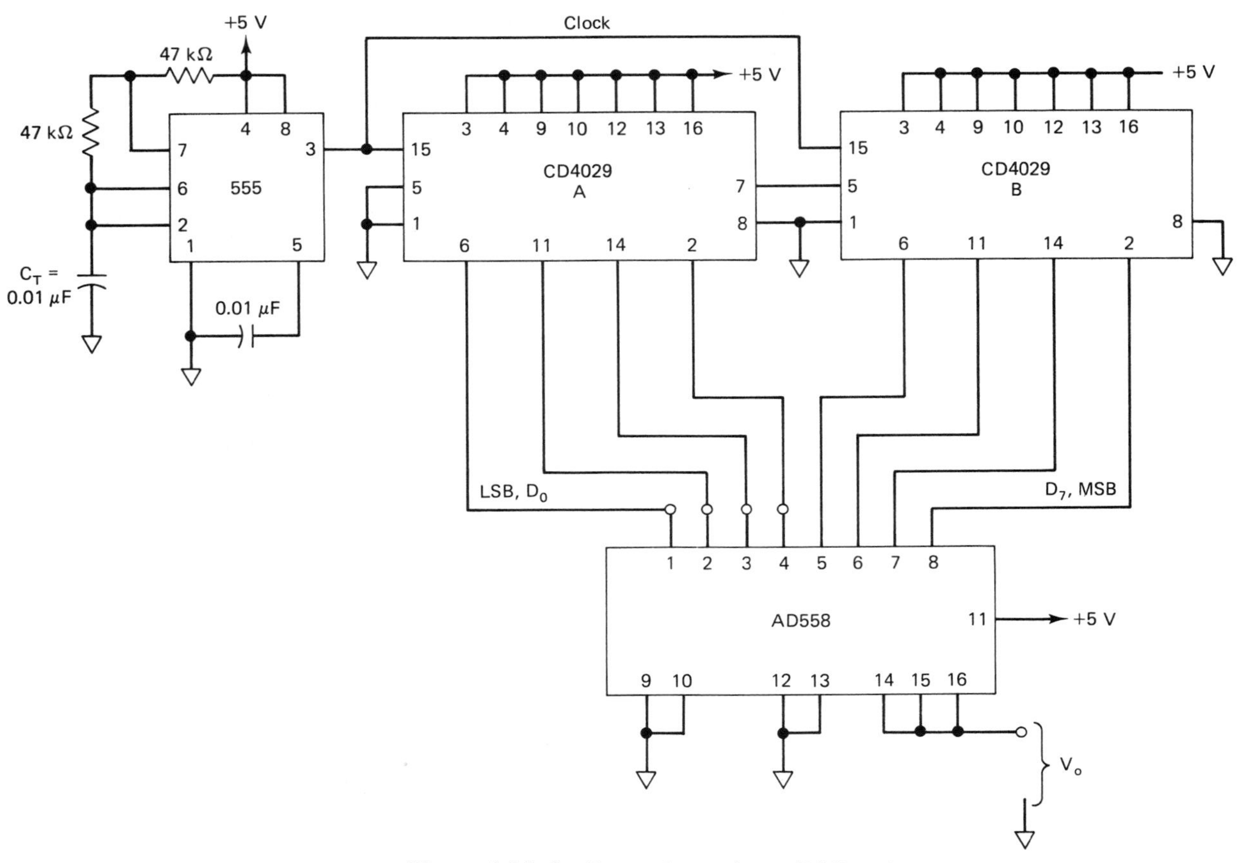

Figure L11-3 Dynamic testing a DAC staircase generator.

at LSB(D_o) pin 6 and pin 2. Pin 6, ________; pin 2, ________. (They should equal $f/2$ and $f/16$.)

10. Add the second B-CD4029 chip and measure the MSB digit frequency at pin 2 (MSB D7). Frequency = ________. (It should equal f/256.)

Procedure E: Dynamic Test of the AD558

11. Wire the 8-bit synchronous counter to the AD558 as in Fig. L11-3 to complete the *staircase generator.* Connect a dual trace CRO to the staircase generator as follows:

- **(a)** Connect the external trigger of the CRO to MSB pin 8 of the AD558. Set to trigger external on the negative edge.
- **(b)** Connect channel 1 of the CRO to measure V_o at pin 16 of the AD558. Zero the trace at the bottom of the CRO screen. The vertical amplifier is set for 0.5 V/div dc-coupled. Time base = 5 ms/div.
- **(c)** Connect channel 2 of the CRO to pin 8 of the AD558 to monitor the MSB. Set the vertical sensitivity for 2 V/div and dc-coupled.
- **(d)** Sketch the waveshape for V_o and also the MSB on Fig. L11-4.

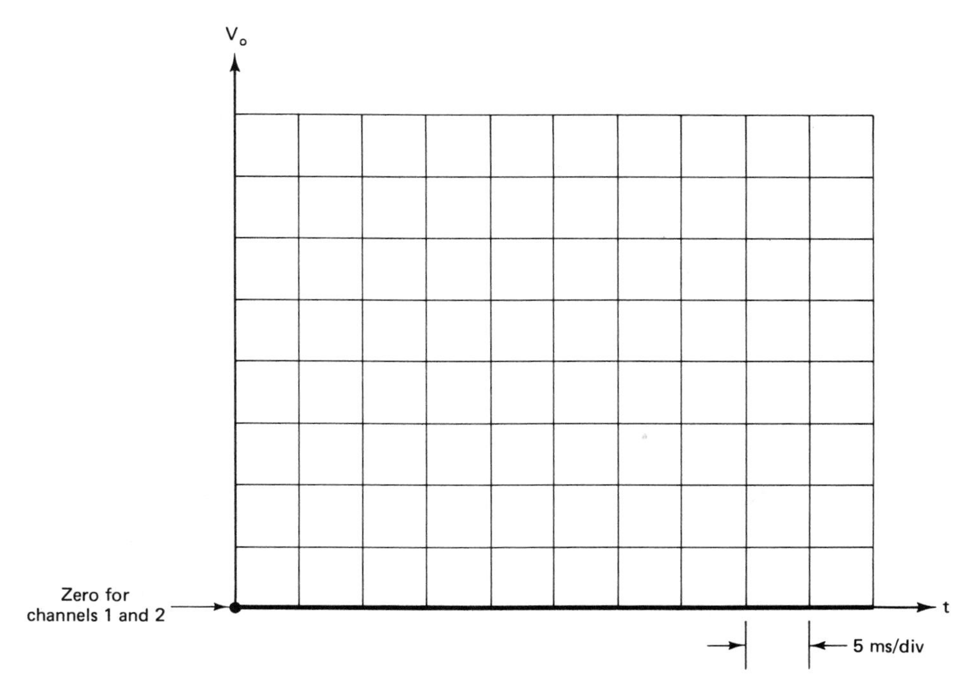

Figure L11-4 Waveshapes of the staircase generator V_O and the most significant bit.

12. Increase the vertical amplifier gain of channel 1 until you can see the voltage increments for V_o. What is the value of ΔV_o? ΔV_o =

__________. This value should correspond to 10 mV and is the resolution or ratio of a change in V_o to a change of 1 bit.

CONCLUSION:

DACPORT
Low-Cost Complete µP-Compatible 8-Bit DAC

AD558*

FEATURES

Complete 8-Bit DAC
Voltage Output — 2 Calibrated Ranges
Internal Precision Band-Gap Reference
Single-Supply Operation: +5V to +15V
Full Microprocessor Interface
Fast: 1µs Voltage Settling to ±1/2LSB
Low Power: 75mW
No User Trims
Guaranteed Monotonic Over Temperature
All Errors Specified T_{min} to T_{max}
Small 16-Pin DIP or PLCC Package
Single Laser-Wafer-Trimmed Chip for Hybrids
Low Cost

PRODUCT DESCRIPTION

The AD558 DACPORT™ is a complete voltage-output 8-bit digital-to-analog converter, including output amplifier, full microprocessor interface and precision voltage reference on a single monolithic chip. No external components or trims are required to interface, with full accuracy, an 8-bit data bus to an analog system.

The performance and versatility of the DACPORT is a result of several recently-developed monolithic bipolar technologies. The complete microprocessor interface and control logic is implemented with integrated injection logic (I^2L), an extremely dense and low-power logic structure that is process-compatible with linear bipolar fabrication. The internal precision voltage reference is the patented low-voltage band-gap circuit which permits full-accuracy performance on a single +5V to +15V power supply. Thin-film silicon-chromium resistors provide the stability required for guaranteed monotonic operation over the entire operating temperature range (all grades), while recent advances in laser-wafer-trimming of these thin-film resistors permit absolute calibration at the factory to within ±1LSB; thus no user-trims for gain or offset are required. A new circuit design provides voltage settling to ±1/2LSB for a full-scale step in 800ns.

The AD558 is available in four performance grades. The AD558J and K are specified for use over the 0 to +70°C temperature range, while the AD558S and T grades are specified for −55°C to +125°C operation. The "J" and "K" grades are available either in 16-pin plastic (N) or hermetic ceramic (D) DIPS. They are also available in 20-pin JEDEC standard PLCC packages. The "S" and "T" grades are available in 16-pin hermetic ceramic DIP packages.

PRODUCT HIGHLIGHTS

1. The 8-bit I^2L input register and fully microprocessor-compatible control logic allow the AD558 to be directly connected to 8- or 16-bit data buses and operated with standard control signals. The latch may be disabled for direct DAC interfacing.
2. The laser-trimmed on-chip SiCr thin-film resistors are calibrated for absolute accuracy and linearity at the factory. Therefore, no user trims are necessary for full rated accuracy over the operating temperature range.
3. The inclusion of a precision low-voltage band-gap reference eliminates the need to specify and apply a separate reference source.
4. The voltage-switching structure of the AD558 DAC section along with a high-speed output amplifier and laser-trimmed resistors give the user a choice of 0V to +2.56V or 0V to +10V output ranges, selectable by pin-strapping. Circuitry is internally compensated for minimum settling time on both ranges; typically settling to ±1/2LSB for a full-scale 2.55 volt step in 800ns.
5. The AD558 is designed and specified to operate from a single +4.5V to +16.5V power supply.
6. Low digital input currents, 100µA max, minimize bus loading. Input thresholds are TTL/low voltage CMOS compatible over the entire operating V_{CC} range.

(continued on page 4)

*Covered by U.S. Patent Nos. 3,887,863; 3,685,045; 4,323795; Patents Pending.
DACPORT is a trademark of Analog Devices, Inc.

One Technology Way; P. O. Box 9106; Norwood, MA 02062-9106 U.S.A.
Tel: 617/329-4700 Twx: 710/394-6577
Telex: 174059 Cables: ANALOG NORWOODMASS

*Courtesy of **Analog Devices, Inc.**

SPECIFICATIONS (@ T_A = +25°C, V_{CC} = +5V to +15V unless otherwise specified)

Model	AD558J Min	AD558J Typ	AD558J Max	AD558K Min	AD558K Typ	AD558K Max	AD558S[1] Min	AD558S[1] Typ	AD558S[1] Max	AD558T[1] Min	AD558T[1] Typ	AD558T[1] Max	Units
RESOLUTION			8			8			8			8	Bits
RELATIVE ACCURACY[2]													
0 to +70°C			±1/2			±1/4			±1/2			±1/4	LSB
−55°C to +125°C									±3/4			±3/8	LSB
OUTPUT													
Ranges[3]		0 to +2.56			0 to +2.56			0 to +2.56			0 to +2.56		V
		0 to +10			0 to +10			0 to +10			0 to +10		V
Current Source	+5			+5			+5			+5			mA
Sink		Internal Passive Pull-Down to Ground[4]			Internal Passive Pull-Down to Ground			Internal Passive Pull-Down to Ground			Internal Passive Pull-Down to Ground		
OUTPUT SETTLING TIME[5]													
0 to 2.56 Volt Range		0.8	1.5		0.8	1.5		0.8	1.5		0.8	1.5	μs
0 to 10 Volt Range[4]		2.0	3.0		2.0	3.0		2.0	3.0		2.0	3.0	μs
FULL SCALE ACCURACY[6]													
@25°C			±1.5			±0.5			±1.5			±0.5	LSB
T_{min} to T_{max}			±2.5			±1			±2.5			±1	LSB
ZERO ERROR													
@25°C			±1			±1/2			±1			±1/2	LSB
T_{min} to T_{max}			±2			±1			±2			±1	LSB
MONOTONICITY[7]													
T_{min} to T_{max}		Guaranteed			Guaranteed			Guaranteed			Guaranteed		
DIGITAL INPUTS													
T_{min} to T_{max}													
Input Current			±100			±100			±100			100	μA
Data Inputs, Voltage													
Bit On – Logic "1"	2.0			2.0			2.0			2.0			V
Bit On – Logic "0"	0		0.8	0			0			0			V
Control Inputs, Voltage													
On – Logic "1"	2.0			2.0			2.0			2.0			V
On – Logic "0"	0		0.8	0		0.8	0		0.8	0		0.8	V
Input Capacitance		4			4			4			4		pF
TIMING[8]													
t_W Strobe Pulse Width	200			200			200			200			ns
T_{min} to T_{max}	270			270			270			270			ns
t_{DH} Data Hold Time	10			10			10			10			ns
T_{min} to T_{max}	10			10			10			10			ns
t_{DS} Data Set-Up Time	200			200			200			200			ns
T_{min} to T_{max}	270			270			270			270			ns
POWER SUPPLY													
Operating Voltage Range (V_{CC})													
2.56 Volt Range	+4.5		+16.5	+4.5		+16.5	+4.5		+16.5	+4.5		+16.5	V
10 Volt Range	+11.4		+16.5	+11.4		+16.5	+11.4		+16.5	+11.4		+16.5	V
Current (I_{CC})		15	25		15	25		15	25		15	25	mA
Rejection Ratio			0.03			0.03			0.03			0.03	%/%
POWER DISSIPATION, V_{CC} = 5V		75	125		75	125		75	125		75	125	mW
V_{CC} = 15V		225	375		225	375		225	375		225	375	mW
OPERATING TEMPERATURE RANGE	0		+70	0		+70	−55		+125	−55		+125	°C

NOTES

[1]The AD558 S & T grades are available processed and screened to MIL-STD-883 Class B. Consult Analog Devices' Military Databook for details.

[2]Relative Accuracy is defined as the deviation of the code transition points from the ideal transfer point on a straight line from the zero to the full scale of the device.

[3]Operation of the 0 to 10 volt output range requires a minimum supply voltage of +11.4 volts.

[4]Passive pull-down resistance is 2kΩ for 2.56 volt range, 10kΩ for 10 volt range.

[5]Settling time is specified for a positive-going full-scale step to ±1/2LSB. Negative-going steps to zero are slower, but can be improved with an external pull-down.

[6]The full range output voltage for the 2.56 range is 2.55V and is guaranteed with a +5V supply, for the 10V range, it is 9.960V guaranteed with a +15V supply.

[7]A monotonic converter has a maximum differential linearity error of ±1LSB.

[8]See Figure 7.

Specifications subject to change without notice.

AD558 METALIZATION PHOTOGRAPH

Dimensions shown in inches and (mm).

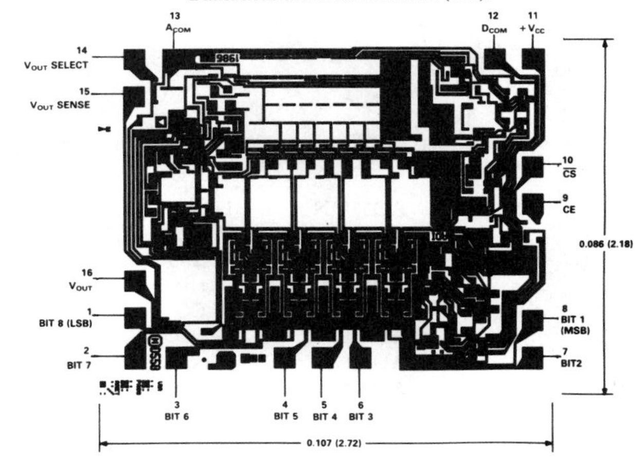

ABSOLUTE MAXIMUM RATINGS

V_{CC} to Ground . . . . .0V to +18V
Digital Inputs (Pins 1-10) . . . . 0 to +7.0V
V_{OUT} . . . . Indefinite Short to Ground
Momentary Short to V_{CC}
Power Dissipation 450mW
Storage Temperature Range
N/P (Plastic) Packages . . . . -25°C to +100°C
D (Ceramic) Package . . . . -55°C to +150°C
Lead Temperature (soldering, 10 second). . . . . 300°C
Thermal Resistance
Junction to Ambient/Junction to Case
D (Ceramic) Package . . . . 100/30°C/W
N/P (Plastic) Packages . . . . 140/55°C/W

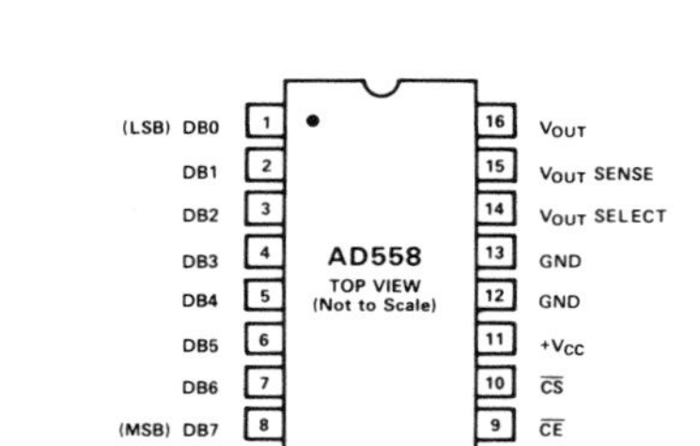

Figure 1a. AD558 Pin Configuration (DIP)

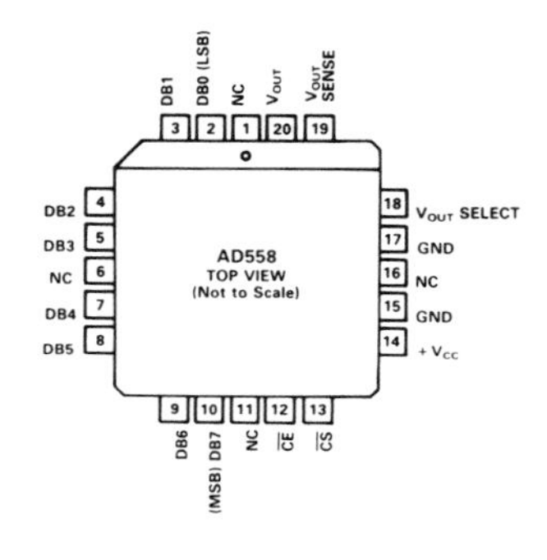

Figure 1b. AD558 Pin Configuration (PLCC)

OUTLINE DIMENSIONS

Dimensions shown in inches and (mm).

N (PLASTIC) PACKAGE

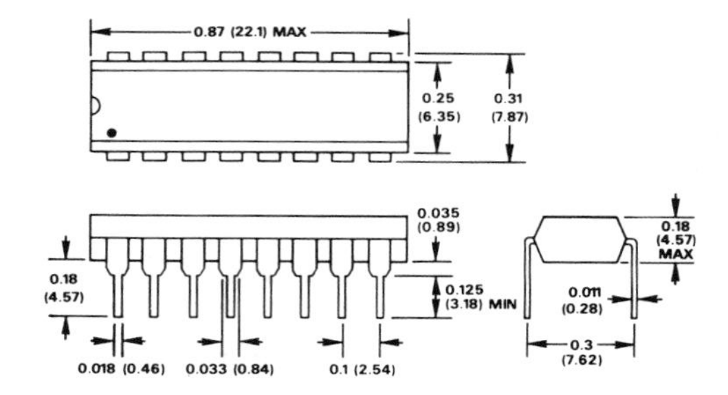

P (PLCC) PACKAGE

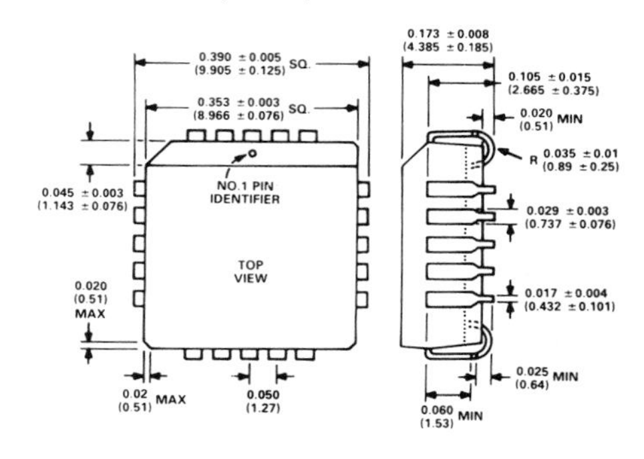

D (CERAMIC) PACKAGE

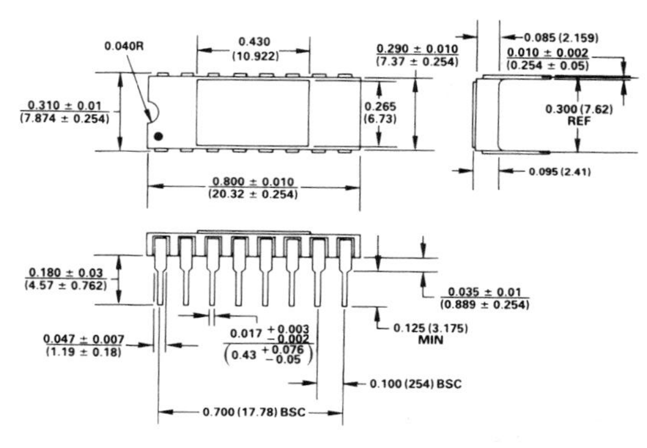

AD558 ORDERING GUIDE

Model	Package	Temperature	Relative Accuracy Error Max T_{min} to T_{max}	Full-Scale Error, Max T_{min} to T_{max}
AD558JN	Plastic DIP	0 to +70°C	±1/2LSB	±2.5LSB
AD558JP	PLCC	0 to +70°C	±1/2LSB	±2.5LSB
AD558JD	Ceramic DIP	0 to +70°C	±1/2LSB	±2.5LSB
AD558KN	Plastic DIP	0 to +70°C	±1/4LSB	±1LSB
AD558KP	PLCC	0 to +70°C	±1/4LSB	±1LSB
AD558KD	Ceramic DIP	0 to +70°C	±1/4LSB	±1LSB
AD558SD	Ceramic DIP	−55°C to +125°C	±3/4LSB	±2.5LSB
AD558TD	Ceramic DIP	−55°C to +125°C	±3/8LSB	±1LSB

(continued from page 1)

7. The single-chip, low power I^2L design of the AD558 is inherently more reliable than hybrid multi-chip or conventional single-chip bipolar designs. The AD558S and T grades which are specified over the −55°C to +125°C temperature range, are available processed to MIL-STD-883, Class B.

8. All AD558 grades are available in chip form with guaranteed specifications from +25°C to T_{max}. MIL-STD-883, Class B visual inspection is standard on Analog Devices bipolar chips. Contact the factory for additional chip information.

CIRCUIT DESCRIPTION

The AD558 consists of four major functional blocks, fabricated on a single monolithic chip (see Figure 2). The main D to A converter section uses eight equally-weighted laser-trimmed current sources switched into a silicon-chromium thin-film R/2R resistor ladder network to give a direct but unbuffered 0mV to 400mV output range. The transistors that form the DAC switches are PNPs; this allows direct positive-voltage logic interface and a zero-based output range.

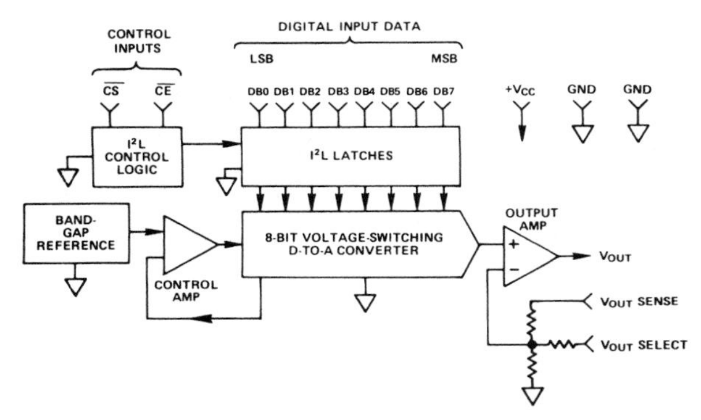

Figure 2. AD558 Functional Block Diagram

The high-speed output buffer amplifier is operated in the non-inverting mode with gain determined by the user-connections at the output range select pin. The gain-setting application resistors are thin-film laser-trimmed to match and track the DAC resistors and to assure precise initial calibration of the two output ranges, 0V to 2.56V and 0V to 10V. The amplifier output stage is an NPN transistor with passive pull-down for zero-based output capability with a single power supply.

The internal precision voltage reference is of the patented band-gap type. This design produces a reference voltage of 1.2 volts and thus, unlike 6.3 volt temperature-compensated zeners, may be operated from a single, low-voltage logic power supply. The microprocessor interface logic consists of an 8-bit data latch and control circuitry. Low-power, small geometry and high-speed are advantages of the I^2L design as applied to this section. I^2L is bipolar process compatible so that the performance of the analog sections need not be compromised to provide on-chip logic capabilities. The control logic allows the latches to be operated from a decoded microprocessor address and write signal. If the application does not involve a μP or data bus, wiring $\overline{CS}$ and $\overline{CE}$ to ground renders the latches "transparent" for direct DAC access.

MIL-STD-883

The rigors of the military/aerospace environment, temperature extremes, humidity, mechanical stress, etc., demand the utmost in electronic circuits. The AD558, with the inherent reliability of integrated circuit construction, was designed with these applications in mind. The hermetically-sealed, low profile DIP package takes up a fraction of the space required by equivalent modular designs and protects the chip from hazardous environments. To further ensure reliability, military-temperature range AD558 grades S and T are available screened to MIL-STD-883. For more complete data sheet information consult the Analog Devices' Military Databook.

CHIP AVAILABILITY

The AD558 is available in laser-trimmed, passivated chip form. AD558J and AD558T chips are available. Consult the factory for details.

Digital Input Code			Output Voltage	
Binary	Hexadecimal	Decimal	2.56V Range	10.00V Range
0000 0000	00	0	0	0
0000 0001	01	1	0.010V	0.039V
0000 0010	02	2	0.020V	0.078V
0000 1111	0F	15	0.150V	0.586V
0001 0000	10	16	0.160V	0.625V
0111 1111	7F	127	1.270V	4.961V
1000 0000	80	128	1.280V	5.000V
1100 0000	C0	192	1.920V	7.500V
1111 1111	FF	255	2.55V	9.961V

Input Logic Coding

CONNECTING THE AD558

The AD558 has been configured for ease of application. All reference, output amplifier and logic connections are made internally. In addition, all calibration trims are performed at the factory assuring specified accuracy without user trims. The only connection decision that must be made by the user is a single jumper to select output voltage range. Clean circuit-board layout is facilitated by isolating all digital bit inputs on one side of the package; analog outputs are on the opposite side.

Figure 3 shows the two alternative output range connections. The 0V to 2.56V range may be selected for use with any power supply between +4.5V and +16.5V. The 0V to 10V range requires a power supply of +11.4V to +16.5V.

Because of its precise factory calibration, the AD558 is intended to be operated without user trims for gain and offset; therefore no provisions have been made for such user-trims. If a small increase in scale is required, however, it may be accomplished by slightly altering the effective gain of the output buffer. A resistor in series with V_{OUT} SENSE will increase the output range.

For example if a 0V to 10.24V output range is desired (40mV = 1LSB), a nominal resistance of 850Ω is required. It must be remembered that, although the internal resistors all ratio-match and track, the *absolute* tolerance of these resistors is typically ±20% and the *absolute* TC is typically −50ppm/°C (0 to −100ppm/°C). That must be considered when re-scaling is performed. Figure 4 shows the recommended circuitry for a full-scale output range of 10.24 volts. Internal resistance values shown are nominal.

NOTE: Decreasing the scale by putting a resistor in series with GND will not work properly due to the code-dependent currents in GND. Adjusting offset by injecting dc at GND is not recommended for the same reason.

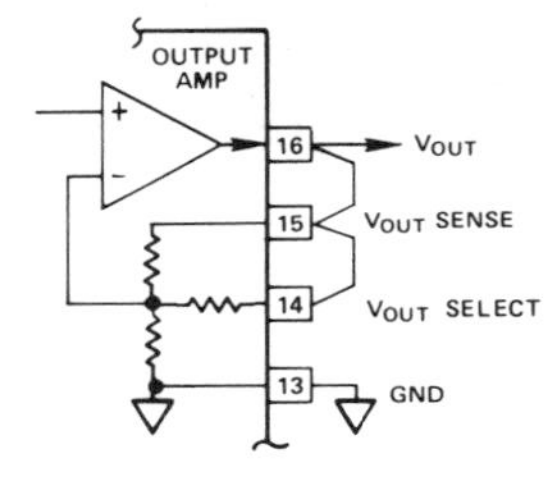

a. 0V to 2.56V Output Range

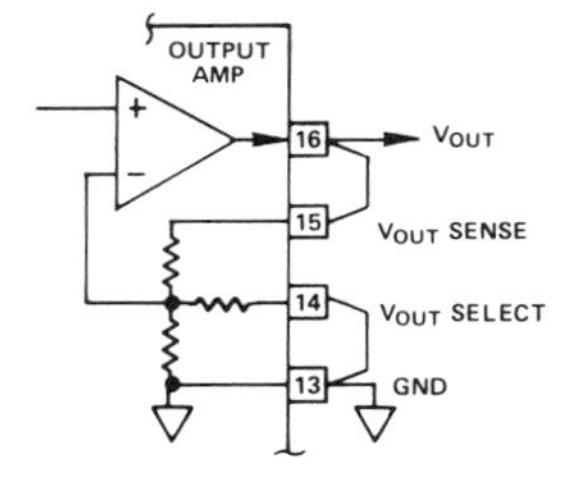

b. 0V to 10V Output Range

Figure 3. Connection Diagrams

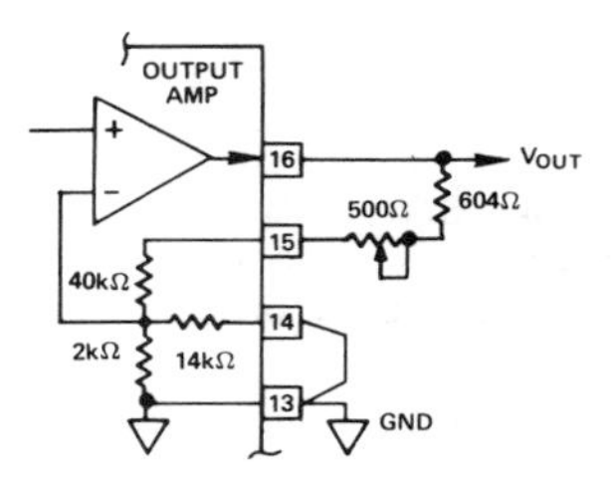

Figure 4. 10.24V Full-Scale Connection

GROUNDING AND BYPASSING*

All precision converter products require careful application of good grounding practices to maintain full rated performance. Because the AD558 is intended for application in microcomputer systems where digital noise is prevalent, special care must be taken to assure that its inherent precision is realized.

The AD558 has two ground (common) pins; this minimizes ground drops and noise in the analog signal path. Figure 5 shows how the ground connections should be made.

It is often advisable to maintain separate analog and digital grounds throughout a complete system, tying them common in one place only. If the common tie-point is remote and accidental disconnection of that one common tie-point occurs due to card removal with power on, a large differential voltage between the two commons could develop. To protect devices that interface to both digital and analog parts of the system, such as the AD558, it is recommended that common ground tie-points should be provided at *each* such device. If only one system ground can be connected directly to the AD558, it is recommended that analog common be selected.

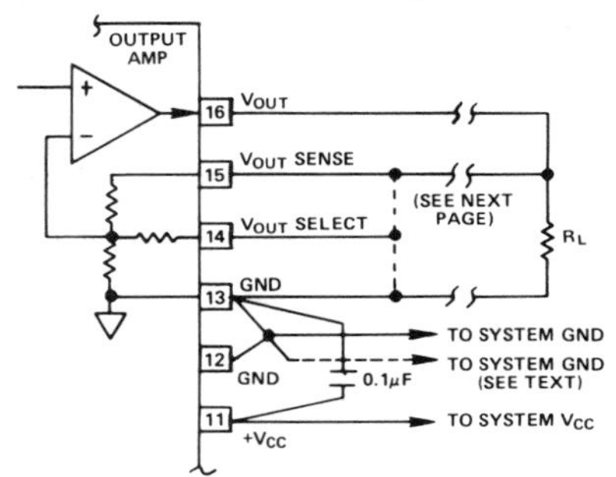

Figure 5. Recommended Grounding and Bypassing

POWER SUPPLY CONSIDERATIONS

The AD558 is designed to operate from a single positive power supply voltage. Specified performance is achieved for any supply voltage between +4.5V and +16.5V. This makes the AD558 ideal for battery-operated, portable, automotive or digital main-frame applications.

The only consideration in selecting a supply voltage is that, in order to be able to use the 0V to 10V output range, the power supply voltage must be between +11.4V and +16.5V. If, however, the 0V to 2.56V range is to be used, power consumption will be minimized by utilizing the lowest available supply voltage (above +4.5V).

*For additional insight, "An IC Amplifier Users' Guide to Decoupling, Grounding and Making Things Go Right For A Change", is available at no charge from any Analog Devices Sales Office.

TIMING AND CONTROL

The AD558 has data input latches that simplify interface to 8- and 16-bit data buses. These latches are controlled by Chip Enable ($\overline{CE}$) and Chip Select ($\overline{CS}$) inputs. $\overline{CE}$ and $\overline{CS}$ are internally "NORed" so that the latches transmit input data to the DAC section when both $\overline{CE}$ and $\overline{CS}$ are at Logic "0". If the application does not involve a data bus, a "00" condition allows for direct operation of the DAC. When either $\overline{CE}$ or $\overline{CS}$ go to Logic "1", the input data is latched into the registers and held until both $\overline{CE}$ and $\overline{CS}$ return to "0". (Unused $\overline{CE}$ or $\overline{CS}$ inputs should be tied to ground.) The truth table is given in Table I. The logic function is also shown in Figure 6.

Input Data	$\overline{CE}$	$\overline{CS}$	DAC Data	Latch Condition
0	0	0	0	"transparent"
1	0	0	1	"transparent"
0	∫	0	0	latching
1	∫	0	1	latching
0	0	∫	0	latching
1	0	∫	1	latching
X	1	X	previous data	latched
X	X	1	previous data	latched

Notes: X = Does not matter
∫ = Logic Threshold at Positive-Going Transition

Table I. AD558 Control Logic Truth Table

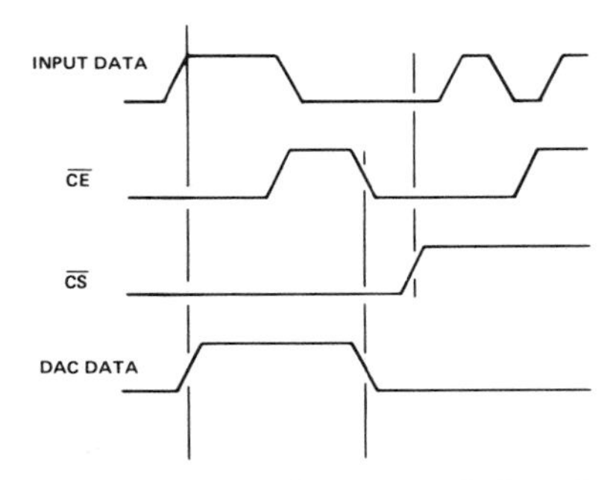

Figure 6. AD558 Control Logic Function

In a level-triggered latch such as that in the AD558 there is an interaction between data setup and hold times and the width of the enable pulse. In an effort to reduce the time required to test all possible combinations in production, the AD558 is tested with $t_{DS} = t_W = 200ns$ at 25°C and 270ns at T_{min} and T_{max}, with $t_{DH} = 10ns$ at all temperatures. Failure to comply with these specifications may result in data not being latched properly.

Figure 7 shows the timing for the data and control signals; $\overline{CE}$ and $\overline{CS}$ are identical in timing as well as in function.

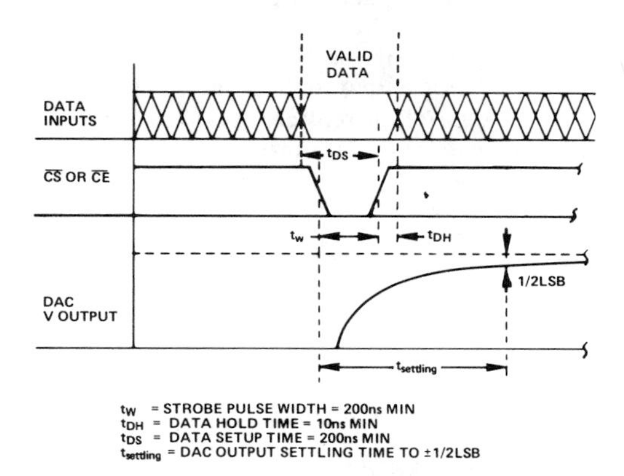

Figure 7. AD558 Timing

USE OF V_{OUT} SENSE

Separate access to the feedback resistor of the output amplifier allows additional application versatility. Figure 8a shows how I × R drops in long lines to remote loads may be cancelled by putting the drops "inside the loop." Figure 8b shows how the separate sense may be used to provide a higher output current by feeding back around a simple current booster.

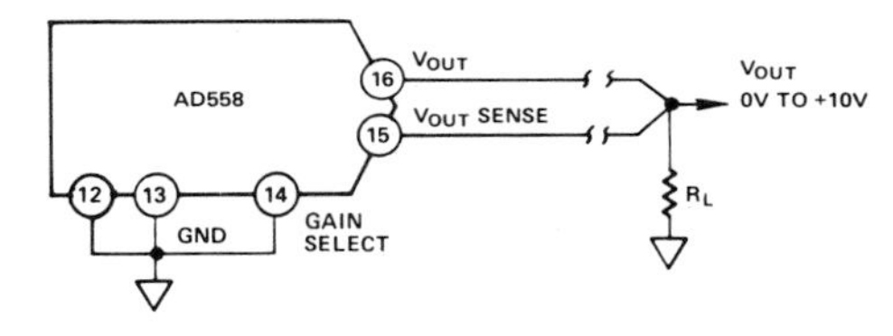

a. Compensation for I x R Drops in Output Lines

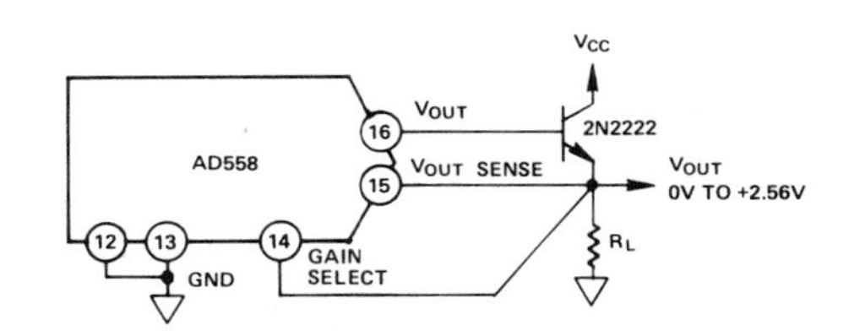

b. Output Current Booster

Figure 8. Use of V_{OUT} Sense

OPTIMIZING SETTLING TIME

In order to provide single-supply operation and zero-based output voltage ranges, the AD558 output stage has a passive "pull-down" to ground. As a result, settling time for negative-going output steps may be longer than for positive-going output steps. The relative difference depends on load resistance and capacitance. If a negative power supply is available, the negative-going settling time may be improved by adding a pull-down resistor from the output to the negative supply as shown in Figure 9. The value of the resistor should be such that, at zero voltage out, current through that resistor is 0.5mA max.

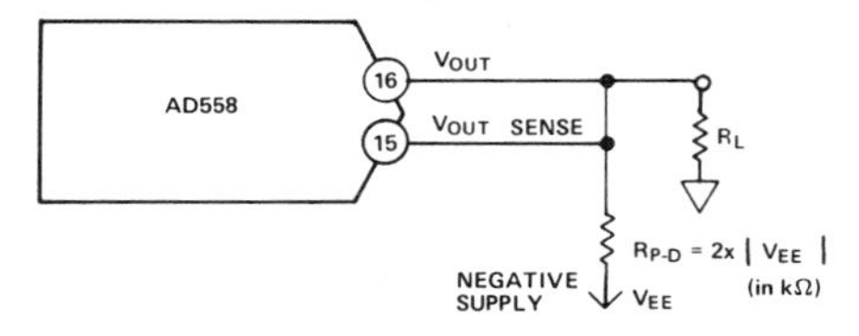

Figure 9. Improved Settling Time

BIPOLAR OUTPUT RANGES

The AD558 was designed for operation from a single power supply and is thus capable of providing only unipolar (0V to +2.56 and 0V to 10V) output ranges. If a negative supply is available, bipolar output ranges may be achieved by suitable output offsetting and scaling. Figure 10 shows how a ±1.28 volt output range may be achieved when a −5 volt power supply is available. The offset is provided by the AD589 precision 1.2 volt reference which will operate from a +5 volt supply. The AD544 output amplifier can provide the necessary ±1.28 volt output swing from ±5 volt supplies. Coding is complementary offset binary.

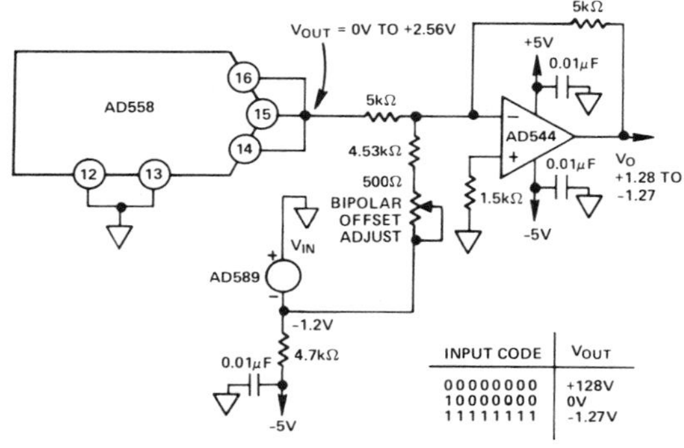

INPUT CODE	VOUT
00000000	+128V
10000000	0V
11111111	-1.27V

Figure 10. Bipolar Operation of AD558 from ±5V Supplies

INTERFACING THE AD558 TO MICROPROCESSOR DATA BUSES*

The AD558 is configured to act like a "write only" location in memory that may be made to coincide with a read only memory location or with a RAM location. The latter case allows data previously written into the DAC to be read back later via the RAM. Address decoding is partially complete for either ROM or RAM. Figure 11 shows interfaces for three popular microprocessor systems.

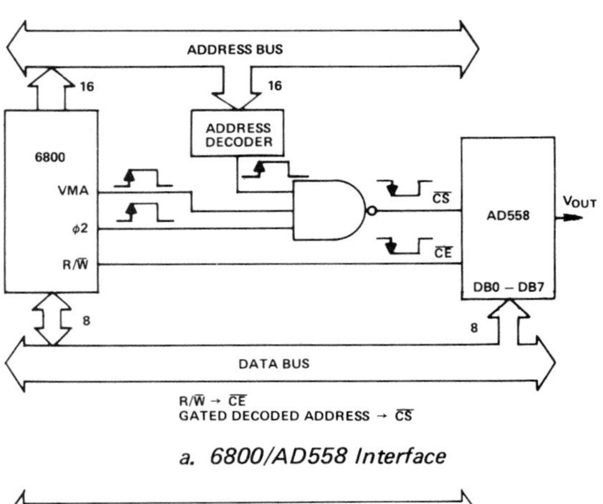

a. 6800/AD558 Interface

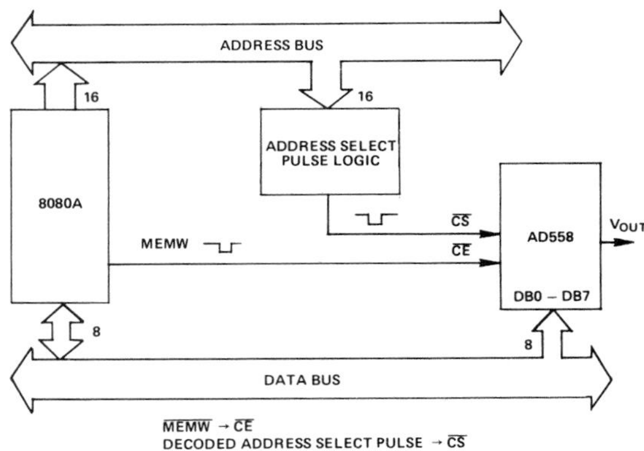

b. 8080A/AD558 Interface

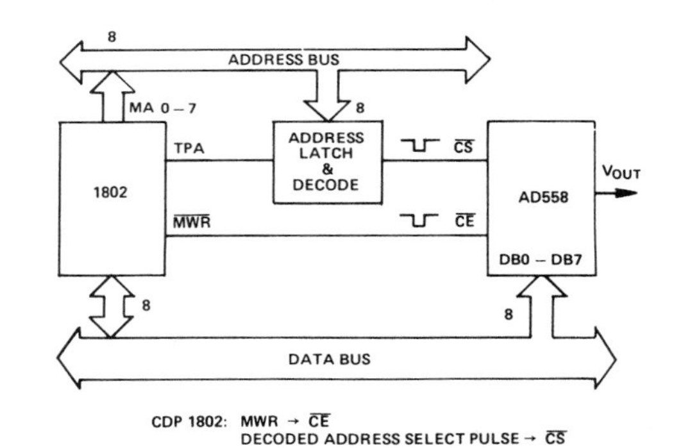

c. 1802/AD558 Interface

Figure 11. Interfacing the AD558 to Microprocessors

*The microprocessor-interface capabilities of the AD558 are extensive. A comprehensive application note, "Interfacing the AD558 DACPORT™ to Microprocessors" is available from any Analog Devices Sales Office upon request, free of charge.

Analog-to-Digital Converters

12

LEARNING OBJECTIVES

Upon completion of this chapter on analog-to-digital converters, you will be able to:

- Draw a typical digital output versus analog input voltage characteristic for a 3- or 4-bit A/D converter, and define their resolution.
- Name the three most common types of A/D converters, define conversion time, and list the converters in order of decreasing conversion time.
- State the need for ADCs.
- Explain the principles of operation for a dual-slope integrating ADC by (1) naming the three operational phases used to make one analog-to-digital conversion, (2) drawing a block diagram, (3) describing the purpose for each operating phase T_1, T_2, and T_z, (4) calculating times required for each phase for a given input voltage, and (5) finding the output versus input equation.
- Draw a block diagram for a successive approximation register and explain its principle of operation.
- Describe operating features of the AD670, an 8-bit microprocessor-compatible ADC by (1) listing the tasks performed by each terminal, (2) showing how it can be pin-programmed for analog input ranges of 0 to 2.55 V or 0 to 255 mV, and (3) pin-programmed for straight binary, offset binary, or 2's-complement output codes.
- Describe the procedure by which a microprocessor tells only one ADC out of many to (1) convert an analog signal to a digital code, and (2) tell the microprocessor when conversion is over.
- Make a stand-alone test circuit to measure output versus input performance of an AD670 without the need of a microprocessor.
- Calculate the frequency response of an ADC if you know its conversion time.
- Describe the operation of a sample-and-hold or follower-and-hold amplifier and show how it can improve frequency response of an ADC.

12-0 INTRODUCTION

12-0.1 Need for A/D Converters

Physical variables and processes are analog in nature. They vary continuously, unlike the digital two-value world of the computer. In modern technology, the powerful capabilities of computers must be applied to the analog world. A compelling need developed for interfaces between analog and digital technologies.

Two types of devices are needed to obtain data about physical processes and send it into a computer. They are transducers and analog-to-digital converters, hereafter called ADCs. Transducers convert changes in physical variables, such as temperature or pressure, into changes of electrical quantities, such as voltage, current, resistance, inductance, or capacitance. These electrical quantities must be converted to an analog signal voltage whose range will match with that required by the ADC. This signal voltage is applied to an analog-to-digital converter. The ADC converts this analog signal voltage into a digital output compatible with microprocessors or computers.

12-0.2 Ideal ADC and Resolution

The output versus input characteristic of an ideal 3-bit ADC is shown in Fig. 12-1. It differs from the ideal DAC characteristics in Fig. 11-1. Suppose that V_{in} varied continuously from a value of 0 to full-scale voltage of $V_{FS} = 7$ V. A 3-bit ADC would have $2^3 = 8$ possible outputs. They are binary 000 to 111. Each additional input volt would increase the digital output by 1 LSB. We have just defined the *resolution* of an ADC. *Resolution equals that change in value of input voltage required to change the digital output by 1 LSB.* Formally, resolution is defined by

$$\text{Resolution} = \frac{\text{value}}{\text{bit}} = \frac{\text{FS output}}{2^n - 1} \tag{12-1a}$$

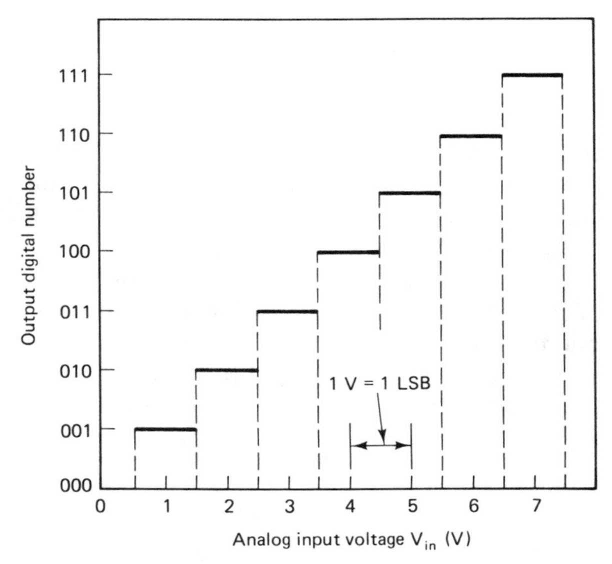

Figure 12-1 Digital output code versus analog input voltage of an ideal 3-bit D/A converter. Resolution is 1 V per bit or $2^3 = 8$.

An alternative definition of resolution is the *number of digital output codes* for V_{in} or

$$\text{Resolution} = 2^n \tag{12-1b}$$

where n is the number of bits.

12-0.3 Quantization Error

Figure 12-1 shows that the ADC's output equals binary 001 for an analog input of 1 V. An output of 001 results also when V_{in} lies between 0.5 and 1.5 V. There is an unavoidable uncertainty about the value of V_{in} when the output is 001. This uncertainty equals $\pm\frac{1}{2}$ LSB and is called *quantization error.* It is inherent in all ADCs. It can be reduced only by dividing the analog input into more partitions by increasing n to obtain finer resolution.

12-0.4 Types of A/D Converters

Dozens of procedures have been devised to perform A/D conversion. However, only three classifications have survived as standards. With respect to conversion time (time to complete one A/D conversion), they are classified as slow, fast, and very fast.

The slow *integrating A/D converter and fast successive approximation converter* services over 80% of the A/D market. *Flash* converters are the fastest. A rough comparison is made between all three types in Table 12-1.

TABLE 12-1
Types of A/D Converters

Type of converter	*Relative speed*	*Conversion time (typical)*	*Typical applications*
Integrating	Slow	Milliseconds	DC voltmeter
Successive approximation	Fast	Microseconds	Audio
Flash	Very fast	Nanoseconds	Video

The integrating ADC is very slow. However, it is superbly fitted to measure dc voltages and ignores line-frequency noise. Its long conversion time allows very high resolutions. Successive approximation converters are fast, accurate, and have conversion times down to about 1 μs. Flash converters are so fast that they can perform essentially continuous conversions. Conversion time can be less than 50 ns. They are costly, particularly if resolution must exceed 8 bits.

12-1 INTEGRATING A/D CONVERTER

The most popular version is the dual-slope integrating A/D converter. These ADCs are complex circuits that combine both analog and digital technology. Fortunately, they are readily available at modest cost in a single integrated circuit, two companion ICs, or in the form of digital panel meters from a variety of manufacturers. We select Intersil's $3\frac{1}{2}$-digit 7106 or 7107 to show the internal operation of a typical integrating ADC.

12-1.1 Principles of Operation

The block diagram of a typical dual-slope integrating A/D converter is shown in Fig. 12-2b. An on-board divide-by-4 counter drives the control logic at a rate of 12 kHz. This frequency is set by the user, via external

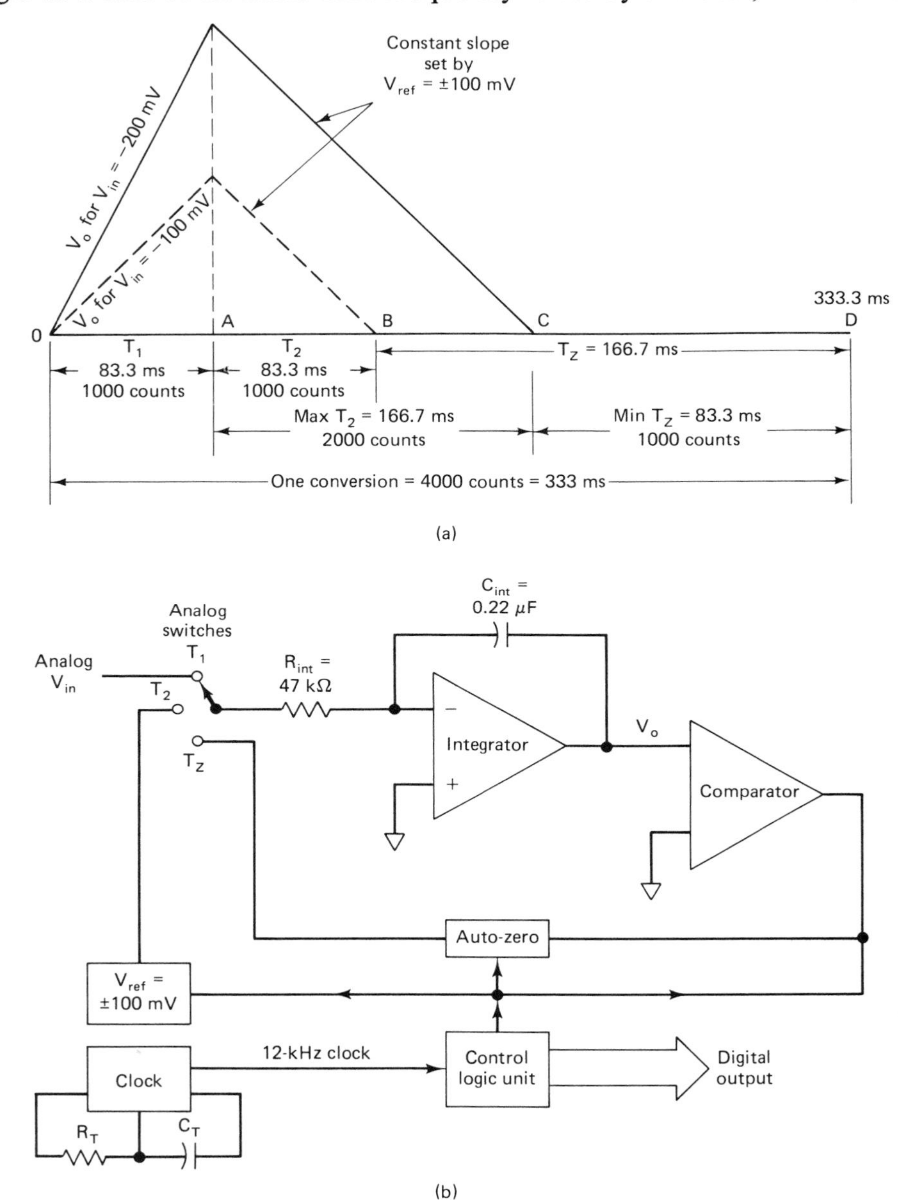

Figure 12-2 (a) Timing diagram of a typical dual-slope integrating ADC; (b) simplified block diagram of a dual-slope integrating ADC typical block diagram. In (a), one A/D conversion takes place in three phases: signal integrating phase T_1, reference T_2, and auto-zero T_z.

timing resistor R_T and C_T. It must be a multiple of the local line frequency (50 or 60 Hz) to render the ADC immune to line-frequency noise.

The control logic unit activates a complex network of logic circuits and analog switches to convert analog input voltage V_{in} into a digital output. The analog-to-digital conversion is performed in three phases and requires about one-third of a second. These operating phases are called *signal integrate phase* T_1, *reference integrate phase* T_2, and *auto-zero phase* T_z. These will be discussed in sequence.

12-1.2 Signal Integrate Phase, T_1

The control logic unit of Fig. 12-2b connects V_{in}, to an integrator to begin phase T_1. The integrator or ramp generator's output V_o ramps up or down depending on the polarity of V_{in} and at a rate set by V_{in}, R_{int}, and C_{int}. If V_{in} is negative, V_o ramps up, as shown in Fig. 12-2a. Time T_1 is set by the logic unit for 1000 clock pulses. Since the 12-kHz clock has a period of 83.3 μs per count, T_1 lasts 83.33 ms.

If $V_{in} = -100$ mV, V_o will ramp from 0 V to 806 mV. The maximum allowed full-scale value of V_{in} is ± 200 mV. When $V_{in} = -200$ mV, V_o will rise to a maximum of 1012 mV. Clearly, V_o is directly proportional to V_{in}. At the end of 1000 counts, the logic unit disconnects V_{in} and connects V_{ref} to the integrator. This action ends T_1 and begins T_2.

12-1.3 Reference Integrate Phase, T_2

During phase T_1, the logic unit determined the polarity of V_{in} and charged a reference capacitor, C_{ref} (not shown), to a reference voltage $V_{ref} = 100$ mV. At the beginning of phase T_2, the logic unit connects C_{ref} to the integrator so that V_{ref} has a polarity opposite to V_{in}. Consequently, V_{ref} will ramp the integrator back toward zero. Since V_{ref} is constant, the integrator's output V_o will ramp down at a *constant* rate, as shown in Fig. 12-2a.

When V_o reaches zero, a comparator tells the logic unit to terminate phase T_2 and begin the next auto-zero phase. T_2 is thus proportional to V_o and consequently, V_{in}. The exact relationship is

$$T_2 = T_1\left(\frac{V_{in}}{V_{ref}}\right) \tag{12-2a}$$

Since $T_1 = 83.33$ ms and $V_{ref} = 100$ mV,

$$T_2 = \left(0.833\,\frac{\text{ms}}{\text{mV}}\right)V_{in} \tag{12-2b}$$

EXAMPLE 12-1

For the ADC of Fig. 12-2, calculate T_2 if (a) $V_{in} = \pm\ 100$ mV; (b) $V_{in} = \pm 200$ mV.

SOLUTION (a) From Eq. (12-2b),

$$T_2 = \left(0.833\,\frac{\text{ms}}{\text{mV}}\right)(100\text{ mV}) = 83.33\text{ ms}$$

$$\text{(b) } T_2 = \left(0.833\,\frac{\text{ms}}{\text{mV}}\right)(200\text{ mV}) = 166.6\text{ ms}$$

12-1.4 A/D Conversion

The actual conversion of analog voltage V_{in} into a digital count occurs during T_2 as follows. The control unit connects the clock to an internal binary-coded-decimal counter at the beginning of phase T_2. The clock is disconnected from the counter at the end of T_2. Thus the counter's content becomes the digital output. This digital output is set by T_2 and the clock frequency:

$$\text{digital output} = \left(\frac{\text{counts}}{\text{second}}\right) T_2 \tag{12-3a}$$

but T_2 is set by V_{in} from Eq. (12-1a) and therefore

$$\text{digital output} = \left(\frac{\text{counts}}{\text{second}}\right)(T_1)\left(\frac{V_{in}}{V_{ref}}\right) \tag{12-3b}$$

Since clock frequency is 12 kHz for the 7106/7107, $T_1 = 83.33$ ms, and $V_{ref} = 100$ mV, the output–input equation is

$$\text{digital output} = \left(12{,}000\ \frac{\text{counts}}{\text{second}}\right)\left(\frac{83.33\ \text{ms}}{\text{mV}}\right) V_{in} \text{ or}$$

$$\text{digital output} = \left(10\ \frac{\text{counts}}{\text{mV}}\right) V_{in} \tag{12-3c}$$

The digital output can be connected to an appropriate $3\frac{1}{2}$-digit display.

EXAMPLE 12-2

V_{in} equals +100 mV in the ADC of Fig. 12-2. Find the digital output.

SOLUTION From Eq. (12-3c),

$$\text{digital output} = \left(10\ \frac{\text{counts}}{\text{mV}}\right)(100\ \text{mV}) = 1000\ \text{counts}$$

Example 12-2 shows the need for some human engineering. The display reads 1000, but it *means* that V_{in} equals 100 mV. *You* must wire in a decimal point to display 100.0 and paste a "mV" sign beside the display.

12-1.5 Automatic Zeroing

The block diagram of Fig. 12-2b contains a section labeled "auto-zero." During the third and final phase of conversion T_z, the logic unit activates several analog switches and connects an auto-zero capacitor C_{AZ} (not shown).

The auto-zero capacitor is connected across the integrating capacitor, C_{int}, and any input offset voltages of both integrating and comparator op amps. C_{AZ} charges to a voltage approximately equal to the average error voltage due to C_{int} and the offset voltages. During the following phases T_1 and T_2, the error voltage stored on C_{AZ} is connected to cancel any error voltage on C_{ref}. Thus the ADC is automatically zeroed for every conversion.

12-1.6 Summary

Refer to the timing diagram in Fig. 12-2a. The logic unit allocates 4000 counts for one conversion. At 83.33 μs per count, the conversion takes 333 ms. The control unit always allocates 1000 counts or 83.3 ms to phase T_1.

The number of counts required for T_2 depends on V_{in}. Zero counts are used for $V_{in} = 0$ V and a maximum of 2000 counts or 166.7 ms are used when V_{in} is at its maximum limit of ± 200 mV.

T_2 and T_z always share a total of 3000 counts for a total of 250 ms. For $V_{in} = 0$ V, $T_2 = 0$ counts and $T_z = 3000$ counts. For $V_{in} = \pm 200$ mV, $T_2 = 2000$ counts and $T_z = 1000$ counts.

Intersil markets a complete $3\frac{1}{2}$-digit digital voltmeter kit. The kit contains a 40-pin dual-slope integrating A/D converter (7106 or 7107), all necessary parts, printed circuit board, and instructions. The instructions make it easy to make, easy to use, and forms an excellent tutorial on integrating ADCs.

12-2 SUCCESSIVE APPROXIMATION ADC

The block diagram of a successive approximation register (ADC) is shown in Fig. 12-3. It consists of a DAC, a comparator, and a *successive approximation register* (SAR). One terminal is required for analog input voltage V_{in}. The digital output is available in either serial or parallel form. A minimum of three control terminals are required. *Start conversion* initiates an A/D conversion sequence and *end of conversion* tells when the conversion is completed. An external clock terminal sets the time to complete each conversion.

12-2.1 Circuit Operation

Refer to Fig. 12-3. An input *start conversion* command initiates one analog-to-digital conversion cycle. The successive approximation register (SAR) connects a sequence of digital numbers, one number for each bit to the inputs of a DAC. This process is explained in Section 12-2.2.

The DAC converts each digital number into an analog output V_o. Analog input voltage, V_{in}, is compared to V_o by a comparator. The comparator tells the SAR whether V_{in} is greater or less than DAC output V_o, once for each bit. For a 3-bit output, three comparisons would be made.

Comparisons are made beginning with the MSB and ending with the LSB, as will be explained. At the end of the LSB comparison, the SAR sends an end-of-conversion signal. The digital equivalent of V_{in} is now present at the SAR's digital output.

12-2.2 Successive Approximation Analogy

Suppose that you had 1-, 2-, and 4-lb weights (SAR) plus a balance scale (comparator and DAC). Think of the 1-lb weight as 1 LSB and the most significant 4-lb weight as 4-LSB. Please refer to Figs. 12-3 and 12-4. V_{in} corresponds to an unknown weight.

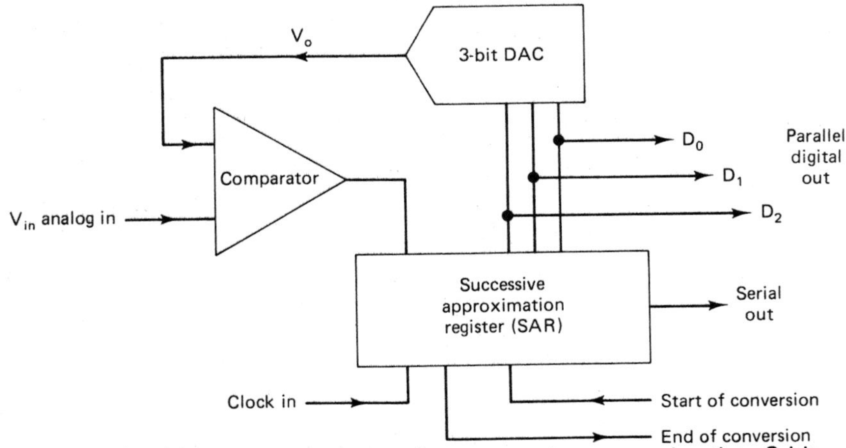

Figure 12-3 Block diagram of a successive approximation 3-bit analog-to-digital converter.

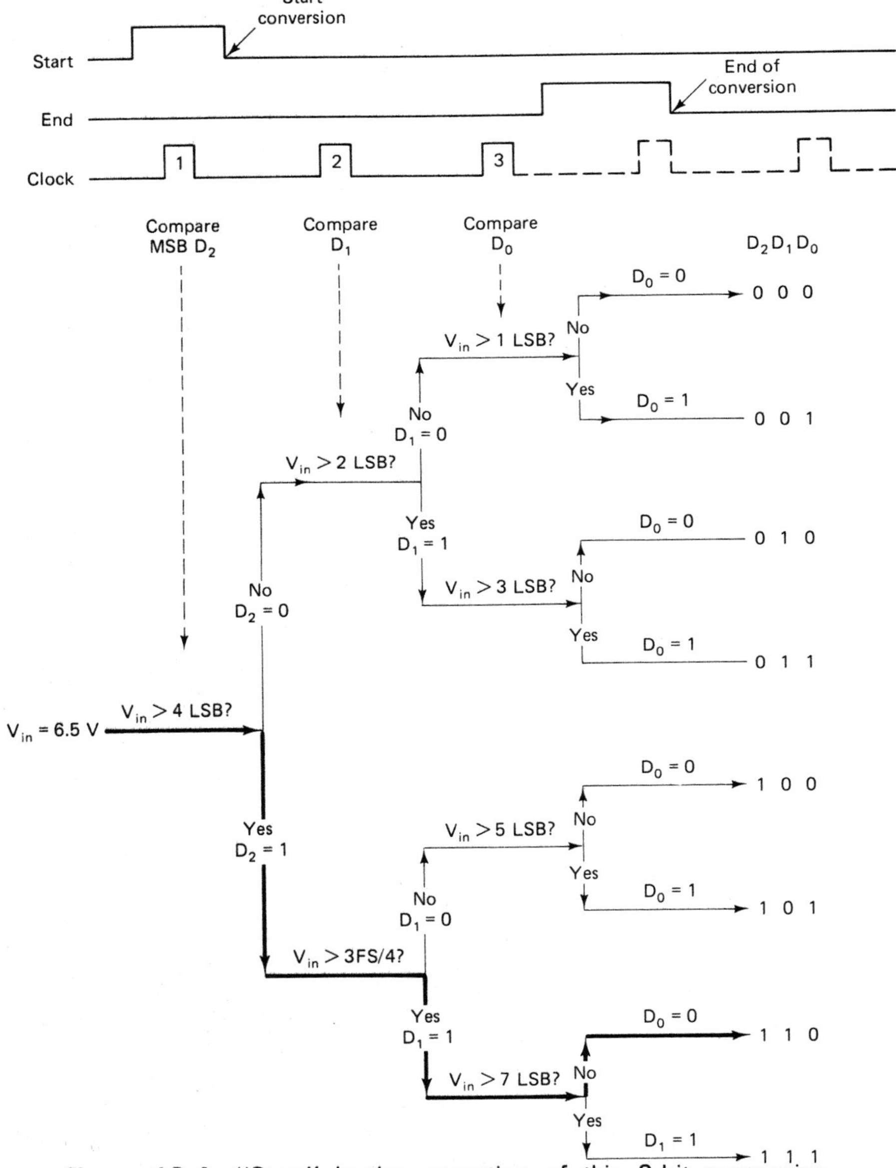

Figure 12-4 "Start" begins operation of this 3-bit successive approximation register. Beginning with the MSB, the weight of each bit is compared with V_{in} by the comparator in Fig. 12-3. If V_{in} is greater, the SAR's output is set to 1, or to 0 if V_{in} is smaller. Heavy lines show the conversion for V_{in} = 6.5 V.

Let us convert $V_{in} = 6.5$ V to a digital output (unknown weight = 6.5 lb). You would place the unknown weight on one platform of the balance, the 4-lb weight on the other, and compare if the unknown weight (V_{in}) exceeded the 4-lb weight. The SAR uses one clock pulse to apply the MSB (100) to the DAC in Fig. 12-3. Its output, $V_o = 4$ V, is compared with V_{in}. The MSB (D_2) in Fig. 12-4 is set to 1 if $V_{in} > V_o$. This is analogous to you, leaving the 4-lb weight on the scale.

The SAR then applies 110 (add a 2-lb weight) to the DAC. D_1 is set to 1 since $V_{in} = 6.5$ V is greater than $V_o = 6$ V. Finally, the SAR applies 111 to the DAC (add 1 lb). Since $V_{in} = 6.5$ V is less than 7 V, D_o is set to zero (1-lb weight removed).

12-2.3 Conversion Time

Figure 12-4 shows that one clock pulse is required for the SAR to compare each bit. However, an additional clock pulse is usually required to reset the SAR prior to performing a conversion. The time for one analog-to-digital concession, T_C, must depend on both the clock's period T and number of bits n. The relationship is

$$T_C = T(n + 1) \tag{12-4}$$

EXAMPLE 12-3

An 8-bit successive approximation ADC is driven by a 1-MHz clock. Find its conversion time.

SOLUTION The time for one clock pulse is 1 μs. From Eq. (12-4),

$$T_C = 1\ \mu\text{s}\ (8 + 1) = 9\ \mu\text{s}$$

12-3 A/D CONVERTERS FOR MICROPROCESSORS

The microprocessor's views a peripheral ADC simply as a "read only" address in the microprocessor's memory map. Refer to Fig. 12-5. The ADC must have a tri-state *memory buffer register* (MBR). In the idle state, the MBR will contain a digital code resulting from the ADC's *last* conversion. Also, the MBR will be disconnected from the data bus.

The microprocessor uses the address bus and decoders to select one ADC out of all the others by bringing its *chip select* terminal low. This process is similar to that shown in Fig. 11-10. A low on the *chip select* terminal in Fig. 12-5 tells the ADC that a command is coming to its read/$\overline{\text{write}}$ terminal. A more descriptive terminal designation is read/$\overline{\text{convert}}$. If read/$\overline{\text{write}}$ is brought low by the microprocessor, the ADC converts V_{in} into a digital code and loads or writes it into its own MBR. When read/$\overline{\text{write}}$ is high *and* chip select is low, the ADC's memory buffer register is connected (transparent) to the data bus.

It is important to look at this operation from the microprocessor's viewpoint. A *read* command means that the microprocessor is going to read data stored in the ADC's memory buffer register. The ADC's digital tri-state outputs must go from hi-Z (high impedance) to transparent and

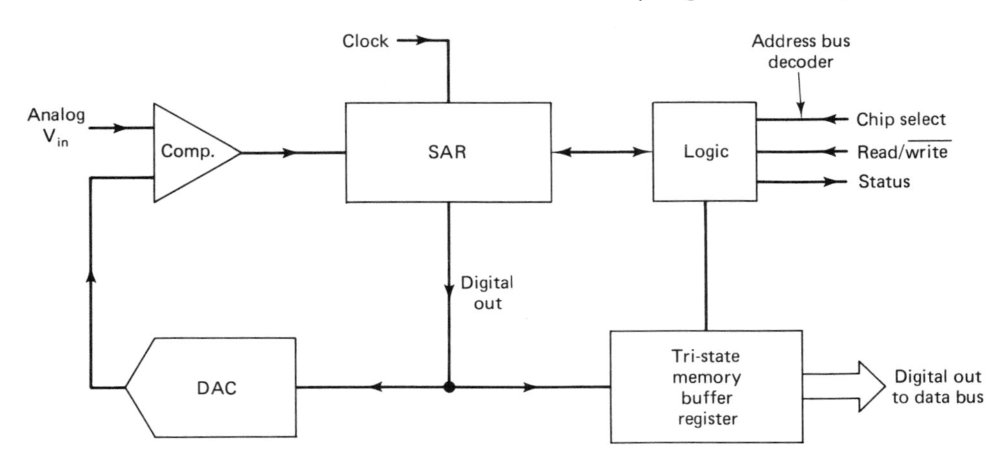

Figure 12-5 An A/D converter for microprocessors.

connect the digital word to the data bus. A *write command is actually a start conversion* command to the ADC. The microprocessor thus tells the ADC: (1) perform a conversion; (2) store (and *write*) it in your memory; and (3) don't tell me the result until I want to *read* it.

Finally, the microprocessor-compatible ADC must tell the microprocessor via its *status* terminal when a conversion is in progress; status goes high. If a conversion is completed, status goes low to signal the microprocessor that data is valid and ready for reading. We select Analog Devices AD670 to learn how all the foregoing features are available in a single 20-pin integrated circuit.

12-4 MICROPROCESSOR-COMPATIBLE A/D CONVERTER

The AD670 is an 8-bit microprocessor-compatible successive approximation analog-to-digital converter. The 20-pin package of Fig. 12-6 contains all the features described in Section 12-3 and Fig. 12-5. In addition, it contains an on-board clock, voltage reference, and instrumentation amplifier, and needs only a single 5-V supply. To understand how the AD670 operates, we examine the tasks performed by each of its terminals and associated circuit blocks.

12-4.1 Analog Input Voltage Terminals

Four analog input terminals are pins 16, 17, 18, and 19. They are inputs to an instrumentation amplifier configured to handle unipolar or bipolar analog input voltages. They are also pin-programmable to make it easy for the user to select resolution. Figure 12-6b shows operation for an analog input of 0 to 2.55 V, resolution = 10 mV/LSB. Figure 12-6c shows operation for 0 to 255 mV or 1 mV/LSB.

12-4.2 Digital Output Terminals

Pins 1 through 8 are tristate, buffered, latching digital outputs for the data bus digits, D_0 through D_7, respectively. When a microprocessor tells the AD670 to perform a conversion (write), the result is latched into its

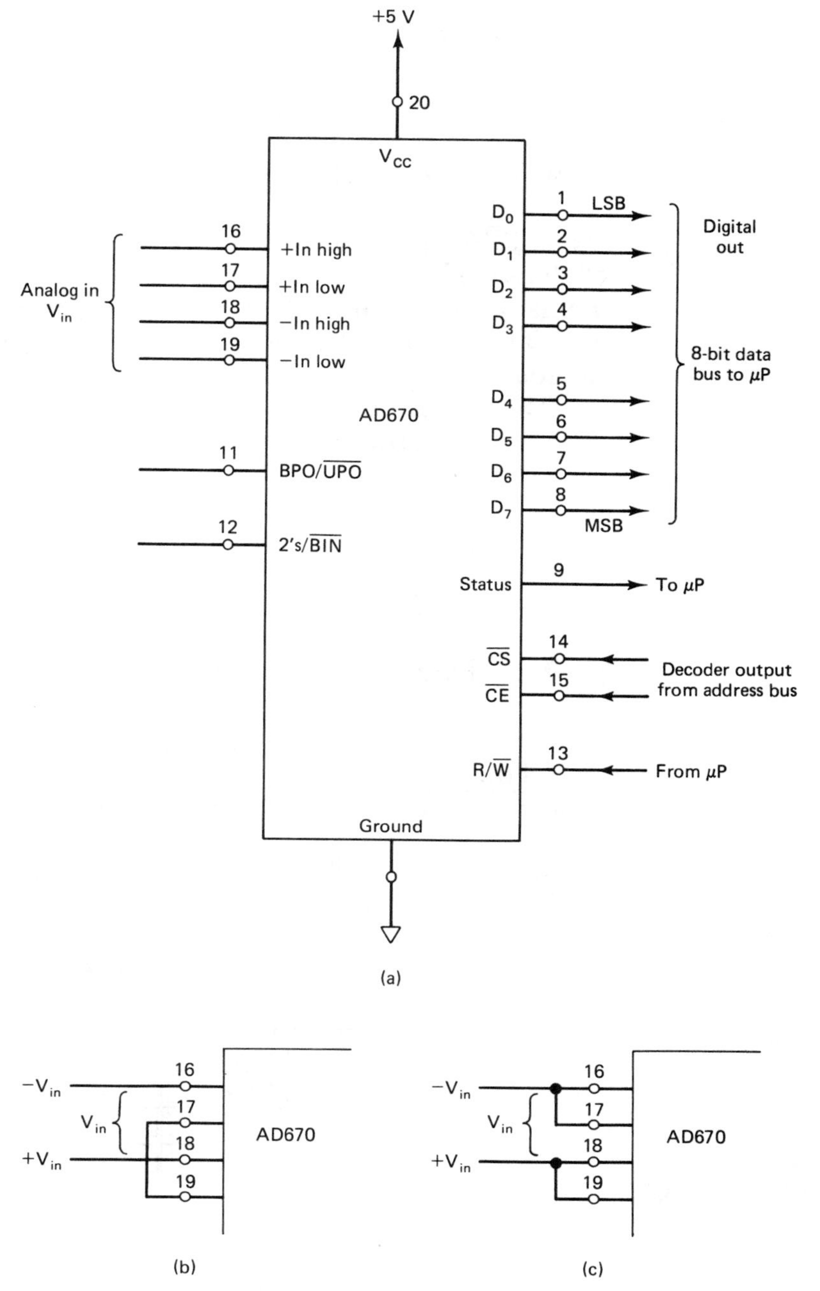

Figure 12-6 (a) AD670 ADC pin connections. Full-scale analog input voltages are 0 to 2.55 V or 0 to ±1.28 V in (b) or 0 to 255 mV or 0 to ±128 mV in (c).

memory buffer register. Tri-state output switches are held in the *high-impedance* (high-Z) state until the microprocessor sends a read command. Thus the ADC's memory register is normally disconnected from the data bus.

12-4.3 Input Option Terminal

Pin 11 is called BPO/$\overline{\text{UPO}}$ and allows the microprocessor to tell the AD670 whether to accept a bipolar analog input voltage range or a unipolar input range. A low on pin 11 selects unipolar operation. A range of 0 to 2.55 V or 0 to 255 mV is set by the user as in Fig. 12-6b and c. A high sent to pin 11 selects bipolar operation. The V_{in} range is then ± 1.28 V (Fig. 12-6b) or ± 128 mV (Fig. 12-6c).

12-4.4 Output Option Terminal

In Fig. 12-6, pin 12 is labeled "2's/$\overline{\text{BIN}}$". It allows the microprocessor to tell the AD670 to present an *output format* in either 2's-complement code, or binary code. A binary output code format will be *straight binary* if V_{in} is unipolar (pin 11 = low) or *offset binary* if V_{in} is bipolar (pin 11 = high). The four possible options are shown in Fig. 12-7a.

The digital output responses to analog input V_{in} are shown in Fig. 12-7b and c. V_{in} is the *differential* input voltage as in Fig. 12-6b and c. It is defined by

$$V_{in} = (+V_{in}) - (-V_{in}) \tag{12-5}$$

where $+V_{in}$ and $-V_{in}$ are measured with respect to ground.

12-4.5 Microprocessor Control Terminals

As shown in Fig. 12-6, pins 13, 14, and 15 are used by a microprocessor to control the AD670. Terminal 14 is called *chip select* ($\overline{\text{CS}}$) and terminal 15 is called *chip enable* ($\overline{\text{CE}}$). Pin 13 is called read/write (R/$\overline{\text{W}}$; or sometimes, write/$\overline{\text{convert}}$).

If $\overline{\text{CS}}$, $\overline{\text{CE}}$, and R/$\overline{\text{W}}$ are all brought low, the ADC converts continuously. It performs one conversion every 10 μs or less. The result of

Pin 11 BPO/$\overline{\text{UPO}}$	Input range	Pin 12 2's/$\overline{\text{BIN}}$	Output format
0	Unipolar	0	Straight binary
1	Bipolar	0	Offset binary
0	Unipolar	1	2's complement
1	Bipolar	1	2's complement

(a)

Differential V_{in}	Unipolar/straight binary, pin 11 = 0, 12 = 0
0	0000 0000
1 mV	0000 0001
128 mV	1000 0000
255 mV	1111 1111

(b)

Differential V_{in}	Bipolar/offset binary, pin 11 = 1, 12 = 0	Bipolar/2's complement, pin 11 = 1, 12 = 1
−128 mV	0000 0000	1000 0000
−1 mV	0111 1111	1111 1111
0	1000 0000	0000 0000
1 mV	1000 0001	0000 0001
127 mV	1111 1111	0111 1111

(c)

Figure 12-7 Input ranges unipolar or bipolar, and output formats are determined by pins 11 and 12 in (a). Output codes are given for unipolar inputs in (b) and bipolar inputs in (c). (a) Input range and output format are controlled by pins 12 and 11, respectively; (b) digital output code for unipolar V_{in} inputs wired as in Fig. 12-6c; (c) digital output codes for bipolar V_{in} inputs wired as in Fig. 12-6c.

each conversion is latched into the output buffer register. However, the digital output code is *not* connected to the data bus because the outputs are high-impedance. This condition is called a *write/$\overline{convert}$* command. That is, the microprocessor tells the AD670 to write converted data into its own buffer register. If $\overline{CS}$ or R/$\overline{W}$ or $\overline{CE}$ is high, the AD670 is unselected (high impedance) and retains the last conversion in its register.

Status terminal, pin 9, stays high during a conversion. When a conversion is completed, pin 9 outputs a low to tell the microprocessor that data is valid in the AD670's buffer register. To read data out of the AD670, the microprocessor brings R/$\overline{W}$ high while status and $\overline{CS}$ *and* $\overline{CE}$ are low. This is a *read* command from the microprocessor.

The AD670's buffer becomes transparent and connects the eight digital outputs (D_7 through D_0) to the data bus. These data will remain on the bus until the AD670 is disconnected by bringing $\overline{CS}$ high, or $\overline{CE}$ high, or R/$\overline{W}$ low.

Summary

1. A low on $\overline{CE}$ and $\overline{CS}$ selects the AD670. What happens next depends on R/$\overline{W}$.
2. If R/$\overline{W}$ is low (for at least 0.3 μs), a conversion is performed and the result is written into the buffer register. Outputs are high-impedance. The conversion requires 10 μs.
3. If R/$\overline{W}$ is high, the last conversion is stored in the buffer and the outputs are transparent. No further conversions are performed. The contents of the register can now be read by the microprocessor via the data bus.
4. Status tells the microprocessor what is going on within the AD670. Status = high means that conversion is being performed. Status = low tells the microprocessor that data are valid. The microprocessor is free to read the selected AD670's data by sending a high to R/$\overline{W}$.

12-5 OPERATING A SUCCESSIVE APPROXIMATION ADC

The 8-bit SAR A/D converter can be tested or operated without the need for a microprocessor. Figure 12-8 shows how to do this. Each data output, D_0 to D_7, is connected to an inverter, resistor, and LED. These components simulate a data bus. An LED lights to signify that a 1 is present on its associated data bus wire.

Pins 14 and 15 are wired so that $\overline{CS}$ and $\overline{CE}$ are low. This causes continuous conversion. The 555 timer drives R/$\overline{W}$ low for 5 μs to simulate a write command. R/$\overline{W}$ thus returns high before a conversion is completed in the 10-μs conversion time. At the end of 10 μs, the high on R/$\overline{W}$ simulates a read command and data are displayed on the LEDs. If R_T = 1.5 MΩ, the AD670 makes one conversion and one readout 1000 times per second. Reduce R_T to 120 kΩ for convert/reads of 10,000 times per second.

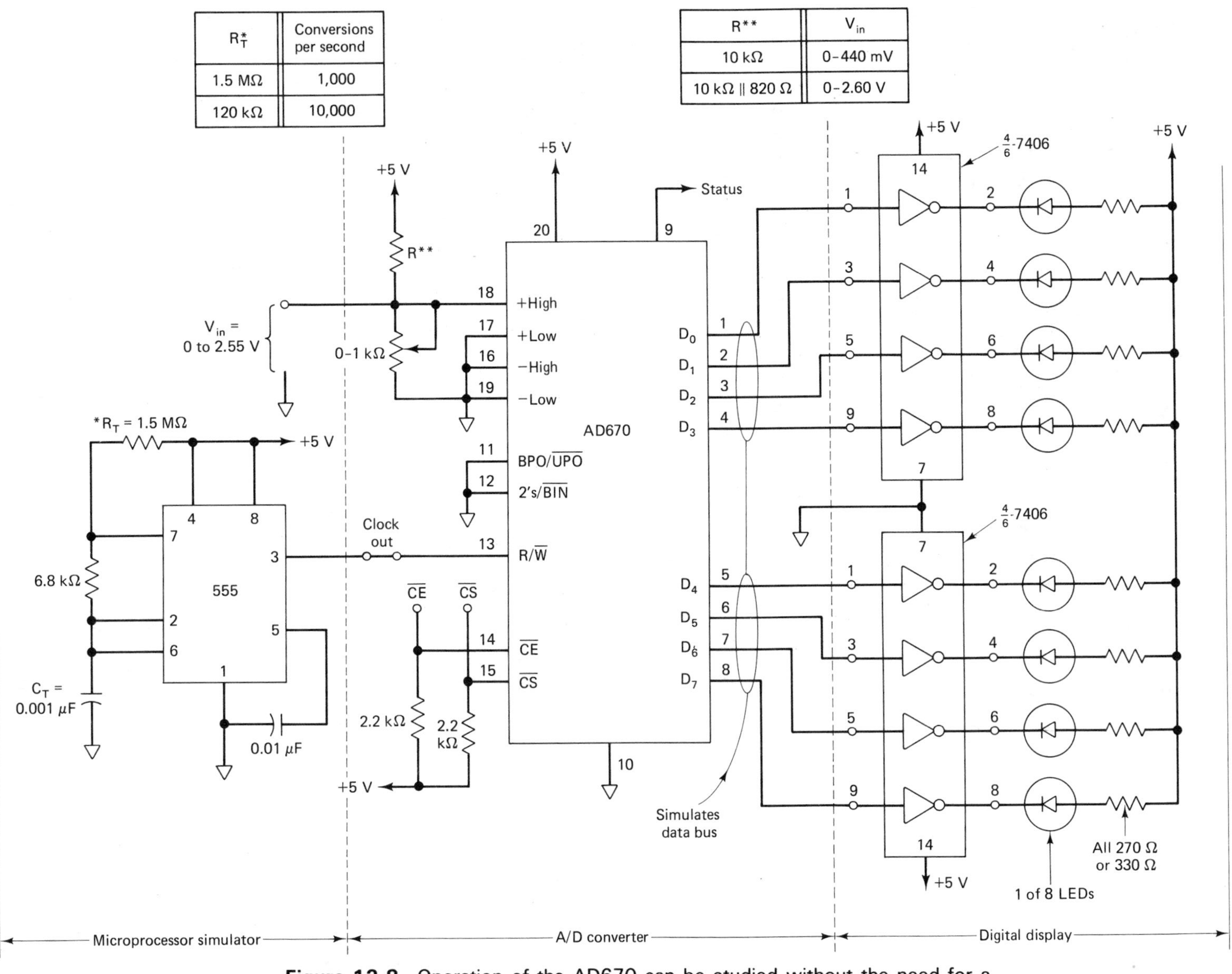

R_T^*	Conversions per second
1.5 MΩ	1,000
120 kΩ	10,000

R**	V_{in}
10 kΩ	0–440 mV
10 kΩ ‖ 820 Ω	0–2.60 V

Figure 12-8 Operation of the AD670 can be studied without the need for a microprocessor. Pins 11 and 12 are grounded to simulate a microprocessor selection via on address bus. The 555 timer simulates continuous read/*write* commands from a microprocessor.

12-6 FLASH CONVERTERS

12-6.1 Principles of Operation

Fastest of all A/D converters is the *flash* converter, shown in Fig. 12-9a. A reference voltage and resistor divider network establishes a resolution of 1 V/LSB. Analog input voltage V_{in} is applied to the + inputs of all comparators. Their outputs drive an 8-line-to-3-line priority encoder. The encoder logic outputs a binary code that represents the analog input.

For example, suppose that $V_{in} = 5.0$ V. The outputs of comparators 1 through 5 would go high and 6 through 8 would go low. As shown in Fig. 12-9b, the digital output would be 101.

12-6.2 Conversion Time

The conversion time of the flash converter is limited only by response time of comparators and logic gates. They can digitize video or radar signals. The flash converter's high speed becomes more expensive as resolution is increased. Figure 12-9 shows that the flash converter requires seven comparators (or $2^3 - 1$) to perform a 3-bit conversion. The number of comparators required for n-bit resolution is

$$\text{number of comparators} = 2^n - 1 \tag{12-6}$$

For example, an 8-bit flash converter requires $(2^8 - 1)$ or 255 comparators. Encoder logic would be more complex requiring a 256-line-to-8-line priority encoder.

12-7 FREQUENCY RESPONSE OF A/D CONVERTERS

12-7.1 Aperture Error

During conversion time, T_C, the analog input voltage must not change by more than $\pm\frac{1}{2}$ LSB (total 1 LSB), or the conversion will be incorrect. This type of inaccuracy is called *aperture error.* The rate of change of V_{in} with respect to time is called *slew rate.* If V_{in} is a sine wave, its slew rate is maximum at its zero crossings. The sine wave's slew rate is determined by both its peak voltage and frequency.

For an A/D converter, the maximum frequency for a sine wave V_{in} to be digitized within an accuracy of $\pm\frac{1}{2}$ LSB is

$$f_{max} \simeq \frac{1}{2\pi(T_C)2^n} \tag{12-7}$$

EXAMPLE 12-4

The AD670 is an 8-bit ADC with a conversion time of 10 μs. Find the maximum frequency of an input sine wave that can be digitized without aperture error.

SOLUTION From Eq. (12-7),

$$f_{max} \simeq \frac{1}{2\pi(2^8)10\ \mu s} = \frac{1}{2\pi(256)10 \times 10^{-6}\ s} = 62\ \text{Hz}$$

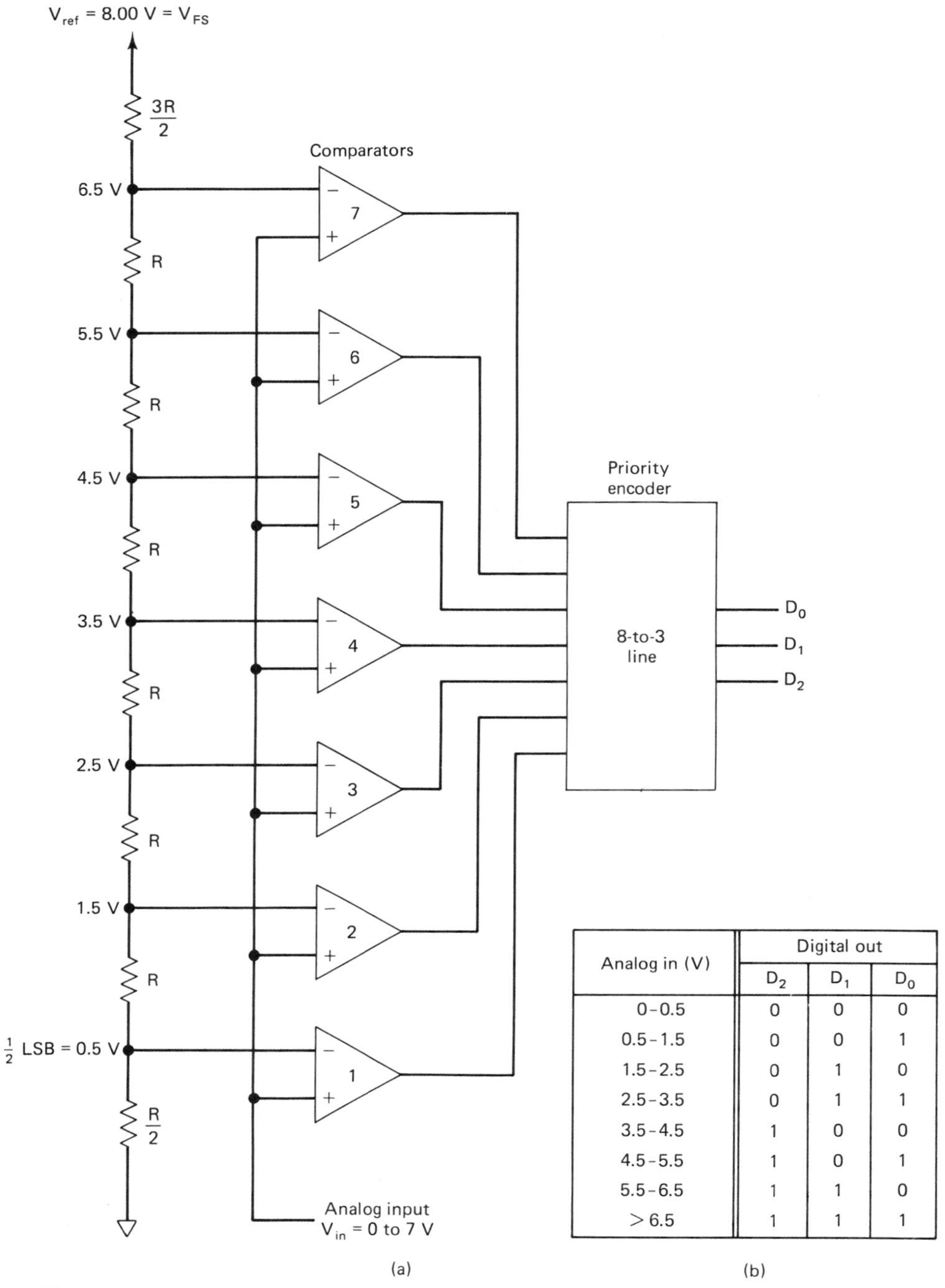

Analog in (V)	Digital out		
	D_2	D_1	D_0
0–0.5	0	0	0
0.5–1.5	0	0	1
1.5–2.5	0	1	0
2.5–3.5	0	1	1
3.5–4.5	1	0	0
4.5–5.5	1	0	1
5.5–6.5	1	1	0
> 6.5	1	1	1

Figure 12-9 (a) Three-bit flash (parallel) A/D converter; (b) output versus input.

Example 12-4 shows that the frequency response of even a fast ADC is surprisingly low. For a 10-bit integrating ADC with a conversion time of $\frac{1}{3}$ s, the highest sine frequency is about 0.5 mHz, or 1 cycle per 2000 s.

Summary: An 8-bit converter with a 10-μs conversion time can theoretically perform $\frac{1}{10\mu s}$ = 100,000 conversions per second. Yet the highest frequency sine wave that can be converted without slew-rate limiting is about 62 cycles per second. To raise the frequency response, we must add another circuit block, the sample-and-hold or follower-and-hold amplifier.

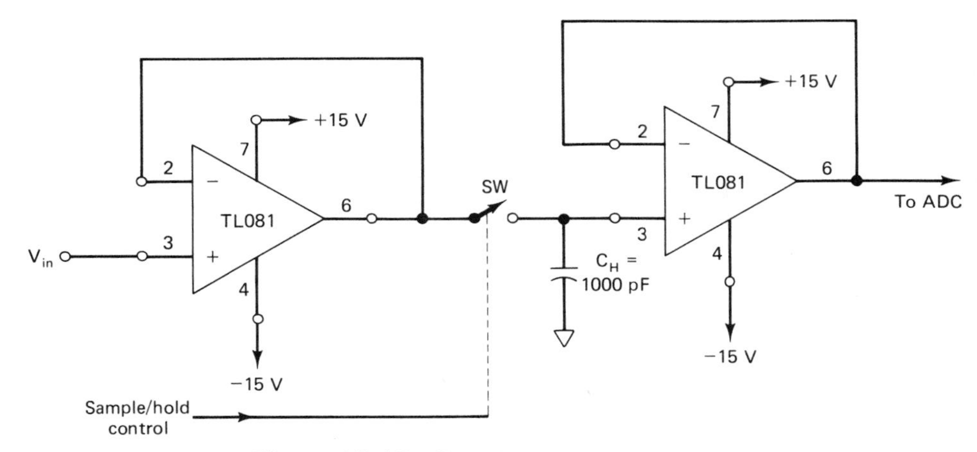

Figure 12-10 Sample-and-hold amplifier.

12-7.2 Sample-and-Hold Amplifier

The sample-and-hold (S/H) amplifier of Fig. 12-10 is made from two op amps, a hold capacitor (C_H), and a high-speed analog switch. This amplifier is connected between an analog input signal and the input to an ADC.

When the S/H amplifier is in the *sample* mode, the switch is closed and hold capacitor (C_H) voltage follows V_{in}. A *hold* command opens the switch and C_H retains a charge equal to V_{in} at the moment of switching. The S/H amplifier thus acts to hold V_{in} (stored on C_H) constant, while the ADC performs a conversion.

Conversion time of the ADC no longer limits frequency response. Instead, the limit is the *aperture time* of the S/H amplifier, which can be made much less than the conversion time. Aperture time is the time elapsed between a hold command and a switch opening. If the hold command is advanced by a time equal to the aperture time, C_H will hold the desired sample of V_{in}. Then the only remaining error is *aperture time uncertainty,* the switch jitter variation for each hold command.

Commercial S/H amplifiers have aperture time uncertainties lower than 50 ns. An example shows the improvement in frequency response due to an added S/H amplifier.

EXAMPLE 12-5

A S/H amplifier with an aperture time uncertainty of 50 ns is connected to an 8-bit ADC. Find the highest-frequency sine wave that can be digitized within an error of 1 LSB.

SOLUTION Replace conversion time by aperture uncertainty time in Eq. (12-7):

$$f_{max} \simeq \frac{1}{2\pi(2^8)50 \times 10^{-9}\text{ s}} = 12.4\text{ kHz}$$

End of Chapter Exercises

Name: ______________________

Date: __________ **Grade:** __

A. FILL-IN THE BLANKS

Fill in the blanks with the best answer.

1. Inherent in all ADCs is the unavoidable uncertainty about the value of input voltage when the output changes. This uncertainty equals $\pm\frac{1}{2}$ LSB and is called ____.
2. Name the three types of ADCs. (a) __________ (b) __________ (c) __________.
3. The actual conversion of an analog voltage into a digital count during the __________ phase of an integrating type ADC.
4. Name the three control terminals required for a successive approximation ADC. __________, __________, __________.
5. The AD670 microprocessor-compatible ADC can be made to perform continuous conversions by forcing the control terminals $\overline{CS}$, $\overline{CE}$ and R/$\overline{W}$ to a __________ voltage.
6. The AD670 tells the microprocessor its status by pin 9 (status). When a conversion is being performed this terminal is __________.
7. During a conversion, the input voltage must not change by more than $\pm\frac{1}{2}$ LSB or the conversion will be incorrect. This type of inaccuracy is called ____.
8. What type of circuit must be added to the input of an ADC to improve its frequency response? __________.

B. TRUE/FALSE

Fill in **T** if the statement is true, and **F** if any part of the statement is false.

1. ______ ADC resolution is defined as the value of input voltage required to change the digital output by 1 LSB.

2. ______ Integrating type A/D converters have the slowest conversion time.

3. ______ The conversion time of a successive approximation ADC is independent of its clock frequency.

4. ______ The AD670 chip is microprocessor-compatible integrating type ADC with its own clock, voltage reference and instrumentation amplifier.

5. ______ The conversion time for a flash converter is limited only by response time of its comparators and logic gates.

6. ______ If a sample-and-hold amplifier is connected between the analog input and the ADC, conversion time no longer limits frequency response.

C. CIRCLE THE CORRECT ANSWER

Circle the correct answer for each statement.

1. Which type of ADC has the fastest conversion time? (Integrating, successive approximation, flash)
2. Which type of ADC has the highest resolution? (integrating, successive approximation, flash)
3. There are (two, four) possible digital output formats for the AD670 microprocessor-compatible ADC.
4. The AD670 can be made to perform continuously if $\overline{CS}$, $\overline{CE}$ and $R/\overline{W}$ are (high, low).
5. Addition of a sample-and-hold amplifier between the analog input signal and the ADC will (increase, decrease) the frequency response of the ADC.

D. MATCHING

Match the name or symbol in column **A** with the statement that matches best in column **B**.

	COLUMN A	COLUMN B
1. ______	Reference Integration Phase	**a.** $\pm\frac{1}{2}$ LSB
2. ______	S/H Amplifier	**b.** T_z

3. ______ Quantization Error c. T_2

4. ______ auto zero d. Tri-state

5. ______ High-Z e. aperture time

PROBLEMS

12-1. Explain the need for analog-to-digital converters.

12-2. Sketch the output–input characteristics of an ideal 4-bit ADC. Assume that full-scale output is 1111 for an input of 15 V.

12-3. Name the ADC most suited to
- **(a)** convert video signals;
- **(b)** convert audio signals;
- **(c)** convert dc signals to digital outputs.

12-4. A dual-slope integrating ADC requires three phases to perform a conversion.
- **(a)** Name these phases.
- **(b)** During which phase is the digital output loaded into the logic unit?

12-5. An analog input voltage, V_{in} = 50 mV, is applied to the IADC of Fig. 12-2. Find
- **(a)** the times required for T_1, T_2, and T_z;
- **(b)** the digital output reading;
- **(c)** the meaning of the digital output word.

12-6. Draw the block diagram for a 4-bit successive approximation ADC. Label the input wires and output wires.

12-7. A 10-bit SAR-ADC is driven by a 5-MHz clock.
- **(a)** How many clock cycles are required for one conversion?
- **(b)** How long does it take to complete one conversion?

12-8. How does the programmer of a microprocessor view an ADC peripheral device?

12-9.
- **(a)** Describe briefly how a microprocessor selects one AD670 ADC out of many.
- **(b)** Does an unselected ADC load (draw current from) the data bus?

12-10. How does a microprocessor tell if an AD670 ADC is ready to send valid data?

12-11. A microprocessor issues a write command to a selected AD670 ADC. Does this tell the ADC to send data to the microprocessor or perform a conversion?

12-12. An analog input voltage with a range of 0 to 2.55 V is applied to an AD670 8-bit ADC.
- **(a)** Which pins should be jumpered to select this range?
- **(b)** If the input signal is connected between pin 18 and ground, what must you do to pin 16?

12-13.
- **(a)** Which pin of the AD670 programs it for a unipolar or bipolar input format? A ground on this pin selects which format?

12-14. How do you pin-program the AD670 for a straight binary output?

12-15. How does the microprocessor tell the AD670 to
(a) perform a conversion;
(b) place the result of a conversion on the data bus?
(c) How does the microprocessor know that these data are valid?

12-16. **(a)** What is the conversion time for an AD670?
(b) How many conversions can it perform in 1 s?
(c) What is the maximum sine frequency that it can convert without a follow-and-hold amplifier?

12-17. What is the maximum sine wave frequency of an AD670 if it is connected to a S/H amplifier that has an aperture uncertainty time of 25 ns?

12-18. How many comparators are required to make an 8-bit flash converter?

Laboratory Exercise 12

Name: ____________________

Date: __________ **Grade:** ___

ANALOG-TO-DIGITAL CONVERSION

OBJECTIVES: Upon completion of this laboratory exercise on A/D conversion, you will be able to (1) operate a microprocessor-compatible ADC; (2) determine what voltage levels occur at the digital outputs; (3) initiate a conversion and measure conversion time; and (4) use the AD670 ADC to communicate with an AD558 DAC.

REFERENCES: Chapters 11 and 12, AD558 and AD670 Data Sheets

PARTS LIST

1	AD670 converter (See data sheet pp 443-54)	1	120-kΩresistor
1	AD558 converter	1	0- to 1-kΩpot
8	LEDs	2	7406 TTL hex inverters
8	330-Ωresistors	1	0.001-uF capacitor
4	2.2-kΩresistors	1	0.01-uF capacitor
1	820-Ωresistor	1	555 timer
	6.8-kΩresistor		

Procedure A: Measuring Conversion Frequency

1. Wire only the 555 timer circuit of Fig. L12-1. Connect a CRO to measure the clock signal at pin 3 of the 555. Set the CRO for dc coupling, time base = 5 μs/div, vertical sensitivity = 2 V/div, triggered on the negative edge. Measure the low time of the clock pulse. T_{LO} = __________. This is the signal that will initiate a conversion. Change the time base to 100 μs/div. Measure the clock frequency. f = __________. This is the number of conversions that will be performed each second (≈10,000 per second).

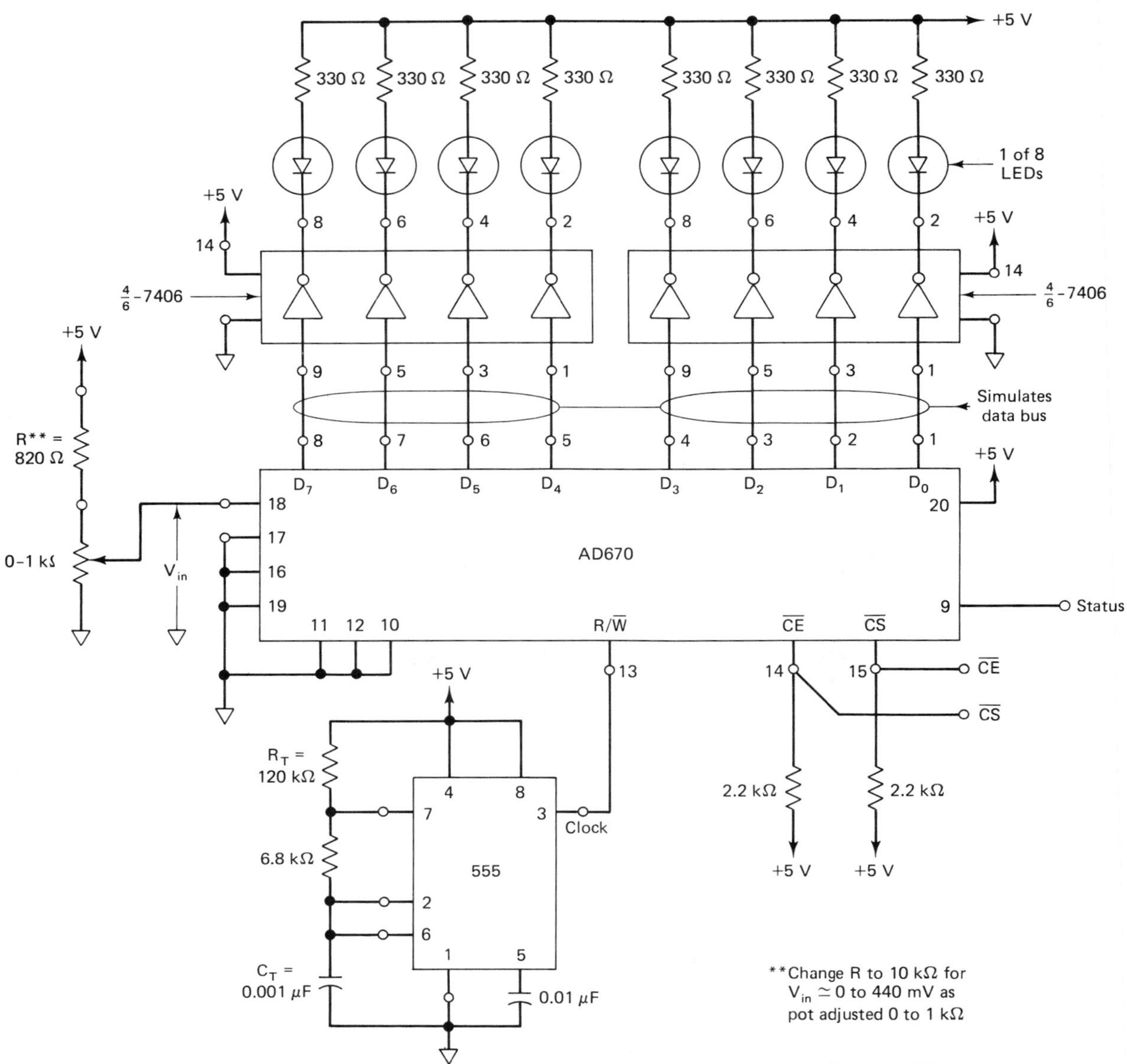

Figure L12-1 ADC test circuit. Grounding *both* $\overline{CE}$ and $\overline{CS}$ connects the AD670's output register to the data bus. The 555 clock circuits applies an 8-μs low to R/$\overline{W}$ to initiate a conversion. With R_1 = 1.20 kΩ as shown, about 10,000 conversions per second are performed.

Procedure B: Setting Up the Digital Display

2. Wire the two 7406 inverters and LEDs as in Fig. L12-1. Do *not* connect to the AD670. Jumper +5 V momentarily to each 7406 input (pins 1, 3, 5, 9) to see if the associated LED lights. Measure the open circuit voltage at input pin 1 of either 7406 with a DMM.

   V_1 = __________. This voltage will tell you when an AD670's digital output is in the high-Z state.

Row	V_{in}	D_7	D_6	D_5	D_4	D_3	D_2	D_1	D_0	$\overline{CS}$ and $\overline{CE}$ jumper
A	2.60 V									Open
B	2.60 V									Ground
C	0 V									Open
D	0 V									Ground

Figure L12-2 Measurements to determine when the AD670 digital outputs indicate a high-*Z* state, "1" state, or "0" state.

Procedure C: Setting Up the AD670 Circuit

3. Wire the AD670 to the 555 clock circuit and digital display circuit as in Fig. L12-1. Use a DMM to adjust V_{in} for 2.60 V with the 1-kΩ pot. Use a DMM to measure and record the voltage at the digital output in row A of Fig. L12-2.

Procedure D: Measuring the AD670's Digital Output Voltages

4. With V_{in} = 2.60 V, ground $\overline{CS}$ and $\overline{CE}$ with a jumper wire. Record the measured voltages at the digital outputs in row B of Fig. L12-2. Jumper a ground to pin 18 of the AD670 so that V_{in} = 0. Record the measured digital output voltages in row D of Fig. L12-2. Finally, remove the grounds on $\overline{CE}$ and $\overline{CS}$. Record the digital output voltages in row C of Fig. L12-2. Remove the short on V_{in}. Which row(s) in the table indicate that the AD670 outputs are in the high -*Z* state? __________. Which rows indicate that the digital outputs represent the digital equivalent of V_{in}? __________.

Procedure E: Measuring Conversion Time and Resolution

5. Adjust V_{in} until the LEDs indicate a code of 10000000 ($V_{in} \approx$ 1.280 V). Measure V_{in}. V_{in} = __________. Calculate resolution in mV/bit from $V_{in} \div 128$. Resolution = __________. Ground $\overline{CE}$ and $\overline{CS}$. Connect channel 1 of a dual-trace CRO to pin 13, R/$\overline{W}$ and channel 2 to pin 9, "Status." Set the CRO controls for 5 V/div vertical sensitivity on both channels, time base to 5 μs/div, trigger from channel 1's negative edge, dc coupled. Zero the traces as in Fig. L12-3. Sketch the waves at Status and R/$\overline{W}$ on Fig. L12-3. Move the CRO's channel 2 to LSB pin 1 at the AD670 and reposition the zero trace. Sketch this LSB waveshape on Fig. L12-3.
6. From your "Status" waveshape, what is the conversion (high) time for T_C in μs? T_C = __________. Indicate when conversion begins

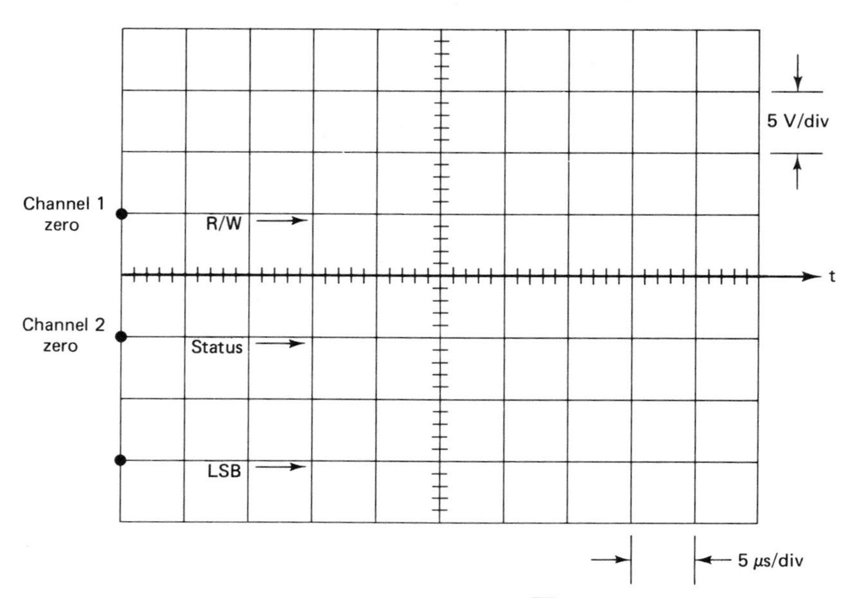

Figure L12-3 Waveshapes at R/$\overline{W}$, status, and D_0.

on your R/$\overline{\text{W}}$ waveshape. Finally, indicate when data is valid on the LSB waveshape.

Procedure F: Communicating from an AD670 ADC to an AD558 DAC

7. Remove the two 7406 inverters, LEDs, and 330-Ω resistors. Add the AD558 and two 2.2-kΩ resistors to complete the circuit shown in Fig. L12-4. With V_{in} = 1.28 V, measure the voltages at each digital input of the AD558. What is the 8-bit digital word in straight binary code? __________. Measure V_o at pin 16 of the AD558. V_o = __________. If $V_o \simeq V_{in}$, the system is working.

8. Connect channel 1 of a dual-trace CRO to monitor V_{in} and channel 2 to monitor V_o. Set the controls and zero as in Fig. L12-5. Remove the 820-Ω resistor and the 0- to 1-kΩ pot. Connect a function generator to V_{in} and adjust V_{in} to be a 0- to 2-V sine wave at 50 Hz as in Fig. L12-5. Sketch the waveshape for V_o on Fig. L12-5 where indicated.

Does V_o essentially follow V_{in}? __________. Keep this circuit intact for the final procedure.

Procedure G: Logic Control Operation of the AD670 and the AD558

9. Remove the ground jumper from $\overline{\text{CS}}$ of the AD670 to force its outputs into the high-Z state. What happened to V_o of the still selected AD558? __________________________________.
Replace the ground jumper on $\overline{\text{CS}}$ of the AD670.

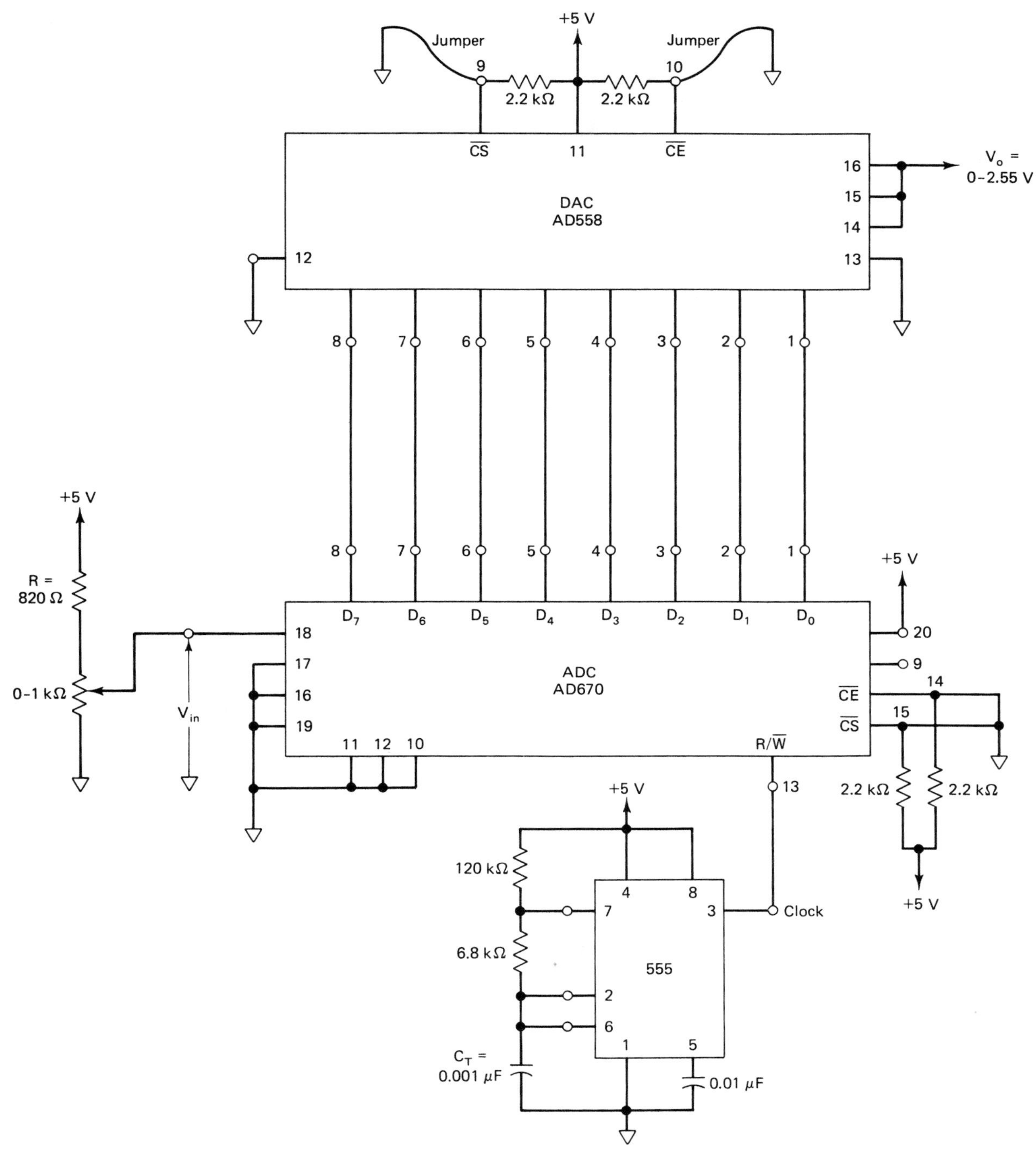

Figure L12-4 The AD670 converts analog voltage V_{in} to a straight binary digital code. Its digital output drives the digital-to-analog converter AD558. Both ICs are pin-programmed so that $V_o = V_{in}$ in this analog-to-digital-to-analog test circuit.

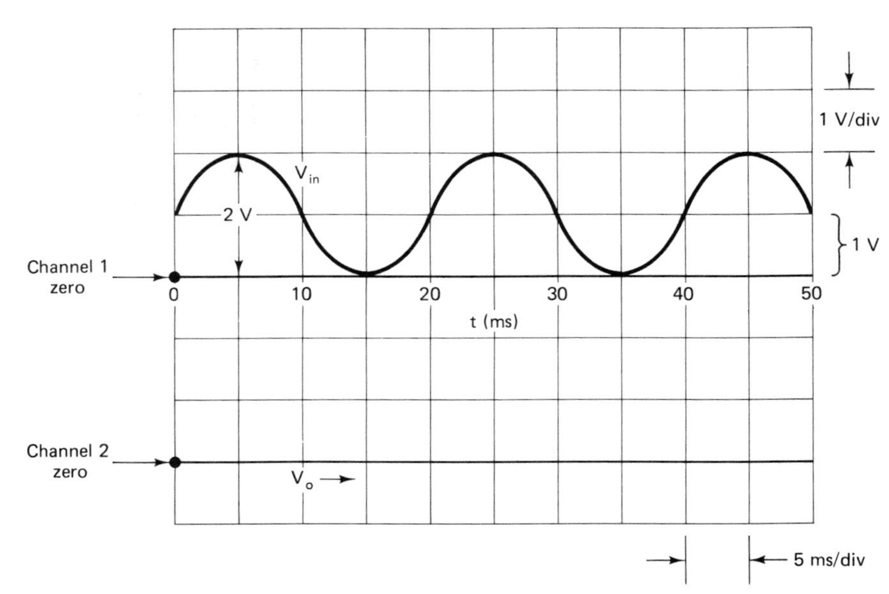

Figure L12-5 Waveshapes for V_{in} and V_O. Dc-coupled vertical amplifiers with both sensitivities at 1 V/div. Time base = 5 ms/d.

10. Remove the ground to $\overline{CS}$ of the DAC so that its digital inputs are no longer transparent. Tap the ground several times to $\overline{CS}$ (DAC). Observe that V_o stabilizes at whatever digital code was present at the instant the ground was removed. In the "Conclusion" section, explain briefly the operation of the logic control terminals.

CONCLUSION:

Low Cost Signal Conditioning 8-Bit ADC

AD670

FEATURES

Complete 8-Bit Signal Conditioning A/D Converter Including Instrumentation Amp, Reference, Comparator
Full Microprocessor Bus Interface
10μs max Conversion Speed
Flexible Input Stage: Instrumentation Amp Front End Provides Differential Inputs and High Common-Mode Rejection
No User Trims Required
No Missing Codes Over Temperature
Single +5V Supply Operation
Convenient Input Ranges
Small 20-Pin Package
Low Cost Monolithic Construction

GENERAL DESCRIPTION

The AD670 analog-to-digital converter is a complete 8-bit signal conditioning successive approximation analog-to-digital converter. It consists of an instrumentation amplifier front end along with a DAC, comparator, successive approximation register (SAR), precision voltage reference, and a three-state output buffer on a single monolithic chip. No external components or user trims are required to interface, with full accuracy, an analog system to an 8-bit data bus. The AD670 will operate on the +5V system supply. The input stage provides differential inputs with excellent common-mode rejection and allows direct interface to a variety of transducers.

The device is configured with input scaling resistors to permit four input ranges (two unipolar and two bipolar): 0 to 255mV (1mV/LSB) and 0 to 2.55V (10mV/LSB). The AD670 can be configured for both unipolar and bipolar inputs over these ranges. The differential inputs and common-mode rejection of this front end are useful in applications such as conversion of transducer signals superimposed on common-mode voltages.

The AD670 incorporates advanced circuit design and proven processing technology. The successive approximation function is implemented with I^2L (integrated injection logic). Thin-film SiCr resistors provide the stability required to prevent missing codes over the entire operating temperature range while laser wafer trimming of the resistor ladder permits calibration of the device to within ±1LSB. Thus, no user trims for gain or offset are required. Conversion time of the device is 10μs.

The AD670 is available in two package types and five grades: the AD670JN and AD670KN, in 20-pin plastic DIP packages are specified over 0 to +70°C while the AD670AD and AD670BD (−25°C to +85°C) are in 20-pin ceramic packages. An extended temperature grade designated as the AD670SD and specified from −55°C to +125°C is available in the 20-pin ceramic package.

PRODUCT HIGHLIGHTS

1. The AD670 is a complete 8-bit A/D including three-state outputs and microprocessor control for direct connection to 8-bit data buses. No external components are required to perform a conversion.
2. The flexible input stage features a differential instrumentation amp input with excellent common-mode rejection. This allows direct interface to a variety of transducers without preamplification.
3. No user trims are required for 8-bit accurate performance.
4. Operation from a single +5V supply allows the AD670 to run off of the microprocessor's supply.
5. Four convenient input ranges (two unipolar and two bipolar) are available through internal scaling resistors: 0 to 255mV (1mV/LSB) and 0 to 2.55V (10mV/LSB).
6. Software control of the output mode is provided. The user can easily select unipolar or bipolar inputs and binary or 2's complement output codes.
7. Maximum conversion time of the AD670 is 10μs.

Route 1 Industrial Park; P.O. Box 280; Norwood, Mass. 02062
Tel: 617/329-4700 **TWX: 710/394-6577**
West Coast 714/641-9391 · **Mid-West** 312/653-5000 · **Texas** 214/231-5094

*Courtesy of **Analog Devices, Inc.**

SPECIFICATIONS (@ V_{CC} = +5V and +25°C unless otherwise noted)

Model	AD670J Min	AD670J Typ	AD670J Max	AD670K Min	AD670K Typ	AD670K Max	Units
OPERATING TEMPERATURE RANGE	0		+70	0		+70	°C
RESOLUTION	8			8			Bit
CONVERSION TIME[1]			**10**			**10**	μs
RELATIVE ACCURACY							
T_{min} to T_{max}			**±1/2**			**±1/4**	LSB
DIFFERENTIAL LINEARITY ERROR							
T_{min} to T_{max}	**GUARANTEED NO MISSING CODES ALL GRADES**						
GAIN ACCURACY[2]							
@ +25°C		1.0	**±1.5**		0.5	**±0.75**	LSB
T_{min} to T_{max}		1.5	**±2.0**		0.75	**±1.0**	LSB
UNIPOLAR ZERO ERROR							
@ +25°C			**±1.5**			**±0.75**	LSB
T_{min} to T_{max}			**±2.0**			**±1.0**	LSB
BIPOLAR ZERO ERROR[3]							
@ +25°C			**±1.5**			**±0.75**	LSB
T_{min} to T_{max}			**±2.0**			**±1.0**	LSB
ANALOG INPUT RANGES							
All Grades		0 to +255			0 to +255		mV
		−128 to +127			−128 to +127		mV
		0 to +2.55			0 to +2.55		V
		−1.28 to +1.27			−1.28 to +1.27		V
BIAS CURRENT (255mV RANGE)							
T_{min} to T_{max}		200	**500**		200	**500**	nA
OFFSET CURRENT (255mV RANGE)							
T_{min} to T_{max}		20	**100**		20	**100**	nA
2.55V RANGE INPUT RESISTANCE	8.0		12.0	8.0		12.0	kΩ
2.55V RANGE FULL SCALE MATCH							
+ AND − INPUT		±1/2			±1/2		LSB
COMMON-MODE RANGE[4]							
@ +25°C	**−0.2**		**$+V_{CC}-3.4$**	**−0.2**		**$+V_{CC}+3.4$**	V
T_{min} to T_{max}	**0**		**$+V_{CC}-3.8$**	**0**		**$+V_{CC}-3.8$**	V
COMMON-MODE REJECTION							
RATIO (255mV RANGE)			**1**			**1**	LSB[4]
COMMON-MODE REJECTION							
RATIO (2.55V RANGE)			**1**			**1**	LSB[4]
POWER SUPPLY							
Operating Range	**4.5**		**5.5**	**4.5**		**5.5**	V
Current I_{CC}		30	**45**		30	**45**	mA
Rejection Ratio T_{min} to T_{max}			**0.015**			**0.015**	% of FS/%
DIGITAL OUTPUTS							
SINK CURRENT (V_{OUT} = 0.4V)							
@ +25°C	**1.6**			**1.6**			mA
T_{min} to T_{max}	**1.6**			**1.6**			mA
SOURCE CURRENT (V_{OUT} = 2.4V)							
@ +25°C	**0.5**			**0.5**			mA
T_{min} to T_{max}	**0.5**			**0.5**			mA
THREE-STATE LEAKAGE CURRENT			**±40**			**±40**	μA
OUTPUT CAPACITANCE		5			5		pF
DIGITAL INPUT VOLTAGE							
V_{INL}			**0.8**			**0.8**	V
V_{INH}	**2.0**			**2.0**			V
DIGITAL INPUT CURRENT							
($0 \leq V_{IN} \leq +5V$)							
I_{INL}	**−100**			**−100**			μA
I_{INH}			**+100**			**+100**	μA
INPUT CAPACITANCE		10			10		pF

NOTES

[1]Detailed timing specifications are given on page 7 of this data sheet.

[2]Maximum specifications apply to 2.55V range only. Typical specifications apply to both the 2.55V range and the 255mV range.

[3]Refer to Figure 5 for characterization.

[4]The digital output code will not change by more than 1LSB with zero or full scale applied to the differential inputs.

Specifications subject to change without notice.

Specifications shown in boldface are tested on all production units at final electrical test. Results from those tests are used to calculate outgoing quality levels. All min and max specifications are guaranteed, although only those shown in boldface are tested on all production units.

Model	AD670A			AD670B			AD670S			
	Min	Typ	Max	Min	Typ	Max	Min	Typ	Max	Units
OPERATING TEMPERATURE RANGE	−25		+85	−25		+85	−55		+125	°C
RESOLUTION	8			8			8			Bit
CONVERSION TIME[1]			**10**			**10**			**10**	µs
RELATIVE ACCURACY										
T_{min} to T_{max}			**±1/2**			**±1/4**		±1/4	**±1/2**	LSB
DIFFERENTIAL LINEARITY ERROR										
T_{min} to T_{max}	**GUARANTEED NO MISSING CODES ALL GRADES**									
GAIN ACCURACY[2]										
@ +25°C		±1.0	**±1.5**		±0.5	**±0.75**		±0.5	**±0.75**	LSB
T_{min} to T_{max}		±2.0	**±2.5**		±1.0	**±1.5**		±2.0	**±2.5**	LSB
UNIPOLAR ZERO ERROR										
@ +25°C			**±1.0**			**±0.5**			**±0.5**	LSB
T_{min} to T_{max}			**±2.0**			**±1.0**			**±1.0**	LSB
BIPOLAR ZERO ERROR[3]										
@ +25°C			**±1.0**			**±0.5**			**±0.5**	LSB
T_{min} to T_{max}			**±2.0**			**±1.0**			**±1.0**	LSB
ANALOG INPUT RANGES										
All Grades		0 to +255			0 to +255			0 to +255		mV
		−128 to +127			−128 to +127			−128 to +127		mV
		0 to +2.55			0 to +2.55			0 to +2.55		V
		−1.28 to +1.27			−1.28 to +1.27			−1.28 to +1.27		V
BIAS CURRENT (255mV RANGE)										
T_{min} to T_{max}		200	**500**		200	**500**		200	**500**	nA
OFFSET CURRENT (255mV RANGE)										
T_{min} to T_{max}		20	**100**		20	**100**		20	**100**	nA
2.55V RANGE INPUT RESISTANCE	8.0		12.0	8.0		12.0	8.0		12.0	kΩ
2.55V RANGE FULL SCALE MATCH										
+ AND − INPUT		±1/2			±1/2			±1/2		LSB
COMMON-MODE RANGE[4]										
@ +25°C	**−0.2**		**$+V_{CC}-3.4$**	**−0.2**		**$+V_{CC}-3.4$**	**−0.2**		**$+V_{CC}-3.4$**	V
T_{min} to T_{max}	**0**		**$+V_{CC}-3.8$**	**0**		**$+V_{CC}-3.8$**	**0**		**$+V_{CC}-3.8$**	V
COMMON-MODE REJECTION										
RATIO (255mV RANGE)			**1**			**1**			**1**	LSB[4]
COMMON-MODE REJECTION										
RATIO (2.55V RANGE)			**1**			**1**			**1**	LSB[4]
POWER SUPPLY										
Operating Range	**4.5**		**5.5**	**4.5**		**5.5**	**4.75**		**5.5**	V
Current I_{CC}		30	**45**		30	**45**		30	**45**	mA
Rejection Ratio T_{min} to T_{max}			**0.015**			**0.015**			**0.015**	% of FS/%
DIGITAL OUTPUTS										
SINK CURRENT (V_{OUT} = 0.4V)	**1.6**			**1.6**			**1.6**			mA
T_{min} to T_{max}	**1.6**			**1.6**			**1.6**			mA
SOURCE CURRENT (V_{OUT} = 2.4V)										
@ +25°C	**0.5**			**0.5**			**0.5**			mA
T_{min} to T_{max}	**0.5**			**0.5**			**0.5**			mA
THREE-STATE LEAKAGE CURRENT			**±40**			**±40**			**±40**	µA
OUTPUT CAPACITANCE		5			5			5		pF
DIGITAL INPUT VOLTAGE										
V_{INL}			**0.8**			**0.8**			**0.7**	V
V_{INH}	**2.0**			**2.0**			**2.0**			V
DIGITAL INPUT CURRENT										
($0 \leq V_{IN} \leq +5V$)										
I_{INL}	**−100**			**−100**			**−100**			µA
I_{INH}			**+100**			**+100**			**+100**	µA
INPUT CAPACITANCE		10			10			10		pF

NOTES
[1]Detailed timing specifications are given on page 7 of this data sheet.
[2]Maximum specifications apply to 2.55V range only. Typical specifications apply to both the 2.55V range and the 255mV range.
[3]Refer to Figure 5 for characterization.
[4]The digital output code will not change by more than 1LSB with zero or full scale applied to the differential inputs.
Specifications subject to change without notice.

Specifications shown in boldface are tested on all production units at final electrical test. Results from those tests are used to calculate outgoing quality levels. All min and max specifications are guaranteed, although only those shown in boldface are tested on all production units.

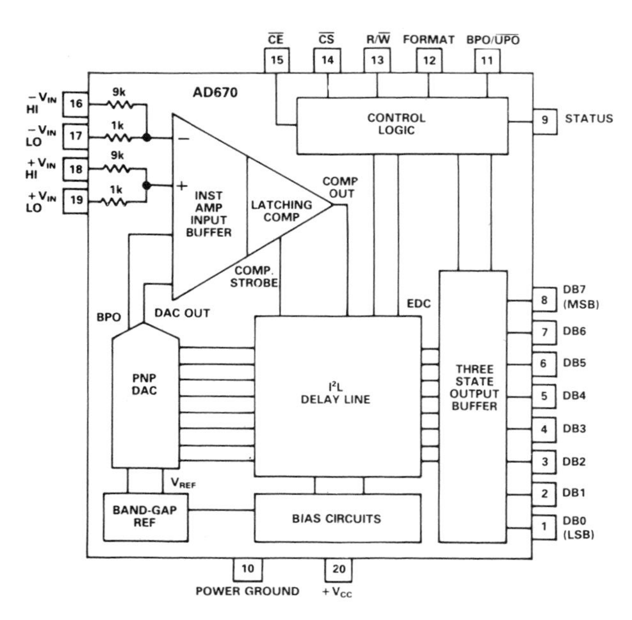

Figure 1a. AD670 Block Diagram

AD670 8-BIT ADC

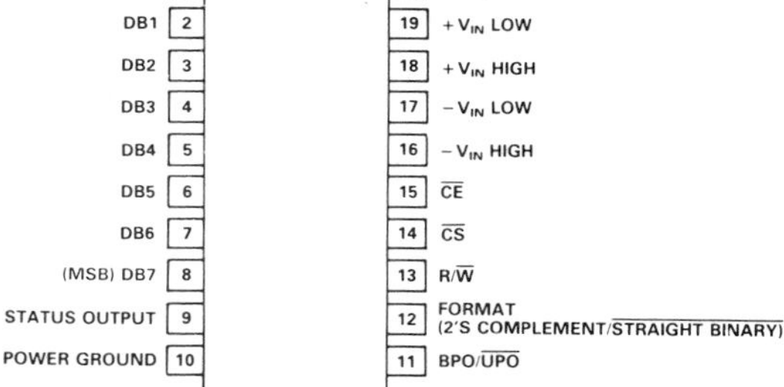

20-PIN DIP

Figure 1b. AD670 Pin Connections

ABSOLUTE MAXIMUM RATINGS

V_{CC} to Ground . . . . 0V to +7.5V
Digital Inputs (Pins 11-15) . . . . −0.5V to V_{CC} +0.5V
Digital Outputs (Pins 1-9) . Momentary Short to V_{CC} or Ground
Analog Inputs (Pins 16-19) . . . . −30V to +30V
Power Dissipation 450mW
Storage Temperature Range . . . . −65°C to +150°C
Lead Temperature (Soldering, 10 Seconds) . . . . . . +300°C

AD670 ORDERING GUIDE

Model	Temperature Range	Relative Accuracy Max (T_{min} to T_{max})	Gain Accuracy (T_{min} to T_{max})	Package Type	Price (100s)
AD670JN	0 to +70°C	±1/2LSB	±2.5LSB	Plastic DIP	$ 6.95
AD670KN	0 to +70°C	±1/4LSB	±1.0LSB	Plastic DIP	$ 8.90
AD670AD	−25°C to +85°C	±1/2LSB	±2.5LSB	Ceramic DIP	$12.25
AD670BD	−25°C to +85°C	±1/4LSB	±1.0LSB	Ceramic DIP	$14.75
AD670SD	−55°C to +125°C	±1/2LSB	±2.0LSB	Ceramic DIP	$34.50

CIRCUIT OPERATION/FUNCTIONAL DESCRIPTION

The AD670 is a functionally complete 8-bit signal conditioning A/D converter with microprocessor compatibility. A block diagram and pin out are shown in Figures 1a. and 1b. The input section uses an instrumentation amplifier to accomplish the voltage to current conversion. This front end provides a high impedance, low bias current differential amplifier. The ground inclusive common-mode range allows the user to directly interface the device to a variety of transducers.

The A/D is signaled to begin a conversion using the three input signals, R/$\overline{W}$, $\overline{CS}$, and $\overline{CE}$. The R/$\overline{W}$ line directs the converter to read or start a conversion. A minimum write/start pulse of 300ns is required on either $\overline{CE}$ or $\overline{CS}$. The conversion thus begun, the internal 8-bit DAC is sequenced from MSB to LSB using a novel successive approximation technique. In conventional designs, the DAC is stepped through the bits by a clock. This can be thought of as a static design since the speed at which the DAC is sequenced is determined solely by the clock. No clock is used in the AD670. Instead, a "dynamic SAR" is created consisting of a string of inverters with taps along the delay line. Sections of the delay line between taps act as one shots. The pulses are used to set and reset the DAC's bits and strobe the comparator. When strobed, the comparator then determines whether the addition of each successively weighted bit current causes the DAC current sum to be greater or less than the input current. If the sum is less, the bit is turned off. After all bits are tested, the SAR holds an 8-bit code representing the input signal to within 1/2LSB accuracy. Ease of implementation and reduced dependence on process related variables make this an attractive approach to a successive approximation design.

The SAR provides an end-of-conversion signal to the control logic which then brings the STATUS line low. Data outputs remain in a high impedance state until edge-triggered by a signal sent to the R/$\overline{W}$ line; R/$\overline{W}$ is then brought high with $\overline{CE}$ and $\overline{CS}$ low and allows the converter to be read. Bringing $\overline{CE}$ or $\overline{CS}$ high during the valid data period ends the read cycle. The output buffers cannot be enabled during a conversion. Any convert start commands will be ignored until the conversion cycle is completed; once a conversion cycle has been started it cannot be stopped or restarted.

The AD670 provides the user with a great deal of flexibility by offering two input spans and formats and a choice of output codes. Input format and input range can each be selected. The BPO/$\overline{\text{UPO}}$ pin controls a switch which injects a bipolar offset current of a value equal to the MSB less 1/2LSB into the summing node of the comparator to offset the DAC output. Two precision 10 to 1 attenuators are included on board to provide input range selection of 0 to 2.55V or 0 to 255mV. Additional ranges of −1.28 to 1.27V and −128 to 127mV are possible if the BPO/$\overline{\text{UPO}}$ switch is high when the conversion is started. Finally, output coding can be chosen using the FORMAT pin when the conversion is started. In the bipolar mode and with a logic 1 on FORMAT, the output is in two's complement; with a logic 0, the output is offset binary.

CONNECTING THE AD670

The AD670 has been designed for ease of use. All active components required to perform a complete A/D conversion are on board and are connected internally. In addition, all calibration trims are performed at the factory assuring specified accuracy without user trims. There are, however, a number of options and connections that should be considered to obtain maximum flexibility from the part.

INPUT CONNECTIONS

Standard connections are shown in the figures that follow. An input range of 0 to 2.55V may be configured as shown in Figure 2a. This will provide a one LSB change for each 10mV of input change. The input range of 0 to 255mV is configured as shown in Figure 2b. In this case, each LSB represents 1mV of input change. As in Figure 2, when unipolar input signals are used, Pin 11, BPO/$\overline{\text{UPO}}$, should be grounded. Pin 11 selects the input format for either unipolar or bipolar signals. Figures 3a and 3b show the input connections for bipolar signals. Pin 11 should be tied to $+V_{CC}$ for bipolar inputs.

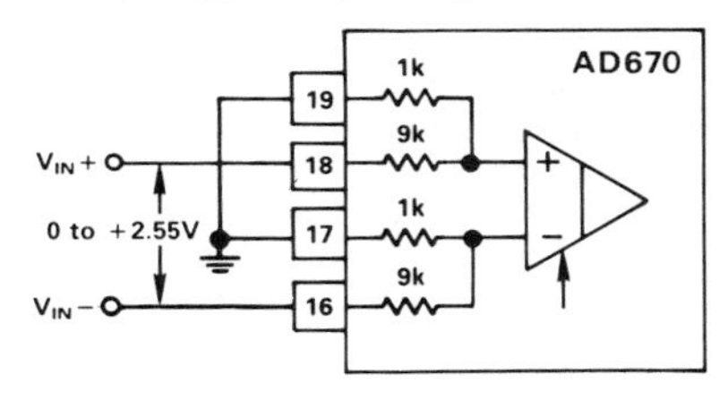

2a. 0 to 2.55V (10mV/LSB)

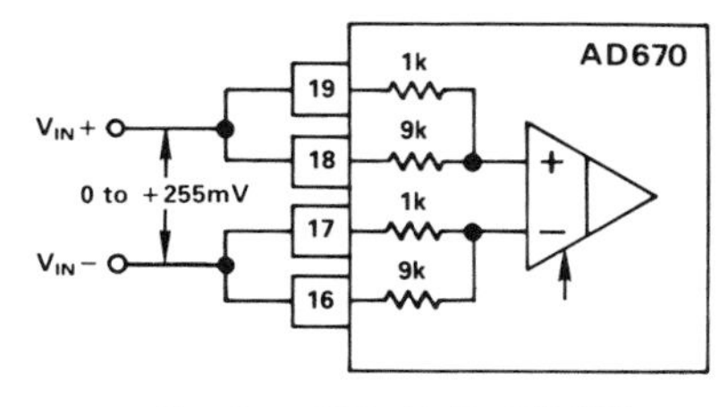

2b. 0 to 255mV (1mV/LSB)

NOTE: PIN 11, BPO/$\overline{\text{UPO}}$ SHOULD BE LOW WHEN CONVERSION IS STARTED.

Figure 2. Unipolar Input Connections

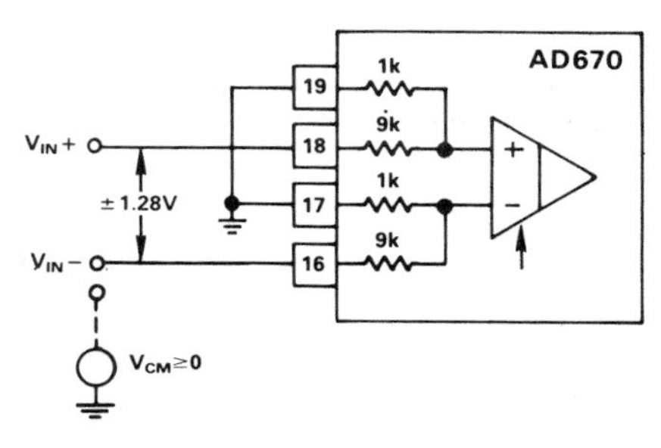

3a. ±1.28V Range

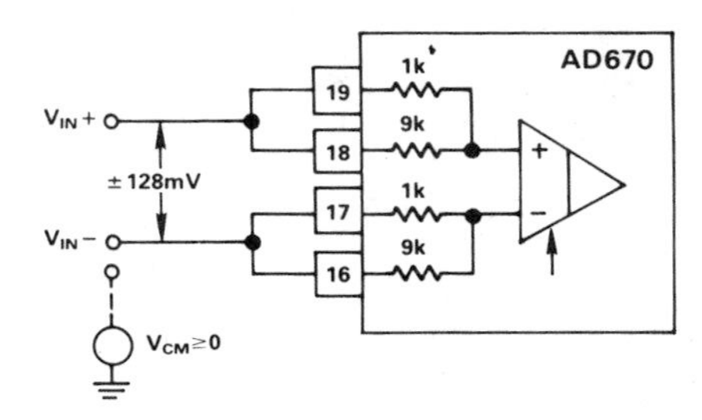

3b. ±128mV Range

NOTE: PIN 11, BPO/$\overline{\text{UPO}}$ SHOULD BE HIGH WHEN CONVERSION IS STARTED.

Figure 3. Bipolar Input Connections

Common-Mode Performance

The AD670 can handle a common-mode voltage of −0.2 to (V_{CC} −3.4V) at 25°C and 0 to (V_{CC} −3.8V) over temperature. The excellent common-mode rejection of the AD670 is due to the differential nature of the instrumentation amplifier front end. The differential signal is maintained until it reaches the output of the comparator. In contrast to a standard operational amplifier, the instrumentation amplifier front end of the AD670 provides significantly improved CMRR over a wide frequency range (Figure 4A).

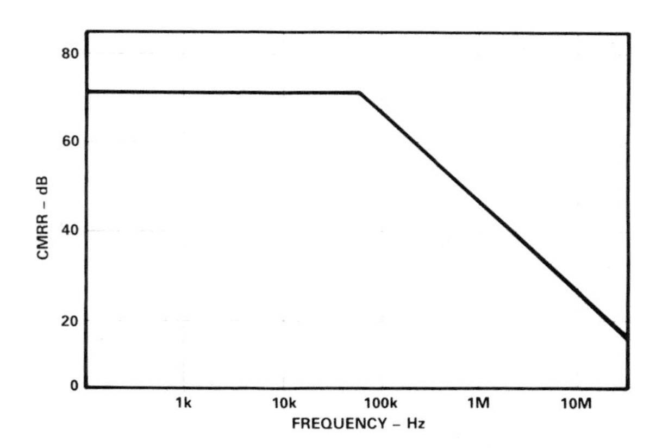

Figure 4a. CMRR over Frequency

Good common-mode performance is particularly useful in a number of situations. In bridge-type transducer applications, such performance facilitates the recovery of differential analog signals in the presence of a dc common-mode or a noisy electrical environment. High-frequency CMRR also becomes important when the analog signal is referred to a noisy, remote digital ground. In each case, the CMRR specification of the AD670 allows the integrity of the input signal to be preserved.

The AD670's common-mode voltage tolerance allows more flexibility in circuit layout. Most other A/D converters require the establishment of one point as the analog reference point. This is necessary in order to minimize the effects of parasitic voltages. The AD670, however, eliminates the need to make the analog ground reference point and ADC analog ground one and the same. Instead, a system such as that shown in Figure 4b is possible as a result of the AD670's common-mode performance. The resistors and inductors in the ground return represent unavoidable system parasitic impedances.

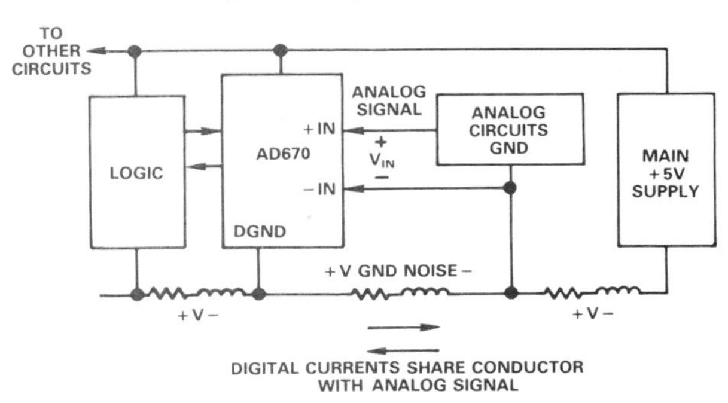

Figure 4b. AD670 Input Rejects Common-Mode Ground Noise

Although the instrumentation amplifier has a differential input, there must be a return path to ground for the bias currents. If it is not provided, these currents will charge stray capacitances and cause internal circuit nodes to drift uncontrollably causing the digital output to change. Such a return path is provided in Figures 2a and 3a (larger input ranges) since the 1k resistor leg is tied to ground. This is not the case for Figures 2b and 3b (the lower input ranges). When connecting the AD670 inputs to floating sources, such as transformers and ac-coupled sources, there must still be a dc path from each input to common. This can be accomplished by connecting a 10kΩ resistor from each input to ground.

Input/Output Options

Data output coding (2's complement vs. straight binary) is selected using the FORMAT pin. The selection of input format (bipolar vs. unipolar) is controlled using Pin 11, BPO/$\overline{\text{UPO}}$. Prior to a write/convert, the state of FORMAT and BPO/$\overline{\text{UPO}}$ should be available to the converter. These lines may be tied to the data bus and may be changed with each conversion if desired. BPO/$\overline{\text{UPO}}$ controls the bipolar offset current. A logic 0 on this input sets up the AD670 for a unipolar input range and a logic 1 sets up the bipolar range. These choices are shown in Table I.

An output signal, STATUS, indicates the status of the conversion. STATUS goes high at the beginning of the conversion and returns low when the conversion cycle has been completed.

Output coding can be selected for the AD670 by using Pin 12, the FORMAT pin. Holding FORMAT high when starting a conversion will give a 2's complement output. Holding FORMAT low will give an offset binary output. Coding for each of these output formats are shown in Figures 5a and 5b.

BPO/$\overline{\text{UPO}}$	FORMAT	INPUT RANGE/OUTPUT FORMAT
0	0	Unipolar/Straight Binary
1	0	Bipolar/Offset Binary
0	1	Unipolar/2's Complement
1	1	Bipolar/2's Complement

Table I. AD670 Input Selection/Output Format Truth Table

$+V_{IN}$	$-V_{IN}$	DIFF V_{IN}	STRAIGHT BINARY (FORMAT = 0, BPO/$\overline{\text{UPO}}$ = 0)
0	0	0	0000 0000
128mV	0	128mV	1000 0000
255mV	0	255mV	1111 1111
255mV	255mV	0	0000 0000
128mV	127mV	1mV	0000 0001

Figure 5a. Unipolar Output Codes

$+V_{IN}$	$-V_{IN}$	DIFF V_{IN}	OFFSET BINARY (FORMAT = 0, BPO/$\overline{\text{UPO}}$ = 1)	2's COMPLEMENT (FORMAT = 1, BPO/$\overline{\text{UPO}}$ = 1)
0	0	0	1000 0000	0000 0000
127mV	0	127mV	1111 1111	0111 1111
1.127V	1.000V	127mV	1111 1111	0111 1111
255mV	255mV	0	1000 0000	0000 0000
128mV	127mV	1mV	1000 0001	0000 0001
127mV	128mV	−1mV	0111 1111	1111 1111
127mV	255mV	−128mV	0000 0000	1000 0000

Figure 5b. Bipolar Output Codes

Calibration

Because of its precise factory calibration, the AD670 is intended to be operated without user trims for gain and offset; therefore, no provisions have been made for such user-trims. Figures 6a, 6b, and 6c show the transfer curves at zero and full scale for the unipolar and bipolar modes. The code transitions are positioned so that the desired value is centered at that code. The first LSB transition for the unipolar mode occurs for an input of +1/2LSB (5mV or 0.5mV). Similarly, the MSB transition for the bipolar mode is set at −1/2LSB (−5mV or −0.5mV). The full scale transition is located at the full scale value −1 1/2LSB. These values are 2.545V and 254.5mV.

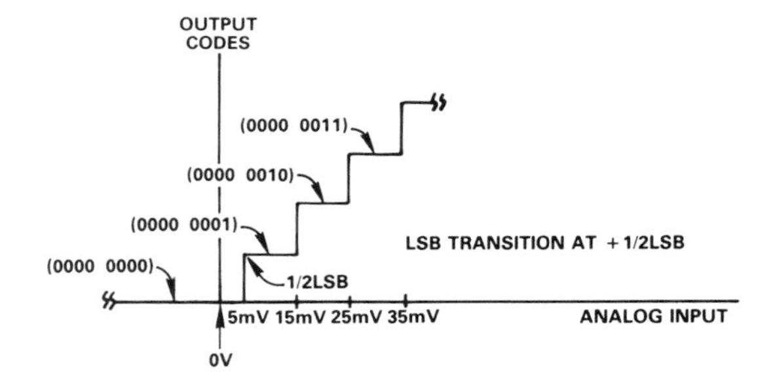

Figure 6a. Unipolar Transfer Curve

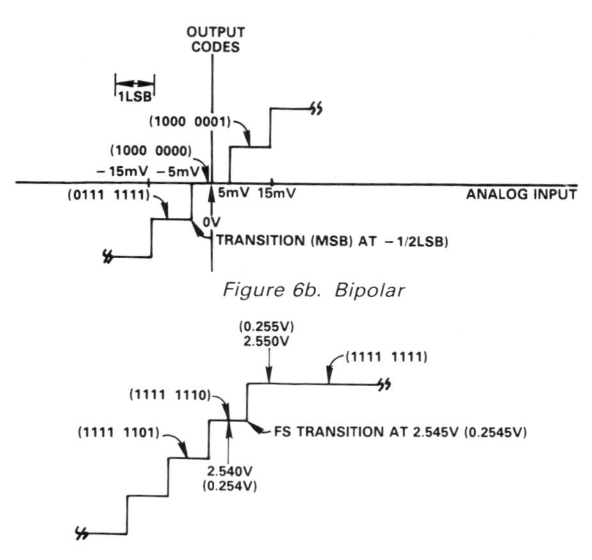

Figure 6b. Bipolar

Figure 6c. Full Scale (Unipolar)

Figure 6. Transfer Curves

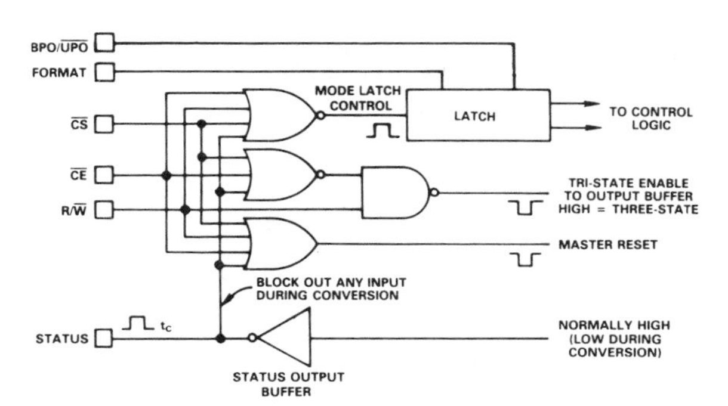

Figure 7. Control Logic Block Diagram

R/$\overline{W}$	$\overline{CS}$	$\overline{CE}$	OPERATION
0	0	0	WRITE/CONVERT
1	0	0	READ
X	X	1	NONE
X	1	X	NONE

Table II. AD670 Control Signal Truth Table

CONTROL AND TIMING OF THE AD670

Control Logic

The AD670 contains on-chip logic to provide conversion initiation and data read operations from signals commonly available in microprocessor systems. Figure 7 shows the internal logic circuitry of the AD670. The control signals, $\overline{CE}$, $\overline{CS}$, and R/$\overline{W}$ control the operation of the converter. The read or write function is determined by R/$\overline{W}$ when both $\overline{CS}$ and $\overline{CE}$ are low as shown in Table II. If all three control inputs are held low longer than the conversion time, the device will continuously convert until one input, $\overline{CE}$, $\overline{CS}$, or R/$\overline{W}$ is brought high. The relative timing of these signals is discussed later in this section.

Timing

The AD670 is easily interfaced to a variety of microprocessors and other digital systems. The following discussion of the timing requirements of the AD670 control signals will provide the designer with useful insight into the operation of the device.

Write/Convert Start Cycle

Figure 8 shows a complete timing diagram for the write/convert start cycle. $\overline{CS}$ (chip select) and $\overline{CE}$ (chip enable) are active low and are interchangeable signals. Both $\overline{CS}$ and $\overline{CE}$ must be low for the converter to read or start a conversion. The minimum pulse width, t_W, on either $\overline{CS}$ or $\overline{CE}$ is 300ns to start a conversion.

Table III. AD670 TIMING SPECIFICATIONS

(Guaranteed Over the Full Operating Temperature Range, Unless Otherwise Noted)
Boldface indicates parameters tested 100% unless otherwise noted. See Specifications page for explanation.

		+25°C				T_{min} to T_{max}			
Symbol	**Parameter**	**Min**	**Typ**	**Max**	**Units**	**Min**	**Typ**	**Max**	**Units**
WRITE/CONVERT START MODE									
t_W	Write/Start Pulse Width	300			ns	360			ns
t_{DS}	Input Data Setup Time	200			ns	240			ns
t_{DH}	Input Data Hold	10			ns	10			ns
t_{RWC}	Read/Write Setup Before Control	0			ns	0			ns
t_{DC}	Delay to Convert Start			700	ns			840	ns
$\mathbf{t_C}$	**Conversion Time (J,K,A,B)/S grades**			**10**	µs			**(11)/13**	µs
READ MODE									
t_R	Read Time	250			ns	250			ns
t_{SD}	Delay from Status Low to Data Read			250	ns			250	ns
$\mathbf{t_{TD}}$	**Bus Access Time**		200	**250**	ns		200	**250**	ns
t_{DH}	Data Hold Time	25			ns	25			ns
$\mathbf{t_{DT}}$	**Output Float Delay**			**150**	ns			**150**	ns
t_{RT}	R/$\overline{W}$ before $\overline{CE}$ or $\overline{CS}$ low	0			ns	0			ns

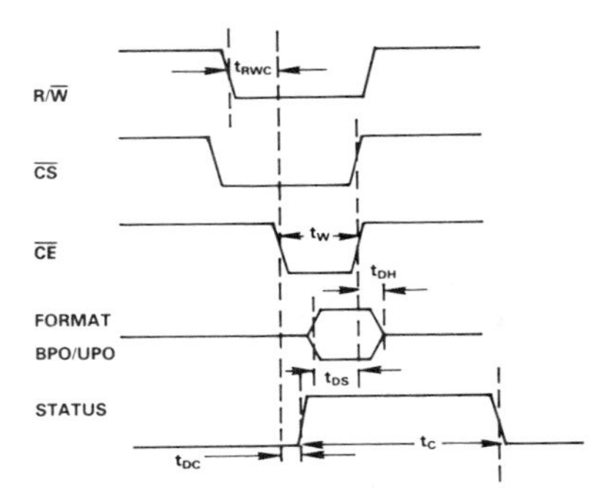

Figure 8. Write/Convert Start Timing

The R/$\overline{W}$ line is used to direct the converter to start a conversion (R/$\overline{W}$ low) or read data (R/$\overline{W}$ high). The relative sequencing of the three control signals (R/$\overline{W}$, $\overline{CE}$, $\overline{CS}$) is unimportant. However, when all three signals remain low for at least 300ns (t_W), STATUS will go high to signal that a conversion is taking place.

Once a conversion is started and the STATUS line goes high, convert start commands will be ignored until the conversion cycle is complete. The output data buffer cannot be enabled during a conversion.

Read Cycle

Figure 9 shows the timing for the data read operation. The data outputs are in a high impedance state until a read cycle is initiated. To begin the read cycle, R/$\overline{W}$ is brought high. During a read cycle, the minimum pulse length for $\overline{CE}$ and $\overline{CS}$ is a function of the length of time required for the output data to be valid. The data becomes valid and is available to the data bus in a maximum of 250ns. This delay between the high impedance state and valid data is the maximum bus access time or t_{TD}. Bringing $\overline{CE}$ or $\overline{CS}$ high during valid data ends the read cycle. The outputs remain valid for a minimum of 25ns (t_{DH}) and return to the high impedance state after a delay, t_{DT}, of 150ns maximum.

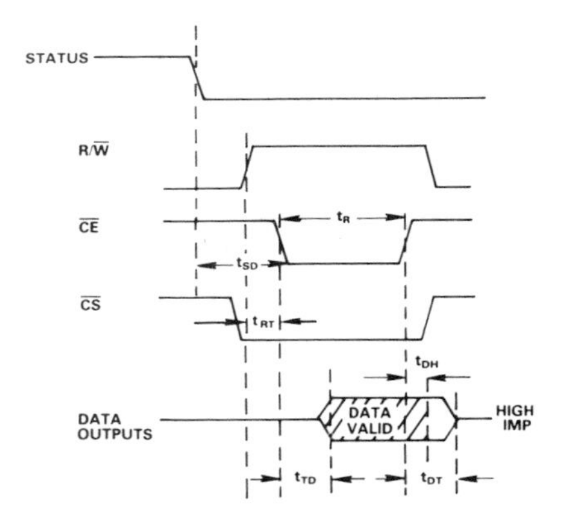

Figure 9. Read Cycle Timing

STAND-ALONE OPERATION

The AD670 can be used in a "stand-alone" mode, which is useful in systems with dedicated input ports available. Two typical conditions are described and illustrated by the timing diagrams which follow.

Single Conversion, Single Read

When the AD670 is used in a stand-alone mode, $\overline{CS}$ and $\overline{CE}$ should be tied together. Conversion will be initiated by bringing R/$\overline{W}$ low. Within 700ns, a conversion will begin. The R/$\overline{W}$ pulse should be brought high again once the conversion has started so that the data will be valid upon completion of the conversion. Data will remain valid until $\overline{CE}$ and $\overline{CS}$ are brought high to indicate the end of the read cycle or R/$\overline{W}$goes low. The timing diagram is shown in Figure 10.

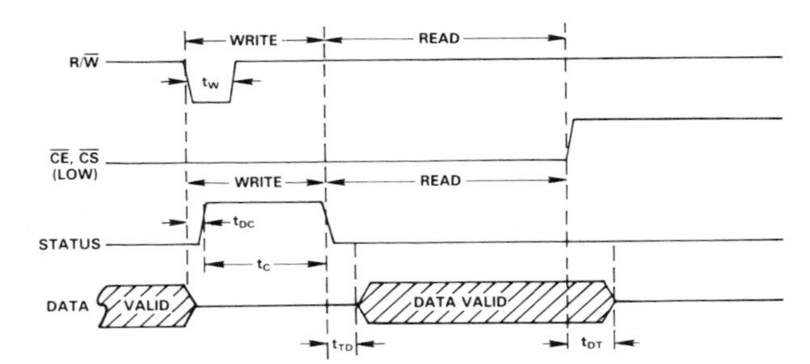

Figure 10. Stand-Alone Mode Single Conversion/ Single Read

Continuous Conversion, Single Read

A variety of applications may call for the A/D to be read after several conversions. In process control systems, this is often the case since a reading from a sensor may only need to be updated every few conversions. Figure 11 shows the timing relationships.

Once again, $\overline{CE}$ and $\overline{CS}$ should be tied together. Conversion will begin when the R/$\overline{W}$ signal is brought low. The device will convert repeatedly as indicated by the status line. A final conversion will take place once the R/$\overline{W}$ line has been brought high. The rising edge of R/$\overline{W}$ must occur while STATUS is high. R/$\overline{W}$ should not return high while STATUS is low since the circuit is in a reset state prior to the next conversion. Since the rising edge of R/$\overline{W}$ must occur while STATUS is high, R/$\overline{W}$'s length must be a minimum of 10.25µs (t_C + t_{TD}). Data becomes valid upon completion of the conversion and will remain so until the $\overline{CE}$ and $\overline{CS}$ lines are brought high indicating the end of the read cycle or R/$\overline{W}$ goes low initiating a new series of conversions.

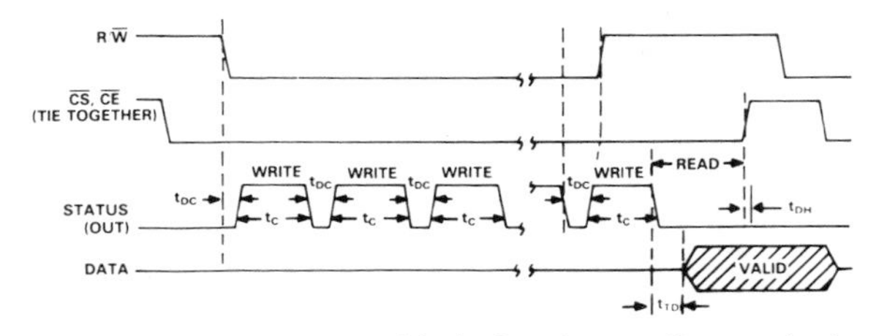

Figure 11. Stand-Alone Mode Continuous Conversion/ Single Read

APPLYING THE AD670

The AD670 has been designed for ease of use, system compatibility, and minimization of external components. Transducer interfaces generally require signal conditioning and preamplification before the signal can be converted. The AD670 will reduce and even eliminate this excess circuitry in many cases. To illustrate the flexibility and superior solution that the AD670 can bring to a transducer interface problem, the following discussions are offered.

Temperature Measurements

Temperature transducers are one of the most common sources of analog signals in data acquisition systems. These sensors require circuitry for excitation and preamplification/buffering. The instrumentation amplifier input of the AD670 eliminates the need for this signal conditioning. The output signals from temperature transducers are generally sufficiently slow that a sample/hold amplifier is not required. Figure 12 shows the AD590 IC temperature transducer interfaced to the AD670. The AD580 voltage reference is used to offset the input for 0°C calibration. The current output of the AD590 is converted into a voltage by R1. The high impedance unbuffered voltage is applied directly to the AD670 configured in the − 128mV to 127mV bipolar range. The digital output will have a resolution of 1°C.

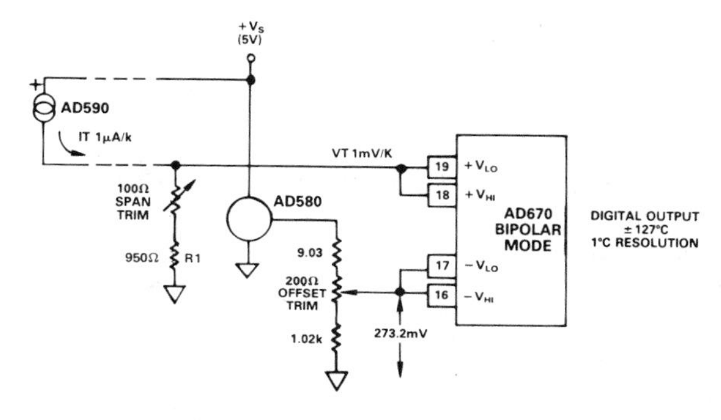

Figure 12. AD670 Temperature Transducer Interface

Platinum RTDs are also a popular, temperature transducer. Typical RTDs have a resistance of 100Ω at 0°C and change resistance 0.4Ω per °C. If a constant excitation current is caused to flow in the RTD, the change in voltage drop will be a measure of the change in temperature. Figure 13 shows such a method and the required connections to the AD670. The AD580 2.5V reference provides the accurate voltage for the excitation current and range offsetting for the RTD. The op-amp is configured to force a constant 2.5mA current through the RTD. The differential inputs of the AD670 measure the difference between a fixed offset voltage and the temperature dependent output of the op-amp which varies with the resistance of the RTD. The RTD change of approximately 0.4Ω/°C results in a 1mV/°C voltage change. With the AD670 in the 1mV/LSB range, temperatures from 0 to 255°C can be measured.

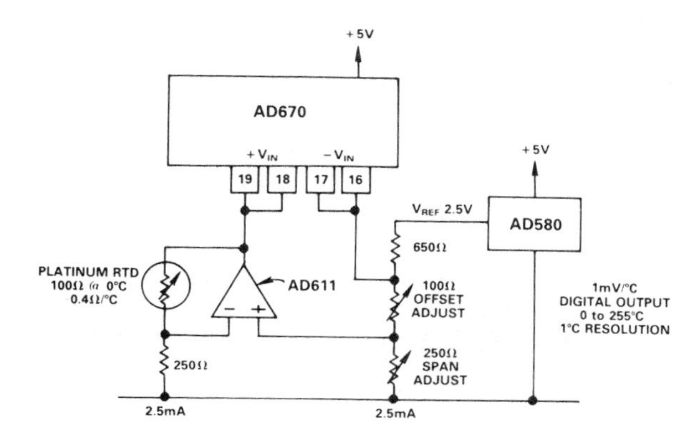

Figure 13. Low Cost RTD Interface

Differential temperature measurements can be made using an AD590 connected to each of the inputs as shown in Figure 14. This configuration will allow the user to measure the relative temperature difference between two points with a 1°C resolution. Although the internal 1k and 9k resistors on the inputs have ± 20% tolerance, trimming the AD590 is unnecessary as most differential temperature applications are concerned with the relative differences between the two. However, the user may see up to a 20% scale factor error in the differential temperature to digital output transfer curve.

This scale factor error can be eliminated through a software correction. Offset corrections can be made by adjusting for any difference that results when both sensors are held at the same temperature. A span adjustment can then be made by immersing one AD590 in an ice bath and one in boiling water and eliminating any deviation from 100°C. For a low cost version of this setup, the plastic AD592 can be substituted for the AD590.

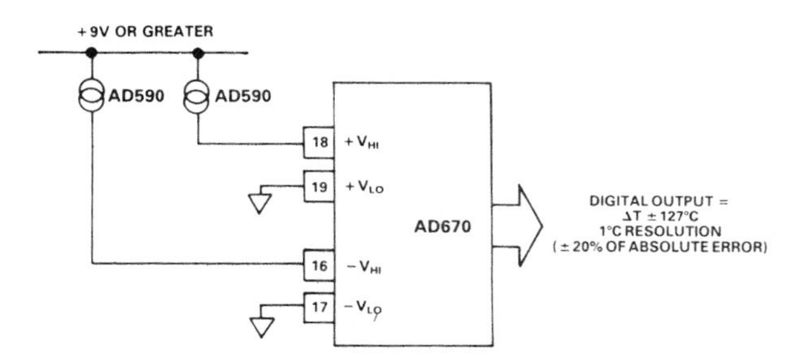

Figure 14. Differential Temperature Measurement Using the AD590

STRAIN GAUGE MEASUREMENTS

Many semiconductor-type strain gauges, pressure transducers, and load cells may also be connected directly to the AD670. These types of transducers typically produce 30 millivolts full-scale per volt of excitation. In the circuit shown in Figure 15, the AD670 is connected directly to a Data Instruments model JP-20 load cell. The AD584 programmable voltage reference is used along with an AD741 op-amp to provide the ±2.5V excitation for the load cell. The output of the transducer will be ±150mV for a force of ±20 pounds. The AD670 is configured for the ±128 millivolt range. The resolution is then approximately 2.1 ounces per LSB over a range of ±17 pounds. Scaling to exactly 2 ounces per LSB can be accomplished by trimming the reference voltage which excites the load cell.

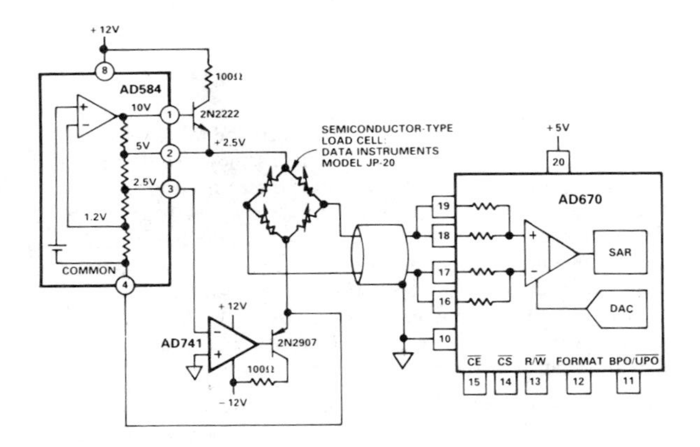

Figure 15. AD670 Load Cell Interface

MULTIPLEXED INPUTS

Most data acquisition systems require the measurement of several analog signals. Multiple A/D converters are often used to digitize these inputs, requiring additional preamplification and buffer stages per channel. Since these signals vary slowly, a differential MUX can multiplex inputs from several transducers into a single AD670. And since the AD670's signal-conditioning capability is preserved, the cost of several ADCs, differential amplifiers, and other support components can be reduced to that of a single AD670, a MUX, and a few digital logic gates.

An AD7502 dual 4-channel MUX appears in Figure 16 multiplexing four differential signals to the AD670. The AD7502's decoded address is gated with the microprocessor's write signal to provice a latching strobe at the flip/flops. A write cycle to the AD7502's address then latches the two LSBs of the data word thereby selecting the input channel for subsequent conversions.

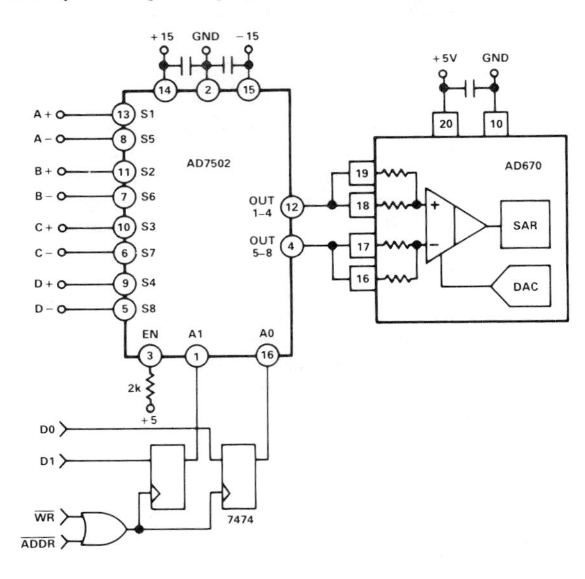

Figure 16. Multiplexed Analog Inputs to AD670

SAMPLE INPUTS

For those applications where the input signal is capable of slewing more than 1/2LSB during the AD670's 10μs conversion cycle, the input should be held constant for the cycle's duration. The circuit shown in Figure 17 uses a CMOS switch and capacitor to sample/hold the input. The AD670's STATUS output, once inverted, supplies the sample/hold (S/$\overline{\text{H}}$) signal.

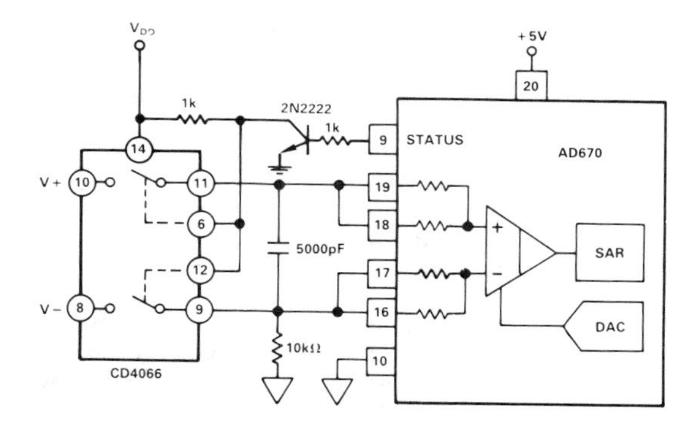

Figure 17. Low Cost Sample-and-Hold Circuit for AD670

The above circuit's performance as a sample/hold strictly depends on the switch characteristics and size of the hold capacitor. While in the hold mode, the AD670's bias current will discharge the hold capacitor yielding a droop rate equal to the bias current divided by the capacitance. The AD670's typical bias current of 200nA will yield a 40μV/μs droop rate with the 5000pF capacitor shown in Figure 17. The input will then droop only 0.4mV during the AD670's 10μs conversion, which meets the 1/2LSB limit for the same 256mV range.

The CD4066 CMOS switch shown in Figure 17 was chosen for its fast transition times and low on resistance. The control input's propagation delay for switch-closed to switch-open should remain less than 150ns to ensure the sample-to-hold transition occurs before the first bit decision. Otherwise, missing codes may arise. The low on resistance, meanwhile, allows for a fast acquisition time. Since settling to 1/2LSB at 8-bits of resolution requires 6.2 RC time constants, the 5000pF hold capacitor and the CD4066's 80Ω on resistance yields an acquisition time of 2.5μs.

Since CMOS switch characteristics vary with supply voltage, timing requirements should be considered when selecting a switch and its supply voltage. The npn transistor and 1k base resistor in Figure 17 invert and level-shift the AD670's STATUS output. If the switch is operated with the +5V supply, a TTL or CMOS inverter may be substituted.

INTERFACING THE AD670 TO MICROPROCESSOR DATA BUSES

The control logic of the AD670 allows it to be interfaced with many popular microcomputers and microprocessors with few or no additional components. Interfaces to two widely used personal computers are discussed in the following text.

Figure 18 shows how the AD670 can be interfaced to an APPLE II microcomputer. The DEVICE SELECT signal available on the APPLE's peripheral connector can serve as the $\overline{CE}$ for the AD670. The $\overline{CS}$ line can be connected to the APPLE's phase 1 clock, and R/$\overline{W}$ can be connected directly to the APPLE's R/$\overline{W}$ line. The BPO/$\overline{UPO}$ and FORMAT controls can be latched into the AD670 from the data bus.

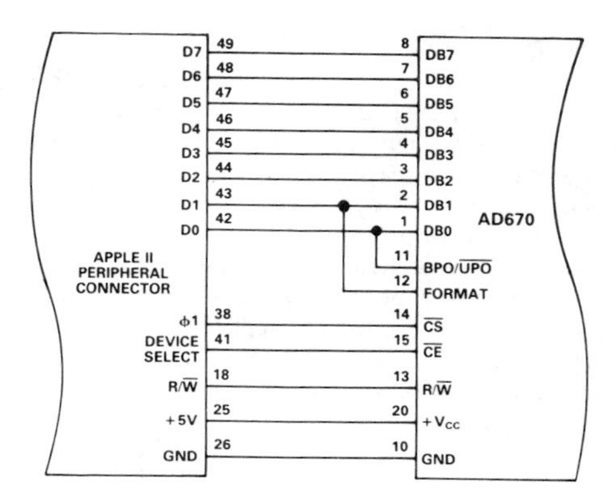

Figure 18. APPLE II Interface

The program shown in Figure 19 is used in interfacing the AD670 to the JP-20 load cell. The load cell output is ±150mV for ±20 pounds and a ±2.5V input to the AD670. This routine determines the weight of the object, averages 100 readings and provides an answer in pounds. The AD670 is interfaced to an APPLE II computer as in Figure 18.

```
1  PR# 1: PRINT  CHR$ (9);"52N": LIST : PR#
     Ø: END : REM  HARDCOPY OF LIST
2  REM   PROGRAM IN APPLESOFT TO INTER-
3  REM   FACE AD67Ø 8-BIT ADC TO DATA
4  REM   INSTRUMENTS MODEL JP-2Ø LOAD
5  REM   CELL WHICH HAS +/-15Ø MV
6  REM   OUTPUT FOR +/-2Ø POUNDS AND
7  REM   +/-2.5 VOLT INPUT.
1Ø  PRINT : PRINT : PRINT : PRINT
2Ø  PRINT "TARE (T) OR WEIGH (W)? ";
3Ø  INPUT A$
4Ø  IF A$ <  > "T" THEN  GOTO 1ØØ
5Ø  PRINT : PRINT "CLEAR SCALE NOW"
6Ø  FOR I = 1 TO 1ØØØ: NEXT I
7Ø TARE = W
8Ø  PRINT "TARE IS ";TARE;" POUNDS"
9Ø  PRINT : PRINT : PRINT
95  GOTO 2Ø
1ØØ  REM   ACTUAL WEIGHT ROUTINE
1Ø5  PRINT "PUT THE OBJECT ON THE SCALE."
11Ø  GOSUB 67Ø
12Ø NETWT = W - TARE
125  PRINT : PRINT : PRINT : PRINT
13Ø  PRINT "NET WEIGHT IS ";NTWT;" POUNDS."
14Ø  PRINT : PRINT : PRINT
15Ø  GOTO 11Ø
67Ø  REM   THIS ROUTINE INTERFACES THE
671  REM   AD67Ø TO THE APPLE AND
672  REM   AVERAGES 1ØØ READINGS, THEN
673  REM   CONVERTS THE ANSWER TO POUNDS.
675 W = Ø
68Ø  FOR I = 1 TO 1ØØ
681  POKE 4936Ø,2 REM   THE '2' SETS THE
682  REM   AD67Ø FOR OFFSET
683  REM   BINARY OPERATION
684  REM  AND BIPOLAR INPUT
69Ø X =  PEEK (4936Ø) REM  TO READ OUTPUT
691  REM  AFTER CONVERSION
695 X = X - 128
7ØØ X = (X / 15Ø) * 2Ø
71Ø W = W + X
72Ø  NEXT I
73Ø W =  INT ((W / 1Ø) + Ø.5)
74Ø W = W / 1Ø
75Ø  RETURN
1ØØØ  END
```

Figure 19. APPLE II Software for AD670 Interface

IBM PC INTERFACE

The AD670 appears in Figure 20 interfaced to the IBM PC. Since the device resides in I/O space, its address is decoded from only the lower ten address lines and must be gated with AEN (active low) to mask out internal (DMA) cycles which use the same I/O address space. This active low signal is applied to $\overline{CS}$. AO, meanwhile, is reserved for the R/$\overline{W}$ input. This places the AD670 in two adjacent addresses; one for starting the conversion and the other for reading the result. The $\overline{IOR}$ and $\overline{IOW}$ signals are then gated and applied to $\overline{CE}$, while the lower two data lines are applied to FORMAT and BPO/$\overline{UPO}$ inputs to provide software programmable input formats and output coding.

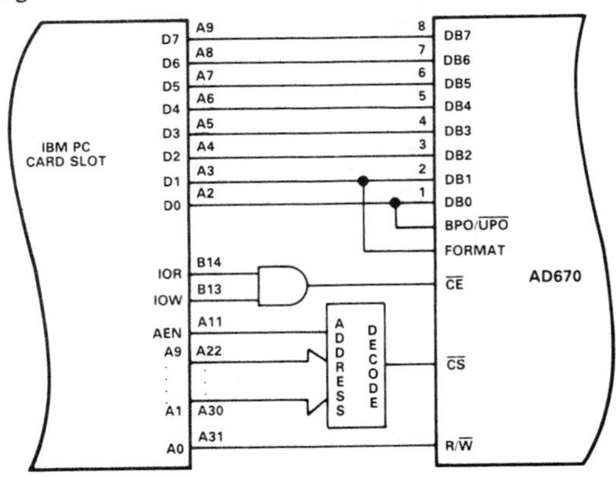

Figure 20. IBM PC Interface to the AD670

In BASIC, a simple OUT ADDR, WORD command initiates a conversion. While the upper six bits of the data WORD are meaningless, the lower two bits define the analog input format and digital output coding according to Table IV. The data is available ten microseconds later (which is negligible in BASIC) and can be read using INP (ADDR + 1). The 3-line subroutine in Figure 21, used in conjunction with the interface of Figure 20, converts an analog input within a bipolar range to an offset binary coded digital word.

NOTE: Due to the large number of options that may be installed in the PC, the I/O bus loading should be limited to one Schottky TTL load. Therefore, a buffer/driver should be used when interfacing more than two AD670's to the I/O bus.

DATA	INPUT FORMAT	OUTPUT CODING
0	Unipolar	Straight Binary
1	Bipolar	Offset Binary
2	Unipolar	2's Complement
3	Bipolar	2's Complement

Table IV.

```
10  OUT &H310,1                    'INITIATE CONVERSION
20  ANALOGIN=INP (&H311)           'READ ANALOG INPUT
30  RETURN
```

Figure 21. Conversion Subroutine

OUTLINE DIMENSIONS

Dimensions shown in inches and (mm).

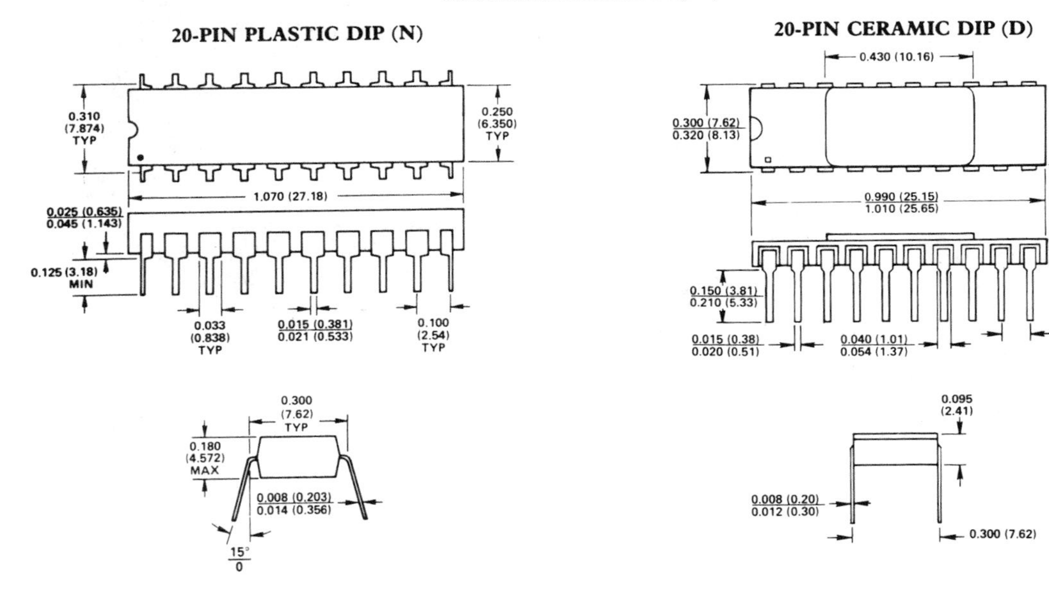

LEAD NO. 1 IDENTIFIED BY DOT OR NOTCH

Index